Genetics and Exploitation of Heterosis in Crops

Editors

James G. Coors and Shivaji Pandey

Associate Editors

Arnel R. Hallauer, Delbert C. Hess, Maarten van Ginkel, Kendall R. Lamkey, Albrecht E. Melchinger, Ganesan Srinivasan, and Charles W. Stuber

Based on the International Symposium on the Genetics and Exploitation of Heterosis in Crops organized and hosted by the International Maize and Wheat Improvement Center [Centro Internacional de Mejoramiento de Maíz y Trigo (CIMMYT)] in Mexico City, 17-22 August 1997

Organizing Committee

Timothy Reeves (Chairman), Ganesan Srinivasan (Executive Chairman), Linda Ainsworth, Shivaji Pandey, Michael Listman, Gregorio Martinez, and Gregory O. Edmeades

American Society of Agronomy, Inc.
Crop Science Society of America, Inc.
Soil Science Society of America, Inc.
Madison, Wisconsin, USA
1999

SPONSORS

CIMMYT wishes to express its sincere thanks to the following sponsors and contributors to the symposium:

Cosponsors (US $10,000 or more)

International Rice Research Institute (IRRI)
Maharashtra Hybrid Seeds Company, Ltd., India
Monsanto Company, USA
Novartis Seeds AG, Switzerland
Peoples Republic of China
Pioneer Hi-Bred International Inc., USA
Sementes Agroceres S.A., Brazil
United States Agency for International Development (USAID)
United States Department of Agriculture (USDA)

Participating Sponsors (U.S. $5,000 – U.S. $10,000)

Asgrow Seed Company, USA
DHL Worldwide Express
Grains Research Development Corporation, Australia
KWS Kleinwanzlebener Saatzucht AG, Germany

Other Sponsors (US $2,000 or less)

Cargill Hybrid Seeds, USA
Ceres Internacional de Semillas, SA de C.V., Mexico
DeKalb Genetics Corporation, USA
Instituto Interamericano de Cooperacion para la Agricultura (IICA)
Maritz de Mexico, SA de C.V.
Plant Genetic Systems Inc. (America), USA

Special thank you to the following:

Rockefeller Foundation for their generosity in enabling the participation of eight researchers from developing countries to the symposium.

American Society of Agronomy for sponsoring the publication of the proceedings of the symposium.

Heterosis Pioneers

the
UNIVERSITY
of
GREENWICH

G.R. Sprague, Jr.
accepting for G.F. Sprague, Sr.

E.J. Wellhausen

Cover art provided by authors

American Society of Agronomy, Inc.
Crop Science Society of America, Inc.
677 South Segoe Road, Madison, Wisconsin 53711 USA

Library of Congress Catalog Card Number: 99-73185

Printed in the United States of America

CONTENTS

SECTION I. OVERVIEW OF HETEROSIS—PAST AND PRESENT

SECTION II. QUANTITATIVE GENETICS OF HETEROSIS

SECTION III. GENETIC DIVERSITY AND HETEROSIS

PREFACE

The chapters for this book are based on 40 plenary presentations by invited speakers and related question-and-answer sessions from the international symposium, *The Genetics and Exploitation of Heterosis in Crops*. The Symposium was organized and hosted by the International Maize and Wheat Improvement Center (Centro Internacional de Mejoramiento de Maíz y Trigo, CIMMYT) in Mexico City, 17–22 August 1997. The chapters basically follow the order of the symposium's nine half-day sessions.

The event focused attention on the momentous contributions of hybrid crop varieties and encouraged experts worldwide to share their knowledge on the phenomenon of hybrid vigor, which is expected to play a key role in meeting humanity's expanding food and feed demands in the future. The symposium drew more than 500 participants—nearly three-quarters of whom came from developing countries—and featured 145 posters in addition to the plenary presentations. The technical program covered actual and potential contributions of heterosis to food security and natural resource conservation through its use in a range of crops—including maize, rice, wheat, sorghum, millets, cotton, vegetables, and oil seeds. Of particular interest were the studies on the genetic, physiological, biochemical, and molecular bases of heterosis that may lead to new strategies for the effective use of hybrid vigor. Among the major conclusions were the following:

- Heterosis is much more widely used in crops than the organizers of the symposium originally envisaged.
- Scientists now generally agree on the genetic mechanisms that underlie heterosis. Dominance, as opposed to overdominance, and epistasis are generally accepted as the principal factors.
- A much clearer understanding of the molecular basis of heterosis emerged; it now seems possible to identify and, eventually, tag the major genes involved in heterosis, allowing their transfer and control by geneticists and breeders.
- Mutations are responsible for a greater proportion of genetic variation than was first thought, and may underlie continued gains in hybrid maize in temperate areas; the importance of genetic diversity in general was emphasized repeatedly.
- Investment in hybrid technology for the cereals appears to make particularly good sense for countries such as China and India to meet domestic demands, and others, such as Brazil, Thailand, Argentina, Myanmar, and South Africa, to produce for exports markets.

Extended abstracts (including methods, results, and selected data) for all presentations and posters were published in a book distributed at the symposium and available through CIMMYT. In a symposium dinner ceremony, five pioneers in research on heterosis—James F. Crow, Yuan Longping, F.W. Schnell, George F. Sprague, and Edwin J. Wellhausen—received special awards for their contributions. Beyond the scientific community itself, extensive media coverage of the symposium brought the issues of food security, sustainable agriculture, and the role of international agricultural research before a global audience.

The Technical Steering Committee and the Editorial Board for this symposium are very grateful to all the authors, reviewers, and sponsors who made the symposium so enjoyable and worthwhile. Special thanks goes to the American Society of Agronomy, the Crop Science Society of America, and the Soil Science Society of America for publishing the book.

Editorial Board:

J.G. Coors	A.R. Hallauer	A.E. Melchinger
S. Pandey	D.C. Hess	G. Srinivasan
M.V. Ginkel	K.R. Lamkey	C.W. Stuber

CONTRIBUTORS

J. Axtell

Professor, Department of Agronomy, Purdue University, 1150 Lilly Hall, West Lafayette, IN 47907-1150

D. Andrews

Professor, Department of Agronomy, 328 Keim Hall, University of Nebraska, Lincoln, NE 68583-0915

R. Bernardo

Assistant Professor, Department of Agronomy, Purdue University, 1150 Lilly Hall of Life Sciences, West Lafayette, IN 47907-1150

N.E. Borlaug

Senior Consultant to the Director General, CIMMYT, Lisboa 27, Apdo. Postal 6-641, 06600 México, D.F., México

J.L. Brewbaker

Professor, Department of Horticulture, University of Hawaii, 3190 Malle Way, Rm. 102, Honolulu, HI 96822

M. Cooper

Senior Lecturer, School of Land and Food, St. Lucia Campus, University of Queensland, Brisbane, Qld. 4072, Australia

J.G. Coors

Professor, Department of Agronomy, University of Wisconsin, Madison, WI 53706

H. Cordova

Leader of Tropical Maize Program, CIMMYT, Lisboa 27, Apdo. Postal 6-641, 06600 México, D.F., México

J.F. Crow

Professor, Department of Genetics, University of Wisconsin, Madison, WI 53706

A. Dhopte

Professor, Department of Botany, Punjabrao Krishi Vidyapeeth, Akola 444 104, Maharashtra, India

D.N. Duvick

Affiliate Professor of Plant Breeding, Department of Agronomy, Iowa State University, P.O. Box 446, 6837, N.W. Beaver Drive, Johnston, IA 50131-0446

J.D. Eastin

Professor, Department of Agronomy, University of Nebraska, P.O. Box 830817, Lincoln, NE 68583-0817

J.W. Edwards

Graduate Research Assistant, Department of Agronomy, Iowa State University, Ames, IA 50011-1010

G.O. Edmeades

Interim Director of Maize Program, CIMMYT, Lisboa 27, Apdo. Postal 6-641, 06600 México, D.F., México

G. Ejeta

Professor, Department of Agronomy, Purdue University, West Lafayette, IN 47907-1150

S.A. Engelbrecht

General Manager of Operations, Sensako, P.O. Box 3295, Brits 9700, South Africa

G.I. Gandoul	Department of Agronomy, University of Nebraska, P.O. Box 830817, Lincoln, NE 68583-0817
T.J. Gerik	Blackland Research Center, 808 E. Blackland Road, Temple, TX 76502
H.H. Geiger	Professor of Population Genetics, University of Hohenheim, 350 Institute of Plant Breeding, Seed Science and Population Genetics, D-70593 Stuttgart, Germany
M.V. Ginkel	Head, Bread Wheat Program, CIMMYT, Lisboa 27, Apdo. Postal 6-641, 06600 México, D.F., México.
I.L. Goldman	Associate Professor, Department of Horticulture, Universit of Wisconsin, 1575 Linden Drive, Madison, WI 53706-1597
J.M. González	Head, Departamento de Coordinación y Desarrollo, Centro de Investigaciones Agrarias de Mabegondo, Apdo. 10, 15080 La Coruña, Spain
M.M. Goodman	William Neal Reynolds and Distinguished University Professor, Crop Science Department, North Caroline State University, Box 7620, Raleigh, NC 27695
C.J. Goodnight	Associate Professor, 115 Marsh Life Science Building, Department of Biology, University of Vermont, Burlington VT 05405-0086
A. Grunst	Consultant, 120 S. 32nd Street, West Des Moines, IA 50265
A.R. Hallauer	C.F. Curtiss Distinguished Professor in Agriculture Professor, Department of Agronomy, Iowa State University Ames, IA 50011-1010
W. Hanna	Research Geneticist, USDA-ARS-SAA, Coastal Plain Exp. Stn., P.O. Box 748, Tifton, GA 31793-0748
V.G. Hernandez	Director, Centro de Genetica, Colegio de Postgraduardos, Montecillo, México 56230
D. Hess	Retired Director, CIMMYT Maize Program, #7 Merion Street, Abilene, TX 79606
K. Hoard	Research Analyst, 13247 NW 121st Place, Madrid, IA50156
M.R.A. Hovney	Agricultural Research Center, Sorghum Research Department, Shandowell Station, Sohag, Egypt
R.B. Hunter	Manager of Product Development, Novartis Seeds, R.R. #1 Plattsville, Ontario N0J 1S0, CANADA
Y. Ibrahim	Graduate Research Assistant, Department of Agronomy, Purdue University, West Lafayette, IN 47907-1150

J. Janick	James Troop Distinguished Professor of Horticulture, Department of Horticulture and Landscape Architecture, Purdue University, West Lafayette, IN 47907-1165
J.P. Jordaan	General Manager of Cereal Grain Research, Sensako, P.O. Box 556, Bethlehem 9700, South Africa
I. Kapran	Sorghum Breeder and INTSORMIL Niger Country Coordinator, Institute National de Recherches Agronomiques du Niger, B. P. 429, Niamey, Niger
M. Kafka	Graduate Research Assistant, Department of Genetics and Plant Breeding, Aristotelian University of Thessaloniki, P.O. Box 261, 540 06 Thessaloniki, Greece
H.A. Knobel	Wheat Breeder, Sensako, P.O. Box 556, Bethlehem 9700, South Africa
J.A. Labate	Research Geneticist, Department of Agronomy, Iowa State University, Ames, IA 50011-1010
K.R. Lamkey	Research Geneticist, USDA-ARS, Department of Agronomy, Iowa State University, Ames, IA 50011-1010
M. Lee	Professor, Department of Agronomy, Iowa State University Ames, IA 50011-1010
A.B. Maunder	Senior Vice President (retired), Dekalb Genetics, 4511 9th Street, Lubbock, TX 79416
J.H. Malan	Wheat Breeder, Sensako, P.O. Box 556, Bethlehem 9700, South Africa
A.E. Melchinger	Professor, Institute of Plant Breeding, Seed Science and Population Genetics, University of Hohenheim, 70593 Stuttgart, Germany
W.R. Meredith, Jr.	Research Geneticist, Crop Genetics and Production Research, Box 314, Stoneville, MS 38776
T. Miedaner	Senior Scientist, University of Hohenheim, Landessaatzuchtanbstalt (720), D-70593 Stuttgart, Germany
J.F. Miller	Research Geneticist, USDA-ARS, Northern Crop Science Laboratory, P.O. Box 5677, Fargo, ND 58102
J.B. Miranda Filho	Professor, Departamento de Genética, Universide de São Paulo/ESALQ, Caixa Postal 83, 13400-970, Piracicaba, Sã Paulo, Brazil
M.L. Munoz	Centro de Genetica, Colegio de Postgraduardos, Montecillo México 56230
V.B. Ogunlea	Professor, Institute for Agricultural Research, Ahmadu Bello University, P.M.B. 1044, Samaru-Zaria, Nigeria

L. M. Onofre   Centro de Genetica, Colegio de Postgraduardos, Montecillo
                            México 56230

P. Ozias-Akins   Associate Professor, Coastal Plain Exp. Stn., P.O. Box 748
                            Tifton, GA 31793-0748

S. Pandey   Maize Program Director, CIMMYT, Lisboa 27, Apdo.
                            Postal 6-641, 06600 México, D.F., México

R. Pandya-Lorch   Coordinator, 2020 Vision for Food, Agriculture and the
                            Environment Initiative, International Food Policy Research
                            Institute, 2033 K Street N.W., Washington, DC 20006

C.L. Petersen   Department of Agronomy, University of Nebraska, P.O.
                            Box 830817, Lincoln, NE 68583-0817

P.A. Peterson   Professor, Departments of Agronomy, Zoology, and
                            Genetics, Iowa State University, Ames, IA 50011-1010

R.L. Phillips   Regents' Professor, Department of Agronomy and Plant
                            Genetics, University of Minnesota, St. Paul, MN 55108

P.L. Pingali   Director, Economics Program, CIMMYT, Lisboa 27, Apdo
                            Postal 6-641, 06600 México, D.F., México

P. Pinstrup-Andersen   Director General, International Food Policy Research
                            Institute, 2033 K Street N.W., Washington, DC 20006

D.W. Podlich   Quantitative Geneticist, School of Land and Food, St.
                            Lucia Campus, University of Queensland, Brisbane, Qld.
                            4072, Australia

A. Polidoros   Postdoctoral Fellow, Department of Genetics and Plant
                            Breeding, Aristotelian University of Thessaloniki, P.O.
                            Box 261, 540 06 Thessaloniki, Greece

T.G. Reeves   Director General, CIMMYT, Lisboa 27,Apdo. Postal 6-
                            641, 06600 México, D.F., México

D. Roche   Research Geneticist, USDA-ARS-SAA, Coastal Plain Exp.
                            Stn., P.O. Box 748, Tifton, GA 31793-0748

F. Shaw   Research Associate, Ecology, Evolution and Behavior
                            Department, 1987 Upper Buford Circle, St. Paul, MN
                            55108

R. Shaw   Associate Professor, Ecology, Evolution and Behavior
                            Department, 1987 Upper Buford Circle, St. Paul, MN
                            55108

J.S.C. Smith   Germplasm Security Coordinator–Research Fellow,
                            Pioneer Hi-Bred International, P.O. Box 1004, 7300 NW,
                            62nd Ave., Johnston, IA 50131-1004

O.S. Smith — Research Fellow, Pioneer Hi-Bred International, P.O. Box 1004, 7300 NW, 62nd Ave., Johnston, IA 50131-1004

C.L. Souza, Jr. — Professor, Departamento de Genética, Universide de São Paulo/ESALQ, Caixa Postal 83, 13400-970, Piracicaba, Sã Paulo, Brazil

G. Srinivasan — Leader of Subtropical Maize Program and Head of International Testing, CIMMYT, Lisboa 27, Apdo. Postal 6-641, 06600 México, D.F., México

C.W. Stuber — Professor (Emeritus), Department of Genetics, North Carolina State University, Raleigh, NC 27695-7614

W.G. Sun — Researcher, Hawaii Agricultural Research Center, 99-193 Aiea Heights Rd., Aiea, HI 96701

E. Tani — Postgraduate Assistant, Department of Genetics and Plant Breeding, Aristotelian University of Thessaloniki, P.O. Bo 261, 540 06 Thessaloniki, Greece

A.S. Tsaftaris — Professor, Department of Genetics and Plant Breeding, Aristotelian University of Thessaloniki, P.O. Box 261, 540 06 Thessaloniki, Greece

S.K Vasal — Distinguished Scientist and Liaison Officer, CIMMYT Asian Regional Maize Program, P.O. Box 9-188, Bangkok 10900, Thailand

P.K. Verma — Proagro Seed Company Ltd., B-1-39 Toli Chowki, Hyderabad AP 500 008, India

S.S. Virmani — Plant Breeder cum Deputy Head, Plant Breeding, Genetics and Biochemistry Division, International Rice Research Institute, P.O. Box 933, Manila, Philippines

S.J. Wall — Senior Research Associtate, Pioneer Hi-Bred International P.O. Box 1004, 7300 NW, 62nd Ave., Johnston, IA 50131 1004

T.C. Wehner — Professor, Department of Horticultural Science, North Carolina State University, Box 7509, Raleigh, NC 27695-7609

M.W. Witt — Kansas Agric. Exp. Stn. Eminence Rt., Garden City Branch Garden City, KS 67846

W.L Woodman — Research Associate, Department of Agronomy, Iowa State University, Ames, IA 50011-1010

F. Zavala-Garcia — Facultad de Agronomia U.A.N.L., Apartado Postal #358, 66450 San Nicolas de los Garza N.L., Ubicacion de la Facultad, Carretera Zuazua-Marin Km. 17, 66700 Marin N.L., México

Chapter 1

Food Security and the Role of Agricultural Research

T. Reeves, P. Pinstrup-Andersen, and R. Pandya-Lorch

INTRODUCTION

Despite impressive food production growth in recent decades, the world is not food secure. Even if available food energy were evenly distributed within each country—which it is not—33 countries would not be able to assure sufficient food energy (2200 calories per person per day) for their populations.

More than 800 million people in the developing world, i.e., 20% of the population, are food insecure, more than 180 million preschool children are malnourished, and many hundreds of millions of people suffer from diseases of hunger and malnutrition.

With two-thirds of the developing world's undernourished, Asia remains the main area of concern. Another emerging area of concern is sub-Saharan Africa, where food security is rapidly deteriorating. The number of undernourished people in that region almost doubled in two decades from 94 million in 1969–1971 to 175 million in 1989–1990, and the proportion of the population that is undernourished rose from 35 to 37%. Between 1988–1990 and 2010, the number of undernourished people is projected to increase by 70% to 296 million, 32% of the region's population. By 2010, almost one-half of the developing world's undernourished will be located in sub-Saharan Africa, up from 10% in 1969–1971.

Hunger is a consequence of poverty. An estimated 1.3 billion people live in households that earn a dollar a day or less per person. About 50% of these absolute poor people live in South Asia, 19% in sub-Saharan Africa, 15% in East Asia, and 10% in Latin America and the Caribbean. Almost one-half of the population of South Asia and sub-Saharan Africa, and one-third of the Middle East and North Africa, live in poverty.

Poverty in the developing world is not expected to diminish much in the near future. The total number of poor people is projected to remain around 1.3 billion in 2000, although regional shifts in the distribution of total poverty are anticipated. Sub-Saharan Africa will increasingly become a new locus of poverty: the number of poor is expected to increase by 40% between 1990 and 2000, and the region's share of the developing world's poor is expected to increase from 19% to 27% during this period. South Asia will continue to be home to one-half the world's poor; the number of poor in that region is expected to decline by only 10% between 1990 and 2000.

Food production growth in recent decades has been impressive. During the period 1961 to 93, cereal production worldwide more than doubled from 877 million tons to 1894 million tons; in developing countries it almost tripled from 396 million tons to 1089 million tons. Since 1950, grain production per person has in-

Copyright © 1999 ASA-CSSA-SSSA, 677 South Segoe Road, Madison, WI 53711, USA. *The Genetics and Exploitation of Heterosis in Crops.*

1

creased about 100 kg per person worldwide and about 80 kg per person in developing countries as a group.

Of note on the food production front is the role of yield increases, which have been the source of 92% of the increased cereal production in the developing world between 1961 and 1990; area expansion contributed only 8%. While cultivated area is still increasing in most developing countries, it is doing so at a low and declining rate. Yield trends in developing countries climbed steadily upward for the three major cereals of rice (*Oryza sativa* L.), maize (*Zea mays* L.), and wheat (*Triticum aestivum* L.) between the 1960s and late 1980s. Yield increases were notably high in Asia: during 1961 to 91 rice yields doubled from 1.7 to 3.6 tons ha^{-1}, wheat yields increased from 0.7 to 2.5 tons ha^{-1}, and maize yields almost tripled from 1.2 to 3.4 tons ha^{-1}.

Yield growth rates in some areas are stagnating and, in a few cases, falling. A slowdown in the rate of increase of yields of major cereals raises concern since increased yields will have to be the source of increased food production in the future. Most cultivable land in Asia, North Africa, and Central America has already been brought under cultivation, and physical and technological constraints as well as environmental considerations are likely to restrain large-scale conversion of potentially cultivable land in sub-Saharan Africa and South America. The option of area expansion as a source of food production increases is rapidly disappearing, and even Africa will have to rely mostly on increased yields to expand food production.

Another cause of concern on the food production front is the leveling off during the 1980s and early 1990s of grain production per person for the world and for the developing countries as a group, after steady increases during the 1950s, 1960s, and 1970s. Since mid-1980, global grain production per person has decreased and grain production for developing countries as a group has been constant. If corrective actions are not taken soon, this trend could turn downward, with potentially adverse repercussions not just because the additional population needs adequate food, but because factors in addition to population growth are pushing up demand for grain. While future demand for grain for direct consumption in developing countries is expected to grow at a rate only slightly above population growth, the expected growth rate in world feedgrain demand is more than twice the expected population growth rate. Once incomes increase beyond a certain level, demand for feedgrain increases rapidly; most developing countries have incomes still below the level where feedgrain use increases rapidly.

THE ROLE OF AGRICULTURAL RESEARCH

Existing technology and knowledge will not permit the necessary expansions in food production to meet needs. Low-income developing countries are grossly under-investing in agricultural research compared with industrialized countries, even though agriculture accounts for a much larger share of their employment and incomes. Their public sector expenditures on agricultural research are typically <0.5% of agricultural gross domestic product, compared with about 2% in higher-income developing countries and 2 to 5% in industrialized countries.

Investment in agricultural research must be accelerated if developing countries are to assure future food security for their citizens at reasonable prices and without irreversible degradation of the natural resource base. Accelerated investment in agricultural research is particularly important and urgent for low-income developing countries, partly because these countries will not achieve reasonable economic growth, poverty alleviation, and improvements in food security without productivity increases in agriculture, and partly because so little research is currently undertaken in these countries. As mentioned above, the negative correlation between investment in agricultural research and the income level of the country is very strong. Poor countries, which depend the most on productivity increases in agriculture, grossly under-invest in agricultural research.

Agricultural research has successfully developed yield-enhancing technology for the majority of crops grown in temperate zones and for several crops grown in the tropics. The dramatic impact of agricultural research and modern technology on wheat and rice yields in Asia and Latin America since the mid-1960s is well known. Less dramatic but significant yield gains have been obtained from research and technological change in other crops, particularly maize.

Large yield gains currently being obtained in experimental varieties of many crops offer great promise for future yield and production increases at the farm level. Many other research results—tolerance or resistance to adverse production factors such as pests and drought, biological and integrated pest control, and improved varieties and hybrids for agroecological zones with less than optimal production conditions—will reduce risks and uncertainty and enhance sustainability in production through better management of natural resources and reduced environmental constraints.

Accelerated agricultural research aimed at more-favored areas will reduce pressures on fragile lands in less-favored areas. Future research for the former must pay much more attention to sustainability than in the past to avoid a continuation of extensive waterlogging, salination, and other forms of land degradation; however, a continuation of past low priority on less-favored agroecological zones is inappropriate and insufficient to achieve the goals of poverty alleviation, improved food security, and appropriate management of natural resources. More research resources must be dedicated to less-favored areas; i.e., areas with agricultural potential, fragile lands, poor rainfall, and high risks of environmental degradation. A large share of the poor and food insecure reside in these agroecological zones.

The low priority given to research to develop appropriate technology for less-favored agroecological zones in the past is a major reason for the current rapid degradation of natural resources, and high levels of population growth, poverty, and food insecurity. Much more research must be directed at the development of appropriate technology for these areas. Outmigration is not a feasible solution for these areas in the foreseeable future simply because of the large numbers of poor people who reside there and the lack of alternative opportunities elsewhere. Strengthening of agriculture and related nonagricultural rural enterprises is urgent and must receive high priority.

Following on the tremendous successes popularly referred to as the Green Revolution, the international agricultural research centers under the auspices of the Consultative Group on International Agricalutural Research (CGIAR) have recognized the importance and urgency of research that fosters sustainability in agricultural intensification through appropriate management of natural resources. Thus, management of natural resources, germplasm conservation, and germplasm enhancement are given high priority in current and future research by the centers.

Declining investment in agricultural research for developing countries since the mid-1980s by both developing-country governments and international foreign assistance agencies is inappropriate and must be reversed. While privatization of agricultural research should be encouraged, much of the agricultural research needed to achieve food security, reduce poverty, and avoid environmental degradation in developing countries is of a public goods nature and will not be undertaken by the private sector. Fortunately, while private rates of return may be insufficient to justify private-sector investment, expected high social rates of returns justify public investment. The major share of such investment should occur in the developing countries' own research institutions (National Agricultural Research Systems, NARS); there is an urgent need to strengthen these institutions to expand research and increase the probability of high pay-offs.

The centers under the auspices of the CGIAR have a well-defined role to play in support of the work by NARS, namely to undertake research of a public goods nature with large international externalities and to strengthen the research capacity of the NARS and networking among NARS, international centers, and research institutions in the industrialized nations. Research institutions in the industrialized nations have played an extremely important role by undertaking basic research required to support strategic, adaptive, and applied research by the international centers and the NARS and by providing training for developing-country researchers. Collaboration among developed-country research institutions, CGIAR centers, and NARS in developing countries is widespread but further strengthening is required to fully use the comparative advantages of each of the three groups for the ultimate benefit of the poor in developing countries.

All appropriate aspects of science, including molecular biology-based research, must be mobilized to solve poor people's problems. Almost all investment in genetic engineering and biotechnology for agriculture during the last 10 to 15 years has focused on problems of temperate-zone agriculture. Products have included herbicide resistance in cotton (*Gossypium hirsutum* L.), longer shelf life for perishable products such as tomatoes (*Lycopersicon esculentum* Mill.), and a variety of others important for industrialized nations. If we are serious about helping poor people, particularly poor women, and if we are serious about assuring sustainability in the use of natural resources, we must use all appropriate tools at our disposal to achieve these goals, including modern science. For example, modern science may help solve such problems as losses due to drought among small farmers in West Africa. Drought-tolerant varieties of maize for poor African farmers could be developed, along with crop varieties that possess tolerance or resistance to a number of other adverse conditions, including certain insects and pests.

While some argue that it is too risky to use genetic engineering to solve poor people's problems because we may be unaware of future side effects, we believe that it is unethical to withhold solutions to problems that cause thousands of children to die from hunger and malnutrition. Clearly, we must seek acceptable levels of biosafety before releasing products from modern science, but it is critical that the risks associated with the solutions be weighed against the ethics of not making every effort to solve the food and nutrition problems.

Effective partnerships between developing-country research systems, international research institutions, and private- and public-sector research institutions in industrialized countries should be forged to bring biotechnology to bear on the agricultural problems of developing countries. Incentives should be provided to the private sector to undertake biotechnology research focused on the problems of developing-country farmers.

Failure to expand agricultural research significantly in and for developing countries will make food security, poverty, and environmental goals elusive. Lack of foresight today will carry a very high cost for the future. As usual, the weak and powerless will carry the major burden, but just as we must all share the blame for inaction or inappropriate action, so will we all suffer the consequences.

ADDITIONAL INFORMATION

Anonymous. 1994. Alleviating poverty, intensifying agriculture, and effectively managing natural resources. Food Agric., and the Environ. Discussion Pap. 1. Int. Food Policy Res. Inst. Washington, DC.

International Food Policy Research Institute. 1995. A 2020 vision for food, agriculture, and the environment: The vision, challenge, and recommended action. Int. Food Policy Res. Inst. Washington, DC.

Pardey, P., J. Roseboom, and N. Beintema. 1994. Agricultural research in Africa: Three decades of development. Briefing Pap. 19. Int. Serv. for Nat. Agric. Res. The Hague, the Netherlands.

Pinstrup-Andersen, P., and R. Pandya-Lorch. 1995. Poverty, food security, and the environment: 2020 Brief 29. Int. Food Policy Res. Inst. Washington, DC.

Rosegrant, M.W., M. Agcaoili-Sombilla, and N. Perez. 1995. Global food supply, demand, trade to 2020: Projections and implications for policy and investment. Int. Food Policy Res. Inst., Washington, DC.

Scherr, S., and S. Yadav. 1996. Land degradation in the developing world: Implications for food, agriculture, and the environment to 2020. Food Agric., and the Environ. Discussion Pap. 14. Int. Food Policy Res. Inst., Washington, DC.

Chapter 2

Inbreeding and Outbreeding in the Development of a Modern Heterosis Concept

I. L. Goldman

HISTORY OF INBREEDING AND OUTBREEDING

Inbreeding and outbreeding, the mating of genetically like and unlike individuals, respectively, form the foundation of the modern heterosis concept. Historical treatments (East, 1908; East & Jones, 1919; Jones, 1918; Zirkle, 1952; Shull, 1952; Hayes, 1952; Stuber, 1994) have reviewed in detail the creation of this concept during the 19th and 20th centuries. Much of the discussion presented herein is therefore concerned principally with those who recognized and exploited the phenomenon to the betterment of agriculture, and in particular those who contributed to our knowledge of inbreeding and outbreeding with respect to hybrid crops.

Substantial evidence exists to support the notion that early agriculturists were aware of the significance of inbreeding and outbreeding in both plant and animal species. Perhaps the first significant example in this regard is the development of the mule. A mule results from a cross between the donkey (*Equus asinus*) and the horse (*E. caballus*). Mules exhibit significant heterosis for size, strength, and endurance; all of which were integral to the development of modern animal agriculture. The Sumerians were producing these crosses at least 3000 years before the Common Era, indicating that hybrid technology may be at least 5000 years old (Clutton-Brock, 1992). In fact, the Sumerians and others in the Middle East and Mediterranean regions were producing donkey × onager (*E. hemionus*) crosses and making donkey × zebra mules (*E. asinus* × *E. caballus*) for the tropics; thus demonstrating that an animal breeding program to exploit heterosis was in place many thousands of years ago (Clutton-Brock, 1992).

Anthropological and genetic evidence also suggest that native Americans may have practiced hybrid technology. Although the cultivation of maize (*Zea mays* L.) by native American peoples has been well documented, little scientific information on their methods of selection and breeding are known. In a remarkable article by G.N. Collins (1909), then Assistant Botanist at the United States Department of Agriculture's Bureau of Plant Industry, the suggestion is made that these early Americans purposefully exploited hybrid technology:

> "Among a number of primitive tribes where the cultivation of corn has reached a high state of development, the injurious effect of this close breeding appears to have been recognized, since they have methods of guarding against it. Thus the Indians in the region of Quezaltenango, in western Guatemala, and the Hopi Indians of Arizona make a regular practice of placing seeds of

more than one local variety in each hill, with the idea that larger yields can be obtained in this way" (Collins, 1909).

The close planting of diverse strains clearly indicates that a recognition of the positive benefits of heterosis was firmly in place many thousands of years ago in the Americas. In addition to this anecdotal comment, some research has been conducted to assess whether ancient maize from these regions fits a genetic pattern of modern day maize hybrids. Helentjaris (1988) described the results of experiments with 700-year-old maize remains obtained from several Anasazi sites in the southwestern USA. The identical restriction fragment length polymorphism (RFLP) patterns for each sample across sites and multiple restriction fragments detected in this study were suggestive of the kind of hybrid uniformity and genetic heterozygosity present in the RFLP patterns of modern day maize hybrids. Although these data do not prove the purposeful construction of maize hybrids by the Anasazi, they do raise the interesting issue that technology designed to exploit heterosis may have been in place for many years before the modern scientific èra would indicate. Archeologists and anthropologists have suggested that the Anasazi probably isolated certain maize populations or varieties for spiritual reasons; thus it is feasible that isolation practices may have led to or been formulated out of observations of heterosis in food crops such as maize.

Ideas about inbreeding and outbreeding have evolved dramatically during human evolution. Investigation of their foundations in human culture provides a glimpse into the power of these two reproductive modes and shows how they were closely connected to the development of religion and society. In the main, ideas about totemism, nature worship, and other forms of idolatry were designed to connect human reproductive behavior with those behaviors observed in nature. This in turn led to a variety of reproductive rites imposed by human populations on themselves and on the natural populations upon which they depended. Expressions of these rites may have included the bringing together of diverse plants and animals for mating rituals and the separation of various plant and animal populations for purification reasons. Epic works such as the Golden Bough (Frazer, 1941) described how early human societies modeled their reproductive behaviors after those observed in nature or those sought in agriculture; for the purposes of good harvests and other forms of positive spiritual and material consequences. Finally, Sigmund Freud and others recognized and wrote about the widespread acceptance of an incest taboo among many of the world's cultures, indicating that humans have known for many thousands of years about the consequences of inbreeding and outbreeding (Freud, 1912).

EARLY PLANT HYBRIDIZERS

Plant hybridizers in the 18th century were among the first to make significant scientific contributions to the development of a modern heterosis concept. Much of their work was concerned more with experimentation of the results of divergent crosses rather than practical plant improvement (Mayr, 1982). Cotton Mather was the first to note and describe the xenia effect; or the immediate effect of the pollen parent on the female seed parent. This casual observation made a significant contribution to pollination biology and was an important conceptual advance in plant hybridization. In 1776, Kolreuter became the first of the plant hybridizers to document in detail the results of his crosses, in this case in the genus *Nicotiana*, and to describe significant heterosis. Knight took these ideas one step further and, in 1799, suggested that the widespread existence of cross-pollination in nature was proof that nature 'intended' this to be the norm. In 1828, Wiegmann described the results of crosses leading to heterosis in the crucifers (Mayr, 1982). Gartner and Focke, in 1849 and 1881, respectively, detailed the results of their crosses, noting heterosis, and encouraging other scientists to think along the lines of enhanced plant growth through such hybridization (Mayr, 1982). Naudin and Mendel, separately in 1865, described not only the fundamental laws of heredity thus founding the science of genetics, but also detailed the results of their crosses in doing so. These workers commented on the luxurious growth observed in hybrids and, because of the eventual influential nature of their work, contributed additional observations to the development of a heterosis concept.

We now turn to the debt we owe Charles Darwin, the contributor of so many important ideas in biology, for his pioneering work in the area of plant hybridization. Darwin drew primary inspiration for natural selection from plant and animal breeders and called the ideas formulated from his visits to the feather clubs of London "the most beautiful part of my theory" (Desmond & Moore, 1991). The continual progress exhibited by the impressive strains the breeders developed during this period pushed Darwin to experiment with inbreeding and outbreeding in domestic plants, where he focused on the results of various crosses. Upon completion of a typically painstaking analysis of reproductive modes in plants and concluded:

> "Nature thus tells us, in the most emphatic manner,
> that she abhors perpetual self-fertilization" (Darwin,
> 1862).

Darwin detailed the results of 37 crosses including maize in which he observed increased height in 24 crosses. He commented on the decrease in height observed in the self-pollinated plants and discussed many natural mechanisms by which plants avoid inbreeding. These observations became critical in developing an understanding of the significance of inbreeding and outbreeding in nature. Interestingly, Darwin noticed that the deleterious effects of inbreeding could be reversed following the crossing of inbred strains. In the cases he tested, the performance of the first cross-pollinated generation was quite vigorous. As such, he suggested that the effects of inbreeding were essentially reversible upon intermating. This hints quite closely at a modern concept of the relationship between inbreeding and outbreeding with respect to heterosis, however Darwin's ideas did not draw a clear path between the effects of inbreeding and those of outbreeding.

EARLY INBREEDING AND OUTBREEDING THEORY

Perhaps Darwin's other major contribution in this area was to serve as an inspiration to the Harvard botanist Asa Gray. Gray was an important figure in the biological sciences at the time in the USA, and he was a correspondent of Darwin. The two had visited in England several times and shared ideas on inbreeding and outbreeding, and this correspondence may have had a very significant impact on the paths taken by Gray in his research. Clearly, correspondence between Darwin and Gray established Darwin's prior claim on the details of natural selection prior to the joint publication of Darwin and Wallace in 1859 (Darwin & Wallace, 1859). Gray in turn served as the research mentor for William James Beal (Fig. 1), who earned his degree at Harvard in 1865 and became one of the first practical contributors to the modern heterosis concept (Wallace & Brown, 1988; Crabb, 1947). Beal's career at Michigan Agricultural College (later Michigan State University) focused on varietal crossing and pollination control in maize; two practices with great influence on early plant hybridizers. Beal then served as the inspiration and mentor of Perry Holden and Eugene Davenport, who made great strides in early inbreeding theory at the University of Illinois at Urbana-Champaign. This latter group also included Cyril Hopkins, the chemist, who began the Illinois Long Term Selection experiment in 1896, and later E.M. East, who was to become perhaps the most important and influential figure in development of modern scientific plant breeding in the USA.

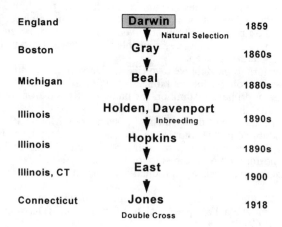

England	**Darwin**	1859
	▼ Natural Selection	
Boston	Gray	1860s
	▼	
Michigan	Beal	1880s
	▼	
Illinois	Holden, Davenport	1890s
	▼ Inbreeding	
Illinois	Hopkins	1890s
	▼	
Illinois, CT	East	1900
	▼	
Connecticut	Jones	1918
	Double Cross	

Fig. 2–1. Academic-influence pedigree chart from Charles Darwin to D.F. Jones showing a flow of ideas about inbreeding and outbreeding that led to a modern heterosis concept. Specific contributions of certain individuals are noted in smaller type under their names. The central column represents the flow of ideas from worker to worker. The left column represents the country or state where the ideas were formulated and the right column represents the year or general era in which the ideas were first discussed.

Beal's contribution to the development of a heterosis concept was an emphasis on pollination control as a means of maize breeding. The idea here was to minimize self-pollination, for its deleterious effects had been carefully noted by many workers beginning with Darwin. Early maize breeding methods, heavily influenced by the work of Beal, thus focused on ways to prevent inbreeding through pollination control rather than exploiting specific matings in the development of cross-bred strains. During this period, Beal spearheaded a widespread effort to examine the deleterious effects of inbreeding. Archibald Shamel and Eugene Funk both began inbreeding programs at the University of Illinois, only to stop them in the early part of the century due to the widespread feeling enough damage had been done by inbreeding already. At the time, the words of Davenport may have summed up the sentiment on inbreeding:

> "The effects of inbreeding appear both pronounced and disastrous; the second generation from inbred seed being less than two-thirds normal size and nearly barren . . . but the second planting from this seed when closely selected after the same plan left almost a full stand, which shows that corn may be brought much nearer a constant type than has ever yet been done" (Davenport; as cited in Fitzgerald, 1990).

A CULTURE OF INBREEDING: MAIZE BREEDING AT THE
UNIVERSITY OF ILLINOIS

Beal was active in training students in maize breeding. Both Eugene Davenport (received Master of Science degree in 1884) and Perry Holden (received Master of Science degree in 1895) learned pollination control theory and techniques from Beal at Michigan State. Following the passage of the Hatch Act establishing agricultural experiment stations at land-grant universities in 1887, much practical maize breeding work began to take place at these institutions. The work conducted by Beal and his students was repeated and confirmed at the University of Illinois Agricultural Experiment Station by G. Morrow and F.D. Gardner. This work demonstrated the positive benefits of hybridization in terms of improved yield (Morrow & Gardner, 1893, 1894). These workers not only emphasized the importance of cross pollination but recommended the alternate planting of varieties with detasseling in order to produce hybrid seed. A section in their 1894 manuscript entitled 'Results from Cross-Bred Corn' (Morrow & Gardner, 1894) is perhaps the first outline of the hybrid maize breeding method still in use today (Troyer, 1996, personal communication). Eugene Davenport was appointed Dean of the College of Agriculture at the University of Illinois and he later appointed Perry Holden as a Professor in the College in 1896. In the same year, Cyril Hopkins initiated the Illinois Long Term Selection experiment, the longest-running crop selection experiment in modern plant breeding history. East was hired by Hopkins to assist with the Long Term Selection project. East had been a student in a botany course with Professor Charles Hottes, who has just returned from European study with the rediscoverers of Mendel's laws. Hottes was therefore attuned to the hereditary foundation by which inbred lines might be developed from segregating populations. It has been suggested that East may have been influenced to think about line uniformity and other such aspects of the genetic foundations of inbreeding through his contact with Hottes. Hottes remarked:

> "I liked East very much. He was a good student. He didn't have to be driven, although he was a rather retiring pleasant sort of fellow who as a youth gave rather little indication of the fine qualities of aggressiveness which he developed later. He always constructed his sentences with great care, speaking in a rather thin high voice. He was many times studious to the point of being preoccupied" (Hottes; as cited in Fitzgerald, 1990).

The Illinois Long Term Selection Experiment was influenced primarily by work conducted in France by early breeders of the fodder beet (*Beta vulgaris* L.; Hopkins, 1899). When the Napoleonic Wars left France without an inexpensive source of sugar, Napoleon offered a prize for a new European source of sugar (Troyer, 1996). The discovery by Andreas Marggraf in 1747 that fodder beets produced sucrose identical to that of sugar cane led to late 18th century mass selection efforts by Franz Karl Achard to increase sugar concentration in this crop (Duvick, 1996). This work was followed by the landmark pedigree breeding program of Louis de Vilmorin, grandson of the founder of the Vilmorin Seed Company. Vilmorin's work focused on progeny testing and thus measured the breeding potential of parental lines by assessing the performance of their crossed progeny. This work necessarily involves accurate pedigree records, and thus both pedigree breeding and progeny testing evolved in 19th century France, supplanting the previous mass selection efforts, in the context of improving a quantitative trait in *Beta vulgaris*. Achard and Vilmorin were therefore important influences on Hopkins, and he based his modification of chemical composition of the maize kernel on these principles.

The change in emphasis at the University of Illinois away from inbreeding research and on to mass selection and progeny testing may have delayed the development of a modern heterosis concept. Hopkins said to East:

> "We know what inbreeding does and I do not propose
> to spend people's money to learn how to reduce corn
> yields" (Hopkins, as cited in Fitzgerald, 1990).

East, who went on to uncover some of the key aspects of the modern heterosis concept, was clearly interested in pursuing inbreeding research at Illinois. It is possible that the shift in emphasis away from this work at Illinois led to East's departure for the Connecticut Agricultural Experiment Station and delayed some of his early inbreeding work. Despite the shift in breeding practices, East's many useful discussions at the University of Illinois with office mates and like-minded geneticists such as H.H. Love (Troyer, 1996) served to sharpen his resolve to try inbreeding. A significant portion of his inbreeding work with maize, which led to key insights into heterosis, was conducted while East was at Connecticut. During this period, intensive research on maize inbreeding was also being carried out by another geneticist on the east coast of the USA, G.H. Shull. Both East and Shull deserve recognition for the formulation of the modern heterosis concept; however, credit is often given to Shull, whose story is described below.

COMPOSITION OF A FIELD OF MAIZE

Turning to a second academic-influence pedigree (Fig. 2–2), I introduce the key progression of thoughts that led to the development of a modern heterosis concept. Both Buffon and Lamarck in France offered key insights into the workings of evolution and natural selection.

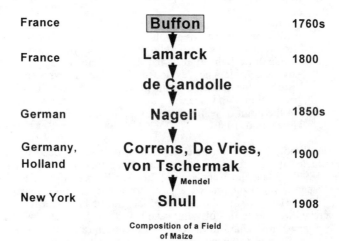

Fig. 2–2. Academic-influence pedigree chart from Buffon to Shull showing a flow of ideas about inbreeding and outbreeding that led to a modern heterosis concept. The central column represents the flow of ideas from worker to worker. Specific contributions of certain individuals are noted in smaller type under their names. The left column represents the country or state where the ideas were formulated and the right column represents the year or general era in which the ideas were first discussed.

These ideas influenced early plant hybridizers such as de Candolle and Na-
geli; both of whom made important contributions to early ideas about the results of
hybridization (Mayr, 1982). These workers in turn served as the inspiration for the
rediscoverers of Mendel: Correns, de Vries, and Von Tschermak. In particular, the
important mutation theories developed by De Vries were of interest to the young
George Harrison Shull, a brilliant young biologist who earned his graduate degree
at the University of Chicago. Shull wished to test some of de Vries' ideas about the
generation of variation by mutation, however his interest was primarily directed at
basic genetic mechanisms and not crop improvement. Interestingly, De Vries and
Pfeffer, both influenced by Sachs and before him, Purkinje (Fig. 2–3) influenced
Johanssen, who made a significant contribution to inbreeding theory and quantitative inheritance by developing the pure-line theory (Johanssen, 1903). Upon assuming his first position in 1904 at Cold Spring Harbor Laboratories in Long Island, New York, Shull began inbreeding a number of crop plants including maize.
The next four years were consumed with inbreeding and outbreeding work in a variety of species. Ultimately, Shull developed a perspective on heterosis that he outlined in a 1908 publication entitled "Composition of a Field of Maize." This article
is considered to be the landmark development in early heterotic theory, as it clearly
states that a maize variety is a complex mixture of genotypes. Shull commented that
since each plant is, in a sense, isolated by inbreeding; each plant is of an essentially
different genotype. He observed and measured a reduction in vigor due to the segregation of these different types into their respective homozygous classes and
showed how the F_1 yield from crosses of these types exceeded the parental varieties
from which they came (Shull, 1908). The significance of this landmark article is
that Shull was able to bring together the key aspects of inbreeding and outbreeding
theory and show how they are related in a coherent heterosis concept (Shull, 1952).

East and Shull are often given credit for developing key aspects of the modern heterosis concept. Although both men clearly contributed crucial ideas to this
field, Shull is thought of as having claim to the first publication of the idea, based
on the following evidence. East had discussed the dangers of inbreeding as late as

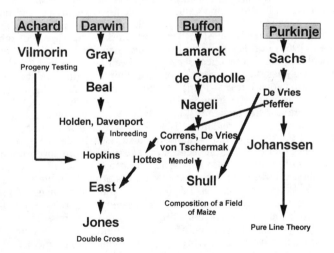

Fig. 2–3. Combined academic-influence pedigree chart showing influences of four
streams of thought on the development of a modern heterosis concept.

the 1907 Connecticut Agricultural Experiment Station Bulletin 158; however, he was heavily influenced by Shull's presentation at the American Breeder's Association Meeting in Chicago in early 1908. In the 1908 Connecticut Agricultural Experiment Station Bulletin, East stated:

> "A recent paper by Dr. G.H. Shull has given, I believe, the correct interpretation of this vexed question. His idea, although clearly and reasonably developed, was supported by few data" (East, 1908).

Finally, in early 1908, East wrote to Shull and said:

> "Since studying your paper, I agree entirely with your conclusion, and wonder why I have been so stupid not to see the fact myself" (Shull, 1952).

East also is to be given a great deal of credit for his extensive training of leaders in the field of plant breeding. Many of the students trained by East at the Bussey Institute at Harvard went on to become influential scientists and plant breeders in the USA and abroad. For example, East trained D.F. Jones, popularizer of hybrid maize by suggesting the double cross in 1918; R.A. Emerson, mentor for many of plant breeding's early pioneers at Cornell University; E.R. Sears, pioneer of chromosome manipulations in plant breeding at Missouri; C.M. Rick, developer of one of the world's most remarkable collections of crop genetic material at the University of California, Davis; R.A. Brink and H.K. Hayes, pioneer teachers and researchers at the University of Wisconsin and University of Minnesota, respectively; and P.C. Mangelsdorf of Texas A&M University and Harvard University, crop evolutionist (Troyer, 1996).

FURTHER VALUE OF INBREEDING

Although Shull and East shed much light on the significance of inbreeding and outbreeding in heterosis, the value of inbreeding extends beyond heterosis to serve as a cornerstone for the success of modern scientific plant breeding. In Appendix C of the second edition of his landmark work, "The Theory of Inbreeding," R.A. Fisher (1965) outlines the practical value of inbreeding to plant and animal improvement. Noting that practical success has been obtained by a 'cycle of operations' involving (i) choice of parental stock; (ii) inbreeding to produce homozygous or nearly homozygous lines; and (iii) crossing of chosen lines; Fisher commented that no change in genotype frequency in a population would occur if foundation individuals were chosen randomly, inbreeding took place without selection, and chosen lines were crossed randomly. Therefore, selection must be taking place at one or more of the stages of the cycle. Fisher dismisses the choice of parental stock as a significant point of selection because the average performance of domesticated varieties does not change very rapidly. The kind of gains made in one cycle of this modern method typically far exceeded any efforts at comparable selection within parental stock. Likewise, Fisher also dismisses the inbreeding stage as a significant point of selection because of arguments that the mutation frequency will come to equilibrium with the loss of deleterious recessives in the population during the inbreeding process. Fisher argues that when many homozygous lines are present, selection is most effective because (i) selection is practiced upon the actual genotype rather than its ancestor; and (ii) seed will be available to test promising crosses over environments with great precision using modern experimental designs; and (iii) any particular advantage or characteristic of the hybrid remains a permanent quality of the hybrid; thus additional work can have a cumulative effect. In this view, a large quantity of inbred lines would serve to increase gain from selection because these lines would well represent the entire germplasm pool in the form of readily-testable and repeatable parent stock.

ACCIDENTS AND AGENDAS IN MODERN HETEROSIS

Some have questioned whether the development of a modern heterosis concept would have come about so rapidly without maize as a model organism. Indeed, maize is thought to be central to the development of heterotic theory because it lent itself so well to experiments involving inbreeding and outbreeding. Simmonds (1979) suggested three reasons why maize was so suitable: (i) The development of hybrid maize came along at a time when new methods of maize breeding were needed. In fact, the early workers at Illinois were charged with the development of new maize breeding methods, and thus there was likely support in both the scientific and agricultural communities for such investigation. (ii) The development of agriculture in the USA was closely tied to the development of maize as a staple crop. The sheer economic weight of maize during the early part of the 20th century was responsible for fueling the development of new and productive varieties. (iii) Finally, and perhaps most important, it has been suggested that the relatively simple emasculation techniques practiced in maize breeding and the biological accident of the simple monoecious reproductive structure of the maize plant allowed for easy inbreeding and outbreeding. This in no small way helped to make maize the model organism for the study of these two key phenomena.

In addition to improved yield performance, many other benefits have been derived from hybrids and the exploitation of heterosis. One of the most often overlooked benefits is uniformity; an element which has certainly allowed for rapid expansion of production in many crop plants such as the vegetables. This characteristic of hybrids has enabled the production of more uniform vegetable crops for fresh market production, contributing to enhanced consumer appeal and greater market value. Furthermore, the benefits of uniformity with respect to maturity have been associated with greater efficiencies during harvest. The development of mechanical harvesting technology for major grain crops such as maize in the 1920's coincided perfectly with the development and widespread use of hybrid maize. These two events are connected in that the added benefit of uniformity obtained from hybrids allowed for widespread dissemination of mechanical harvesting technologies. Additional benefits may include stress tolerance and pest resistance and other performance characteristics. Severe droughts in the Cornbelt during 1934 and 1936 resulted in poor maize crops; however hybrids often out-performed their open-pollinated counterparts under these conditions.

In his landmark 1908 article, Shull discussed the non-renewability of hybrids, and East and Jones (1919) had commented that the advancement of hybrid technology would provide less incentive for individual breeders to improve open-pollinated strains. The dramatic shift toward hybrid technology revolutionized many sectors of the agricultural economy. One sector that was affected greatly was the seed industry, because the adoption of hybrids meant being able to sell seed to farmers year after year. Berlan and Lewontin (1986) have criticized the move toward hybrids as merely an expression of class interest. They further suggest that improvement of open-pollinated varieties through careful selection should have resulted in agronomic performance on a par with hybrid varieties. Kloppenburg (1988) has even called hybrid maize agriculture's Manhattan project. The intriguing perspective that genetic gain in open-pollinated populations might have rivaled those of hybrids has, in some cases, been supported experimentally (see Duvick, 1977, 1992, 1996; Hallauer & Miranda, 1981). Despite this contention, others have suggested that many of these arguments against hybrid technology reveal their political agenda by expressing surprise that clever individuals learned to profit from agriculture (Goodman, 1989). One of the major agricultural changes to emerge from hybrid technology, therefore, has been a flow of germplasm and control of germplasm from the public to the private sector; a situation that has aroused a great deal of interest in both the scientific community and the general population.

N.W. Simmonds (1979) remarked about certain aspects of plant improvement that "exact knowledge is inessential-otherwise plant breeding would be impossible . . . (and) there are situations in which it is perfectly reasonable to disregard formal genetics as such and talk in terms of a workable statistical abstraction.." The development of inbreeding and outbreeding methods in exploiting the poorly-understood phenomenon of heterosis serves as an excellent historical example of a pragmatic merger between science and technology in the service of agriculture and humankind.

The pioneering work of E.M. East and G.H. Shull is built on foundations only several academic generations removed from the pioneers Darwin and Buffon. Today, we work only several generations removed from E.M. East and G.H. Shull, hopeful that our insight may be as keen and our sense of purpose as positive as the founders of the modern heterosis concept.

ACKNOWLEDGMENTS

I am indebted to my colleague William F. Tracy for many helpful discussions and for bringing to my attention the article by Collins. I am also grateful to Dr. Forest Troyer, Cargill Hybrid Seeds, for many important contributions on the history of maize breeding at the University of Illinois and for helpful discussions during the preparation of this manuscript. A portion of this manuscript was presented at the Heterosis Workshop in Indianapolis, IN, as part of the joint ASHS–CSSA Plant Breeding Symposium, November, 1996.

REFERENCES

Berlan, J.P, and R. Lewontin. 1986. The political economy of hybrid corn. Monthly Rev. 38:35–47.

Clutton-Brock, J. 1992. Horse power. Harvard Univ. Press, Cambridge, MA.

Collins, G.N. 1909. The importance of broad breeding in corn. USDA Bureau of Plant Industry Bull. 141. Part 4: 33–42.

Crabb, A.R. 1947. The hybrid-corn makers: Prophets of plenty. Rutgers Univ. Press, New Brunswick, NJ.

Darwin, C., and A. Wallace. 1859. On the tendency of species to form varieties; and on the perpetuation of varieties and species by natural means of selection. J. Linn. Soc. London. (Zool.) 3:45–62.

Darwin, C. 1862. On the various contrivances by which British and foreign orchids are fertilised by insects, and on the good effects of intercrossing. Murray, London.

Desmond, A., and J. Moore. 1991. Darwin: The life of a tormented evolutionist. Norton. London.

Duvick, D. 1977. Genetic rates of gain in hybrid maize yields during the past 40 years. Maydica. 22:187–196.

Duvick, D. 1992. Genetic contributions to advances in yield of U.S. maize. Maydica. 37:69–79.

Duvick, D. 1996. Plant breeding, an evolutionary concept. Crop Sci. 36:539–548.

East, E.M. 1908. Inbreeding in corn. Rep. Conn. Agric. Exp. Stn. for 1907. p. 419–428.

East, E.M., and D.F. Jones. 1919. Inbreeding and outbreeding. J.B. Lippincott Co., Philadelphia, PA.

Fisher, R.A. 1965. The theory of inbreeding. 2nd ed. Academic Press, London.

Fitzgerald, D. 1990. The business of breeding: hybrid corn in Illinois, 1890–1940. Cornell Univ. Press, Ithaca, NY.

Frazer, J.G. 1941. The golden bough. Macmillan, New York.

Freud, S. 1912. Totem and taboo. In P. Gay (ed.) The Freud reader. 1989. W.W. Norton and Co., New York.

Goodman, M.M. 1989. Diversity. 33–35.

Hallauer, A.R., and J.B. Miranda. 1981. Quantitative genetics in maize breeding. Iowa State Univ. Press, Ames.

Hayes, H.K. 1952. Development of the heterosis concept. p. 49–65. In J.W. Gowen (ed.) Heterosis. Iowa State College Press, Ames.

Helentjaris, T. 1988. Does RFLP analysis of ancient Anasazi samples suggest that they utilized hybrid maize? Maize Newsletter. 62:104–105.

Hopkins, C.G. 1899. Improvement in the chemical composition of the corn kernel. Illinois Agric. Exp. Stn. Bull. 55:205–240.

Johannsen, W. 1903. Ueber Erblichkeit in Populationen und in reinen Linien. Gustav Fischer, Jena. Translated version (Heredity in populations and pure lines) appears in Classic papers in genetics. 1959. J.A. Peters, ed. Prentice-Hall, Englewood Cliffs, NJ.

Jones, D.F. 1918. The effects of inbreeding and crossbreeding upon development. Conn. Agric. Expt. Sta. Bull. 207:419–428.

Kloppenburg, J.R., Jr. 1988. First the seed. Cambridge University Press. New York.

Mayr, E. 1982. The growth of biological thought. Belknap Press, Cambridge, MA.

Morrow, G.E., and F.D. Gardner. 1893. Field experiments with corn, 1892. Univ. of Illinois Agric. Exp. Stn. Bull. 25:173–203.

Morrow, G.E., and F.D. Gardner. 1894. Field experiments with corn, 1893. Univ. of Illinois Agric. Exp. Stn. Bull. 31:333–360.

Shull, G.F. 1908. The composition of a field of maize. Rep. Am. Breed. Assoc. 5:51–9.

Shull, G.F. 1952. Beginnings of the heterosis concept. p. 14–48. In J.W Gowen (ed.) Heterosis. Iowa State College Press, Ames.

Simmonds, N.W. 1979. Principles of crop improvement. Longman, London.

Stuber, C.W. 1994. Heterosis in plant breeding. Plant Breeding Reviews. John Wiley and Sons, New York, NY.

Troyer, F. 1996. Early Illini corn breeders: Their quest for quality and quantity. American Seed Trade Assoc. Hybrid Corn-Sorghum Res. Conf. 50:56–67.

Wallace, H.A., and W.L. Brown. 1988. Corn and its early fathers. Revised ed. Iowa State Univ. Press, Ames.

Zirkle, C. 1952. Early ideas on inbreeding and crossbreeding. p. 1–13. *In* J.W. Gowen (ed.) Heterosis. Iowa State College Press, Ames.

Chapter 3

Heterosis: Feeding People and Protecting Natural Resources

D. N. Duvick

INTRODUCTION

Hybrid vigor, or heterosis, usually refers to the increase in size or rate of growth of offspring over parents; for example, hybrid vigor in crop plants can be observed as in increase in yield of grain, or reduction in number of days to flower. Heterosis in plants has been used on large scale for the past 75 years, as carefully selected and reproduced hybrid cultivars. Field crops such as maize (*Zea mays* L.) sorghum [*Sorghum bicolor* (L.) Moench.] and sunflower (*Helianthus annus* L.) are produced as hybrids in all of the industrialized world; they also are grown as hybrids in increasing amounts in the developing world. Hybrid rice (*Oryza sativa* L.) is grown extensively in China, and increasingly in India (Virmani, 1994). Many commercial vegetable and flower crops are grown almost entirely as hybrids. Heterosis is credited for large increases in production per unit area, thus sparing large amounts of land for other uses such as environmentally benign nature preserves.

Examination of the historical record suggests that the major gift of heterosis was its stimulation of interest in the entire system of breeding and use of hybrid crops, rather than a simple exploitation of hybrid vigor per se. Development and use of hybrid seeds can enhance crop yields and performance in ways that are different from and not necessarily dependent on heterosis by itself. This essay will examine the historical record on use of hybrids to exploit heterosis in some of the major field crops, discuss the present use of heterosis and hybrids, and attempt to predict how heterosis and hybrids may be used in the future. The purpose of this analysis will be to show how hybrids and heterosis can help to supply food for burgeoning populations and also help to improve environmental health of the global food production system.

Hybrid Maize

Hybrid maize was first bred and produced in the USA. The first hybrids, in the 1920s, yielded about 15% more than the better open pollinated varieties (OPVs; Iowa State Dept. of Agric., 1934). Starting in the mid-1930s, the area planted to hybrid maize began to increase rapidly and after 15 years 95% of the land in the USA Corn Belt was planted to hybrid maize (USDA, 1953). Concomitantly with the rise in plantings of hybrid maize, on-farm maize yields began to climb. They continued to climb after USA maize plantings were essentially 100% hybrid (about 1965) and they are still rising (Fig. 3–1). Several studies have shown that about 40 to 50% of USA maize yield gain since the 1930s is due to changes in management

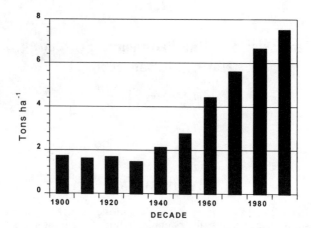

Fig. 3-1. USA maize yields 1900 to 1996, decade means. Hybrid maize was introduced in about 1930 and was used on 100% of USA maize plantings by
about 1965. (USDA NASS, Agricultural Statistics)

such as increases in nitrogen fertilizer and higher plant densities, while the other 50
to 60% is due to changes in maize genotype (Duvick, 1992, Russell, 1991).

If one assumes that present-day maize hybrids will yield on average 15%
more than OPVs, as they did in the USA in the 1920s, one can conclude that use of
hybrids to produce a given quantity of grain requires only 85% as much land as if
the production were made with OPVs (Table 3–1). One can say, further, that present-day maize yields, as a worldwide average, are about 10% higher than they
would be if all maize were open pollinated. Assumptions are as follows: 65% of
maize area, worldwide, is planted to hybrids; 65% of the maize area times the hybrid yield advantage of 15% gives an estimated worldwide average yield increase
of 10%.

Similar assumptions can be used to support claims that (i) the hybrid yield
advantage can be credited for 55 million of the current total world maize production of about 550 million tons, and (ii) use of hybrids has spared cultivation of
about 13 million hectares of land for maize production.

Table 3–1. Estimates of the annual global contributions of hybridization to production of maize, sorghum, sunflower, and rice.

Crop	Area planted to hybrids†	Hybrid yield ad-vantage‡	Annual added yield	Annual added yield	Annual land savings
	%	%	%	$t \times 10^6$	$ha \times 10^6$
Maize	65	15	10	55	13
Sorghum	48	40	19	13	9
Sunflower	60	50	30	7	6
Rice	12	30	4	15	6

† Production data from USDA National Agricultural Statistics Service.
‡ Estimated gain in yield of hybrids over superior open pollinated varieties at time
of hybrid introduction.

These exercises in arithmetic indicate that use of hybrids no doubt saves land by increasing yield per unit area, but one should not give large credence to the precision of these numbers. For example, if as much effort had been put into improvement of OPVs as has been devoted to hybrid improvement over the years, the gap between the best hybrids and the best OPVs might be < 15%, or it might be > 15%, depending on relative effectiveness of breeding methods to produce the two kinds of product. Some authors say that OPVs would be superior to hybrids if as much effort had been expended on OPVs as went into hybrid development (Lewontin & Berlan, 1990), but their assumption is not backed up by data. No experiments have been designed and conducted to test the theory that OPVs could be improved at the same rate as hybrids (or at a higher rate), if equal effort were expended on each kind of breeding.

Minimal data indicate, however, that some improved populations (they are more or less equivalent to elite OPVs) were improved in yield at rates equivalent to those achieved in a time series of commercial hybrids (Duvick, 1992). The populations, however, were not subjected to selection for the full panoply of traits that were improved in the hybrids, thus stronger selection could be made for yield in the populations than in the hybrids, other things being equal. The question remains unanswered, as to whether or not OPVs can be improved in all needed traits at the same rate as hybrids, given equal effort to both kinds of breeding.

More to the point, any advantage of hybrids over OPVs is not necessarily due to an increase in heterosis per se. OPVs are themselves a collection of hybrid plants, each one exhibiting heterosis and other yield traits to a greater or lesser degree. The yield of the OPV is an average of the yield of all its hybrid combinations. The highest-yielding hybrid genotype in an OPV by definition will outyield the OPV as a whole.

The goal of hybrid breeding is to identify and then reliably reproduce superior hybrid genotypes. Virtually all commercial maize hybrids are made from crosses of inbred lines. The inbreds are low yielding but their hybrids exhibit a high degree of heterosis for yield as well as for other traits such as maturity and plant height. Maize hybrids typically yield two to three times as much as their inbred parents.

But superior hybrid genotypes, from the farmer's point of view, are not necessarily genotypes with high heterosis. A cross of two extremely low yielding inbreds can give a hybrid with high heterosis but comparatively low yield, whereas a cross of two high yielding inbreds might exhibit less heterosis but nevertheless produce a high yielding hybrid. High yielding hybrids owe their yield not only to heterosis but also to other heritable factors that are not necessarily influenced by heterosis. One needs to know the relative importance of each genetic contribution —of heterosis and nonheterosis—in individual hybrids.

Furthermore, when examining yield trends in a time series of successively released hybrids, breeders need to know what portion of the genetic yield gains (if gains are made) is due to increases in heterosis, and what portion to increases in nonheterosis. This knowledge can help them as they plan future breeding operations.

The record shows clearly that the yielding ability of maize hybrids worldwide has improved steadily over the years (Castleberry et al., 1983; Derieux et al., 1987; Duvick, 1992; Eyhérabide et al., 1994; Ivanovic & Kojic, 1990; Russell, 1991). Rates of improvement in the USA have averaged about 100 kg ha^{-1} yr^{-1}. Although maize breeders informally often credit the steady increase in genetic yielding ability of maize hybrids to improvements in heterosis, they also know that hybrids continually are improved in yield limiting traits (often called defensive traits) such as disease and insect resistance, stronger roots, resistance to stalk lodging, tolerance to heat and drought, and tolerance to abnormally cool and wet growing conditions. Many of the defensive traits exhibit little or no heterosis. Inheritance of most of these traits is quantitative, their interaction with environment is high, and so they are not easily subjected to genetic analysis. Perhaps for this reason, few studies have attempted to sort out the relative importance of defensive traits for provision of high and reliable yields. Even fewer studies have been designed and carried out to compare the importance of sequential improvements in defensive traits, relative to sequential improvements in heterosis, over the years.

Fifty years ago, Frederick Richey (Richey, 1946) observed that, "there seem to be no comprehensive data showing a negative relation between parents and hybrids. On the contrary, the higher yielding inbreds have consistently tended to produce the higher yielding hybrids."

But early attempts to correlate inbred yields with yields of their hybrids showed disappointingly low positive correlations, and subsequent studies have confirmed the early findings (Hallauer et al., 1988); however, when some of the early analyses compared correlations of (i) inbreds with their specific single cross hybrids, and (ii) inbreds with the mean of all their hybrid progeny, the correlations for method ii were appreciably higher than those for method i (Sprague, 1964). The results showed that inbred yields predicted general combining ability more accurately than they predicted specific combining ability. But the correlations still were not high enough to warrant selecting inbreds on the basis of their yield per se; performance in crosses was and still is essential for evaluating the worth of an inbred for yield in hybrids, as well as for other traits affected by yield, such as standability.

Despite their low values, the inbred-hybrid yield correlations were positive. They indicated a tendency for high yielding inbreds to produce high yielding hybrids. The question was still open: how important is heterosis, as a cause of increases in yielding ability of maize hybrids over the years? Does heterosis account for all of the increase, part of it, or none of it?

To answer this question one must compare a time series of inbred lines with their single cross hybrids. If heterosis has increased over time, single crosses will yield increasingly more than the mean of their parents. But the question still can be asked, "Has the proportionate contribution of heterosis to hybrid yield changed over the years?". Thus, two methods of analysis are needed: (1) calculation of heterosis as grain yield per unit area (i.e., yield of a single cross minus average yield of its inbred parents), and (2) calculation of heterosis as proportion of hybrid yield (i.e., yield of single cross minus yield of midparents, divided by yield of single cross). Alternatively, one can calculate rate of genetic gain for single cross yield and compare it to rate of gain for heterosis.

A few experiments have been designed to compare yields of a time series of important inbreds with yields of their single crosses. Two are reported in the literature (Duvick, 1984, Meghi, et al., 1984), and unpublished data are available for a third experiment (Duvick, unpublished data). All experiments were conducted in the USA Corn Belt, with inbreds that were highly successful in the Corn Belt at some period of time during the past 60 years. Some of the results of the three experiments are summarized in Fig. 3–2, and Tables 3–2 and 3–3. All three experiments show that:

1. Yields of inbreds, and of their single crosses, have risen continually since the 1930s.

2. Rates of yield gain were higher for single crosses than for their midparents in all experiments except at the low density (typical of the 1930s) in Experiment 3.

3. Heterosis, calculated as yield of single cross minus yield of midparent, rose continually in all experiments, except at the lowest density in Experiment 3.

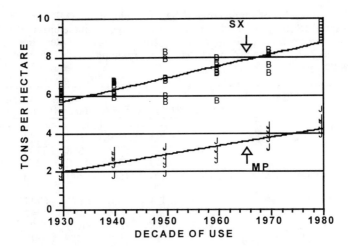

Fig. 3–2. Yields of maize single crosses (SX) and means of their parent inbreds
(MP). Widely used pedigrees in Iowa, seven SX per decade, 1930s through
1980s. Means of three densities, two years. (Duvick, unpublished data)

Table 3–2. Yields of single crosses (SX) and midparents (MP), and values for SX
minus MP (Het) and for heterosis as percentage of single cross yield (Het ÷ SX
(%). b = rate of gain for yield category, in kg ha^{-1} yr^{-1}. Data from three experiments.†

Expt.	Yield Category	1930s	1940s	1950s	1960s	1970s	1980s	b
				--- kg ha^{-1} ---				
1	SX	7097		7407		9538		61
	MP	2985		2969		3400		10
	Het	4112		4438		6138		51
	Het ÷ SX (%)	(58)		(60)		(64)		
2	SX	4600	5300	6900	7000	7900		83
	MP	1900	2100	2800	3400	3600		47
	Het	2700	3200	4100	3600	4300		36
	Het ÷ SX (%)	(59)	(60)	(59)	(51)	(54)		
3	SX	5941	6371	6865	7174	7929	9164	60
	MP	2154	2618	2722	2985	3827	4506	45
	Het	3787	3754	4143	4188	4102	4658	16
	Het ÷ SX (%)	(64)	(59)	(60)	(58)	(52)	(51)	

† Exp. 1: Recalculated data (Meghi, et al., 1984), means of two densities, 31 500
and 58 800 plants ha^{-1}. Exp. 2: (Duvick, 1984), means of three densities, 30 000,
47 000, and 64 000 plants ha^{-1}. Exp. 3: (Duvick, unpublished data), means of three
densities, 30 000, 54 000, and 79 000 plants ha^{-1}

Table 3–3. Yields of single crosses (SX) and midparents (MP), and values for SX minus MP (Het) and for heterosis as percentage of single cross yield (Het ÷ SX (%) in Experiment 3. Means of 7 SX per decade at 3 densities: L = 30 000 plants ha^{-1}, M = 54 000 plants ha^{-1}, H = 79 000 plants ha^{-1}. b = rate of gain for yield category, in kg ha^{-1} yr^{-1}.

Density	Yield cate- gory	1930s	1940s	1950s	1960s	1970s	1980s	b
		------------------------------------ kg ha^{-1} ------------------------------------						
L	SX	6717	6703	7099	7387	7187	8174	26
	MP	2062	2373	2395	2658	3309	3875	35
	Het	4655	4330	4703	4730	3877	4298	-8
	Het ÷ SX (%)	(69)	(65)	(66)	(64)	(54)	(52)	
M	SX	6569	7033	7960	8171	8747	10024	64
	MP	2337	3065	3174	3493	4352	4969	50
	Het	4232	3968	4786	4679	4396	5055	15
	Het ÷ SX (%)	(64)	(56)	(59)	(57)	(50)	(50)	
H	SX	5708	6648	6823	7192	9098	10492	90
	MP	2308	3003	3063	3390	4463	5476	59
	Het	3400	3645	3760	3802	4635	5016	32
	Het ÷ SX (%)	(59)	(55)	(53)	(53)	(51)	(48)	

  4. Heterosis, calculated as percentage of single cross yield, ranged from about 50 to 65%. Heterosis as percentage of single cross yield increased in the most recent decade (1970s) in Experiment 1, but it declined in the most recent decades (1960s through 1980s), in Experiments 2 and 3.
  5. Gains in heterosis did not account for all of the gains in hybrid yield over time; gains in nonheterosis (measured as midparent yield) made important contributions, also. Presumably this means that improvements in the contribution of additive genes was an important factor in improvement of hybrid yield.
  Further study of these three experiments shows that yield gains in the hybrids always were accompanied by improvements in tolerance to abiotic and biotic stress, and that the improvements occurred in parental inbreds as well as in their hybrid progeny. Numerous additional comparisons of commercially important maize hybrids also show that all important defensive traits have shown strong improvement over the years (Cardwell, 1982; Castleberry et al., 1983; Derieux et al., 1987; Duvick, 1992; Eyhérabide et al., 1994; Russell, 1991). Most of these defensive traits do not exhibit heterosis.
  One can conclude, therefore, that:
  1. Heterosis plays an important role in maize hybrid yields.
  2. Heterosis has increased in absolute amounts (e.g., kg ha^{-1}) over the years.
  3. Heterosis probably will contribute increasingly smaller proportions to total yield gains in years to come, because of proportionately higher rates of improvement in inbred yield. (An interesting observation: Table 3–3 shows that modern inbreds, grown at today's high densities, can yield nearly as much as hybrids of the 1930s.)
  This proportionate lack of increase in heterosis, as hybrid yields increase, was foretold quite clearly in a summary of data from experiments designed for other purposes but which included yields for inbreds and their single cross hybrids (Schnell, 1974). Schnell summarized his analysis by saying, "there was only a modest increase in heterosis as compared to the large simultaneous increase in the

yields of inbreds, . . . the . . . non-heterotic part of the yields of corresponding hybrids."

Hybrid Sorghum

Grain sorghum has been grown as hybrids for about 40 years, starting in the USA (Doggett, 1988). About 48% of grain sorghum plantings, worldwide, are now hybrid (Table 3–1). The first hybrids yielded at least 40% more than the local varieties; the advantage was much higher under severe drought stress. Grain sorghum varieties naturally are quite highly inbred and presumably have low numbers of deleterious recessive genes. Perhaps for this reason, sorghum hybrids exhibit much less heterosis for yield than is typical for crosses of maize inbreds. Using assumptions similar to those for the maize calculations, one can state that (i) worldwide sorghum yields are about 19% higher than they would be without use of hybrids, (ii) the hybrid yield advantage accounts for 13 MMT of the total annual worldwide production of about 66 MMT, and (iii) use of hybrids spares about 9 million hectares of land that otherwise would be planted to grain sorghum. These figures are no more reliable than those for the maize calculations, since they, too, depend on specific assumptions about the advantage of current hybrids over hypothetical improved varieties.

Grain sorghum yields have risen over the years, but it is not a simple matter to calculate the genetic contribution to yield gains, even in the USA with 100% hybrids. In part, this is because the relative amounts of land planted to irrigated and rain-fed sorghum have varied widely over the years. The area planted to irrigated sorghum increased for several years after hybrids were introduced, then declined as water supplies declined, and pumping costs increased (Miller & Kebede, 1984). Onsets of severe insect and disease problems also have caused high variability in yields, despite breeders' success in countering new problems. Furthermore, yield increases sometimes are stepwise, as superior new breeding lines come into use. Breeders have estimated, nonetheless, that genetic gains have been made over the years, at rates of about 1% per year. They estimate that genetic contributions have provided 35 to 40% of the total yield gains in grain sorghum in the USA, since the advent of hybrids. Improvements in cultural practices, such as added nitrogen fertilizer and irrigation, are responsible for 60 to 65% of the gains in yield.

Measurements of sequential changes in heterosis for yield have not been made for sorghum, but it seems likely that heterosis is greater in present-day hybrids than in the first hybrids. Breeders say that inbred yields per se are not useful indicators of hybrid yields, and they also say that inbred yields have not improved over the years to any large degree, certainly not at rates comparable to those for maize inbreds. Major yield gains have come from discovery of well-balanced heterotic combinations of superior new germplasm families. Nevertheless, new inbreds do tend to be more vigorous than their predecessors (Doggett, 1988). In particular, they are better equipped with defensive traits such as disease and insect resistance, and tolerance to abiotic stresses such as heat and drought or mineral imbalance. It seems likely, therefore, that increased yield in grain sorghum hybrids gradually will depend less on heterosis per se and more on other kinds of gene action, particularly for defensive traits that confer yield stability. As any breeder knows, high average yield depends on stability of performance year in and year out under all expected stresses, as well as on ability to make top yields in highly favorable environments. Selection for specific combining ability often means balancing non-heterotic defensive traits as much as or more than balancing theoretical dominant or epistatic gene combinations for yield.

Hybrid Sunflower

Oil sunflower has been grown as hybrids for about 20 years (Miller, 1987), starting in the USA. Sunflower hybrids now are planted in all parts of the world where sunflower is grown commercially as an oil crop. About 60% of the crop is now hybrid, worldwide (USDA, 1995). Sunflower hybrids yield about 50% more than the better OPVs (Miller, 1987). Sunflower is cross-pollinated, and OPVs, like those of maize, are themselves a collection of hybrid plants, each exhibiting heterosis to some degree; however, most commercial sunflower OPVs are somewhat inbred, as a consequence of strong selection for uniformity for desired traits; thus, oil sunflower hybrids show a relatively high yield advantage compared to OPVs. (The estimated sunflower hybrid yield advantage over OPVs is more than three times as large as that for maize.) Making assumptions like those already described for maize and sorghum, one can say that (i) worldwide sunflower yields are 30% higher than they would be if all were OPVs, (ii) the hybrid yield advantage is responsible for about 7 million tons of the total global production of 24 million tons, and (iii) about 6 million hectares of land are spared from planting to sunflower, because of the hybrid yield advantage. And as with maize and grain sorghum, these calculations are based on unproved assumptions about advantage of hybrids over OPVs.

Experiments designed to measure genetic contribution to yield gains in hybrid sunflower have not been made, nor have experiments been conducted to measure changes in amount of heterosis. New heterotic groups involving new germplasm lines show promise of giving hybrids with even higher yield, but no data are on hand, to show whether or not the higher yields are due to gains in heterosis. As with both maize and sorghum, inbred yield per se is not a good predictor of hybrid yield (Miller, 1987); however, as with sorghum and maize, some of the best new inbred parents (at least the females) are clearly more vigorous and higher yielding than their predecessors. Demands of low cost seed production also dictate that female seed yields be increased if at all possible, so it seems likely that yields per se of female inbreds gradually will increase in years to come; however, breeders will not try directly to increase yields of male inbreds. Sunflower is unique among the hybrid crops in that females are single-headed but males have multiple heads, a recessive trait. Presence of multiple heads in the male ensures a long period of pollen availability, and better seed yields on the female, but it also hinders visual estimates of yield of the line per se.

Performance gains in hybrid sunflower primarily are due to improvements in traits that confer stability of performance, e.g., in disease and insect resistance, and resistance to lodging. High oil percentage of course also is important in this oil crop. Parents as well as hybrids have acquired these improvements; they are not the unique product of heterosis. Thus, changes in nonheterotic traits that could just as well have been improved in the OPVs (but with more difficulty) are responsible for much of the improved hybrid performance in sunflower, as with maize and sorghum. It seems likely, therefore, that increased yield in oil sunflower gradually will depend less on heterosis per se and more on nonheterotic traits, for gains in yield and yield stability. In this regard, sunflower will be like maize and sorghum — gains in average yield (across years and in many locations) are more important than high yield under ideal conditions, and gains in average yield owe much to improvements in defensive traits, many of which do not exhibit heterosis per se.

DISCUSSION

Heterosis is an important cause of the increasingly high yields of maize, grain sorghum, and oil sunflower but it is not the only cause. Improvements in general combining ability as well as in specific combining ability, in additive genes as well as in dominant, over-dominant or epistatic gene combinations, have been crucial to improvement of hybrids in all three crops. For each of these crops, one can theorize that similar gains in yield and performance might have been made if more attention had been devoted to improvement of OPVs, starting with the then existing superior varieties. The facts are, however, that impressive gains in yield and performance have been made since hybrid breeding began for each crop, but minimal or even no breeding progress had been made before the advent of hybrids to those crops. Why did this happen? What was the unique contribution of heterosis? And what can one say about the future importance of hybrid breeding in these crops, or in other crops?

Perhaps the greatest gift of heterosis has been its indirect effect, in forcing attention on hybrids as the medium for genetic improvement of the crop. The hybrid seed industry has made several important contributions to crop improvement that are not due directly to heterosis. Some of the beneficial indirect effects of hybrids are:

1. Precise genotype identification and multiplication. Instead of a random collection of hybrid and/or inbred plants in an OPV, the most superior hybrid combinations can be identified and reproduced at will, in unlimited quantity. Despite concerns about dangers of genetic uniformity, experience shows that stability in performance can be most easily identified and used in uniform genotypes, including hybrids.

2. Breeders of hybrid crops can react faster and with more options to meet changing times and changing demands, as compared with breeders of either inbred crops or OPVs. New hybrids with needed new traits can be made and put out to test within one or two seasons, given a broad-based pool of inbred lines.

3. Hybrids facilitate combination of multiple traits into one cultivar, e.g., one hybrid can carry several dominant genes for disease resistance, some coming from one parent, some from the other, or one hybrid may derive its drought tolerance from one parent and its lodging resistance from the other parent.

4. Farmers can easily identify hybrids as a class, and in some cases they can identify specific hybrids. They expect more from hybrids, they are more likely to provide extra agronomic inputs to hybrids, and they are more likely to press breeders to make continuing and rapid improvements in hybrids. Commercial breeders are especially vulnerable to this farmer pressure for continuing improvement.

5. The prospect of annual seed sales at profitable prices attracts private capital to hybrid breeding and sales. Hybrid breeding, and the associated seed production and distribution technologies, are doubly supported, by both public and private funds.

Within limits, it seems likely that some of the benefits of the inbred–hybrid method also can be applied to improvement of highly self-pollinated crops such as rice and wheat, but the likelihood of high seeding rates in these crops introduces an economic problem: seed production costs must be low enough and yield of hybrids in the farmers' fields must be high enough that farmers can profit from purchase and use of hybrid seed and companies can profit from production and sale of hybrid seed. Hybrids have been successful with sorghum, an inbred crop, but seed yields are high, and seeding rates are low, compared to those for wheat and direct-seeded rice. For some parts of the world, this seed-yield–seeding-rate problem may be insurmountable. But one should never discount the ability of farmers and breeders to come up with ingenious solutions to the problem.

CONCLUSIONS

Gains in yield and yield stability offered by heterosis have prompted use of hybrids in several crops. As a result, substantially increased amounts of breeding effort have been devoted to these crops. Genetic yielding ability has been increased greatly, and thus total production has been increased, with minimal dependence on chemical inputs and maximum use of biological power. Enthusiasm and funds have been directed to hybrid breeding, in part because of the proven efficiency of the inbred–hybrid method for producing products that farmers need and want, and in part because private capital was attracted to the profit potential of hybrid breeding and sales. The inbred–hybrid method has given breeders greater precision in developing, identifying, and multiplying the best hybrid genotypes in cross-pollinated crops. The pace of genetic improvement in these crops has been greatly accelerated thereby, and shows no sign of slackening. Inbred crops have benefited from use of the inbred–hybrid method as well. Initial improvements in the inbred crops were in yield gains caused by heterosis, but continuing gains will depend on breeders' facility in using the inbred–hybrid method to make superior new hybrid genotypes with greater speed and precision than is possible with homozygous inbred varieties.

Evidence to date indicates that improvements in inbreds per se will play an increasingly large role in improving the performance of hybrids. Improvements in nonheterotic traits that confer stability of performance (defensive traits) will enhance hybrid yields as well as their overall performance.

But heterosis itself will continue to be a highly important cause of hybrid superiority in yield and yield stability. Specific combining ability—specific combinations of inbred lines with good general combining ability—will remain as the final and always essential requirement for production of superior new hybrids.

REFERENCES

Cardwell, V.B. 1982. Fifty years of Minnesota corn production: Sources of yield increase. Agron. J. 74:984–990.

Castleberry, R.M., C.W. Crum, and C.F. Krull. 1983. Genetic yield improvement of U.S. maize cultivars under varying fertility and climatic environments. Crop Sci. 24:33–36.

Derieux, M., M. Darrigrand, A. Gallais, Y. Barriere, D. Bloc, and Y. Montalant. 1987. Estimation du progrès genétique réalisé chez le maïs grain en France entre 1950 et 1985. Agronomie 7:1–11.

Doggett, H. 1988. Sorghum. Longman Scientific & Technical, Harlow, Engand.

Duvick, D.N. 1984. Genetic contributions to yield gains of U.S. hybrid maize, 1930 to 1980. p. 15–47. In W.R. Fehr (ed.) Genetic contributions to yield gains of five major crop plants. CSSA Spec. Publ. 7. CSSA and ASA, Madison, WI.

Duvick, D.N. 1992. Genetic contributions to advances in yield of U.S. maize. Maydica 37:69–79.

Eyhérabide, G.H., A.L. Damilano, and J.C. Colazo. 1994. Genetic gain for grain yield of maize in Argentina. Maydica 39:207–211.

Hallauer, A.R., W.A. Russell, and K.R. Lamkey. 1988. Corn breeding. p. 469–565. In G.F. Sprague and J.W. Dudley (ed.) Corn and corn improvement. ASA, CSSA, and SSSA, Madison, WI.

Iowa State Department of Agriculture. 1934. Iowa corn and small grain growers' association. p. 135–148. In Thirty-fifth Annual Iowa Year Book of Agriculture, Iowa State Dep. of Agric. Des Moines.

Ivanovic, M., and L. Kojic. 1990. Grain yield of maize hybrids in different periods of breeding. Informatsionnyi Byulleten po Kukuruza 8 8:93–101.

Lewontin, R.C., and J.-P. Berlan. 1990. The political economy of agricultural research: The case of hybrid corn. p. 613–628. *In* C.R. Carroll et al. (ed.) Agroecology. McGraw Hill, New York.

Meghi, M.R., J.W. Dudley, R.J. Lambert, and G.F. Sprague. 1984. Inbreeding depression, inbred and hybrid grain yields, and other traits of maize genotypes representing three eras. Crop Sci. 24:545–549.

Miller, F.R., and Y. Kebede. 1984. Genetic contributions to yield gains in sorghum, 1950 to 1980. p. 1–14. *In* W.R. Fehr (ed.) Genetic contributions to yield gains of five major crop plants. CSSA Spec. Publ. 7. CSSA and ASA, Madison, WI.

Miller, J.F. 1987. Sunflower. p. 626–668. *In* W. R. Fehr (ed.) Principles of cultivar development. MacMillan Publ. Company, New York.

Richey, F.D. 1946. Hybrid vigor and corn breeding. J. Am. Soc. Agron. 38:833–841.

Russell, W.A. 1991. Genetic improvement of maize yields. Adv. Agron. 46:245–298.

Schnell, F.W. 1974. Trends and problems in breeding methods for hybrid corn. p. 86–98. *In* XVI British Poultry Breeders Roundtable. Birmingham, England.

Sprague, G.F. 1964. Early testing and recurrent selection. p. 400–417. *In* J. W. Gowen (ed.) Heterosis. Hafner Publ. Company, New York.

USDA. 1953. Agricultural statistics 1953. USDA, Washington, DC.

USDA. 1995. Historical world wide sunflower: Production, supply and disposition. USDA, Washington, DC.

Virmani, S.S. 1994. Hybrid Rice Technology: New Developments and Future Prospects. *In* S.S. Virmani (ed.) Selected papers from the Int. Rice Res. Conf. IRRI, Philippines.

Chapter 4

Quantitative Genetics of Heterosis

K. R. Lamkey and J. W. Edwards

INTRODUCTION

Nearly 50 years have elapsed since the seminal heterosis conference was held at Iowa State College (Gowen, 1952). That conference undoubtedly grew out of the obvious importance of maize (*Zea mays* L.) hybrids in the agricultural economy of Iowa and the USA as well as the lack of understanding of the phenomenon of heterosis. Farmers in Iowa rapidly adopted maize hybrids. In just 15 years, Iowa went from 0 to 100% of the maize acreage being planted to hybrids. Gowen (1952) stated the following about hybrid maize "It seems likely that in no other period of like years has there been such an increase in food produced over so many acres of land. The return from hybrid corn has been phenomenal, but it is now evidently approaching an asymptotic value." If only Gowen could have looked ahead 50 years, because the best was yet to come (Fig. 4–1).

The term heterosis was coined by Shull (1952). He defined the heterosis concept as ". . . the interpretation of increased vigor, size, fruitfulness, speed of development, resistance to disease and to insect pests, or to climatic rigors of any kind, manifested by crossbred organisms as compared with corresponding inbreds, as the specific results of unlikeness in the constitutions of the uniting parental gametes." This definition is often interpreted as not implying a genetic basis for heterosis, because the definition basically describes the phenotype that results from crossing two different inbred lines.

For our purposes, we will define heterosis or hybrid vigor as the difference between the hybrid and the mean of the two parents (Falconer & Mackay, 1996). This definition is usually called midparent heterosis. Midparent heterosis is often expressed as a percentage of the midparent in the literature. It is important to note, however, that percentage midparent heterosis is difficult to interpret from a quantitative genetic point of view, and statistical tests of percentage midparent heterosis are nearly impossible. High parent heterosis is preferred in some circumstances, particularly in self-pollinated crops, for which the goal is to find a better hybrid than either of the parents.

To some, the terms hybrids and heterosis are synonymous. This is misleading, however, because there are hybrids that do not exhibit heterosis, but there cannot be heterosis without hybrids. In some species, hybrids are sold commercially because crossing of two varieties brings together complementary traits controlled by additive gene action. Distinguishing between hybrids and heterosis is important, because hybrids bring factors other than heterosis per se, i.e., uniformity, reproducibility, etc., to crop production. Often these factors are

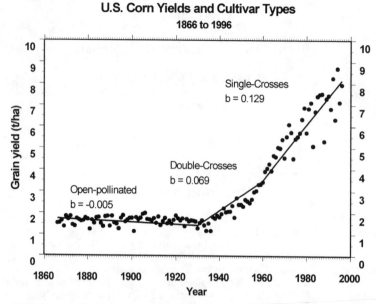

Fig. 4–1. Average U. S. maize yields from 1866 to 1996. Regression lines were splined together at 1930 and 1960, which corresponds roughly to when double-cross and single-cross hybrids started to become important. Regression coefficients are in t ha^{-1} yr^{-1}. Data are from the USDA, National Agricultural Statistics Service.

confounded and difficult to separate. For example, uniformity may result in higher yields. Is uniformity a genetic or nongenetic cause of increased yield? Is uniformity a factor in heterosis? In other species (such as wheat), hybrids are being sought as a means to prevent farmers from saving and planting their own seed and as a means of protecting research investments in transgenes.

In this manuscript we will review basic quantitative genetic concepts in heterosis, introduce the concept of baseline heterosis, review the results of over 50 years of gene action studies, and suggest needs for future research. We feel that the quantitative genetics of heterosis must be tied to plant improvement and that any theory of heterosis must explain and be consistent with the increase in yield of hybrid maize since 1930 (Fig. 4–1).

BASIC QUANTITATIVE GENETIC CONCEPTS OF HETEROSIS

Much of the quantitative genetic theory of heterosis is based on single locus theory. Single locus heterosis theory assumes the absence of epistasis, which considerably simplifies the mathematics and interpretations of the theory. Single locus theory will be reviewed in detail, because basic properties of heterosis are derived from this theory. Willham and Pollak (1985) developed single locus heterosis theory for predicting the performance of the F_1, F_2, parents, and the backcross to the parents. They used the random mated F_1 (the F_2 generation) as the base population in which all genetic effects are defined. Willham and Pollak (1985) were interested in applying this theory to animals, for which inbreeding the parental populations is rare. Therefore, we have extended this theory to include any level of inbreeding of the F_1, F_2, P_1, and P_2 generations. A pedigree showing

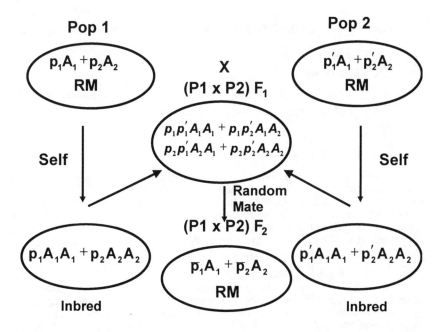

Fig. 4–2. Mating scheme for populations described in text. Populations 1 and 2 start with the gametic arrays shown and are in random mating equilibrium (panmixia). Crossing the panmictic populations together forms the F_1 shown. Random mating the F_1 gives rise to the F_2 generation, the population in which genetic effects are described in Willham and Pollak (1985). Inbreeding populations 1 and 2 to complete homozygosity generates the populations with the genotypic arrays shown. Crossing these two inbred populations produces the same F_1 as crossing the two panmictic populations.

how each generation is obtained is given in Fig. 4–2. Assuming two alleles per locus, the generation means are

$$\overline{F}_{1(f)} = (1 - f)(\overline{F}_2 + 2\Delta^2 d) + fa(\overline{p}_1 - \overline{p}_2),$$

$$\overline{F}_{2(f)} = (1 - f)(\overline{F}_2) + fa(\overline{p}_1 - \overline{p}_2),$$

$$\overline{P}_{1(f)} = (1 - f)(\overline{F}_2 + 2\Delta\alpha - 2\Delta^2 d) + fa(\overline{p}_1 - \overline{p}_2 + 2\Delta), \text{ and}$$

$$\overline{P}_{2(f)} = (1 - f)(\overline{F}_2 - 2\Delta\alpha - 2\Delta^2 d) + fa(\overline{p}_1 - \overline{p}_2 - 2\Delta),$$

where

f = inbreeding coefficient of a generation;

p_i = frequency of the ith allele in population 1;

p_i' = frequency of the ith allele in population 2;

$\overline{p}_i = \dfrac{p_i + p_i'}{2}$ = average allele frequency in the cross of population 1 and 2; and

$\delta_i = \dfrac{p_i - p'_i}{2}$ = one half the difference in allele frequency between populations.

In the two allele case,

$\delta_1 = -\delta_2 = \Delta$; and

d = the deviation of the heterozygote from the homozygote midparent;

a = half of the difference between homozygotes;

$\alpha = a + d(\bar{p}_2 - \bar{p}_1)$ = average effect of an allele substition;

$\overline{F}_2 = a(\bar{p}_1 - \bar{p}_2) + 2\bar{p}_1\bar{p}_2 d$ = mean of F$_2$ generation.

Panmictic-midparent heterosis is the heterosis observed when two random mating populations are crossed to form an F$_1$ hybrid. The strict definition of panmictic-midparent heterosis is the difference between the mean of the F$_1$ and the average of the two random mated parent populations (midparent value). F$_2$ heterosis is defined as the difference between the mean of the F$_2$ generation and the midparent value. Algebraically, these heterosis values are:

Panmictic-midparent heterosis = $4\Delta^2 d$, and

F$_2$ heterosis = $2\Delta^2 d$.

Four conclusions can be drawn from these expressions: (i) heterosis is dependent on directional dominance; (ii) heterosis is a function of the square of the difference in allelic frequency between two populations and therefore, heterosis is specific to a particular cross; (iii) if two inbred lines are crossed, Δ can only be 0 or 1, therefore, heterosis in a cross of two inbred lines is a function of dominance at those loci that carry different alleles in the inbred lines (Falconer & Mackay, 1996); and (iv) randomly mating the F$_1$ reduces heterosis by 50%. Although genetic divergence (difference in allelic frequency) and dominance are necessary for there to be heterosis, they are not sufficient in the case of multiple alleles. Cress (1966) showed that with multiple alleles segregating in a population the lack of heterosis cannot be used to infer a lack of genetic divergence between the parental populations. This result has important implications when breeders are screening populations to establish new heterotic groups.

Falconer and Mackay (1996) refer to heterosis as the converse of inbreeding depression. This is sometimes an overlooked fact and represents one of the breakthroughs in the discovery of heterosis. This fact is particularly relevant to the inbred–hybrid system of breeding in which heterosis is generally calculated by using the mean of inbred parents, as opposed to the mean of random mated populations, as in Falconer and Mackay (1996). We define inbred-midparent heterosis as the difference between the mean of the F$_1$ and the mean of the parent populations when inbred to homozygosity. The vigor lost during inbreeding of the parent populations is restored in the F$_1$. This gives rise to the concept of *baseline heterosis*. *Baseline heterosis* is simply the restoration of what was lost because of inbreeding depression. Inbred-midparent heterosis is equal to baseline heterosis plus *panmictic-midparent heterosis*. Thus, baseline heterosis is equal to inbred-midparent heterosis minus panmictic-midparent heterosis, which is equal to the difference between the panmictic midparent value and the inbred midparent value, or simply the average inbreeding depression observed in the two panmictic parent populations. Algebraically,

Inbred-midparent heterosis = $2\bar{p}_1\bar{p}_2 d + 2\Delta^2 d$, and

baseline heterosis = $2\bar{p}_1\bar{p}_2 d - 2\Delta^2 d$.

Note that inbred-midparent heterosis is a function of inbreeding depression, genetic divergence, and dominance whereas panmictic-midparent heterosis is a function only of genetic divergence and dominance.

Surprisingly little attention has been given to epistasis and heterosis. In the 1950 conference on heterosis (Gowen, 1952), epistasis as a cause of heterosis was not directly addressed. Willham and Pollak (1985) presented heterosis theory for the case of two linked loci with epistasis. Although the equations are too complicated to model, a couple of important points emerge concerning epistasis and heterosis. Panmictic-midparent heterosis is a function of dominance and unlinked additive x additive epistasis at loci with genetic divergence. The performance of the F_1 hybrid, however, is a function of dominance and unlinked dominance x dominance epistasis at those loci showing genetic divergence. These observations have important implications for the genetic interpretation of the midparent heterosis observed in self-pollinated crops that show little inbreeding depression. The heterosis observed may be due primarily to additive x additive epistasis, which does not contribute to inbreeding depression.

GENE ACTION AND HETEROSIS

Theory

Single locus heterosis theory coupled with the detrimental effect of recessiveness led to two prominent theories of heterosis called the dominance and overdominance hypotheses (Crow, 1952). Heterosis under the dominance hypothesis is produced by the masking of deleterious recessives in one strain by dominant or partially dominant alleles in the second strain. Heterosis under the overdominance hypothesis is due to heterozygote superiority and, therefore, increased vigor is proportional to the amount of heterozygosity. Supporters of the overdominance hypothesis put forth two main objections to the dominance hypothesis. First, it should be possible to accumulate by selection all the favorable dominance alleles into one homozygous strain and obtain inbreds that are as vigorous as hybrids. Second, F_2 distributions should be skewed because of the ¾ dominants to ¼ recessives segregation. Jones (1917) showed with linkage and Collins (1921) showed with large numbers of loci that the overdominance and dominance hypotheses were essentially indistinguishable. Crow (1948) using a mutation-selection equilibrium argument felt that dominance was insufficient to explain heterosis in maize. Hull (1952) presented eight reasons why he felt overdominance was the cause of heterosis in maize. The debate over the type of gene action controlling heterosis has gone on for more that 80 years. As we will see later in the section entitled gene action, this debate had a major influence on the development of breeding methodology.

We have applied our extension of the theory presented by Willham and Pollak (1985) to the cases of dominance and overdominance to further illustrate some key principles. In the case of complete dominance (Fig. 4–3), several key points are obvious. (i) F_1 performance is maximized when the favorable allele is fixed in one of the populations. (ii) With one locus the best inbred is as good as the best hybrid. (iii) Panmictic-midparent heterosis is maximized as the midparent values are declining. (iv) When inbred populations are crossed, heterosis exists even in the absence of genetic divergence. Pure overdominance is similar to complete dominance (Fig. 4–4), but the major exception is that it is not possible to obtain an inbred line as good as a hybrid with overdominance.

Baseline and panmictic heterosis for dominance and overdominance are plotted in Fig. 4–5. The graphs are very similar for the two types of gene action with the major difference being the magnitude; more heterosis is observed with overdominance than with dominance. The second and most important point is that

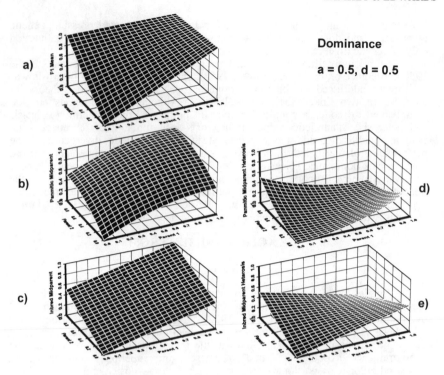

Fig. 4–3. Plots of mean performance and heterosis vs. frequencies of the dominant allele in parent populations for the case of complete dominance. (a) F_1 hybrid performance, (b) midparent mean for the cross of two panmictic populations, (c) midparent mean for the cross of two populations that have been inbred to $f = 1$, (d) panmitic-midparent heterosis, and (e) inbred-midparent heterosis.

panmictic-midparent heterosis only exceeds baseline heterosis when allelic frequencies are at the extremes. This is an important point to keep in mind when studying heterosis among inbred lines. A significant portion of the heterosis among inbred lines is due simply to recovery of what was lost during inbreeding and in some instances little of the observed heterosis may actually be due to genetic divergence.

Empirical Studies

Gene action and gene effects have been extensively studied in many crop species. Gene action is important in determining cultivar type (e.g., hybrid, pure line, synthetic), breeding methodology used to develop cultivars, and in the interpretation of quantitative genetic experiments. The study of gene action has been approached in two ways (Sprague, 1966). One characterizes the predominant types of genetic variance (additive vs. dominant) in populations, an activity that lead to development and analysis of mating designs, including the North Carolina mating designs (for a review see Hallauer & Miranda, 1988). Because of the difficulties in artificial hybridization, the variance component approach is not used frequently in self-pollinated crops; instead generation mean analysis has been the

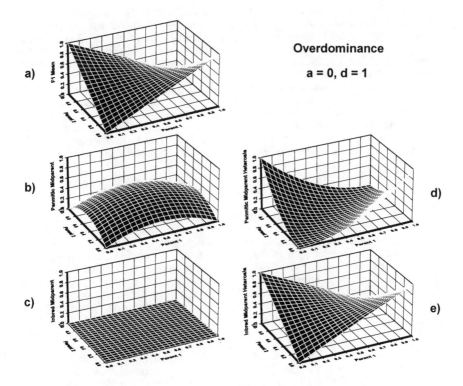

Overdominance

a = 0, d = 1

Fig. 4–4. Plots of mean performance and heterosis vs. allele frequency in the parent populations for the case of pure over dominance. (a) F1 hybrid performance, (b) midparent mean for the cross of two panmictic populations, (c) midparent mean for the cross of two populations that have been inbred to f = 1, (d) panmitic-midparent heterosis, and (e) inbred-midparent heterosis.

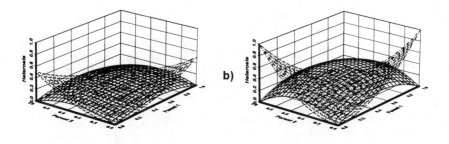

Fig. 4–5. Plots of baseline (solid lines) and functional heterosis (dashed lines) vs. allelic frequencies in the parents. Inbred-midparent heterosis is the sum of panmictic-midparent heterosis and baseline heterosis. (a) is for the case of complete dominance, and (b) is for the case of pure overdominance.

most prominent approach to determining gene action in these species. The results of these studies lead to the proposal of many breeding methods that capitalize on different types of gene action, including recurrent selection for general combining ability and inbred per se selection (additive effects), recurrent selection for specific combining ability (dominance effects), and reciprocal recurrent selection (both additive and dominance effects).

Of the major crop species, gene action has been most extensively studied in maize. A review of gene action studies in maize is therefore appropriate. Four lines of evidence will be reviewed: variance component estimation, generation means analysis, recurrent selection, inbreeding depression, and measured genotypes studies. Sprague and Eberhart (1977), Gardner (1963), and Hallauer and Miranda (1988) have excellent reviews of gene action studies in maize. Our review will be confined to grain yield.

Variance Component and Generation Means Studies

Numerous variance component estimation studies have been conducted in maize. Hallauer and Miranda (1988) reviewed and summarized variance component studies in maize conducted through the mid 1980s. The general conclusion from these studies is that in most maize populations, additive genetic variance for grain yield is usually two to four times larger than dominance variance. Dominance variance is important in maize populations and often is significant, but it is usually much smaller than additive variance. These results are often interpreted as implying that additive effects are of primary importance for grain yield of maize and that grain yield is controlled by genes with partial to complete dominance. The Design III mating design was developed specifically to estimate the degree of dominance (Comstock & Robinson, 1952) by using F_2 populations, in which the allele frequency is 0.5. Several Design III experiments have been conducted. Estimates of degree of dominance from F_2 populations were usually in the overdominant range. These scientists realized from the outset that repulsion phase linkage would bias degree of dominance upward, so experiments were developed to reduce the linkage bias by random mating the F_2 populations. Estimates of average degree of dominance estimated from random mated F_2 populations were always smaller than the estimates from nonrandom mated F_2 populations and usually in the partial to complete dominance range. These results convinced all but the most adamant overdominance supporters that much of the observed overdominance was probably due to linkage bias.

The early variance component studies assumed that epistasis was unimportant for grain yield of maize. This assumption was required because the number of covariances of relatives were not available to estimate epistasis and because epistatic models are difficult to handle mathematically.

From both a breeding methodology and statistical point of view, epistasis is difficult to estimate. Studies estimating epistasis in maize are too numerous for comprehensive review (for review see Hallauer & Miranda, 1988), but a few interesting conclusions can be drawn. Studies estimating epistasis by generation means analysis generally have reported significant epistatic effects. Estimates made by the analysis of variance (covariance of relatives) approach generally have reported nonsignificant epistatic effects. Studies with open-pollinated varieties generally have shown additive effects to be more important than dominance or epistatic effects, and studies with elite inbred lines generally have reported dominance and epistatic effects to be more important than additive effects.

These results are interesting and ambiguous at best. The lack of detection of epistasis with variance component studies suggests either a lack of statistical power or that epistasis is relatively unimportant. The ability to detect epistasis with generation means studies is indicative of the greater statistical power of using means, but these studies usually have a narrow inference base. Generally, it has been accepted that epistasis is relatively unimportant, but that there may be specific hybrid combinations in which epistasis is important. These conclusions are interesting considering the findings from molecular biology during the past 15 years. It is well known now that genes at the molecular level interact with each other or exhibit epistasis (Coe et al., 1988). The question is: why have we been relatively unsuccessful at detecting epistasis at the phenotypic level? The difficulty in detecting epistasis in populations at the phenotypic level, despite its ubiquitous presence at the molecular level, may be related mostly to an inadequate understanding of epistasis at the population level. Geneticists have long known about epistasis, but their concept of epistasis (physiological epistasis) is different from a quantitative geneticist's statistical or population epistasis. Physiological epistasis occurs when phenotypic differences among individuals with various genotypes at one locus depends on their genotypes at another locus (Cheverud & Routman, 1995). Statistical epistasis is a deviation of multilocus genotypic values from the additive combination of their single locus components (Cheverud & Routman, 1995). The main distinction between these two definitions is that statistical epistasis is a population phenomenon **dependent** on allelic frequencies in a specific population, whereas physiological epistasis is a genotypic phenomenon **independent** of allelic frequencies at the loci in question (Cheverud & Routman, 1995),

Cheverud and Routman (1995) demonstrated that additivity (a), dominance (d), and epistasis (e) all contribute to the average effects of alleles and the additive genetic variance. Only dominance and epistasis contribute to dominance deviations and variance, and epistasis alone contributes to the epistatic interaction deviations and variance. This means that physiological epistasis makes important contributions to additive and dominance variance and only the remainder contributes to statistical epistasis. It is also important to note that this concept is different from the confounding of statistical epistasis with additive and dominance genetic variances as often happens in one and two factor mating designs.

Cheverud and Routman (1995) were able to show that physiological epistasis can either suppress or enhance additive and dominance genetic variance. In some instances, depending on allelic frequencies, genetic variances and in particular dominance variance, can be suppressed or enhanced up to 50%. Using this same two locus approach and varying allelic frequencies at the two loci, Cheverud and Routman (1996) set up models with only additive-by-additive, additive-by-dominance, and dominance-by-dominance epistasis. Additive (a) and dominance (d) genotypic values for these models were zero. They were able to show that under certain allele frequencies that additive and dominance genetic variance exists in these populations. In essence, they along with others have shown that with finite populations, epistasis can contribute to the additive genetic variance.

Results from generation means analyses have been more ambiguous. Hallauer and Miranda (1988) reviewed the advantages and disadvantages of generation means models. Generally, two types of generation means studies have been conducted. One type involves a diallel among a group of inbred lines or populations. For diallel studies, models of Griffing (1956), Eberhart and Gardner (1966), and Gardner and Eberhart (1966) are often used. With these studies, the reference or inference population is restricted to the set of lines or populations included in the study. Typically, only general and specific combining ability effects can be estimated, although in the more advanced models of Eberhart and Gardner (1966) epistatic effects can be estimated as well. The second type of generation means analysis involves the cross between two inbred lines and generations derived from this cross (e.g., F_2, backcrosses to the parents). These studies are even more restricted in their inference base and have been used mostly for studying the inheritance of specific traits. Several methods are available for analyzing these types of studies (for a review see Hallauer & Miranda, 1988).

Classical generation means studies involving inbred lines and derived generations typically have the F_2 as the inference population. This is a disadvantage in crops exhibiting heterosis, because the inference population is not reflective of elite maize hybrids. Melchinger (1987) proposed a generation means model to analyze testcrosses of generations derived from two inbred lines. The reference population for this model is the F_2 testcross population in gametic phase equilibrium, which is more directly applicable to elite maize hybrids. He developed models for means and variances that included linkage and epistasis.

The typical genetic design for Melchinger's model involves choosing two inbreds from the same heterotic group (P1 and P2) and an inbred from the opposite heterotic group (PT). P1 and P2 are used to generate F_1, F_2, BCP1, and BCP2 generations. In addition, to enhance the power of the model, an F_∞ generation can be developed by selfing the F_2 to homozygosity or the F_2 generation can be random mated for several generations (Melchinger, 1987; Lamkey et al., 1995). Each of these generations is testcrossed to PT and generation means analysis is calculated by using the testcross generation means. Variances of each of the segregating generations can also be analyzed by crossing individual plants from each of the generations onto the tester, PT. Because linkage has no effect on the means in the absence of epistasis, only two models need to be fit to the data. Model 1 allows for linkage, but not epistasis:

$$Y = m^T + x(d^T),$$

where

Y = generation testcross mean;

M^T = testcross mean of the F_2 population in gametic equilibrium,

$(d^T) = \sum_j \theta_j d_j^T$,

θ_j = +1 if P1 carries the favorable allele at locus j and -1 otherwise,

d_j^T = one-half the average effect of a gene substitution at locus j in the F_2 testcross

　　　population, and

x = coefficient that is generation dependent.

T denotes parameters that are intrinsic to the tester used in the study. Model 2 allows for epistasis, but not linkage:

$$Y = m^T + x(d^T) + x^2(i^T),$$

where

$(i^T) = \sum_{j<k} \theta_j \theta_k i_{jk}^T$ and

i_{jk}^T = additive x additive epistatic effect between loci j and k.

Lamkey et al. (1995) fit Melchinger's model to the P1, P2, F_2, F_2-Syn 8, BCP1, and BCP2 generations derived from the inbreds B73 (P1) and B84 (P2). B73 and B84 are from the same heterotic group and are related to the extent that they were both developed from Iowa Stiff Stalk Synthetic (BSSS). Inbred Mo17 from the Lancaster Sure Crop heterotic group was used as the tester. The results of fitting Models 1 and 2 to the testcross means are shown in Fig. 4–6. Model 1, which allows linkage, but not epistasis, explained 48% of the variation among

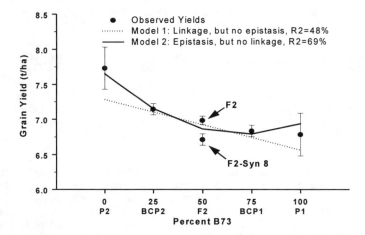

Fig. 4-6. The results of fitting Models 1 and 2 to the six testcross generation means. Testcross performance is plotted against the percentage of B73 germplasm in the population (Lamkey et al., 1995).

testcross means, but had a highly significant lack of fit. Model 2, which allows epistasis, but not linkage, explained 69% of the variation among testcross means and detected significant epistatic effects. These results indicated that unlinked epistatic effects accounted for 21% of the variation among generation means. In addition, the lack of fit for Model 2 was significant as well, indicating that other epistatic effects, both linked and unlinked, are important in this population. Favorable epistatic gene combinations have been accumulated in B73 and B84. Lamkey et al. (1995) found that the genetic variance among BCP2 progenies was not significant. Melchinger et al. (1988) also reported that backcrossing to the higher yielding parent resulted in a nonsignificant genetic variance component. These results are further evidence of the importance of epistasis and suggest that it may not be possible to accumulate favorable alleles for grain yield into one parent in an additive fashion as predicted by the dominance theory of heterosis (Lamkey et al., 1995). This result has important implications for marker assisted selection and backcrossing programs, which rely on the additive accumulation of favorable alleles into a parent.

Recurrent Selection

Patterns of response to recurrent selection also are indicative of the type of gene action controlling a trait. Sprague and Miller (1950) proposed a selection experiment to test what type of gene action was important for a trait. The premise of their method was that selection for general combining ability is made on the assumption that dominant favorable genes are important in heterosis and selection for specific combining ability is made on the assumption that overdominance and epistasis are mainly responsible for heterosis. With selection for general combining ability, the average allele frequency for genes affecting a trait will approach 1.0 as a limit. With selection for specific combining ability, the allele frequency in the population undergoing selection would approach $1-q$ if the average allele frequency in the homozygous tester is q. The experiment involves choosing two populations A and B, in which selection will be practiced and an inbred line C

Table 4–1. Selection response in Alph and the F_2 of (WF9 x B7) after five cycles of selection using B14 as the tester (data from Russell et al., 1973).

Tester	Alph Cn	(WF9 x B7)Cn
Population per se	2.06 ± 0.44	1.55 ± 0.44
B14	3.09 ± 0.50	1.32 ± 0.50
BSBB	3.63 ± 0.50	1.51 ± 0.50
C0 of population per se	2.29 ± 0.44	2.38 ± 0.44
C0 of other population	2.63 ± 0.18	1.46 ± 0.18
Interpopulation cross	4.09 ± 0.18	

as the tester parent. Standard half-sib selection is conducted in A and B for a number of cycles by using C as the tester. Improved cycles of A and B are designated as A', A'', and so on. If selection has been primarily fixing dominant alleles in A and B, the crosses between A' x B', A'' x B'', . . . should exhibit an increase in yield relative to A x B. Similarly, A', A'', ... should be higher in yield that the original A. If selection has been to primarily fix recessive alleles for those loci where tester C carries dominants and dominant alleles where tester C carries recessive alleles, then the crosses A' x B', A'' x B'', . . . should exhibit a downward trend relative to the original cross A x B. Trends in A', A'', . . . relative to A would depend on allele frequency in the tester C.

Russell et al. (1973) reported on an experiment to compare the importance of dominance and overdominance for yield heterosis in maize by using the procedure of Sprague and Miller (1950). They conducted five cycles of selection for specific combining ability in the open-pollinated variety 'Alph' and the F_2 of WF9 x B7. Responses for yield for six different testers are shown in Table 4–1. They found significant increases in grain yield in both the populations per se and in the interpopulation crosses suggesting that overdominance was not important for grain yield in these two populations. A significant result of this study was that selection for specific combining ability was effective for improving general combining ability as well as evidenced by the performance of the populations when crossed to BSBB (a broad based population) as well as the C0 of the other population. The implication of this study was that using a single tester would also give improvement with other testers as well. This was further evidence in support of the concept of early testing (Jenkins, 1935; Sprague, 1946) that is commonly used in maize breeding today.

The finding that additive effects were of primary importance for grain yield of maize and that overdominance was relatively unimportant increased interest in inbred progeny selection methods and the search for high yielding inbred lines. Comstock (1964) demonstrated that in the absence of overdominance S_1- or S_2-progeny selection was expected to be superior to other methods of recurrent selection for population improvement. Inbred progeny selection has had variable levels of success. Lamkey (1992) and others found that results from S_2-progeny selection were discouraging. Several reasons could account for the lack of response including lack of genetic variance, overdominance, and random genetic drift. Horner et al. (1989) compared S_2-progeny selection with half-sib selection by using an inbred tester in two maize populations. They found greater rates of gain for half-sib selection and concluded that nonadditive gene action in the overdominant range was important for grain yield in these populations.

Long-term reciprocal recurrent selection (RRS) studies also provide evidence on the type of gene action for heterosis. Cress (1967) conducted simulation studies of RRS by using both dominant and overdominant gene action models. With complete dominance, the mean of the interpopulation cross (hybrid), and the mean of the two populations are expected to increase, except by chance.

But, with overdominance, the change in population mean depends on the equilibrium gene frequency. It is possible to get short term increases in the population per se means, but in the long-term they always decrease.

Keeratinijakal and Lamkey (1993a) evaluated response to 11 cycles of RRS in BSSS and Iowa Corn Borer Synthetic #1 (BSCB1). They reported gains of 7% per cycle in the interpopulation cross, no change in BSSS per se, and a small significant increase in BSCB1. The small genetic gains in the populations per se did not resemble the response patterns that Cress (1967) predicted for overdominance and was attributed to random genetic drift due to small effective population size (Keeratinijakal & Lamkey, 1993b). Inbreeding depression in BSSS decreased over cycles of selection and showed no change in BSCB1, whereas inbreeding depression in the interpopulation cross doubled from C0 to C11. Heterosis of the interpopulation cross increased from 0.86 to 2.92 Mg ha^{-1}. These changes in inbreeding depression and heterosis suggest that selection has been for alleles at complementary loci in each population, such that the interpopulation cross is becoming more heterozygous with selection. More recent molecular data seem to support this conclusion [Labate et al., 1997, 1999 (this publication)]. An analysis of genetic divergence of dominance-associated distances also indicated that overdominance was not important in these populations (Keeratinijakal & Lamkey, 1993b).

Inbreeding Depression

Inbreeding depression is the converse of heterosis. The mean of a population with inbreeding coefficient f is:

$$M_f = M_0 - 2f \sum d\bar{p}\bar{q}$$

where summation is over all loci controlling a trait and p and q are the allele frequencies in the whole population (Falconer & Mackay, 1996). Several conclusions can be drawn from this equation, like the one for heterosis. First of all, a locus will not contribute to inbreeding depression if $d = 0$ or there is no dominance. Second, the direction of the change in mean is toward the value of the recessive allele. Third, inbreeding depression is maximized when $p = q = 0.5$, which also is where the number of heterozygotes are maximized. Fourth, in the absence of epistasis, inbreeding is a linear function of f. Fifth, if there is epistasis, but no dominance, there will not be any inbreeding depression (Crow & Kimura, 1970). Sixth, if there is epistasis and dominance, then inbreeding depression will be a quadratic or higher function of f (Crow & Kimura, 1970).

These basic results have several important implications regarding heterosis and hybrids in self-pollinated crops. Several studies with soybean [*Glycine max* (L.) Merr.] have reported significant heterosis (Burton, 1987). Across all the heterosis studies that Burton reviewed, 85% of the F₁ crosses showed midparent heterosis and 62% showed high parent heterosis. In soybean, there obviously is heterosis, but the genetic cause of the heterosis remains obscure. Heterosis alone, is not good evidence for dominance; however, heterosis studies conducted in conjunction with inbreeding depression studies should give a clear picture of the types of gene action involved in heterosis in soybean. For example, if there is midparent heterosis, but no inbreeding depression then there would be good evidence for the existence of additive x additive epistasis.

Numerous inbreeding depression studies have been conducted in maize. The majority of the studies have reported a linear relationship between inbreeding depression and f, and have concluded that epistasis is unimportant for grain yield (Hallauer & Miranda, 1988). It should be realized, however, that these studies all measured population bulks, and hence were looking at the average over the whole population. To our knowledge, there is no published data in maize on the variation in inbreeding depression. Pray and Goodnight (1995) reported that inbreeding depression can be genetically variable among lineages within a single population of flour beetle (*Tribolium castaneum*). Variation in inbreeding depression can be due to variation in the actual level of inbreeding, past history of inbreeding (whether inbreeding is due to the expression of deleterious recessives or overdominance), genetic drift and fixation of different alleles in different lines (Pray & Goodnight, 1995). Pray and Goodnight found evidence for nonlinearity in inbreeding depression suggesting that epistasis may be important for some traits. They concluded that the genetic variation present for inbreeding depression suggests that inbreeding depression may be a heritable trait.

Measured Genotypes

Measured genotypes refers to the situation in which a phenotype is scored on all possible genotypes of a two or three locus system in an otherwise homogeneous genetic background. These types of studies provide considerable power in estimating genetic effects. The disadvantage of these studies, however, is that only two or three loci can be studied at a time. Measured genotype studies may offer us the best opportunity for doing detailed studies of gene effects.

Conducting a measured genotype study requires two features. First, you need genes controlling traits that you are interested in and second, you need a method of creating the appropriate genotypes in an isogenic background. The only technique for doing this in plants is backcrossing. Backcrossing has the usual problem of linkage drag, but if the drag is the same for all gene combinations, then the bias may not be too severe. In *Drosophilia* for example, the appropriate genotypes can be created in identical genetic backgrounds without backcrossing (Clark & Wang, 1997). More recently, in plants, data from quantitative trait loci (QTL) mapping studies have been used like a measured genotype study primarily to estimate epistasis.

For the sake of simplicity, only two measured genotype studies will be discussed. One from maize and one from *Drosophilia*. Together, these two studies bring out the salient features of the analyses. Russell (1976) developed B14 isolines of the 27 genotypes possible for three loci with two alleles. The experiment was grown for three years at one location and data were collected for 10 traits. The standard Cockerham model was fit to the data to estimate additive, dominance, and epistatic effects. Russell (1976) found that 87, 27, 47, 15, 23, and 30% of the additive, dominance, additive x additive, additive x dominance, and dominance x dominance effects were significant at the 5% level, indicating that even at the population level, loci were interacting fairly frequently. This study also demonstrates the pleiotropic effect of loci.

Clark and Wang (1997) reported on a measured genotypes study in *Drosophila*. They constructed all possible two locus genotypes for each of eight pairs of *P*-element insertions. They found significant epistatic effects in 27% of their comparisons. They applied the method of Cheverud and Routman (1995) and found significant physiological epistasis in 15% of the comparisons. Clark and Wang (1997) reported epistatic effects on the same order of magnitude as main effects.

These studies clearly demonstrate that genes interact. Measured genotypes using a combination of Cockerham's analysis and Cheverud and Routman's analysis, may be one of the best tools for understanding epistasis and its contribution to heterosis.

NEEDS FOR FUTURE RESEARCH

The data clearly indicate a need for future empirical and theoretical research into heterosis; however, we need to be very careful about how future experiments are designed and analyzed. Enfield (1977) was critical of empirical quantitative genetic experiments indicating that the literature was either cluttered with experiments that were meaningless because of standard errors that were too large or because standard errors were absent altogether. Despite the tremendous developments that have recently occurred in molecular biology, quantitative genetics is still the only theory linking genotype to phenotype. It is imperative that we design better experiments to test the adequacy and validity of quantitative genetic models.

We would like to propose several areas of research that are needed to better understand heterosis. We will not present experimental approaches, because we do not have all of the answers.

1. Gene action and effects are key to understanding the inheritance of quantitative traits. For maize at least, it seems that from average population estimates, there is no evidence for overdominance. Although this is useful information, what would be even better is to gain insight into the distribution of gene effects and gene action for individual traits. Are there lots of loci with equal and small effects or is there some type of distribution? The conventional approach to gene action studies will not answer this question and new approaches will be needed.

2. Selection experiments in plants need to be better designed. Most of our current recurrent selection experiments are not adequately designed to separate the effects of selection from drift, so it becomes nearly impossible to reliably interpret the results of these experiments. It is clear that recurrent selection works, and future experiments to demonstrate the effectiveness of recurrent selection are probably not needed. But we do need well-designed experiments with adequate controls and replication. New selection experiments should either be replicated or include an unselected, replicated control population of the same effective size.

3. The view of random genetic drift in agricultural is the one that drift leads to a loss of heterozygosity and eventual erosion of genetic variance. Although this is true in the additive gene action case, with dominance and epistasis, drift may reduce heterozygosity with a corresponding increase in additive genetic variance. We need to incorporate this new information from evolutionary biology into the design of our breeding programs.

4. Epistasis has long been ignored in breeding programs and is generally assumed to be absent or unimportant. Evidence from molecular biology clearly shows that genes interact. Recent theory from evolutionary biology that distinguishes physiological epistasis from population epistasis may indicate why population level epistasis may be undetectable. More theoretical work is needed to optimally design breeding programs to select for epistatic effects.

5. Despite several classic studies, we know very little about inbreeding depression in plants. Because much of the observed heterosis among inbred lines may be due to the recovery of inbreeding depression, the genetics of heterosis may be best elucidated by studying the genetics of inbreeding depression. Our preoccupation with heterosis has caused us to overlook the importance of inbreeding depression.

There are of course several problems in designing and conducting good quantitative genetics experiments. First, they are often large and consume considerable physical resources, even for laboratory species. Second, there seems to be little funding available in agricultural species to do quantitative genetics and plant breeding related research. Third, there are few scientists being trained to do this type of research. Lack of funding and support in quantitative genetics may in the long-term severely limit future genetic gains.

ACKNOWLEDGMENTS

Joint contribution from the Corn Insects and Crop Genetics Research Unit, Agricultural Research Service, USDA, Dep. of Agronomy, Iowa State Univ. and Journal Paper J-17717 of the Iowa Agric. and Home Economics Exp. Stn. Project 3495, and supported by Hatch Act and State of Iowa.

REFERENCES

Burton, J.W. 1987. Quantitative genetics: Results relevant to soybean breeding. p. 211-247. *In* J.R. Wilcox (ed.), Soybeans: Improvement, production and uses. ASA, CSSA, and SSSA, Madison, WI.

Cheverud, J.M., and E.J. Routman. 1995. Epistasis and its contribution to genetic variance components. Genetics 139:1455–1461.

Cheverud, J.M., and E.J. Routman. 1996. Epistasis as a source of increased additive genetic variance at population bottlenecks. Evolution 50:1042–1051.

Clark, A.G., and L. Wang. 1997. Epistasis in measured genotypes: Drosophila *P*-element insertions. Genetics 147:157–163.

Coe, E.H. Jr., M.G. Neuffer, and D.A. Hoisington. 1988. The genetics of corn. p. 83–258. *In* G.F. Sprague and J.W. Dudley (ed.) Corn and corn improvement. Agron. Monogr. 18. ASA, CSSA, and SSSA, Madison, WI.

Comstock, R.E., 1964. Selection procedures in corn improvement. p. 87–94. *In* W. Heckendorn and J.I. Sutherland (ed.) Report 19th Hybrid Corn Industry-Res. Conf., Chicago, IL, 9–10 Dec. 1964. Amer. Seed Trade Association, Washington, DC.

Comstock, R.E., and H.F. Robinson. 1952. Estimation of the average dominance of genes. p. 494–516. *In* J. W. Gowen (ed.) Heterosis. Iowa State College Press, Ames.

Collins, G.N. 1921. Dominance and vigor of first generation hybrids. Am. Nat. 55:116–133.

Cress, C.E. 1966. Heterosis of the hybrid related to gene frequency differences between two populations. Genetics 53:269–74.

Cress, C.E. 1967. Reciprocal recurrent selection and modifications in simulated populations. Crop Sci. 7:561–67.

Crow, J.F. 1948. Alternative hypothesis of hybrid vigor. Genetics 33:478–487.

Crow, J.F. 1952. Dominance and overdominance. p. 282–297. *In* J. W. Gowen (ed.) Heterosis. Iowa State College Press, Ames.

Crow, J. F., and M. Kimura. 1970. An introduction to population genetics theory. Burgess, Minneapolis, MN.

Eberhart, S.A., and C.O. Gardner. 1966. A general model for genetic effects. Biometrics 22:864–81.

Enfield, F.D. 1977. Selection experiments in Tribolium designed to look at gene action issues. p. 177–190. *In* E. Pollak et al. (ed.) Proc. of the Int. Conf. on Quantitative Genetics, Ames, IA. 16–21 Aug. 1976. Iowa State Univ. Press, Ames.

Falconer, D.S., and T.F.C. Mackay. 1996. Introduction to quantitative genetics. 4th ed. Longman, Essex, England

Gardner, C.O. 1963. Estimates of genetic parameters in cross-fertilizing plants and their implications in plant breeding. p. 225–252. *In* W.D. Hanson and H.F. Robinson (ed.) Statistical genetics and plant breeding. NAS-NRC Publ. 982.

Gardner, C.O., and S.A. Eberhart. 1966. Analysis and interpretation of the variety cross diallel and related populations. Biometrics 22:439–452.

Griffing, B. 1956. Concept of general and specific combining ability in relation to diallel crossing systems. Aust. J. Biol. Sci. 9:463–493.

Gowen, J.W. (ed.) 1952. Heterosis. Iowa State College Press, Ames.

Hallauer, A.R., and J.B. Miranda, Fo. 1988. Quantitative genetics in maize breeding. 2nd ed., Iowa State Univ. Press, Ames.

Horner, E.S., E. Magloire, and J.A. Morera. 1989. Comparison of selection for S_2 progeny vs. testcross performance for population improvement in maize. Crop Sci. 29: 868–874.

Hull, H.F. 1952. Recurrent selection and overdominance. p. 451–473. *In* J. W. Gowen (ed.) Heterosis. Iowa State College Press, Ames.

Jenkins, M.T. 1935. The effect of inbreeding and of selection within inbred lines of maize upon the hybrids made after successive generations of selfing. Iowa State Coll. J. Sci. 9:429–450.

Jones, D.F. 1917. Dominance of linked factors as a means of accounting for heterosis. Genetics 2:466–479.

Keeratinijakal, V., and K.R. Lamkey. 1993a. Responses to reciprocal recurrent selection in BSSS and BSCB1 maize populations. Crop Sci. 33:73–77.

Keeratinijakal, V., and K.R. Lamkey. 1993b. Genetic effects associated with reciprocal recurrent selection in BSSS and BSCB1 maize populations. Crop Sci. 33:78–82.

Labate, J.A., K.R. Lamkey, M. Lee, and W.L. Woodman. 1997. Genetic diversity after reciprocal recurrent selection in BSSS and BSCB1 maize populations. Crop Sci. 37:416–423.

Lamkey, K.R. 1992. Fifty years of recurrent selection in the Iowa stiff stalk synthetic maize population. Maydica 37:19–28.

Lamkey, K.R., B.S. Schnicker, and A.E. Melchinger. 1995. Epistasis in an elite maize hybrid and choice of generation for inbred line development. Crop Sci. 35:1272–1281.

Melchinger, A.E. 1987. Expectation of means and variances of testcrosses produced from F_2 and backcross individuals and their selfed progenies. Heredity 59:105–115.

Melchinger, A.E., W. Schmidt, and H.H. Geiger. 1988. Comparison of testcrosses produced from F_2 and first backcross populations in maize. Crop Sci. 28:743–749.

Pray, L.A., and C.J. Goodnight. 1995. Genetic variation in inbreeding depression in the red flour beetle *Tribolium castaneum*. Evolution 49:176–188.

Russell, W.A. 1976. Genetic effects and genetic effect x year interactions at three gene loci in sublines of a maize inbred line. Can. J. Genet. Cytol. 18:23–33.

Russell, W.A., S.A. Eberhart, and O.U.A. Vega. 1973. Recurrent selection for specific combining ability for yield in two maize populations. Crop Sci. 13:257–61.

Shull, G.H. 1952. Beginnings of the heterosis concept. p. 14–48. *In* J. W. Gowen (ed.) Heterosis. Iowa State College Press, Ames.

Sprague, G.F. 1946. Early testing of inbred lines of corn. J. Am. Soc. Agron. 38:108–117.

Sprague, G.F. 1966. Quantitative genetics in plant improvement. p. 315–343. *In* K. J. Frey (ed.) Plant Breeding. Iowa State University Press, Ames.

Sprague, G.F., and S.A. Eberhart. 1977. Corn Breeding. p. 305–362 *In* G.F. Sprague and J.W. Dudley (ed.) Corn and corn improvement. Agron. Monogr. 18. ASA, CSSA, and SSSA, Madison, WI.

Sprague, G.F., and P.A. Miller. 1950. A suggestion for evaluating current concepts of the genetic mechanism of heterosis of corn. Agron. J. 42:161–62.

Willham, R.L., and E. Pollak. 1985. Theory of heterosis. J. Dairy Sci. 68:2411–2417.

Chapter 5

Dominance and Overdominance

J. F. Crow

INTRODUCTION

It has been exactly 50 years since I wrote my first article on heterosis, published in the Genetics Society Records of 1947 with a fuller exposition a few months later. I argued then and in the Heterosis Conference at Iowa State College in the summer of 1950 that heterosis is largely due to overdominance (Crow, 1948, 1952). But shortly afterward, additional data rendered my argument ineffective and more evidence for dominance began to appear. So, within a few years I changed my mind, and have retained the view that heterosis is mainly due to loci that are dominant or partially so. I propose here to discuss the reasons for this about-face, which roughly parallels the swing of general opinion on the subject, as best I can judge. But first, a bit of historical background.

EARLY HISTORY

G. H. Shull is responsible for coining the word heterosis. He regarded this as a descriptive synonym for hybrid vigor, not intended to imply any mechanism. The early history is thoroughly reviewed in Gowen (1952), so I shall not give specific references to any of the pioneers. The book also includes a historical article by Shull himself.

The phenomenon of hybrid vigor was well documented in the 19th century; Darwin in particular gave a very detailed, lucid, and convincing account. Mendel himself observed that the hybrids between his tall and short pea varieties were taller than the tall parent. Some examples of hybrid luxuriance are spectacular. Shull notes the large size of hybrids between wild and cultivated sunflowers (*Helianthus* sp.). Perhaps the most extreme example is the hybrid between radish (*Raphanus sativus* L.) and cabbage (*Brassica oleracea* L.), which not only grew to the top of the greenhouse but out through an opening in the roof and down the sides. Yet, although luxuriant, it was sterile and became fertile only on polyploidization. But we are here concerned with heterosis as it applies to crosses within a species, and especially between inbred lines.

Almost from the time that Mendelian inheritance first came to be generally accepted, there have been two alternative theories. In 1908, both Shull and East suggested that different germ plasms stimulate development and that the stimulus increases with the diversity of these germ plasms; in Mendelian terms, this means heterozygosity. This is the *overdominance hypothesis*. I personally prefer Fisher's term *super-dominance*, but this has never caught on.

Alternatively, heterosis can be the result of the masking of deleterious recessives by dominant or partially dominant alleles, each strain bringing to the hybrid a somewhat different collection of favorable dominants. The generally deleterious nature of recessives was first emphasized by Davenport as early as 1908. This hypothesis, the *dominance hypothesis*, was first stated explicitly by Bruce in 1910, although curiously Bruce's algebraic argument could have been used just as effectively in support of overdominance. Experimental evidence came from Keeble and Pellew the same year. They found that pea hybrids exceeded their parents in height, and noted that two different dominant factors were involved, one lengthening the internodes and the other increasing their number. They mentioned that the same principle might apply to more complex crosses.

There were two early objections to the dominance hypothesis. First was the absence of a skewed F_2 distribution, which would be expected from the expansion of $(3/4 + 1/4)^n$. Second was the failure of selection to produce inbreds as good as hybrids. But in 1916, Jones noted that with linkage the two hypotheses became indistinguishable and Collins pointed out in 1921 that with a large n the F_2 distribution is essentially symmetrical.

Three quarters of a century ago, Wright (1922) presented a particularly, detailed and thoughtful analysis of extensive inbreeding and crossbreeding experiments with guinea pigs. He astutely realized that such studies could not distinguish among partial-, complete-, or overdominance. His articles could be used as a reference today without changing a word. In particular he showed that the results of inbreeding of the hybrids could be predicted *quantitatively*. And of course he told us how to measure inbreeding in complex pedigrees.

During the 1920s, 1930s, and early 1940s the dominance hypothesis held sway. It was generally preferred because it depended only on the common observation that recessives are deleterious and did not invoke a form of gene action for which examples were scarce.

THE HEYDAY OF OVERDOMINANCE

In the late 1940s there was a resurgence of interest in overdominance, largely through the influence among plant breeders of Fred Hull (1945). Hull was a relative and close friend of J.L. Lush, the leading American livestock geneticist, so the idea also spread rapidly through the animal breeding community. Several other geneticists also supported the idea. The idea was in the air and several people invented breeding systems that would capitalize on overdominance.

I published the mutation load argument for overdominance in 1948 and presented it at the 1950 Heterosis Conference (Crow, 1952). Using Haldane's (1937) genetic load principle, I noted that a randomly mating population would have its fitness reduced by u, the recessive mutation rate, relative to a population with all deleterious recessives at that locus removed. If the deleterious effects are additive over loci the overall reduction is Σu, the mutation rate per gamete. As long as the total reduction is small, the value is about the same if the effects are multiplicative, which seemed more likely.

This algebra deals with fitness, whereas my concern is yield. But, because of the long history of selection for yield, it seemed reasonable to equate yield to fitness. At least I proceeded on this assumption. Then the amount by which the yield would be increased if all homozygous recessives were removed, and hence the maximum increase in yield of hybrids over that of an open pollinated strain, is Σu.

In 1950 the estimates of total mutation rate, mainly from *Drosophila*, were about 0.05 per haploid genome. This argued that the dominance hypothesis could account for an increased performance of 5% or less, but not the 15 to 20% that was observed. So, I suggested overdominance as the most likely explanation for the discrepancy.

In a similar statement, Fisher (1949) said: ". . . it would appear that the total elimination of deleterious recessives would make less difference to the yield of cross-bred commercial crops than the total mutation rate would suggest. Perhaps no more than a 1% improvement could be looked for from this cause. Differences on the order of 20% remain to be explained."

My treatment when recessiveness was incomplete was confused. I noted that the mutation load is about twice as great with partial dominance, but wasn't sure of the relevance, although I mentioned this later (Crow, 1993) as an upper limit on the increase in hybrids. There is a tighter limit, however, obtained by assuming that the inbred is mated to an inbred free of mutations at this locus. Consider the standard model. The inbred is derived from the random-mating population and the prime allows for possible allele frequency change during inbreeding.

	Fitness or frequency			Mean fitness
Genotype	AA	Aa	aa	
Fitness	1	$1-hs$	$1-s$	
Random mating	p^2	$2pq$	q^2	$W_R = 1 - 2pqhs - q^2s$
Inbred	p'	0	q'	$W_I = 1 - q's$
Other inbred	1	0	0	
Hybrid	p'	q'	0	$W_H = 1 - q'hs$

First, assume that no change occurred during inbreeding ($q' = q$). The difference between the hybrid and random mating fitnesses is

$$\Delta = W_H - W_R = qhs + q^2(1 - 2h)s .$$ [1]

Let the mutation rate from A to a be u. With random mating the equilibrium allele frequency is given by solving for q in the expression

$$q^2(1 - 2h)s + qhs(1 + u) - u = 0 .$$ [2]

Substituting [2] into [1] leads to

$$\Delta = u(1 - qhs) .$$ [3]

Since q, h, and s are all positive quantities, $qhs > 0$. Also, from [2], noting that if there is to be heterosis, $h < 1/2$,

$$qhs = (u - q^2s(1 - 2h))/(1 + u) < u/(1 + u) < u .$$ [4]

Thus, from [3] we see that

$$u(1 - u) \leq \Delta \leq u ,$$ [5]

so the increase in fitness is equal to or less than the haploid mutation rate for either complete or partial recessiveness.

But there may have been selection during inbreeding so that $q' < q$. If this were completely effective, so that $q' = 0$, the increase in hybrid performance is u if dominance is complete and $2u$ if partial. It is unlikely, however, that selection during inbreeding has caused a substantial change, so allowance for this may raise the 5% limit to more like 6 or 7. In any case, it is not enough to account for the observed 15 to 20% increase.

Overdominance was the Zeitgeist of the Heterosis Conference and the main proponents were present. Hull presented his arguments. They were: (i) The yield of a hybrid frequently exceeds the sum of the inbred parents; with dominance and complete additivity, this shouldn't happen. (ii) Mass selection and especially selection in second and third cycle lines usually failed. (iii) There was a negative regression of the F_1 on parents when the other parent was held constant; this was consistent with overdominance. Hull had a large influence.

Comstock and Robinson presented their analyses of several mating schemes pointing to overdominance for yield. On a scale where 0 is no dominance (heterozygote half way between the homozygotes) and 1 is complete dominance, they found values above 1, such as 1.5, especially for yield. Dickerson introduced selection schemes that would improve performance with overdominance. A particularly popular suggestion from Comstock et al. (1949) was a breeding scheme, reciprocal recurrent selection, which was expected to improve performance with partial-, complete-, or overdominance. Many experiments based on this idea were started in both plants and animals. For a discussion of some of the later results, after many generations of selection, see Sprague (1983) and Stuber (1994).

Although overdominance appeared to be an important factor in the greater performance of hybrids than randomly mated populations, it did not appear to be an important factor in inbreeding depression. Consider a standard model.

Genotype	AA	Aa	aa	Mean or total
Relative fitness	1-t	1	1-s	W
Zygotic frequency	p^2	$2pq$	q^2	1

Ignoring mutation, the equilibrium allele frequencies are $p = s/(s + t)$ and $q = t/(s + t)$. The mean fitness of a randomly mating population is $W_R = 1 - st/(s + t)$ and of an inbred population with the same allele frequencies is $W_I = 1 - 2st/(s + t)$. Thus the inbreeding decline relative to a randomly mating population is the same as the increase if all individuals were heterozygous. Since increased yields from hybrids were some 15 to 20% whereas inbreeding decline was 50% or more, I concluded that overdominance is not the main explanation for inbreeding decline.

On the contrary, no such limitation applies to dominance. A recessive allele with a frequency q^2 in a randomly mating population becomes the much larger value q after inbreeding. For example, if $q^2 = 0.0001$, $q = 0.01$. Genes that are rare and make little contribution to the mean or variance make a much larger contribution after inbreeding. A corollary is that those alleles that are making the largest contribution to the outbred population are not usually the same as those contributing heavily to the inbred population.

Conclusions in 1952

The conclusions that I reached at the time (Crow, 1952), which were generally consonant with the views of many at the Heterosis Conference, were:

1. The dominance hypothesis can explain the deterioration from inbreeding and the recovery on outcrossing.

2. It is inadequate to explain how hybrids can greatly exceed the randomly mating populations from which the hybrids were derived.

3. The overdominance hypothesis demands a kind of gene action that is rare, but even if only a small minority of loci are of this type, they may be a major factor in population variance and heterosis.

INCREASING DOUBTS ABOUT OVERDOMINANCE IN MAIZE

Not long after the publication of "Heterosis" in 1952, I began to have doubts because of new experimental data. The estimates of mutation rates became larger as more studies were done, particularly rates for mutations with very small effects. This became particularly striking with the work of Mukai, first published in 1964. His *minimum* estimates of mutation rates for genes with very small effects on viability were an order of magnitude higher than the mutation rate of lethals. Not surprisingly, his work was greeted with some skepticism. I was both interested and skeptical and invited Mukai to repeat the experiment in my laboratory, which he did, with concordant results (Mukai et al., 1972).

At about the same time, Sprague reported high mutation rates for phenotypic characters in inbreds and doubled haploids (see Sprague, 1983). So, I had to conclude that my earlier argument for overdominance was weak. Rather than being limited to about 5%, the increase could be 25% or more.

In the 1960s other evidence also began to appear. One of the most impressive studies was done by Gardner (1963). He carried out one of the Comstock and Robinson mating schemes, one that allowed the chromosomes to randomize before each selection cycle. The results were striking. On the scale where 0 is no dominance and 1 is complete dominance, the values in successive cycles dropped from the overdominant range, 1.4, to about 0.5, partial dominance. Similar results were reported by Moll et al. (1963). The clear conclusion was that the statistical overdominance in the early generation was a consequence of linkage disequilibrium, i. e., favorable dominants linked to deleterious recessives.

There was other evidence, too. Mass selection experiments were effective when the experimental design and statistical controls were adequate to minimize environmental effects (Gardner, 1969). Inbred lines were much improved. Although not as good as the hybrids, they were better than hybrids of earlier years. Another source of evidence was an experimental test devised by Sprague. The results are summarized in Sprague (1983). In this experiment two maize (*Zea mays* L.) populations were each selected for five generations for improved performance of crosses with an inbred tester. With overdominance the two selected populations should become similar to each other, each having alleles complementary to the tester leading to increasing heterozygosity in the cross. The results were otherwise. Each of the populations showed increased yield and crosses between the two also gave increased yield in successive generations. Furthermore, the selected populations performed well with different inbred testers. These experiments indicated that additive and dominant gene effects were mainly responsible for the increase in heterosis.

Despite these average effects, there are individual examples of what appears to be true overdominance in several plant species. Although unusual, they could well be important if they could be utilized. Likewise, there is evidence for some epistasis. Stringfield (1950) found that in many cases backcrosses showed consistently higher yields than the F_2. This suggests interaction, as if the gene combinations selected during inbreeding are broken up in the F_2, but partially retained in backcrosses. Similarly, double crosses usually yield somewhat less than two-way crosses (Troyer, 1991).

Conclusions Regarding Maize

In summary, I can do no better than to quote George Sprague (1983):

"Studies have shown that additive and dominance gene effects are generally much greater than other types of gene effects. Additive effects are precisely those which respond to selection. Specially designed experiments have shown that both overdominance and epistasis exist, but neither has been shown to be important at the population level. . . . Thus, as far as the maize breeder is concerned, a pragmatic solution to the dominance-overdominance controversy has been reached. Additive and dominance effects provide a satisfactory model for the heterosis and for the rather remarkable progress achieved through breeding. Genetic variance estimates for populations under selection indicated little decrease in variability thus giving assurance of further substantial progress."

EVIDENCE FROM DROSOPHILA

The 1960s and 1970s saw a number of *Drosophila* studies bearing on this question. One is the set of experiments of Mukai and others (1964, 1972), mentioned above, and Ohnishi (1977). The experiments consisted of sequestering a chromosome for many generations, keeping it heterozygous and in males so there was no crossing over. To accomplish this, a male, heterozygous for this chromosome, was mated to a standard stock containing identifying mutant markers. A single son was mated to the same standard stock, and this backcrossing process was continued for many generations. Thus, except for dominant effects, which were minimized by using a single male each generation and growing the progeny under optimum conditions for fly-husbandry, there was no selection on the sequestered chromosome. During this period, mutations accumulated, largely unopposed by selection. Then after a suitable time (about every 10 generations) the chromosome was made homozygous and its viability measured by comparison to its sibs.

As expected, the accumulated mutations lowered the viability. The decrease in homozygous viability for this chromosome was about 0.0035 per generation (0.0017 to 0.0049 in different replications); this amounts to about 0.01 per haploid genome. The decrease was essentially linear over the time studied, about 40 generations. These experiments were also used to estimate the mutation rate of mild viability-reducing alleles, but I shall not discuss them here except to note that they were much more frequent than lethals and were part of the evidence for the mutation improvement being much higher than I had assumed in 1950 (for a summary, see Crow, 1993).

Keightley (1996) has suggested that the effect of mutation accumulation on the sheltered chromosomes is an artifact, possibly because of improved viability of the tester chromosomes with which the sheltered chromosomes were compared. This seems unlikely because the testers were long-time laboratory strains that had had many generations to equilibrate. Furthermore Kondrashov (personal communication) has used frozen flies as a control with similar results. So I think the estimates of mutational decline are essentially correct.

In a similar experimental technique, chromosomes can be extracted from a natural population and made homozygous. After exclusion of lethals, the reduction in viability due to mutations with individually minor effects was about 0.12 per chromosome (Temin, 1966). We should expect that there will be as many mutations as occur in a single generation multiplied by the number of generations that a mutation remains in the population before being eliminated by selection. I call this number the *persistence*. Thus the ratio of the decreased homozygous viability of chromosomes from natural populations to the decrease per generation from mutation accumulation is a rough estimate of the mean persistence. The values (0.12/0.0017 and 0.12/0.0049) are 71 and 24. There are reasons why these may be underestimates. Other experiments, based on accumulated mutational variance give values between 50 and 100. A reasonable value is 50 to 80, with considerable confidence that it is 100 or less.

If the mutations were completely recessive they would persist until eliminated by inbreeding or by chance homozygosity. In *Drosophila* this occurrence is probably not oftener than 1/1000, so the persistence is too short for complete recessivity. *A fortiori*, it is much too short for overdominance. Overdominance has been invoked as an explanation for polymorphisms that persist for very long periods, such as antigenic and self-incompatibility factors. I emphasize, however, that such studies based on allele frequency kinetics cannot decide between overdominance and frequency-dependent selection (Denniston & Crow, 1990).

Note that, although the measurements are on viability alone, the persistence is a measure of fitness, since persistence of a mutant depends on viability, fertility, and all components of fitness acting on that chromosome. The selection is almost always on heterozygotes, since homozygotes are so rare.

Finally, direct measurements of genetic variability of viability (Mukai et al. 1974) showed a large additive component. The dominance variance was too small, relative to additive variance, to leave much room for overdominance.

Conclusions from Drosophila

I conclude from these and other *Drosophila* studies that mutations affecting homozygous viability have appreciable deleterious effects on fitness as heterozygotes. There is no evidence for any large number of complete recessives, or of overdominants. Nor does there seem to be much epistasis. Individual experiments indicate that overdominant loci do exist, but they must be a small minority.

OTHER SPECIES

I have discussed only maize and Drosophila. How far can we generalize from these two? I suspect that the conclusions concerning the large amounts of additive and dominance variance, and relatively small contributions from overdominance and epistasis will turn out to be general among cereals. Self-fertilizing species may have differences because of the purging of deleterious recessives by long time inbreeding. In any case, whether the detailed mechanisms are the same, heterosis is ubiquitous and can be of great practical value.

Research on other species is proceeding rapidly, often following the patterns set by earlier maize studies. Some indication of the diversity of species studied (as well as diversity of countries and languages) is suggested by the following books: Frankel (1983), Fischer (1978), Rai (1979), Turbin et al. (1967), and Virmani (1994); and of course by other papers in this conference. For a recent summary of selection for improvement in tropical maize, see Pandey and Gardner (1992).

MOLECULAR STUDIES

Although population experiments, such as I have been describing, argue that the great bulk of variance in fitness traits and yield is additive and dominance, there may well be relatively rare, but perhaps important, instances of overdominant loci and specific epistatic interactions.

We can expect sequence information to be increasingly valuable as it becomes more extensive. The maize genome is very large, some 2 billion base pairs, comparable to the human. In contrast, rice, which presumably has the same number of genes, has much less DNA, 430 million base pairs. Whether to proceed directly with maize sequencing or start with rice is, if I read the news correctly, still being debated. But regardless of the strategy adopted, we shall eventually have a complete sequence for all the common cereals. Sequence information will surely add to our fundamental core of knowledge and should be of great value in identifying genes affecting qualitative traits. How useful it will be for complex traits like yield remains to be seen. But we needn't wait for full sequence information. Much useful information can come from expressed sequence tags. These should clearly provide information useful for qualitative, simply inherited traits. And many of these are of great practicality.

Quantitative trait loci are being actively sought. Stuber et al. (1992) carried out extensive studies on two elite inbred lines and their hybrids, along with F_3 and backcrosses. These were replicated in six environments in three states. They had 76 markers available, which covered 90 to 95% of the genome. Almost every chromosome had at least one QTL affecting grain yield. Some of these had large effects on yield, and even more on some other traits. The experiments were designed to maximize the ability to detect QTLs affecting heterosis. They were successful, for collectively the identified QTLs contributed substantially to the heterosis of the cross.

For the purposes of this discussion, two points are of special interest. First, there was very little genotype x environment interaction. Although perhaps surprising, in view of other experiments, this suggests that detection of major QTLs may not require tests in many environments. Second, almost all the QTLs seemingly showed overdominance. Yet the size of the linked units is such that it is not possible to distinguish between overdominance and pseudo-overdominance due to repulsion linkages, as the authors note. From what has been discovered from population experiments, I would expect most if not all of these to turn out to be pseudo-overdominance. The same conclusion was reached by Cockerham and Zeng (1996). It may be difficult to determine the answer, for extensive recombination tests would be required. Possibly direct cloning can provide the answer. Meanwhile, this information can be put to practical use without answering the dominance-overdominance question. For a discussion of combining molecular markers with standard breeding methods, see Stuber (1994). As individual genetic components are discovered, they can be used to enhance the performance of future hybrids.

If important overdominant loci are discovered, they can be incorporated into appropriate inbreds. There also is a fascinating possibility that if two alleles interact to produce a highly favorable combination, the individual components can be gotten into the same chromosome by using molecular trickery to produce unequal crossovers. The question was raised by Haldane in connection with sickle-cell anemia many years ago, but now with gene cloning it seems like a practical possibility. And we need not be content with just two copies. Perhaps three or more can be built in without the necessity for polysomy or polyploidy.

The large amount of genetic variability that is available (Sprague, 1983) means that rapid progress can be made by standard breeding methods. Grafted on to this will be the new findings that are sure to come from molecular analysis (Stuber, 1994) There are exciting times ahead.

ACKNOWLEDGMENTS

I thank Carter Denniston and the reviewers for some very helpful comments.

REFERENCES

Cockerham, C.C., and Z.-B. Zeng. 1996. Design III with marker loci. Genetics 143:1437-1456.

Comstock, R.E., H.F. Robinson, and P.H. Harvey. 1949. A breeding procedure designed to make maximum use of both general and specific combining ability. Agron. J. 41:360-367.

Crow, J.F. 1948. Alternative hypotheses of hybrid vigor. Genetics 33:477-487.

Crow, J.F. 1952. Dominance and overdominance. p. 282-297. In J.W. Gowen (ed.) Heterosis. Iowa State College Press, Ames.

Crow, J.F. 1993. Mutation, mean fitness, and genetic load. Evol. Biol. 9:3-42.

Denniston, C., and J.F. Crow. 1990. Alternative fitness models with the same allele frequency dynamics. Genetics 125:201-205.

Fischer, H.E. 1978. Heterosis. Grundlagen, Ergebnisse und Probleme in Einzel-darstellungen. Genetik 9:9–163.

Fisher, R.A. 1949. The theory of inbreeding. Oliver and Boyd, Edinburgh, England.

Frankel, R. (ed.). 1983. Heterosis. Springer-Verlag, Berlin.

Gardner, C.O. 1963. Estimates of genetic parameters in cross fertilizing plants and their implications in plant breeding. p. 225–252. *In* W.D. Hanson and H.F. Robinson (ed.) Statistical genetics and plant breeding. NAS–NRC, Washington, DC.

Gardner, C.O. 1969. Genetic variation in irradiated and control populations of corn after 10 cycles of mass selection for high grain yield. p. 469–477. *In* Induced mutations in plants. AEA, Vienna.

Gowen, J.W. (ed.). 1952. Heterosis. Iowa State Col. Press, Ames.

Haldane, J.B.S. 1937. The effect of variation on fitness. Am. Nat. 71:337–349.

Hull, F.H. 1945. Recurrent selection for specific combining ability in corn. J. Am. Soc. Agron. 37:134–145.

Keeble, F., and C. Pellew. 1910. The mode of inheritance of stature and time of flowering in peas (*Pisum sativum*). J. Genet. 1:47–56.

Keightley, P.D. 1996. The nature of deleterious mutation load in *Drosophila*. Genetics 144:1993–1999.

Moll, R. H., M. F. Lindsey, and H. F. Robinson. 1963. Estimates of genetic variances and level of dominance in maize. Genetics 49:411–423.

Mukai, T. 1964. Spontaneous mutation rate of polygenes controlling viability. Genetics 50:1–19.

Mukai, T., S.I. Chigusa, L.E. Mettler, and J.F. Crow. 1972. Mutation rate and dominance of genes affecting viability in *Drosophila melanogaster*. Genetics 72:335–355.

Mukai, T., R.A. Cardellino, T.K. Watanabe, and J.F. Crow. 1974. The genetic variance for viability and its components in a local population of Drosophila melanogaster. Genetics 78:1195–1208.

Ohnishi, O. 1977. Spontaneous and ethyl methanesulfonate-induced mutations controlling viability in Drosophila melanogaster. II. Homozygous effects of polygenic mutations. Genetics 87:529–545.

Pandey, S., and C.O. Gardner, 1992. Recurrent selection for population, variety, and hybrid improvement in tropical maize. Adv. Agron. 48:1–87.

Rai, B. 1979. Heterosis Breeding. Agro-Biological Pub., Delhi, India.

Sprague, G.F. 1983. Heterosis in maize: Theory and practice. p. 48–70. *In* R. Frankel (ed.) Heterosis. Springer-Verlag, Berlin.

Stringfield, G.H. 1950.   Heterozygosis and hybrid vigor in maize. Agron. J. 42:145–152.

Stuber, C.W. 1994. Heterosis in plant breeding. Plant Breed. Rev. 12:227–251.

Stuber, C.W., S.E. Lincoln, D.W. Wolff, T. Helentjaris, and E.S. Lander. 1992. Identification of genetic factors contributing to heterosis in a hybrid from two elite maize inbred lines using molecular markers. Genetics 132:823–839.

Temin, R.G. 1966. Homozygous viability and fertility loads in Drosophila melanogaster. Genetics 53:27–46.

Troyer, A.F. 1991. Breeding corn for the export market. p. 165–176. In D.. Wilkinson (ed.) Proc. 46th Ann. Corn and Sorghum Res. Conf., 11–12 Dec. 1991. Chicago, IL. Amer. Seed Trade Assn., Washington, DC.

Turbin, N.V., V.G. Volodin,. A.N. Palilova, and L.V. Khotyleva. 1967. Genetics of Heterosis. Israel Program for Scientific Translation, Jerusalem.

Virmani, S. S. 1994. Heterosis and hybrid rice breeding. Springer-Verlag, Berlin.

Wright, S. 1922. The effects of inbreeding and crossbreeding on guinea pigs. U.S. Dep. Agric. Tech. Bull. 1090, 1121.

Chapter 6

Epistasis and Heterosis

C. J. Goodnight

INTRODUCTION

Traditionally heterosis has been attributed to dominance interactions, particularly overdominance and the masking of deleterious alleles (Falconer, 1989; Hartl & Clark, 1989); however there are several other possible causes of heterosis. Early on it was recognized that when a complex trait is due to several underlying multiplicatively acting traits heterosis may occur even in the absence of dominance. For example, if plant height is the product of internode length and number of nodes heterosis can result from a cross between a strain with a large number nodes and a strain with large internode length (Richey, 1942; Williams, 1959, 1960; Grafius, 1959; Schnell & Cockerham, 1992). Schnell and Cockerham (1992) modeled the more general case of multiplicatively acting gene effects, and found that as with interactions among traits, additive by additive epistasis resulting from multiplicatively acting gene effects can lead to heterosis.

Inbreeding depression in systems with only additive and dominance effects result in a linear decline in fitness with increased inbreeding; however, experimental studies of inbreeding depression in *Tribolium* flour beetles finds that in many lines there is a significant nonlinear decline in fitness with increases in Wright's inbreeding coefficient, F (Pray & Goodnight, 1995), a result that is consistent with models including epistasis, but at odds with models including only additive and dominance effects. Although in *Tribolium* there is evidence that epistasis does influence traits in populations subjected to inbreeding and interline crosses, other studies have failed to find evidence of two locus epistasis contributing to heterosis (e.g., Xiao et al., 1995).

In this chapter I will discuss one of the consequences of inbreeding and hybridization in systems with epistasis. That is, the effect of an allele on the phenotype changes with inbreeding, and as a result heterosis in an interline cross will not be predictable either from the characteristics of the individual lines, or from crosses between the individual lines and other lines.

PHYSIOLOGICAL AND STATISTICAL EPISTASIS

Cheverud and Routman (1995) identify two concepts of gene interaction, physiological epistasis and statistical epistasis. They define physiological epistasis to occur when genotypic values at one locus vary depending on the genotype present at other loci. Perhaps understandably physiological epistasis is the concept of epistasis with which most biologists are familiar. Following Cheverud and Routman (1995), and for simplicity assuming two loci each with two alleles per locus, the two locus (physiological) genotypic values, G_{ijkl}, are the average phenotype of individuals with the **ij**th genotype at the first locus and the **kl**th genotype at the second locus. The one locus genotype is defined as the unweighted average across the genotypes at the second locus:

$$Gij.. = \frac{(G_{ij11} + G_{ij12} + G_{ij22})}{3}$$

and

$$G..kl = \frac{(G_{11kl} + G_{12kl} + G_{22kl})}{3}$$

where the subscripts 1 and 2 refer to the two alleles at the interacting locus. Cheverud and Routman define the *nonepistatic genotypic value* to be:

$$ne_{ijkl} = G_{ij..} + G_{..kl} - G_{....}$$

and the *epistatic genotypic value* to be:

$$e_{ijkl} = G_{ijkl} - ne_{ijkl}$$

A value of e_{ijkl} different from zero indicates that physiological epistasis is present (see Cheverud & Routman [1995] for a more complete mathematical description of physiological epistasis.) Physiological epistasis has the conceptual advantage that it is a property of genotypes that can be measured independently of population context. Most importantly physiological epistasis does not change as gene frequencies change.

The general concept physiological epistasis is developed for two interacting diallelic loci, but it is easily extended to multiple interacting loci, and multiple alleles per locus.

Statistical epistasis occurs when there are quantitative genetic variance components that can be attributed to interactions among genes. Whereas physiological epistasis is a property of a genotype that is unaffected by population gene frequencies, statistical epistasis is a property of populations that does change with gene frequency.

In standard quantitative genetics the phenotype of an individual can be divided into genetic and environmental effects (assuming no genotype-environment interactions):

$$z_j = g_j + e_j$$

where z_j, g_j and e_j are the phenotype, genotypic value, and environmental deviation respectively for the *j*th individual. In the absence of genotype by environment interactions g_j is a constant and corresponds to the G_{ijkl} defined for physiological epistasis. Environmental deviations are measured from the mean of the genotypic value. Thus environmental deviations have a mean of zero.

The genotypic value, g_j, can be further broken up into the mean, μ, additive, a, dominance, d, and interactive, i, effects:

$$z_j = \mu + a_j + d_j + i_j + e_j$$

Unlike the genotypic value the partitioning of the genotypic value into the mean, additive, dominance, and epistatic components is not constant, rather, it changes as gene frequencies change. This occurs because the components of the genotype are defined statistically, and they are a function not only of the genotype, but also the population in which they are measured. This is best illustrated using only dominance, however, a similar three dimensional graph can be developed for a two locus epistatic system. Following Falconer (1989) consider a single locus with two alleles, A_1 and A_2. The three genotypes are A_1A_1, A_1A_2, and A_2A_2 with (physiological) genotypic values of G_{11}, G_{12}, and G_{22} respectively (Fig. 6–1). The arbitrary scale on which the physiological genotypic values are measured is shown on the left side of the graph. The quantitative genetic effects are measured as a deviation from the population mean (+ in Fig. 6–1). Obviously a change in the population mean does not change the genotypic value, but it does change the partitioning of the genotypic value into quantitative genetic components. The additive genetic value (or breeding value) of the genotype can be defined by the regression of phenotypic value on genotypic value (solid line in Fig. 6–1). This regression is weighted by the genotype frequencies, and it changes as gene frequencies change. The dominance deviation is defined as the difference between the additive effect and the single locus genotypic value, as a result the dominance deviation changes whenever the additive effect changes. By similar reasoning epistatic effects will also change with gene frequency.

The additive effects hold a special place in quantitative genetics and they are normally referred to as breeding values (Falconer, 1989). Breeding values are measured experimentally as twice the deviation of the offspring of an individual

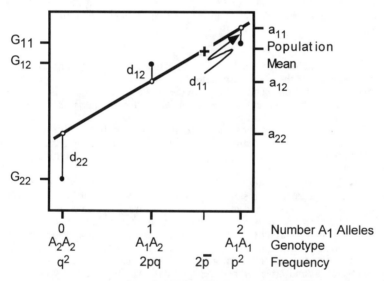

Fig. 6–1. A schematic of the relationship between physiological genotypic values and statistical genetic components for a system with only additive and dominance effects. Closed circles are genotypic values measured on an arbitrary scale on the left side of the graph (G_{11} etc.). The genotypic values are divided into the mean, additive genetic effects (breeding value, a_{11} etc.) and dominance deviations (d_{11} etc.) by performing a regression of phenotype (G_{11} etc.) on genotype (number of A_1 alleles). This regression is weighted by gene frequency, thus the slope will change as gene frequencies change. Note that the regression line will always go through the bivariate mean phenotype and genotype. A similar, although more complicated, three dimensional graph can be drawn for two locus epistatic systems (redrawn from Falconer, 1989).

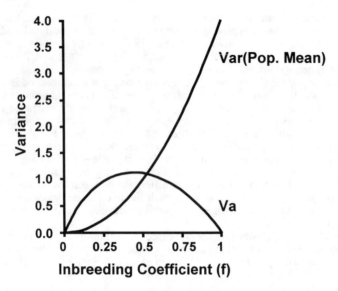

Fig. 6–2. The additive genetic variance (Va), variance among lineages [Var(Pop. Mean] as a function of Wright's inbreeding coefficient for additive by additive epistasis.

measured as a deviation from the population mean. The within population variance in breeding values is the additive genetic variance that is directly proportional to the response to selection. The power provided by a statistical concept of genetic effects, including epistasis, is that the variances, and particularly the additive genetic variance, provide the best predictors of evolutionary potential within a population. Thus, while there is some cost in conceptual simplicity in moving from a physiological concept of genetic effects to a statistical concept, the predictive power provided by the statistical concept of genetic effects justifies this more complicated view.

Standard models of quantitative genetics consider only additive genetic effects with no dominance or epistasis (e.g., Lande, 1979). Under the assumption of additive effects, inbreeding and genetic drift on average decrease the additive genetic variance, whereas migration and crossing between populations will increase the additive genetic variance. When there is gene interaction (dominance and epistasis) this simple relationship will not necessarily be maintained. That is, inbreeding may increase the additive genetic variance (Goodnight, 1988) and conversely, migration and crossing between populations can potentially decrease the additive genetic variance (Whitlock et al., 1993) (Fig. 6–2).

EPISTASIS AND THE AVERAGE EFFECT OF AN ALLELE

The increase in additive genetic variance described above raises some interesting questions. The additive effects or breeding values are the sum of the average effects of alleles. The average effect of an allele is the effect of an allele on the phenotype measured as a deviation from the population mean. For diploid organisms the additive genetic variance is two times the variance in the average effects of alleles. Because of the simple definition of additive genetic variance there are only two ways that it can increase. One of these is to increase the number of alleles, and the second is to increase the range of alleles. Inbreeding inevitably de-

creases the effective number of alleles, thus the increase in additive genetic variance associated with inbreeding in epistatic systems must be due to a spreading of
the range of the alleles (Goodnight, 1995). However, this spreading of average effects is not a simple change of scale, rather it also involves a change in the rank
order of alleles (Fig. 6–3). This change in rank order of alleles is of particular importance during inbreeding and the crossing of populations. Typically discussions
of inbreeding and population crosses focus on inbreeding depression and heterosis.
This shifting in the average effects of alleles is a very different effect. It means that
the effect of an allele on the phenotype may change as the genetic background
changes. In general any increase in additive genetic variance associated with inbreeding will be associated with shifts in the average effects of alleles. From a
practical standpoint what this shifting in average effects means is that when there is
gene interaction the performance of an allele in one population is not a reliably
predictor of its performance in other populations. It also means that crossing inbred lines may cause a sufficient shift in the genetic background to cause marked
changes in the average effects of alleles. Clearly in any attempt to predict the results of inter line crosses it is helpful to have some idea of the role of gene interaction in determining the expression of a trait.

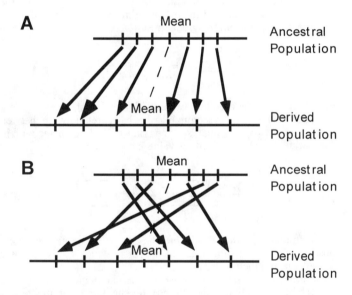

Fig. 6–3. Schematic of the increase in additive genetic variance associated with
    inbreeding. The horizontal lines represent the range of average effects in the
    ancestral and derived populations, the vertical lines represent particular alleles,
    and the lines connecting them track there movement. (A) Hypothetical situa-
    tion where the increase in additive genetic variance is simply due to a change in
    scale. (B) Actual situation where an increase in additive genetic variance is as-
    sociated with a change in the rank order of alleles. Note that the population
    mean is expected to change with inbreeding.

MEASURING THE SHIFT IN AVERAGE EFFECTS OF ALLELES

Recently (Goodnight, 1995; Wade & Goodnight, 1997) methods for measuring this shift in average effects have been developed. Consider first the concept of breeding value in a single randomly mating population. In this population the phenotype of the lth offspring of the kth dam and jth sire can be represented as Z_{jkl}. The breeding value of the jth sire is:

$$Aj = \frac{\sum_k \sum_l Z_{jkl}}{KL} - \overline{Z}_{...} = \overline{Z}_{j..} - \overline{Z}_{...}$$

A dot subscript indicates that an average has been taken. Fisher (1958) defined the variance in A_j to be the additive genetic variance (noting that $\overline{A}_. = 0$):

$$V_A = \frac{\sum_j A_j^2}{J}$$

Now consider a set of sires mated in several genetically differentiated strains. It is now necessary to add an additional summation, and define the *local breeding value*, or the mean value of the offspring of the jth sire mated in ith deme and measured as a deviation from the global mean:

$$A_{ij} = \frac{\sum_k \sum_l Z_{ijkl}}{KL} - \overline{Z}_{....} = \overline{Z}_{ij..} - \overline{Z}_{....}$$

Note that the variance in local breeding values no longer equals the additive genetic variance (Falconer, 1985). By recognizing the basic relationship between this breeding design and a two way analysis of variance (and again recognizing that $\overline{A}_{..} = 0$) the variance in local breeding value can be divided into three components:

$$V_{LBV} = \frac{\sum_i \sum_j A_{ij}^2}{IJ} = \frac{\sum_i A_{i\bullet}^2}{I} + \frac{\sum_j A_{\bullet j}^2}{J} + \frac{\sum_i \sum_j A_{i*j}^2}{IJ}$$

where $A_{i\bullet}$ is the among sires mean local breeding value in the ith deme, $A_{\bullet j}$ is the among demes mean local breeding value of the jth sire, and A_{i*j} is the interaction between the ith deme and the jth sire.

The variance in the interaction between sires and demes, A_{i*j}, is of particular interest. It can be shown that in the absence of gene interaction this term will be zero, whereas when dominance or epistasis is present it will be non zero. Thus, it is the size of this parameter that will determine the stability of an allele across an array of strains. If the variance in A_{i*j}, $\mathrm{Var}(A_{i*j})$, is approximately 0 then the effect of an allele is highly predictable in interstrain crosses. If $\mathrm{Var}(A_{i*j}) > 0$ then its effect will be less predictable.

In practical terms A_{i*j} is no more difficult to measure than any other quantitative genetic parameter. The mean within sire among deme variance in the local breeding value measures

$$V_{sire} = \frac{\sum_i A_{i\bullet}^2}{I} + \frac{\sum_i \sum_j A_{i\bullet j}^2}{IJ}$$

while the among deme variance in the mean local breeding value measures

$$V_{mean(sire)} = \frac{\sum_i A_{i\bullet}^2}{I}$$

thus, the variance in the sire by deme interaction is given by the residual variance in local breeding values, $V_{sire} - V_{mean(sire)}$.

Finally, it is convenient to express the stability of the local breeding value for a trait as the correlation in local breeding values:

$$Corr(LBV) = \frac{V_{mean(sire)}}{V_{sire}} = \frac{V_{i\bullet}}{V_{i\bullet} + V_{i\bullet j}}$$

In the absence of gene interaction this correlation will equal 1. When gene interaction is contributing to a trait it will be < 1. Values for Corr(LBV) are shown in Fig. 6–4 for the standard forms of two locus effects (additive, dominance, additive by additive epistasis, additive by dominance epistasis, dominance by additive epistasis and dominance by dominance epistasis (Cockerham, 1954).

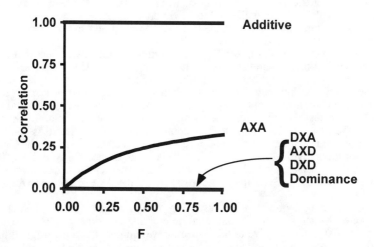

Fig. 6–4. The correlation in local breeding values for the six standard forms of two locus genetic effects (Cockerham, 1954) as a function of Wright's inbreeding coefficient, F. Additive = additive effects, Dominance = dominance effects , AXA = additive by additive epistasis, DXA = dominance by additive epistasis, AXD = additive by dominance epistasis, DXD = dominance by dominance epistasis. Note that dominance as well as epistasis can cause a shift in the average effects of alleles.

The effects of genetic drift on systems has been developed here in terms of standard quantitative genetics and breeding values, it can equally well be developed for quantitative trait loci (QTL) and the average effects of alleles.

EPISTASIS AND HETEROSIS

The above discussion makes it apparent that one of the major effects of epistasis on inbreeding and line crossing is to shift the average effects of alleles; however, the hallmark of line crossing is heterosis, or a generalized increase in vigor in hybrids. Traditionally this has been attributed to the effects of within locus dominance. There are two models of how dominance affects heterosis. These are the "overdominance" model in which there is a generalized advantage to heterozygosity and the "deleterious recessive" model in which deleterious recessive alleles are exposed by the increased homozygosity of inbred lines and hidden by the complete heterozygosity of hybrids. With epistasis there is a third possible model in which inbreeding in epistatic systems causes a change in the genetic background shifting the average effects of alleles.

I will use dominance by dominance epistasis (Fig. 6–5A) to illustrate how epistatic interactions can cause heterosis. Dominance by dominance epistasis is an interaction between the dominance effects at two loci. Inbreeding increases the frequency of double homozygote genotypes ($A_1A_1B_1B_1$ etc.). For dominance by dominance epistasis it leads to inbreeding depression, even though there is no evidence of dominance in the outbred population ($p = 0.5$ at both loci). Some crosses between inbred lines (e.g., $A_1A_1B_1B_1 \times A_2A_2B_1B_1$) will lead to heterosis, whereas others (e.g., $A_1A_1B_1B_1 \times A_2A_2B_2B_2$) will show no heterosis.

It is interesting to note that (statistical) epistatic interactions will often be reduced to a simpler form of genetic effect. In the first cross ($A_1A_1B_1B_1 \times A_2A_2B_1B_1$) the B locus is not segregating since both lines are fixed for the B_1 allele. Thus, a genetic analysis of this cross would reveal that the heterosis was due to simple overdominance at the A locus. Even though the underlying cause is dominance by dominance epistasis the two locus interaction can not be detected in this particular cross.

This epistatic model is similar to the dominance models, except that it is a multilocus effect rather than the effect of loci acting singly. This has several interesting consequences. First, the loci responsible for heterosis may not be apparent in some genetic backgrounds. In the dominance by dominance epistasis example both loci are neutral in an outbred population with a gene frequency of 0.5 at both loci. Nevertheless these loci lose their neutrality as fixation occurs at the interacting locus and they do contribute to inbreeding depression and heterosis. Second, the observed level of heterosis is not necessarily predictable based on the characteristics of the individual lines. Note that the $A_2A_2B_1B_1$ and the $A_2A_2B_2B_2$ lines have identical phenotypes, and they show heterosis when crossed. However when crossed with the $A_1A_1B_1B_1$ line only crosses with the $A_2A_2B_1B_1$ line will show heterosis.

Dominance by additive epistasis (Fig. 6–5B) can be used to illustrate another point of interest. That is that interline crossing can dramatically change the dynamics of selection relative to how selection would operate in the original lines. For example, a dominant by additive locus fixed for the B_2 allele by inbreeding will be overdominant at the A locus. An $A_1A_1B_2B_2 \times A_2A_2B_2B_2$ cross will show heterosis apparently due to overdominance, whereas an $A_1A_1B_2B_2 \times A_2A_2B_1B_1$ cross will initially show apparent additivity. However selecting on the population resulting from the second cross will potentially result in two populations with the same phenotype ($A_1A_1B_1B_1$ and $A_2A_2B_1B_1$), but showing hybrid breakdown when crossed.

A:
Dominance by Dominance Epistasis

	A_1A_1	A_1A_2	A_2A_2
B_1B_1	−1	1	−1
B_1B_2	1	−1	1
B_2B_2	−1	1	−1

B:
Dominance by Additive Epistasis

	A_1A_1	A_1A_2	A_2A_2
B_1B_1	1	−1	1
B_1B_2	0	0	0
B_2B_2	−1	1	−1

Fig. 6–5. Two "pure" forms of two locus epistasis. Two locus genotypes are listed on the margins, and the phenotypic value (standardized to 1 or −1) is listed in the table. **(A)** Dominance by dominance epistasis showing overdominance or underdominance at one locus depending on the genotype at the interacting locus. **(B)** Dominance by additive epistasis. The dominant (A) locus is overdominant, neutral or underdominant depending on the genotype at the additive (B) locus. The direction of increased phenotype at the B locus depends on the genotype at the A locus. At a gene frequency of 0.5 at both loci there is no additive genetic variance or dominance variance for either form of epistasis.

CONCLUSIONS

Inbreeding depression and heterosis inevitably require some form of gene interaction. Traditionally only within locus interactions have been considered. However, experimental evidence suggests that interlocus genetic interactions or epistasis should also be considered (e.g., Pray and Goodnight, 1995). Inbreeding and line crossing in systems with epistasis are best considered in terms of their effects on the correlation in local breeding values. In additive systems this correlation is 1. This means that the effect of an allele on the expression of a trait (relative to other alleles at the same locus) in any line or cross can be predicted from its effect in a single line. When there is epistasis this correlation drops below 1, and for epistatic interactions involving dominance it falls to 0. This means that the relative performance of an allele in one line is not predictive of its performance in other lines or crosses. When this occurs heterosis at a particular locus is a function of the genetic background, and an allele of mediocre performance in one line or cross may lead to substantial heterosis when moved to another genetic background.

REFERENCES

Cheverud, J.M., and E.J. Routman 1995. Epistasis and its contribution to genetic variance components. Genetics 139:1455–1461.

Cockerham, C.C. 1954. An extension of the concept of partitioning heredity variance for analysis of covariance among relatives when epistasis is present. Genetics 39:859–882.

Falconer, D.S. 1985. A note on Fisher's "average effect" and average excess". Genetical Research Cambridge 46:337–347.

Falconer, D. S. 1989. Introduction to quantitative genetics. London, Longman.

Fisher, R. A. 1958. The genetical theory of natural selection. 2nd ed. Dover, New York.

Goodnight, C.J. 1988. Epistasis and the effect of founder events on the additive genetic variance. Evolution 42: 441–454.

Goodnight, C.J. 1995. Epistasis and the increase in additive genetic variance: Implications for Phase 1 of Wright's shifting balance process. Evolution 49:502–511.

Goodnight, C.J. 1997. Quantitative trait loci and gene interaction: The quantitative genetics of metapolulations. (in preparation??)

Grafius, J.E. 1959. Heterosis in barley. Agron. J. 51:551–554.

Hartl, D., and A.G. Clark 1989. Principles of population genetics. Sinauer Associates. Sunderland, MA.

Lande, R. 1979. Quantitative genetic analysis of brain: Body size allometry. Evolution 33:402–416.

Pray, L.A., and C.J. Goodnight 1995. Genetic variability in inbreeding depression in the flour beetle, *T. castaneum*. Evolution 49:176–188.

Richey, F.D. 1942. Mock-dominance and hybrid vigour. Science(Washington, DC) 96:280–281.

Schnell, F.W., and C.C. Cockerham 1992. Multiplicative vs. arbitrary gene action in heterosis. Genetics 131:461–469

Whitlock, M.C., P.C. Phillips, and M.J. Wade 1993. Gene interaction affects the additive genetic variance in subdivided populations with migration and extinction. Evolution 47:1758–1769.

Wade, M.J., and C.J. Goodnight. 1997. Genetics and Adaptation in Metapopulations: When Nature Does Small Experiments. Evolution (submitted).

Williams, W. 1959. Heterosis and the genetics of complex characters. Nature(London) 184:527–530

Williams, W. 1960. Heterosis and the genetics of complex characters. Heredity 15:327–328.

Xiao, J.H., J. Li, L.P. Yuan, and S.D. Tanksley. 1995. Dominance is the major genetic basis of heterosis in rice as revealed by QTL analysis using molecular markers. Genetics 140:745–754.

Chapter 7

Inbreeding and Heterosis

J. B. Miranda Filho

INTRODUCTION

The inbred-hybrid system outlined by Shull (1908, 1909) and East (1908) still remains as the most important breeding scheme for the commercial production of hybrid seeds. To accomplish the purpose of hybrid seed production two phenomena are inevitably involved as the genetic framework toward the exploitation of genetic effects: inbreeding and heterosis. The inevitable consequence of inbreeding is the increase in the level of homozygosity that leads to a depressive effect in the expression of traits, known as *inbreeding depression*. Heterosis refers to the increase in the expression of quantitative traits in the cross between divergent parents, so that different alleles existing separately in the parents appear in the hybrid in heterozygous condition; the high proportion of heterozygosity in the hybrid is then a natural consequence of hybridization and the basis for heterosis expression. Despite the magnificence of the inbred-hybrid system in the context of grain production all over the world, as evidenced by the hundreds of thousands of new inbred lines and hybrids produced and tested each year, the genetic, physiological, and biochemical bases of inbreeding and heterosis still remain largely unexplained (see Stuber, 1994). Also, despite the multitude of investigations for detecting and quantifying inbreeding and heterosis effects, consistent results attached to strong scientific arguments are not available at the desired levels. A review on inbreeding and heterosis concepts and their values for a better understanding of the inbred–hybrid system for hybrid seed production are the objectives of this work.

INBREEDING

Inbreeding is a mating system in which matings occur between relatives (consanguineous mating). The negative effect of inbreeding, inbreeding depression, has been known since ancient civilizations and is due to recessive deleterious alleles in the homozygous state. In quantitative genetics inbreeding depression is quantified by the reduction in the mean that occurs only under nonadditive genetic effects. The increase of homozygosity or the decrease in the frequency of heterozygous genotypes is a natural consequence of inbreeding and reaches its maximum rate through selfing, the most extreme form of inbreeding. Other less severe forms of inbreeding lead to slower increases in homozygosity and the rate of increase depends on the degree of relationship between the mating parents. The inbreeding coefficient (F) is the probability of randomly uniting gametes having alleles identical by descent. So F is the probability of homozygosis where the homozygotes carry alleles that are identical copies of the same allele in a common ancestor. The inbreeding due to random drift, as a consequence of small population size, leads to an increase in the frequency of homozygotes at the rate of $F = 1/2N_e$ per generation; N_e is the effective population size. At any generation (t) of reduced size $F_t = [1/(2N_e)] + [1 - 1/(2N_e)]F_{t-1}$. The reduction of the population size also causes genetic drift by which gene frequency (p) in a subpopulation departs from the original gene frequency as a consequence of the dispersive process. After a large number of generations of reduced size the subpopulation tends to become homoallelic for all genes, meaning that some genes will be fixed ($p = 1$) and others will be lost ($p = 0$). However, genes are also fixed or lost in a few generations of inbreeding. The fixation of gene frequencies ($p = 0$ or $p = 1$) is the base for the strong differentiation among races with their own peculiar characteristics (Sprague, 1983). The inbreeding coefficient is always defined in relation to a generation or reference population in which the alleles in the homozygotes are considered nonidentical by descent ($F = 0$). In this sense, a random mating population may be viewed as a noninbred population in equilibrium, in spite of its history in previous generations; however, it can be considered as an inbred population ($F > 0$) in relation to a reference population in a previous generation. The history of a population is not always available but, even if it is considered noninbred one must be cognizant that it can exhibit a high proportion of homozygotes as a consequence of genetic drift (fixation of alleles) during its lifetime.

Selfing is the most common system of inbreeding used in breeding programs and is equivalent to a reduced population size with $N_e = 1$. Either for selfing or any population size, one must be aware that F is the expected proportion of homozygosity on the average among subpopulations; the proportion of homozygosity will differ among subpopulations as a consequence of genetic drift and the difference between contiguous classes is $1/2N$ for each locus. For $N_e = 1$ and initial gene frequency $p = \frac{1}{2}$, the new gene frequencies in the subpopulations will be 0, $\frac{1}{2}$ and 1 with probabilities $\frac{1}{4}$, $\frac{1}{2}$ and $\frac{1}{4}$, respectively. The probability for gene frequencies in the distribution of subpopulations also is the expected proportion of genes with that frequency within subpopulations. For example, within a given subpopulation it is expected that $\frac{1}{4}$ of its genes occur at frequency $p = 1$. The dispersion of the Y allele, controlling kernel color in maize (*Zea mays* L.), was reported by Paterniani and Miranda Filho (1995) in subpopulations derived from two base populations (A and B: F_2 generations of single crosses between inbred lines), after two generations under reduced size ($N = N_e = 5$). The genotypic classes (YY:Yy:yy) in the subpopulations follow the trinomial distribution, while the dispersion of the allele frequency within subpopulations follows the binomial distribution. For example, the classes (labeled by the number of the respective genotypes) **050** (five Yy genotypes) and **014** (one Yy and four yy genotypes) occur with probabilities 0.0313 and 0.0098 and the gene frequencies within classes are 0.5 and 0.1, respectively. The range for the dispersion of the allele Y in the first generation of reduced size was wider in population A (0.1 to 0.9) and narrower in population B (0.2 to 0.7). The inbreeding coefficients were $F_A = 0.105$ and $F_B = 0.091$, and were very close to the expected value ($F = 1/2N_e = 0.10$). The average gene frequency was 0.51 and 0.50, respectively. After two generations of reduced size, the ranges of dispersion was the same (0.1 to 0.9) in A and 0.3 to 1 in B, averaging 0.58 and 0.66, respectively. A flatter distribution was observed in A and B, than in the first generation. The consequence of the genetic drift as measured by the inbreeding effect was also studied in ten quantitative traits (Paterniani, 1995) after one generation of reduced size (Table 7–1). The dispersion of subpopulation means was wider in population B for practically all traits. For grain yield, the averages over subpopulations were 131 g plant^{-1} and 108 g plant^{-1}, within the ranges of 113 to 145 g plant^{-1} and 87 to 131 g plant^{-1}, for populations A and B, respectively. Results indicate a higher level of heterozygosity in B for most of the traits and that, in spite of being a little more depressive for grain yield, its wider dispersion can result in subpopulations with high

Table 7–1. Means of noninbred (m_o) and inbred (m_1) populations; inbreeding depression (I% for $F = \frac{1}{2}$) and ranges (I_L: low and I_H: high) for subpopulations.

Traits	Population A					Population B				
	m_o	m_1	I%	I_L	I_H	m_o	m_1	I%	I_L	I_H
Plant height, cm	201	198	-1.7	-10	8	215	195	-9.7	-23	5
Ear height, cm	116	110	-5.6	-18	9	128	117	-8.8	-20	7
Tassel branches	18.0	15.7	-9.8	-24	0	25.4	22.4	-10.2	-30	0
Ear length, cm	16.2	15.9	-1.6	-8	6	16.2	15.2	-6.3	-16	6
Ear diameter, cm	4.2	4.5	7.1	-2	16	4.2	4.1	-0.9	-7	14
Row number	13.8	13.6	-1.1	-7	7	12.9	13.5	4.1	-6	15
Kernels/row	36.2	35.6	-1.8	-11	8	37.8	35.9	-5.1	-13	3
Kernel density	1.01	1.03	-2.4	-2	5	1.08	1.05	-2.2	-7	2
300 grain weight, g	86.1	81.1	-5.8	-19	9	82.8	71.9	-13.2	-27	5
Total grain weight, g	131	125	-5.0	-14	10	108	102	-6.2	-20	21

concentration of useful alleles. After two generations of reduced size, the means of seven quantitative traits were estimated in the A subpopulations (m_s) and their testcrosses (m_t) with six testers in two locations (L_1: Piracicaba, SP; L_2: Anhembi, SP) (Blandon, 1996). For yield m_s ranged from 3.22 to 6.13 t ha^{-1} and 1.90 to 4.65 t ha^{-1}, averaging 4.40 t ha^{-1} in L_1 and 3.05 t ha^{-1} in L_2; the means for the base population (A) were 5.34 t ha^{-1} in L_1 and 3.55 t ha^{-1} in L_2. The ranges for m_t were 5.35 to 6.71 t ha^{-1} in L_1 and 4.25 to 5.60 t ha^{-1} in L_2; topcrosses of the base population averaged 5.16 t ha^{-1} in L_1 and 4.49 t ha^{-1} in L_2. The correlation coefficients between m_s and m_t were -0.07 and 0.33, and the estimates of the inbreeding depression (average of subpopulations), were 17.6% and 14.3%. The most commonly used system to produce homozygous lines is continuous selfing, which is also a dispersive process ($N_e = 1$). Selfing gives the most rapid increase in homozygosity and also the most rapid decrease in fitness due to inbreeding depression. In this sense, a less severe form of inbreeding, such as that caused by reduced population size, may give greater opportunities for selection toward the development of less inbreeding depression, since the fixation of deleterious genes would be slower (Hallauer & Miranda Filho, 1995) and the dispersive process would give more opportunity for concentration of desirable alleles in differentiated subpopulations.

For a better understanding of the level of homozygosity one can visualize the distribution of gene frequencies in a population. Genes controlling a quantitative trait will be distributed with frequencies varying from 0 to 1 in the population The completely inbred lines will have genes at frequencies 0 or 1. Partially inbred populations have a proportion of fixed alleles at frequencies of 0 or 1 and a proportion in the range $0 < p < 1$, following a nonpredictable distribution. Synthetics obtained from single crosses and double crosses will have gene frequencies in the classes 0, $\frac{1}{2}$, 1 and 0, $\frac{1}{4}$, $\frac{1}{2}$, $\frac{3}{4}$, 1; respectively. Synthetics of many inbred lines, composite varieties and broad base open-pollinated varieties follow a distribution with a low probability of genes at the extremes (0 and 1) and most of their genes with intermediate frequencies. For populations in equilibrium the maximum level of heterozygosity is 50% for one locus with two alleles and occurs for $p = \frac{1}{2}$; however, with multiple alleles heterozygosity may reach levels higher than 50%.

Under a simple model for one locus with two alleles (**B** and **b**) controlling a quantitative trait, the mean for a random mating population in equilibrium is given by $m_o = (p - q)a + 2pqd$; p and q are the allele frequencies for **B** and **b**,

Table 7–2. Estimates of the means and its componeñts in for some traits in five original and inbred ($F = \frac{1}{2}$) maize populations.

Population	Character	m_o	m_1	$I\%$	$\mu + a^*$	d^*	Ref.†
ESALQ-PB1	Plant height, cm	178.0	161.0	9.6	144.0	34.0	[1]
	Ear height, cm	92.0	81.0	12.0	70.0	22.0	[1]
	Tassel branches	19.3	16.3	15.3	13.3	6.0	[1]
ESALQ-PB2	Grain yield, g pl^{-1}	155.6	83.5	46.3	11.4	144.2	[2]
ESALQ-PB3	Grain yield, g pl^{-1}	143.6	78.1	45.6	12.6	131.0	[2]
EE1	Grain yield, g pl^{-1}	76.8	41.3	46.2	5.8	71.0	[3]
	Tassel branches	36.2	29.1	19.4	22.1	14.1	[3]
EC4	Grain yield, g pl^{-1}	109.4	58.3	46.7	7.1	102.3	[3]
	Tassel branches	28.7	25.1	12.7	21.4	7.3	[3]
ESALQ-PB2	Grain yield, g pl^{-1}	174.8	105.7	40.5	36.7	138.1	[4]
	Plant height, cm	234.8	216.0	8.1	197.1	37.7	[4]
	Ear height, cm	136.0	124.2	8.9	112.4	23.6	[4]
ESALQ-PB3	Grain yield, g pl^{-1}	160.6	94.7	44.2	28.8	131.8	[4]
	Plant height, cm	225.2	207.1	8.0	189.0	36.3	[4]
	Ear height, cm	127.0	116.7	8.2	106.3	20.6	[4]
ESALQ-PB1	Ear yield, g pl^{-1}	115.2	68.0	41.0	20.8	94.4	[5]

† [1] Kassouf & Miranda Filho (1984);[2] Miranda Filho & Meirelles (1986); [3] Nass & Miranda Filho (1995); [4] Terazawa (1993). [5] Packer et al. (1996).

respectively; a is half the difference between the homozygotes and d is a deviation due to dominance. Under inbreeding ($F > 0$) the population mean is $m_1 = (p - q)a + 2pq(1-F)d$.

Both m_o and m_1 are here expressed as deviations from the mean of the two homozygotes (μ). The inbreeding depression is $I = 2Fpqd$; the expected means are $M_o = \mu_o + a^* + d^*$ and $M_1 = \mu_o + a^* + (1 - F)d^*$; a^* and d^* are the overall contributions of homozygotes and heterozygotes to the mean; and $\mu_o = \Sigma_i \mu_i$ (summation over all loci). The quantity $\mu_o + a^*$ is the expected mean of a random sample of completely homozygous lines extracted from the base population. Inbreeding depression is therefore an important feature in choosing populations to be used as source of inbred lines (Eagles & Hardacre,1993). In fact, strong inbreeding depression may impose limitations in the hybrid seed production. Table 7–2 shows estimates of means and their components for some traits in maize populations. For traits under lower levels of dominance, the estimates of $\mu_o + a^*$ and d^* seem to be realistic in the sense that a lower proportion of the mean is attributed to dominance effects. On the other hand, for complex traits such as yield the strong inbreeding depression leads $\mu_o + a^*$ to be underestimated, because a higher proportion of heterozygous effects would not be expected. Negative estimates of $\mu_o + a^*$ for yield were reported by Lima et al. (1984) in four out of 32 Brazilian maize populations; populations derived from inbred lines (synthetics) had less inbreeding depression, in agreement with the hypothesis that highly heterozygous populations that were not under inbreeding have a larger hidden genetic load. Significant populations x inbreeding interaction was detected. In fact, deleterious recessive genes with major effects and qualitative inheritance not only may cause strong inbreeding depression but also may act as epistatic genes, in the sense that they can prevent important physiological pathways thus precluding the expression of other quantitative genes of minor effects (Lima et al., 1984). An extreme example of such an epistatic effect is the recessive allele for albinism, which when homozygous causes seedling death in a few days, thus precluding the expression of all other genes. Although many studies have reported on a linear response to inbreeding (Hallauer & Miranda Filho, 1995), the absence of a nonlinear regression does not provide sufficient evidence to reject epistasis (Sing et al., 1967).

Earlier works attributed the inbreeding effect mainly to deleterious recessive genes of larger effect and easily detected visually. Jones (1917) emphasized: "By inbreeding, strains of maize are isolated which are dwarf; some are sterile; some have contorted stems; some fasciated ears. Some are more susceptible to the bacterial wilt disease, and still others have brace roots so poorly developed that they cannot stand upright when the plants become heavy . . . All the characters cited are recessive, either completely or to a large degree, to the normal condition. More than one of these unfavorable characters may be present together in one inbred strain." Nevertheless, Jones (1952) recognized that "most of the deleterious genes of large effects are eliminated when a cross fertilized species, such as maize, are artificially self-pollinated" and that "there are many genes of small effects without visible morphological changes that are not eliminated by natural or artificial selection either in the wild or under domestication, and that these deficiencies or degenerative mutants do have a large part in bringing about reduced growth". Crow (1952) pointed out that detrimental recessive genes include the lethals and semilethals (such as chlorophyll deficiencies) that show up during inbreeding, but more important are the large number of factors, not individually detectable, which collectively result in the loss of vigor with inbreeding despite rigorous selection. Recurrent selection with inbred families should be recommended for eliminating deleterious recessives toward the development of more vigorous inbred lines (Hallauer, 1980; Miranda Filho, 1981; Hallauer & Miranda Filho, 1995) and results have shown the efficiency of selection to increase the average performance of S_1 lines (Genter & Alexander, 1966; Burton et al., 1971; Mulamba et al., 1983; Tanner & Smith, 1987; Odhiambo & Compton, 1989). Marques (1988) reported on the decrease in yield of S_1 lines in the range of 15.2 to 76.3% in ESALQ-PB2 (C1) and 22.5 to 76.1% in ESALQ-PB3(C1), averaging of 43.7 and 49.4%, respectively. After three cycles of selection in the same populations, the decrease in yield of S_1 lines varied from 1.5 to 65.8% and from 12.4 to 69.1%, averaging 40.5% and 44.2%, respectively, on the average of two locations (Terazawa, 1993). The difference of inbreeding depression when comparing two and three cycles of recurrent selection was thus apparent. In spite of the increase in the vigor of S_1 lines from the second to the third cycle, the overall inbreeding depression remained too high, corroborating the hypothesis on the importance of recessive genes of minor and not visually detected effects. The range of distribution of lines from random mating populations provides important information on their potential as source of superior lines.

Some problems and/or limitations for studying inbreeding and its effects can be outlined as follows: (i) Yield trials for evaluating inbred progenies result in lower experimental precision (Kassouf & Miranda Filho, 1984). (ii)] The inbreeding depression is not always fully detected because lethal recessives do not contribute to the performance of progenies, as when seeds are planted in excess and the lethals will not appear in the harvest (Fisher, 1965). (iii) Planting in excess followed by thinning may favor the less severely handicapped types, because a control for random elimination is not always done (Lima et al., 1984). (iv) Any correction of the mean for stand variation, as a consequence of death and injuries of lethals and semilethals, introduces a bias in the comparisons for measuring inbreeding effects (Lima et al., 1984). (v) The use of inbred lines could result in a biased estimate of the inbreeding effect because a set of random inbred lines cannot be a random sample of subvital genes and certainly lethal genes (Sing et al., 1967). (vi) The real number of defective genes with large deleterious effects and their specific contribution (e. g., sensitivity to aluminum toxicity, susceptibility to pests and diseases) to inbreeding depression are not known at all, respective to their direct effects and interactions. (vii) Despite the evidences of a proportionally high contribution of deleterious genes of minor and not visually detected effects, a better-knowledge of their actual proportion on the net inbreeding depression would be helpful for understanding the phenomenon and planning breeding strategies.

HETEROSIS

Heterosis is the genetic expression of the superiority of a hybrid in relation to its parents. It also depends on dominance and dominance types of epistatic effects; in this sense no heterosis can be detected for a quantitative trait in a hybrid, in relation to the average of the parents, if genes controlling the trait act in a strictly additive way (no dominance). The phenomenon of heterosis is the opposite of inbreeding depression in the sense that the vigor lost as a consequence of inbreeding is recovered by crossing; however, although inbreeding depression and heterosis depends on dominance genetic effects, they are not merely the same phenomenon in opposite directions. In fact, when dealing with populations, inbreeding depression is an *intra*population effect, while heterosis is expressed at the *inter*population level. Earlier studies on heterosis associated the phenomenon as merely the recovery of inbreeding depression caused by major genes visually detected (Jones, 1917); however, under a quantitative viewpoint, heterosis may occur whenever there is genetic divergence (difference in gene frequencies) between parents and some level of dominance. The recovery of vigor, lost because of the action of deleterious genes of large effects, can be seen partly as a nullification of the epistatic effects (preventing other genes of minor effects of full expression) of the recessive major genes. The contribution of detrimental polygenes of minor and not visually detected effects to heterosis, come as a consequence of increasing heterozygosis provided that there are differences in gene frequencies between the parents. If dominance exists controlling the trait, heterosis will be a function of the distribution of gene frequencies. The maximum heterosis would be attained by crossing completely inbred lines fully contrasting in their allele frequencies, i.e. alleles with $p = 0$ in one line and $p = 1$ in the opposite line. The same principle holds for crossing between populations, with the difference that there is a distribution of gene frequencies in the range $0 \leq p \leq 1$. When crossing two populations of this kind, heterosis will be expressed if their genes have some difference in gene frequencies; for example, genes with frequencies at the left side of the distribution ($p < 0.5$) in one population and at the right side ($p > 0.5$) in the opposite population. Skewed distributions are also expected theoretically. Old races of maize possibly have a large amount of genes in a fixed allelic state ($p = 1$ or $p = 0$), as a consequence of continuous genetic drift and this should be the cause of relatively high heterosis in crosses.

The simplest model for estimating heterosis is based on the contrast between the hybrid (F_1 generation) mean and the parental (P_1 and P_2) means ($h = m_{F1} - m_p$). When advanced generations (F_2 and backcrosses for both parents, B_1 and B_2) are available , besides P_1, P_2, and F_1, the least square procedure is recommended for estimation of heterosis. The model is $Y_i = m_o + \theta_{1i} b + \theta_{2i} h + \bar{e}_i$, where Y_i is the mean for a quantitative trait in a set of parental populations and derived generations; m_o is the mean between the parental populations; b is a measure of divergence between parents or the selection response when both parents come from divergent selection; h is the mid-parent heterosis; $\bar{e}_i$ is the error term; θ_{1i} and θ_{2i} are the coefficients defined according to the expected genetic structure of the different generations The vectors, represented by $[P_1, P_2, F_1, F_2, B_1, B_2]$ for θ_1 and θ_2 are $[1,-1,0 ,0, \frac{1}{2}, -\frac{1}{2}]$ and $[0, 0, 1, \frac{1}{2}, \frac{1}{2}, \frac{1}{2}]$. (Miranda Filho, 1991). Essentially, the model is the same as that given by Mather and Jinks (1971) and Jinks (1983), extended for crosses between random mating populations. Procedures for estimation of effects and their standard deviations, and for the analysis of variance are given by Miranda Filho (1991). The vector for θ_2 indicates that one-half of the heterosis expressed in

F_1 is maintained in the advanced generations (F_2, B_1 and $\underline{B}_2$). In the same way, in the synthesis of composites part of the average heterosis ($\overline{h}$) of all crosses is maintained in the newly formed population, i.e., $Y_{co} = m_V + [(n-1)/n]\overline{h}$, where Y_{co} is the expected mean of the composite formed by intercrossing n parental varieties with mean m_V .

For diallel crosses in a fixed set of varieties (random mating populations in equilibrium), the model of Gardner and Eberhart (1966) is appropriate and has been extensively used for the study of heterosis ($h_{ii'}$) and its components: $\overline{h}$ (average heterosis of all crosses), h_i (variety heterosis), and $s_{ii'}$ (specific heterosis). Eberhart and Gardner (1966) extended the model to include the contribution of homozygotes and heterozygotes within varieties and nonallelic interaction (epistasis) in the crosses. Examples of application of diallel analysis are given by Miranda Filho and Vencovsky (1984) for three maize traits using the complete model (Gardner, 1967); and by Pereira and Miranda Filho (1996), using the model with additive x additive epistasis, in the study of the resistance to stalk rot caused by *Colletotrichum graminicola* in maize.

Miranda Filho and Geraldi (1984) adapted the complete model of Gardner and Eberhart (1966) for the analysis of partial diallel crosses between two groups of varieties. The analysis include I varieties of Group 1, J varieties of Group 2, and the IJ crosses. The effects in the model are defined for each group of varieties and have similar meaning as in the complete model; the effect d is included as a measure of the difference between the two groups of varieties. Variations of the basic model for diallel crosses include: analysis with reciprocal crosses (Miranda Filho, 1995), analysis with interaction of effects (Miranda Filho & Vencovsky, 1995) and analysis with F_2 generations (Miranda Filho & Chaves, 1996).

Chaves and Miranda Filho (1997) suggested the analysis of topcrosses (intragroup) instead of diallel crosses for estimating heterosis and predicting variety composite means, based on the reduced model of Gardner (1967) in which specific heterosis is not included. Yield trials comprise the set of n varieties and n topcrosses using the whole set of varieties as tester. The means over replications for varieties (V_i) and topcrosses (T_i) are used for estimating the mean (μ), the average heterosis ($\overline{h}$), and variety heterosis (h_i). A procedure designated as intergroup topcross (IT) was suggested to replace the partial diallel by the analysis of crosses between two groups of varieties (Chaves, personal communication). The estimates of heterosis and its components are explained at the intergroup level.

Studies with heterosis refer to the mid-parent heterosis ($h = m_{F1} - m_p$). Using the additive-dominant model the heterosis is expressed by $h = \Sigma_i (p_i - r_i)^2 d_i$, where p_i is the frequency of the favorable allele at the ith locus in one parent and r_i the frequency of the same allele in the other parent; d_i is the deviation due to dominance. Genetic divergence between parents ($p \neq r$) and nonadditive genetic effects are then required for heterosis expression. The high-parent heterosis (h_H) is sometimes mentioned, referring to the superiority of the hybrid in relation to the best parent. Souza and Zinsly (1985) showed that $h_H = \Sigma_i (p_i - r_i) \alpha_i$ where p and r are the allele frequencies in the low and high parent, respectively; and α is the average effect of gene substitution. In the same way, one could define the low-parent heterosis (h_L). Actually the definitions for h_H and h_L hold only for one locus so that one can specify the lower and higher gene frequencies. For a quantitative trait, the low and high parent will depend on the proportional quantities of genes at low and high frequencies, the net effect resulting from summation over all loci. Therefore, we will refer to h_1 and h_2 as deviations from the respective parent means. Definitions are: $h = \Sigma_i (p_i - r_i)^2 d_i$, $h_1 = \Sigma_i (r_i - p_i) [a_i + (1 - 2p_i)d_i] = \Sigma_i (r_i - p_i)\alpha_{1i} = \Sigma_i \Delta_i\alpha_{1i}$ and $h_2 = \Sigma_i(p_i - r_i)[a_i + (1-2r_i)d_i] = \Sigma_i (p_i - r_i)\alpha_{2i} = -\Sigma_i \Delta_i\alpha_{2i}$. Hence, h_1 and h_2 are functions of the difference in gene frequencies between parents and the respective average effects of gene substitution. It can be shown that $h = (\frac{1}{2})(h_1 + h_2)$ and $h_1 - h_2 = 2\Sigma_i(r_i - p_i)\alpha_{12i} = 2\Sigma_i \Delta_i\alpha_{12i}$, where α_{12} is the average effect of gene substitution defined for the F_2 generation (Miranda Filho, 1991). A similar model was used by

Jinks (1983) to define heterosis between completely homozygous lines. If the population means are such that $m_1 < m_2$ then $h_L = h_1$ and $h_H = h_2$. If $h > 0$ then h_L is always positive and h_H may be positive or negative depending on the genetic structure of the parents. Negative estimates of h_H have been reported (Hallauer & Miranda Filho, 1995; Miranda Filho & Vencovsky, 1984; Souza & Zinsly, 1985).

In studying heterosis and its application in breeding programs, the answers to some questions or inquiries have remained obscure: (i) The amount and net contribution of major genes, that are deleterious in the recessive state, to the heterosis expression. (ii) The rationale for combining complementary traits (e.g., disease resistance) irrespective of the heterotic expression. (iii) Rational strategies for the use of genetic divergence, particularly for the use of exotic germplasm. (iv) Models and methods for translating interpopulation heterosis into outstanding hybrids of inbred lines. (v) The appropriate use of biometrical concepts and methodologies for the interpretation and use of heterosis in plant breeding.

INBREEDING AND HETEROSIS

Because inbreeding is defined and/or quantified in relation to a reference population, it should also be convenient to specify reference populations when dealing with heterosis. One should use h_o as the heterosis expressed in crosses between noninbred populations. Any effect of inbreeding (I) in the base populations will increase the heterosis in the same amount. Under inbreeding the population mean is expressed by $m_I = m_o - Fd^*$; for two different populations (1 and 2), it follows: $d^*_1 = \Sigma_i \, p_i q_i d_i$ and $d^*_2 = \Sigma_i r_i s_i d_i$. It can be shown that in the cross between inbred populations, with inbreeding coefficient F, the heterosis is $h_I = h_o + F\Sigma_i \, (p_i q_i + r_i s_i) d_i$ or $h_I = h_o + \bar{I}$, where $\bar{I}$ is the average inbreeding depression. If the parents are in different levels of inbreeding (say $F_{(1)}$ and $F_{(2)}$), then $h_I = h_o + F_{(1)} d^*_1 + F_{(2)} d^*_2$ or $h_F = h_o + (\frac{1}{2})(I_1 + I_2) = h_o + \bar{I}$. The high-parent and low-parent heterosis are also affected by the inbreeding depression; i.e., $h_1 = \Sigma_i (r_i - p_i) \, \alpha_{1i} + I_1$ and $h_2 = \Sigma_i (p_i - r_i) \, \alpha_{2i} + I_2$, which can be written as $h_{I1} = h_{01} + I_1$ and $h_{I2} = h_{02} + I_2$. The effect I is part of the heterosis merely due to the recovery of the inbreeding depression of the parents; and h_o is the true heterosis referred to the noninbred reference populations. Crossing completely inbred lines $(F = 1)$ the resulting heterosis is largely due to I, thus limiting the interpretation of heterosis as a measure of genetic divergence between parents. Defining reference populations is not always possible because the history of breeding populations is not always available, but one must be cognizant that the expected heterosis will vary considerably between inbred and noninbred parents. Genetic drift not only leads to an increase in homozygosity but also to an unpredictable dispersion of alleles and differentiation among subpopulations for heterosis expression. Evidences for those effects are the high differentiation among topcrosses of subpopulations of reduced size (Blandon, 1996).

Paterniani and Lonnquist (1963) reported on mid-parent heterosis varying from -11 to 101% (average of 33%) in crosses among 12 tropical races of maize; the high parent heterosis varied from -19 to 84% (average of 14%). Grouping the parental varieties as high (H) and low (L) yielding, Paterniani and Lonnquist (1963) found average heterosis of 19, 31, and 46% in the crosses H×H, H×L, and L×L, respectively, supporting the hypothesis of high level of homozygosity in the L class, which also revealed a rather poor adaptation to the conditions of the study. The less heterotic crosses involved the races Guarany Yellow, Caingang, and Cristal Paraguay and were attributed to lack of genetic divergence. The classification of the races for endosperm types (Dent, Flint, and Floury) resulted in higher heterosis within types than between types, averaging 36 and 30%, respectively. Paterniani (1980) reported on diallel crosses with six local populations; the average mid-parent heterosis was 18.8%, varying from 5.6 to 36.8%. The highest heterosis was in the cross of Piracar (a narrow base synthetic) with Cristal (a low yielding flint type from Paraguai). Evidences indicate that at least one parent does exhibit a large amount of homozygosity within population and sufficient genetic divergence for the expression of heterosis. Vasal et al. (1992) studied seven late Mexican populations in diallel crosses. Mid-parent and high-parent heterosis averaged 8.5 and 4.5%, ranging from 2.2 to 17.7% and -3.1 to 12.7%, respectively. Similar heterotic patterns were reported in studies with early and intermediate types (Beck et al., 1990; Crossa et al., 1990), indicating that populations and pools representing tropical germplasm have probably a large amount of heterozygosity and a low inbreeding depression. Mid-parent heterosis for grain yield in maize could be expected up to 20 or 25% without a great influence of inbreeding in the parents; however, low heterosis expression does not necessarily mean absence of inbreeding depression, as for example when two partially inbred subpopulations are derived from the same base population. On the other hand, strong inbreeding depression will result in a high heterosis expression. Hypothetically, if two nonallelic recessive semi-lethal genes are one in each parent, and their effects are so drastic as to reduce yield to near zero in the parents, then heterosis will be proportionally too high and not meaningful.

REFERENCES

Beck, D.L., S.K. Vasal, and J. Crossa. 1990. Heterosis and combining ability of CIMMYT's tropical early and intermediate maturity maize germplasm. Maydica 35:279–285.

Blandon, S.C. 1996. Efeito da deriva genética sobre caracteres quantitativos em uma população de milho (Zea mays L.). Ph.D. diss. Escola Superior de Agricultura "Luiz de Queiroz", Universidade de São Paulo, Piracicaba, Brazil.

Burton, J.W., L.H. Penny, A.R. Hallauer, and S.A. Eberhart. 1971. Evaluation of synthetic populations developed from a maize variety (BSK) by two methods of recurrent selection. Crop Sci. 11:361–365.

Chaves, L.J., and J.B. Miranda Filho. 1997. Predicting variety composite means without diallel crossing. Braz. J. Genetics 20:501–506.

Crossa, J., S.K. Vasal, and D.L. Beck. 1990. Combining ability study in diallel crosses of CIMMYT's tropical late yellow maize germplasm. Maydica 35:273–278.

Crow, J.F. 1952. Dominance and overdominance. p. 282–297. In J.W. Gowen (ed.) Heterosis. Iowa State College Press, Ames.

Eagles, H.A., and A.K. Hardacre. 1993. Inbreeding depression and other genetic effects in populations of maize containing highland tropical germplasm. Plant Breed. 110:230–236.

East, E.M. 1908. Inbreeding in corn. p. 419–428. Report for 1907. Connecticut Agric. Expt. Sta.

Eberhart, S.A., and C.O. Gardner. 1966. A general model for genetic effects. Biometrics 22:864–881.

Fisher, R.A. 1965. The theory of inbreeding. Academic Press, New York.

Gardner, C.O. 1967. Simplified methods for estimating constants and computing sums of squares for a diallel cross analysis. Fitotec. Latinoamer. 4:1–12.

Gardner, C.O., and S.A. Eberhart. 1966. Analysis and interpretation of the variety cross diallel and related populations. Biometrics 22:439–452.

Genter, C.F., and M.W. Alexander. 1966. Development and selection of productive S_1 inbred lines of corn (*Zea mays* L.). Crop Sci. 6:429–431.

Hallauer, A.R. and J.B. Miranda Filho, 1995. Quantitative genetics in maize breeding. 2nd ed. Iowa State Univ. Press, Ames.

Hallauer, A.R. 1980. Relation of quantitative genetics to applied maize breeding. Braz. J. Genetics 3:207–233.

Jinks, J.L. 1983. Biometrical genetics of heterosis. p. 1–46. *In* R. Frankel (ed.) Heterosis: Reappraisal of theory and practice. Springer-Verlag Berlin Heidelberg, Germany.

Jones, D.F. 1917. Dominance and linked factors as a means of accounting for heterosis. Genetics 2:466–479.

Jones, D.F. 1952. Plasmagenes and chromogenes in heterosis. p. 224–235. *In* J.W. Gowen (ed.) Heterosis. Iowa State College Press, Ames.

Kassouf, A.L., and J.B. Miranda Filho. 1984. Variabilidade e endogamia na população de milho ESALQ-PB1. p. 119–131. *In* Proc. XV Congr. Nac. Milho e Sorgo, Maceió (AL).

Lima, M., J.B. Miranda Filho, and P.B. Gallo. 1984. Inbreeding depression in Brazilian populations of maize (*Zea mays* L.). Maydica 29:213–215.

Marques, J.R.B. 1988. Seleção recorrente com endogamia em duas populações de milho (*Zea mays* L.). M.S. thesis ESALQ-USP, Piracicaba, Brazil.

Mather, K., and J.L. Jinks. 1971. Biometrical genetics. 2nd ed. Chapman & Hall, London.

Miranda Filho, J.B. 1981. Seleção e métodos de melhoramento. p. 145–150. *In* Proc. Reunião Bras. Genética, UNESP, Jaboticabal.

Miranda Filho, J.B. 1991. Quantitative analysis of a cross between populations and their derived generations. Braz. J. Genetics 14:547–561.

Miranda Filho, J.B. 1995. Analysis of diallel tables with reciprocal crosses. Braz. J. Genetics 18:633–637.

Miranda Filho, J.B., and L.J. Chaves. 1996. Analysis of diallel crosses with F_2 generations. Braz. J. Genetics 19:127–132.

Miranda Filho, J.B. and I.O. Geraldi. 1984. An adapted model for the analysis of partial diallel crosses. Braz. J. Genetics 7:117–126.

Miranda Filho, J.B., and W.F. Meirelles. 1986. Depressão por endogamia em progênies S_1 de duas populações de milho. p 320–327. *In* Proc. XVI Congr. Nac. Milho e Sorgo, Belo Horizonte (MG).

Miranda Filho, J.B., and R. Vencovsky .1984. Analysis of diallel crosses among open-pollinated varieties of maize (*Zea mays* L.). Maydica 24:217–234.

Miranda Filho, J.B., and R. Vencovsky. 1995. Analysis of variance with interaction of effects. Braz. J. Genetics 18:129–134.

Mulamba, N.N., A.R. Hallauer, and O.S. Smith. 1983. Recurrent selection for grain yield in a maize population. Crop Sci. 23:536–540.

Nass, L.L., and J.B. Miranda Filho. 1995. Inbreeding depression rates of semi-exotic maize (*Zea mays* L.) populations. Braz. J. Genetics 18:585–592.

Odhiambo, M.O., and W.A. Compton. 1989. Five cycles of replicated S_1 vs. reciprocal full-sib index selection in maize. Crop Sci. 29:314–319.

Packer, D., A. Regitano Neto, and J.B. Miranda Filho. 1996. Estudo de progênies endogâmicas da população ESALQ-PB1 de milho. p. 99. XXI Congr. Nac. Milho e Sorgo, Abstracts.

Paterniani, E. 1980. Heterosis in intervarietal crosses of maize (*Zea mays* L.) and their advanced generations. Braz. J. Genetics 3:235–249.

Paterniani, E., and J.H. Lonnquist. 1963. Heterosis in interracial crosses of corn (*Zea mays* L.). Crop Sci. 3:504–507.

Paterniani, M.E.A.G.Z., and J.B. Miranda Filho. 1995. Oscilação gênica em populações de tamanho reduzido de milho. 41th Congr. Bras. Genética. Braz. J. Genetics 18 (Supplement):95.

Paterniani, M.E.A.G.Z.1995. Efeito da redução do tamanho de populações de milho (*Zea mays* L.). Ph.D. diss. ESALQ-USP, Piracicaba, Brazil.

Pereira, O.A.P., and J.B. Miranda Filho. 1996. Base genética da resistência à podridão do colmo causada por Colletotrichum graminicola. p. 105. XXI Congr. Nac. Milho e Sorgo, Abstracts.

Shull, G.H. 1908. The composition of a field of maize. Rep. Am. Breeders Assoc. 4:296–301.

Shull, G.H. 1909. A pure line method of corn breeding. Rep. Am. Breeders Assoc. 5:51–59.

Sing, C.F., R.H. Moll, and W.D. Hanson. 1967. Inbreeding in two populations of (*Zea mays* L.). Crop Sci. 7:631–636.

Souza Jr., C.L., and J.R. Zinsly. 1985. Relative genetic potential of brachytic maize (*Zea mays* L.) varieties as breeding populations. Braz. J. Genetics 8:523–533.

Sprague, G.F. 1983. Heterosis in maize: theory and practice. p. 47–70. *In* R. Frankel (ed.) Heterosis: Reappraisal of theory and practice. Springer-Verlag Berlin Heidelberg, Germany.

Stuber, C.W. 1994. Heterosis in plant breeding. Plant Breed. Rev. 12:227–251.

Tanner, A.H., and O. S. Smith. 1987. Comparison of half-sib and S_1 recurrent selection in Krug Yellow Dent maize population. Crop Sci. 27:509–513.

Terazawa Jr., F. 1993. Seleção recorrente com endogamia em duas populações de milho: avaliação quantitativa e perspectiva para seleção de híbridos. MS thesis, ESALQ-USP, Piracicaba, Brazil.

Vasal, S.K., G. Srinivasan, D.L. Beck , J. Crossa, S. Pandey, and S. De Leon. 1992. Heterosis and combining ability of CIMMYT's tropical late white maize germplasm. Maydica 37:217–223.

Chapter 8

Genotype × Environment Interactions, Selection Response and Heterosis

M. Cooper and D. W. Podlich

INTRODUCTION

Selection strategies are applied to improve the yield and quality performance of genotypes for a target population of environments (TPE). This is often considered in terms of improving broad adaptation. Genotype × environment (G × E) interactions that result in a change in the rank of genotypes (cross-over interactions) complicate selection for broad adaptation. Interactions can also be distinguished on the basis of whether they are repeatable or non-repeatable within the target genotype–environment system. The presence of repeatable G × E interactions identifies cases of specific adaptations to types of environments that are repeatedly encountered in the TPE. When the bases of these are adequately understood this source of genetic variation can be exploited by selection for positive interactions. Examples that are relevant to the northeastern region of Australia include: breeding for midge resistance (*Stenodiplosis sorghicola* Coquillet) and staygreen in sorghum [*Sorghum bicolor* (L.) Moench.], and breeding for root lesion nematode (*Pratylenchus thornei*) resistance and a range of maturity types for different sowing opportunities in wheat (*Triticum aestivum* L.). Non-repeatable G × E interactions are a source of error that interfere with selection for both broad and specific adaptation. Recent publications dealing with G × E interactions in plant breeding have considered their incidence, analysis methodology, and biophysical–genetic causes. Here we consider their implications for selection response and heterosis in relation to a hybrid improvement program based on the half-sib reciprocal recurrent selection breeding strategy.

Hybrid breeding programs operate for a number of crops in the northeastern region of Australia, including sorghum, sunflower (*Helianthus annus* L.), maize (*Zea mays* L.), canola (*Brassica napus* L.), and wheat. In this chapter we will concentrate on issues that are associated with strategies to deal with G × E interactions in relation to work we are involved with on wheat and sorghum. For both crops the TPE in northeastern Australia is highly heterogeneous and G × E interactions for grain yield are large. Cross-over interactions account for a large proportion of the observed G × E interaction (Cooper et al., 1996). Research on wheat (Sheppard et al., 1996) and sorghum (Cooper & Chapman, 1996) has identified repeatable G × E interactions. Some of the regional elements of these interactions appear to be common to both crops, even though wheat is grown in winter and sorghum in summer. The work on wheat is focused on population improvement by recurrent selection for grain yield and protein content (Fabrizius et al., 1996), with the objective of developing parental lines for pedigree breeding. The work on sorghum is strategic, with the objective of developing selection strategies that accommodate the effects of G × E interactions for use in both germplasm development and hybrid selection (Cooper & Chapman, 1996).

Features of G × E Interaction Studies

G × E interactions have been examined from a number of perspectives in plant breeding. Three areas of activity are: (i) understanding the causes of interactions as a basis to select for specific attributes, (ii) developing statistical methodology to describe and characterize interactions, and (iii) examining gene action by environment interactions. Many of these studies have concentrated on picking winners from sets of genotypes. Less attention has been given to the implications of the different models of G × E interactions on selection response in context with breeding strategy. Classically the interactions are treated as a source of error in prediction of response to selection. This treatment is suboptimal where repeatable interactions contribute to specific adaptations to the types of environments encountered in the TPE. An interesting example of the improvement of maize populations by recurrent selection for drought tolerance was summarized by Chapman et al. (1996). A difficulty associated with experimental evaluation of selection strategies is the long time frame involved. Computer simulation methodology can be used to complement experimental methods (e.g., Fabrizius et al., 1996, Podlich & Cooper, 1997).

The Yoyo Effect and Predicting Response to Selection

Rathjen (1994) coined the term *yoyo effect* to dramatize the difficulty that is often associated with achieving response to selection in a complex TPE based on the results from multi-environment trials (METs). The METs represent a sample of environments from the TPE and are subject to sampling variation. Samples of environments that deviate from the TPE provide a sub-optimal basis for selection. Cycles of selection that are conducted for contrasting adaptations in METs that deviate from the TPE can result in disruptive selection and thus reduce the rate of genetic gain. Cooper et al. (1996) quantified this effect in terms of the principles of direct and indirect response to selection. Response in the TPE from selection on the results from a MET can be considered as a case of direct response to selection (ΔG_T), where the sample of environments in the MET is assumed to be a random sample from the TPE, or a case of indirect response to selection ($\Delta G_{T/M}$), where response in the TPE is evaluated as an indirect response to selection on the results of the MET. The prediction of direct response to selection is given by

$$\Delta G_T = i_M h_M^2 \sigma_{P(M)}, \tag{1}$$

where i_M is the standardized selection differential, h_M^2 is the appropriate line mean heritability and $\sigma_{p(M)}$ is the square root of the line mean phenotypic variance, each estimated from the results of the MET. The prediction of indirect response to selection is given by

$$\Delta G_T = i_M h_M h_T r_{g(MT)} \sigma_{P(M)}, \tag{2}$$

where, h_M and h_T are the square roots of the line mean heritability in the MET and the TPE, respectively, $r_{g(MT)}$ is the genetic correlation between the results of the MET and the TPE, $\sigma_{p(M)}$ and i_M were defined above. Equations [1] and [2] can be shown to have the same expectation for response in the TPE by recognizing the relationship between the genetic component of variance used in the heritability term of the direct response equation and the genetic component of covariance used in the genetic correlation of the indirect response equation (Cooper et al., 1996). Equation [1] is widely used to predict genetic gain for breeding strategies. Equation [2] is useful when investigating the influence of sampling variance on the expected response to selection in the TPE, because it emphasizes the point that as the sample of environments obtained in the

MET deviates from that expected for the TPE there can be a reduction in the genetic correlation between the results of the MET and the TPE, and therefore a reduction in the response in the TPE. The variation in the $r_{g(MT)}$ associated with this sampling variation and G × E interactions among types of environments in the TPE can often explain the yoyo effect. We are investigating selection strategies that deal with the complexities that result from the joint influences of the sampling variation associated with METs and G × E interactions in complex genotype–environment systems.

This paper concentrates on some of our findings from computer simulation studies that were conducted to investigate the concept of weighting data from the environments sampled in METs to match the expectations for the TPE. This is referred to as weighted selection. It is being considered as a methodology to adjust for the influence of the sampling variation associated with METs when the genes controlling variation for traits show G × E interactions. The objective of the weighted selection strategy is to improve the realized response to selection in the TPE achieved by breeding strategies, when working with finite samples of environments in METs.

COMPUTER SIMULATION MODEL

The computer simulation experiments were conducted using the QU-GENE simulation software (Podlich & Cooper, 1997). The QU-GENE simulation platform enables the design of $E(N:K)$ models for genotype–environment systems; E is the number of different types of environments in the TPE, N is the number of genes and K is a measure of the level of epistasis. Using the $E(N:K)$ notation identifies that different $N:K$ genetic models (Kauffman, 1993) are nested within the different types of environments encountered in the TPE. Therefore, a type of environment is defined to correspond with a specific $N:K$ genetic model. For consistency with expression of the model in terms of numerals, we denote the $E(N:K)$ model with a colon between N and K as distinct from Kauffman who represents his models without the colon (NK). Where information is available on the gene action associated with variation for a trait, this can be used to specify relevant $N:K$ models. The specification of different $N:K$ models for the types of environments in a TPE generates the G × E interaction within the genotype–environment system. The factors that can be manipulated in QU-GENE to define a genotype–environment system include: the number and frequency of types of environments in the TPE, the number of traits, heritability, number of genes, intra-locus gene action based on specification of midpoint (m), additive (a) and dominance (d) factors, inter-locus gene action by specification of alternative epistatic models, linkage groups based on specification of recombination frequencies, the initial frequencies of the alleles at each locus, the sets of genes that interact with the types of environments, and the form of G × E interaction. These variables are manipulated to generate families of genotype–environment systems of varying complexity.

The operation of the QU-GENE software is as follows. The engine generates a reference population of genotypes and defines the TPE to which improvement is directed. Simulation modules are designed to represent breeding strategies. These modules operate within a recurrent selection framework and manipulate the reference population of genotypes. Recurrent selection proceeds by sampling individuals from the base population, evaluating these in a MET, identifying a select group on performance in the MET, and intermating the select group to generate the base population for the next cycle of selection. This process is repeated for a number of cycles. The weighted selection strategy considered in this study operates at the point of identifying the select group of individuals from the results of the MET. A module was developed to simulate a half-sib reciprocal recurrent selection strategy (HSRRS). The breeding objective was improvement of interpopulation single cross hybrids. For this module two reference populations (A and B) are generated by the engine. The same $E(N:K)$ genotype–environment system model applies to both populations, however the frequencies of the alleles at the N loci can be manipulated to generate different starting conditions in populations A and B.

Materials and Methods: Simulation Experiment 1

Simulation Experiment 1 used the HSRRS simulation module to investigate the impact of an unweighted and weighted selection strategy on the response to selection in the TPE for a half-sib reciprocal recurrent selection breeding strategy. The unweighted and weighted selection strategies were applied to the performance of the half-sib families from populations A and B in the MET. For the unweighted selection strategy the merit of the half-sib families was assessed as the arithmetic average of the performance values for the half-sib families across the environments sampled in the MET. Thus, the unweighted selection strategy is the procedure that is commonly used to identify the select group of individuals in plant breeding. For the weighted selection strategy the merit of the half-sib families was assessed as a weighted average of the performance values for the half-sib families across the environments sampled in the MET. The weights used were the frequencies of occurrence of the environment types in the TPE. To obtain a weighted family mean, the environments sampled in the MET were first classified on environment type, then the family mean was computed for each environment type represented. The weights were applied by multiplying the family means for each environment type by their respective weights and then summing across environment types.

The joint influence of G × E interaction and variation in the environmental composition of METs was examined in terms of genetic gain in the TPE. A set of $E(N:K) = 2(20:0)$ genotype–environment system models was considered. Therefore, there were two types of environments (E1 and E2), 20 independent genes, with no epistatic interactions. Two forms of G × E interaction were introduced: (i) G × E interaction due to heterogeneity of genetic variance between the two environment types, and (ii) G × E interaction due to cross-over interaction between the two environment types. For each of the forms of G × E interaction three levels of dominance were considered: (i) partial dominance ($d/a = 0.75$), (ii) complete dominance ($d/a = 1$), and (iii) overdominance ($d/a = 1.25$). The level of dominance was defined to be the same for all 20 genes. The combination of two types of interaction and three levels of dominance generated the six genetic models examined. The genotype values at a locus (aa, Aa, AA) for the two forms of G × E interaction for each level of dominance in the two environment types are shown in Fig. 8–1 (a, b).

Two classes of environment were considered, the environments sampled in the MET and those that defined the TPE. A MET of ten environments was simulated in all cases. For each of the six genetic models (Fig. 8–1), the joint influence of G × E interactions and variation in the frequency of occurrence of types of environment was investigated by analyzing the relationship between genotype performance in the MET and the TPE as the frequency of both environment types changed in the MET and TPE. To examine this relationship the frequency of occurrence of E1 and E2 was quantified as a percentage of the total number of environments. The mixture of E1 and E2 was varied by changing the frequency of occurrence of the two types of environments from 0% to 100%, in steps of 20% to give a total of six states. This was done independently for both the sample of environments in the MET and the TPE to give a total of 36 pairwise comparisons for each of the genetic models. For each pairwise comparison twenty independent runs of the unweighted and weighted selection strategies were conducted for 15 cycles by the HSRRS module.

For all runs a broad sense heritability of 0.9 on a single plant basis was used for the reference population. In population A, 15 of the loci started with a frequency of 0.3 for the favorable allele. The favorable alleles for the other five loci were not present in population A. For the five loci where the favorable allele was absent in population A, the favorable allele was present in population B with an initial frequency of 0.3. For another five of the loci the favorable allele was not present in population B. The remaining ten loci started with a frequency of 0.7 for the favorable allele in population B. This genetic structure of the reference populations A and B was used for all comparisons.

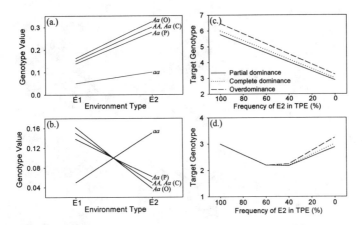

Fig. 8–1. Genotype values at a locus (a, b) and the target genotype responses (c, d) for three levels of dominance (P = partial, C = complete, O = overdominance) for two forms of G × E interactions, both based on two types of environment (E1 and E2); G × E interactions due to heterogeneity of variance (a, c) and cross-over interaction (b, d).

The unweighted and weighted selection strategies were evaluated as the average of the 20 runs after 15 cycles of the HSRRS module. For each comparison the following was plotted at cycle 15: (i) inter-population hybrid mean response in the TPE, (ii) inter-population mid-parent heterosis value, and (iii) number of favorable alleles fixed in either of the populations. The unweighted and weighted selection strategies were compared to the target genotype performance, where the target genotype was defined as being the best possible genotype that could be obtained from the initial populations A and B, with no mutation or movement of alleles between populations. The target genotype mean performance for the six genetic models, for varying percentages of E2 in the TPE, is shown in Fig. 8–1 (c, d).

Results and Discussion

There was no specific influence of G × E interactions due to heterogeneity of variance (Fig. 8–1a) on response to selection or heterosis. Therefore, only the results for the three genetic models incorporating G × E interactions due to cross-over interactions (Fig. 8–1b) are discussed.

Variation in the environmental composition of the MET influenced the mean performance of the hybrids (Fig. 8–2) and the level of heterosis (Fig. 8–3) achieved in the TPE after 15 cycles of selection. The impact of the variation depended on the environmental composition of the TPE. When the TPE contained similar proportions of E1 and E2, variation in the environmental composition of the MET samples had a smaller influence on the hybrid mean (Fig. 8–2) and heterosis (Fig. 8–3) in the TPE than for those situations where one of the types of environments occurred with a high frequency and the other with a low frequency in the TPE. In all cases the mean in the TPE was closest to that of the target genotype (Fig. 8–1d) when the environmental composition of the MET matched that of the TPE (Fig. 8–2), as is expected following the principles embodied in prediction equations [1] and [2]. When the MET was correctly matched with the TPE the available heterosis was effectively exploited (Fig. 8–3).

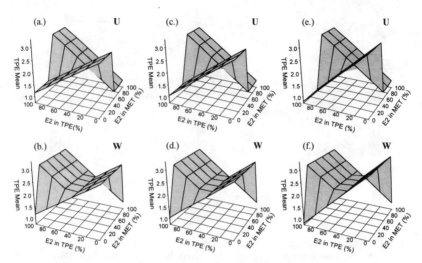

Fig. 8–2. Mean performance of interpopulation hybrids in the TPE after 15 cycles of
the HSRRS simulation module for combinations of the two environment types (E1
and E2) in the MET and TPE, and selection based on the unweighted (U) and
weighted (W) selection strategies applied to genetic models based on cross-over
G × E interactions and three levels of dominance; partial (a, b), complete (c, d),
and overdominance (e, f).

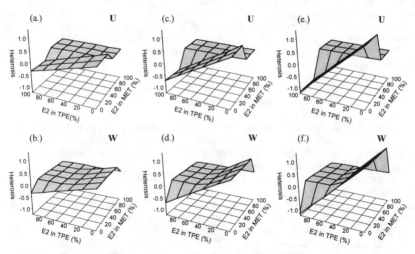

Fig. 8–3. Mid-parent heterosis of interpopulation hybrids in the TPE after 15 cycles
of the HSRRS simulation module for combinations of the two environment types
(E1 and E2) in the MET and TPE, and selection based on the unweighted (U) and
weighted (W) selection strategies applied to genetic models based on cross-over
G × E interactions and three levels of dominance; partial (a, b), complete (c, d),
and overdominance (e, f).

Comparison of the results for the unweighted and weighted selection strategies indicated that the mean response in the TPE was more consistently matched with the target mean response for the weighted selection strategy (Fig. 8–2). The weighted selection strategy compensated for poor matches between the MET and the TPE by adjusting the contribution of performance information from the two environment types in the MET to more accurately reflect the expectations for the TPE. This positive effect on response to selection was observed for the three levels of dominance examined (Fig. 8–2) and contributed to more consistent exploitation of the available heterosis in each case (Fig. 8–3). The instances where the weighted selection strategy was not effective were those where both of the environment types occurred in the TPE but one of them was not represented in the MET.

The influence of the weighted selection strategy was examined in terms of the number of favorable genes fixed in both populations A and B, relative to that required to achieve the target genotype response (Fig. 8–4). For the three levels of dominance the weighted selection strategy fixed genes in the target combination more consistently than the unweighted selection strategy. The comparisons between the unweighted and weighted selection strategies for the G × E interaction model based on cross-over interactions indicate that the weighted selection strategy can compensate for the sampling variation in composition of the MET. Throughout the simulation experiment, when the weighted selection strategy was applied to the results of the MET a more accurate assessment of the merit of the half-sib families in relation to the TPE was achieved than for the unweighted strategy. This enabled the weighted selection strategy to more consistently fix the genes from the reference populations in the configuration that was necessary to achieve the target genotype.

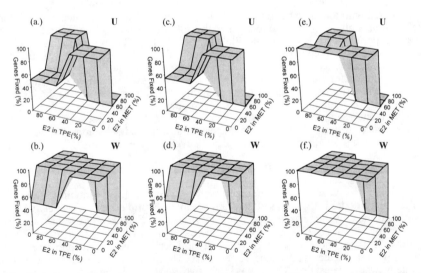

Fig. 8–4. Percentage of genes fixed in populations A and B, relative to that required to achieve the target genotype, after 15 cycles of the HSRRS simulation module for combinations of the two environment types (E1 and E2) in the MET and TPE, and selection based on the unweighted (U) and weighted (W) selection strategies applied to genetic models based on cross-over G × E interactions and three levels of dominance; partial (a, b), complete (c, d), and overdominance (e, f).

Materials and Methods: Simulation Experiment 2

Simulation Experiment 2 was conducted to compare the unweighted and weighted selection strategies when different numbers of environments were sampled in METs and was examined for four genotype–environment systems with different levels of cross-over G × E interactions. The selection procedures for the unweighted and weighted strategies were as in simulation Experiment 1 and were applied to the performance of the half-sib families in the MET. A range of $E(N:K) = 10(20:0)$ genotype–environment system models was considered. Four levels of cross-over G × E interaction were introduced with an increasing amount of G × E interaction defined from Model 1 to Model 4. As a measure of the amount of G × E interaction, the ratio of the G × E interaction variance component on the genetic variance component for each model was computed, assuming a base population in Hardy-Weinberg equilibrium; Model 1 = 0.47, Model 2 = 1.30, Model 3 = 1.71, and Model 4 = 3.77. For each level of G × E interaction, three dominance models were considered: (i) partial dominance ($d/a = 0.75$), (ii) complete dominance ($d/a = 1$), and (iii) overdominance ($d/a = 1.25$). The level of dominance was defined to be the same for all 20 genes. The combination of four levels of interaction and three levels of dominance generated the 12 genetic models examined. Each model was examined for interpopulation hybrid mean response in the TPE across eight cycles of the HSRRS module when different numbers of environments were sampled in the METs. The numbers of environments sampled were: 2, 3, 4, 5, 8 and 10. For each of the runs, the environment types in the METs were sampled at random from the TPE. The frequency of occurrence of the 10 environment types in the TPE was varied and ranged from 0.01 to 0.28.

For each of the twelve genetic models and six MET sample sizes, 80 independent runs of the unweighted and weighted selection strategies were conducted for eight cycles of the HSRRS module. A broad sense heritability of 0.9 on a single plant basis in the reference population was used in all cases. In population A, 10 of the loci started with a frequency of 0.3 with the remaining 10 with a frequency of 0.7 for the favorable allele. For the 10 loci where the initial frequency of the favorable allele was 0.3, population B had an initial allele frequency of 0.7. The remaining 10 loci had an initial allele frequency of 0.3 for the favorable allele in population B. This genetic structure of reference populations A and B was used for all runs.

The performance of the unweighted and weighted selection strategies was evaluated in terms of the average interpopulation hybrid performance for the 80 runs of the HSRRS module. For each of the models with the varying numbers of environments sampled in the METs, the mean difference between the hybrid performance in the TPE for the weighted and unweighted selection strategies was computed and plotted after the third and sixth cycles of the program.

Results and Discussion

For the genotype-environment system with the least amount of G × E interaction (model 1) there was little difference in the hybrid performance between the weighted and unweighted selection strategies (Fig. 8–5). For all cases examined, as the amount of cross-over G × E interaction increased from Model 1 to 4, there was a positive increase in the difference in mean hybrid performance between the weighted and unweighted selection strategies. The observed advantage of the weighted selection strategy was greatest for METs based on two to six environments. However, an advantage was still observed for METs based on a sample size of ten environments. The positive effect of the weighted selection strategy was observed for the three levels of dominance considered. The advantage of the weighted selection strategy was attributed to a more rapid rate of change in gene frequencies towards those required to achieve the target genotype and a reduced chance of losing favorable alleles from populations A and B.

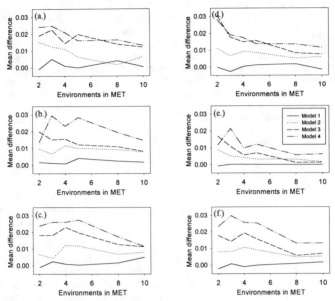

Fig. 8–5. Mean difference between weighted and unweighted selection strategies for interpopulation hybrid performance from 80 runs of the HSRRS simulation module after cycles 3 (a, b, c) and 6 (d, e, f), based on selection on performance in METs comprising of different numbers of environments. Genotype–environment systems based on four G × E interaction models (Models 1 to 4; 1 = the least and 4 = the most cross-over G × E interaction) and three levels of dominance; partial (a, d), complete (b, e) and overdominance (c, f) were examined.

DISCUSSION

For the two simulation experiments considered, different levels of heterosis were generated by manipulating: (i) the level of dominance, and (ii) establishing a difference in allele frequencies between the two base populations. G × E interactions were generated by: (i) specifying different levels of gene expression in the different types of environment in the TPE, and (ii) changing the frequency of occurrence of the different types of environment in the TPE. The presence of cross-over G × E interactions reduced progress from selection in both simulation experiments. The influence of the G × E interactions on response to selection was observed as a reduced mean performance and mid-parent heterosis of the inter-population hybrids. These effects were interpreted as a result of the increased difficulty associated with selecting genotype combinations required to exploit heterosis and achieve high levels of hybrid performance in the presence of complex G × E interactions.

The selection strategy commonly used to improve crop performance in a TPE is to select for average performance across a sample of environments taken to represent the TPE. The classical prediction equations for response to selection treat G × E interactions as a source of errors. These errors interfere with the determination of the genetic merit of individuals, and are assumed to be averaged out by evaluating genotypes across environments. Therefore, all of the variance associated with G × E interactions is allocated to the denominator of the heritability expressions within the genetic gain equations, and response is predicted for average performance across the environments of the TPE. This selection strategy focuses on improving broad adaptation in relation to the environmental challenges presented by the TPE and tends to foster an attitude of ignoring opportunities for improving specific adaptation. As our understanding of G × E interactions within a genotype–environment system improves, and when these interactions account for a large component of the variation among genotypes, it becomes necessary to consider selection strategies that provide scope for improvement of both broad and specific adaptation. An important component of such a selection strategy is establishing a MET protocol that ensures a sample of environments that is relevant to the TPE, so that any opportunities for improvements in both broad and specific adaptation can be detected and selected.

Responses to the incidence of large G × E interactions depend on the nature of the interactions. The results of simulation Experiment 1 emphasized what is known from theory, that it is the cross-over types of G × E interactions, and not those that result from heterogeneity in the magnitude of genetic variance across environments, that complicate selection decisions. Therefore, when evaluating the impact of G × E interactions on response to selection it is important to distinguish between these two forms of interaction. Where G × E interactions are large and predominantly of the heterogeneity of variance type, the reduction in the estimate of heritability from the G × E interaction component of variance may be considered to contribute to an underestimate of the potential for response to selection; however, where the interactions are predominantly of the cross-over type, the reduction in the predicted response to selection will more accurately reflect the realized response to selection. It is common to partition the G × E interaction into components of interaction associated with locations, years and location–year combinations. Where there is a regional structure to the G × E interactions there may be interest in subdividing the region into subregions, with a breeding program for each subregion. Where interactions with years are large, testing across years is necessary. Often the interaction with year–location combinations (G × L × Y) is large. The G × Y and G × L × Y interactions are often assumed to be nonrepeatable. This is not necessarily so. The repeatability of these interactions is more a function of the environmental conditions responsible for their generation (which we have referred to as an environment type) and the frequency of occurrence of these conditions with the TPE. The allocation of G × E interactions into components associated with the year–location cross-classification, while useful in structuring the MET systems of our breeding programs, may be more a function of the way we sample the genotype–environment system rather than a strongly structured feature of the system itself. The incidence of large G × L × Y interactions may be an indication of this. Under this situation, and in the absence of a comprehensive characterization of the TPE and/or understanding of the adaptation of our germplasm, we can compensate by testing genotypes in more environments, to more reliably approach the expectations of the TPE; however, resources place an obvious limit on this strategy.

Breeding programs test genotypes across small samples of environments relative to the size of the TPE. For plant breeding programs that operate within a complex TPE, where cross-over G × E interactions are common, the sampling variation associated with finite samples of environments in METs can result in disruptive selection and can be a major factor reducing realized response in the TPE. Similarly a structured MET that is subject to little sampling variation but is poorly matched with the TPE will reduce realized response to selection in the TPE. An understanding of the environmental composition of the TPE provides the basis for design and implementation of selection strategies to overcome these constraints on progress from selection. Our current work on wheat and sorghum in the northeastern region of Australia is providing an evolving understanding of this complex TPE.

The weighted selection discussed in this paper provides a strategy for buffering progress in plant breeding against the disruptive selection influences that result from the joint influences of cross-over G × E interactions and sampling variation associated with finite samples of environments in METs, i.e., the yoyo effect. To apply the weighted selection strategy two pieces of information are required: (i) a reliable characterization of the TPE, and (ii) a characterization of the environments sampled in METs. The TPE characterization provides a quantitative description of the major types of environment that generate G × E interactions, and their frequency of occurrence over time and space within the target region of the breeding program. The MET characterization provides a specification of the types of environments that were sampled in a MET and how well the composition of these environments matches the TPE. The characterization of the MET needs to be achieved in real time if the information is to be used to assist selection decisions. Implementation of the weighted selection strategy is being evaluated in our wheat recurrent selection program. Capable personnel, pattern analysis methodology, a set of reference genotypes, strategic measurements linked to a biophysical crop growth model, and considerable computer capacity underlie the implementation of the concept into an applied selection strategy. In addition, the deployment of trials to sites and their agronomic management is implemented to increase the chances of obtaining a MET that matches the expectation of the TPE. The weighted selection strategy is viewed as a safety net that increases the chances of matching the results from METs with the expectations for the TPE. If this can be more consistently achieved by the weighted selection strategy than the current unweighted selection strategy, this will increase the chances of realizing the potential rate of genetic gain achievable from the germplasm.

Any characterization of the TPE must recognize and be responsive to the possibility of changes in the mixture of types of environments and their expected frequencies of occurrence. Such changes may occur with the introduction of new agronomic packages in the target region. An example of this, that is relevant to grain production in Australia, is the replacement of intensive tillage systems with forms of reduced and minimum tillage practice. In addition, as the adaptation of the germplasm used within a breeding program improves over time, the types of environments and the form of G × E interaction that impacts on selection response may change. Therefore, any characterization of the TPE must be systematically reviewed to ensure that it is relevant to the objectives of the breeding program. For both wheat and sorghum in northeastern Australia, we are currently characterizing environments in terms of the timing of significant water-stress events in relation to the critical developmental stage of flowering. Environments are categorized based on the incidence of moisture conditions that generate pre-flowering stress, post-flowering stress, stress coincident with flowering, combinations of these stress events or low-stress throughout development. These types of environments have been shown to explain a significant proportion of the G × E interactions for grain yield of wheat (Cooper et al., 1997) and they are repeatedly encountered in the northeastern region TPE, however their frequencies of occurrence are unknown. More accurate knowledge of their frequency of occurrence in the TPE, and the amount of yield G × E interaction for which they are responsible, will provide a basis for implementing the weighted selection strategy. It is possible that the frequencies of these types of environments will change over time and also with locality across the northeastern region. Such changes in the environmental composition of the TPE can be accommodated in the weighted selection strategy by updating the estimates of the frequency of occurrence of the types of environment as necessary.

The results of the second simulation experiment indicated that the weighted selection strategy had a larger advantage over the unweighted strategy when the MET comprised of fewer environments, and as the complexity of the G × E interactions increased. These conditions may be particularly relevant to the early generation stages of breeding programs, where it is common for a large number of breeding lines to be tested across a few environments. Therefore, investment of resources to implement the weighted selection strategy for the early generation testing stages would appear to offer opportunities for improving the overall effectiveness of a breeding program.

We are currently examining the influence of heritability on the effectiveness of the weighted selection strategy. This paper concentrated on systems with a high broad sense heritability in the reference population. With lower levels of heritability (e.g.,

0.05 to 0.20) the influence of weighting of errors needs to be considered. The results from our investigations (not presented in this chapter) indicate that this weighting of errors can act to reduce the realized response to selection. Therefore, in assessing the merit of the weighted selection strategy, the positive effect achieved from providing protection against the disruptive influence of METs that deviate from the expectations of the TPE need to be weighed up against the negative effects of weighting errors. We have found that when cross-over G × E interactions are common, the benefits of the weighted selection strategy generally outweigh the costs and there is a net improvement in response to selection.

REFERENCES

Chapman, S.C., G.O. Edmeades, and J. Crossa. 1996. Pattern analysis of gains from selection for drought tolerance in tropical maize populations. p. 513–527. *In* M. Cooper, and G.L. Hammer (ed.) Plant adaptation and crop improvement. CAB Int., IRRI and ICRISAT, Wallingford.

Cooper, M., and S.C. Chapman. 1996. Breeding sorghum for target environments in Australia. p. 173–187. *In* M.A. Foale et al. (ed.) Proc. of the 3rd Australian Sorghum Conf. AIAS Occ. Pub. No. 93.

Cooper, M., I.H. DeLacy, and K.E. Basford. 1996. Relationships among analytical methods used to analyse genotypic adaptation in multi-environment trials. p. 193–224. *In* M. Cooper, and G.L. Hammer (ed.) Plant adaptation and crop improvement. CAB Int., IRRI and ICRISAT, Wallingford.

Cooper, M., R.E. Stucker, I.H. DeLacy, and B.D. Harch. 1997. Wheat breeding nurseries, target environments, and indirect selection for grain yield. Crop Sci. 37:1168–1176.

Fabrizius, M.A., M. Cooper, D.W. Podlich, P.S. Brennan, F.W. Ellison, and I.H. DeLacy. 1996. Design and simulation of a recurrent selection program to improve yield and protein in spring wheat. p. P8–P11. *In* R.A. Richards et al. (ed.) Proc. of the 8th Assembly of the Wheat Breeding Society of Australia, Canberra.

Kauffman, S.A. 1993. The origins of order, self-organization and selection in evolution. Oxford Univ. Press, New York.

Podlich, D.W., and M. Cooper. 1997. QU-GENE: A platform for quantitative analysis of genetic models. Centre for Statistics Res. Rep. 83. Univ. of Queensland, Brisbane, Queensland.

Rathjen, A.J. 1994. The biological basis of genotype × environment interaction - its definition and management. p. 13–17. *In* J. Paull et al.(ed.). Proc. of the 7th Assembly of the Wheat Breeding Society of Australia, Adelaide.

Sheppard, J.A., W.G.A. Ratnasiri, I.H. DeLacy, D.G. Butler, M. Cooper, and P.S. Brennan. 1996. Classification of Queensland test sites into zones for yield evaluation using multi-environment trials from 1972–1994. p. O213–O217. *In* R.A. Richardset al. (ed.) Proc. of the 8th Assembly of the Wheat Breeding Society of Australia, Canberra.

Chapter 9

Quantitative Genetics and Heterosis

Discussion Session

QUESTIONS FOR K.R. LAMKEY

J. Suárez, Argentina: I would like you to comment on heterosis for species that don't show inbreeding depression. I think that I have heard you say that heterosis in those crops arises from epistatic effects alone. Am I right?

Response: I am not very optimistic about the prospects for heterosis in crops that do not show inbreeding depression. The absence of inbreeding depression means there is no baseline level of heterosis in the species, unless there is something going on during the inbreeding process that we do not understand. But it appears to me that you have to have dominance to have heterosis, and if you have dominance you should have inbreeding depression. I think that studying inbreeding depression is the key. For example, with soybean there have been lots of heterosis studies, but no inbreeding depression studies. I do not know whether there is inbreeding depression in soybean or not. I am not sure if soybean breeders know whether there is inbreeding depression in soybean. The only way to have heterosis without having inbreeding depression is with additive x additive epistasis.

A. Melchinger, University of Hohenheim, Germany: You mentioned the importance of knowing something about the distribution of gene effects. Do you think that QTL studies can give us a good answer to this problem? What is your opinion on that?

Response: QTL studies certainly have the potential to answer the question. I'm really concerned about over estimation of gene effects in QTL models. QTL mapping is just a massive statistical problem that we need more solutions to. But if we can get reasonable and unbiased estimates of the effects from QTL models, and I am not sure we are there yet, I think QTL studies can help answer this question.

A. Grunst, USA: I have been sitting here studying panmictic and base line heterosis. As corn breeder, could you explain to me how I could visualize the difference between what panmictic heterosis is and what I think of as specific combining ability. You know that you often find that one genotype performs in one particular combination but does not perform in any other combination.

Response: In the context of a pedigree breeding program, you start with two F_2 populations (one for each heterotic group), and you theoretically inbreed these populations to infinity. The difference between the mean of inbred populations and the noninbred F_2 population is baseline heterosis. Baseline heterosis is essentially the average inbreeding depression observed in your F_2 source populations. Any specific inbred will deviate from that average to some extent, which we have not accounted for in these models, but it could be done. So we have to recover that baseline and then get above that to get your good hybrid. Any heterosis above the baseline is functional heterosis and is analogous to specific combining ability. I think the data will show, if we start collecting data on genetic gain over time, that you've been good at selecting for panmictic heterosis, and that is giving rise to the elite performing hybrids. The data will also show that we have been effective for improving parental performance, but I am not convinced yet that parental performance has any direct relationship to hybrid performance. But I think we need to look at that question in more detail.

H. Geiger, University of Hohenheim, Germany: I would like to comment on the term functional heterosis. Why was it called that? Functional heterosis, I think it's just an expression of divergence and is misleading. I agree to the term baseline, but I would call functional heterosis divergence heterosis.

Response: I called it functional heterosis because it's the only meaningful part of the heterosis, and if you're going to have a functional hybrid, you've got to have a functional heterosis. Albrecht Melchinger and I have talked about this, and maybe we can come to terms between calling it a heterotic deviation or panmictic heterosis. (Note: The term functional heterosis was changed to panmictic heterosis to agree with current usage in the literature.)

A. Gallais, INRA, France: You have not mentioned the use of molecular markers to study the quantitative genetics of heterosis. Do you think it is possible to use molecular markers to develop models to explain heterosis?

Response: We have not looked at using molecular markers to explain heterosis, but maybe others have.

J. González, Centro de Investigaciones Agrarias de Mabegondo, Spain: If I understood correctly, you said that hybrid performance is a function of dominance and unlinked dominant by dominant epistasis. There are some studies, however, demonstrating that the dominant by dominant epistasis is negative. Is there any solution to reduce the negative effect of dominance by dominance epistatic effects?

Response: No, I do not know of any solution. Of course if epistasis is negative, it is going to decrease F_1 performance, but I do not think we are going to select out those types myself.

P. Sun, Dairyland Research, Int., USA: You mentioned that S_2 selection is the most effective method; however, Cooper says that S_1 is the most effective. Can you tell us the difference?

Response: No, I can not. We have several problems in interpreting selection responses in selection experiments that have been done on agronomic species. And I mentioned that for some of them we cannot separate drift from selection. The ability of any particular method to work for a population, despite theoretical predictions, really boils down to the population structure or the genetic architecture of populations. Since we know so little about that, I think it's hard to predict response. The population that S_2 selection responded in well for us has high levels of additive variance. Another population with additive and dominance variance nearly equal did not respond to S_2 selection. I think that's the reason why. But I can't tell for sure. I wish I could.

B. Dhillon, Punjab Agricultural University, India: How do you explain the quantitative genetics of heterosis in the graph of inbreds versus hybrids that Don Duvick showed this morning.

Response: I do not know how to answer that question. I think Don and I need to talk a little more before I even attempt to answer that, because we need to really think about how to interpret that kind data in light of the inbreeding depression that is going on. The performance of inbreds in the private sector, I think, is really indicative of how much selection pressure there is within a company on inbred per se performance, and this varies a lot from company to company.

QUESTIONS FOR J.F. CROW

Anonymous: Dr. Crow, why do you prefer the term superdominance to overdominance?

Response: Well, I guess for classical reasons. I like Superman rather than Overman. It's just closer to the usual meaning of the word.

A. Dogra, University of Missouri, USA: Dr. Crow do you know of any instance where someone has tried to look into the possible role of codominance in heterosis?

Response: Well we certainly see it with additive genetic traits, but whether there are very many yield or performance traits that have codominance, in fact I don't know.

O. Smith, Pioneer Hi-Bred, Int., USA: With the higher mutation rates per gamete from the Mukai experiments, and the persistence of these mutations, might you not expect inbreeding depression to persist a long time in populations undergoing selection?

Response: Let me say a little about the Mukai experiments. They have been questioned recently, by Peter Keightley from Edinburgh, who has done a likelihood surface analysis. He ends up with much smaller estimate of the mutation rate. He offers two explanations for the Mukai results, which we reported in 1972. One is that the strain had an active transposable element in it. I am pretty sure it didn't, because most such elements produce lethal as well as mild mutations, and the lethal rate didn't change. Furthermore, when Mukai did an experiment with a known active element involved, the mutation rate was much higher. Keightley's second suggestion was that the control strain improved during the course of the experiment. This, I think, is unlikely because this was an old laboratory strain that should long before have reached equilibrium. So, I think the early results are probably correct, but I eagerly await the outcome of several experiments that are now going on to settle the question.

In answer to the question, yes I agree that a higher mutation rate would cause inbreeding depression to persist longer in populations undergoing selection.

QUESTIONS FOR C. J. GOODNIGHT

Arudja, India: I would like to know how much importance do you give to epistasis for the expression of heterosis in comparison to dominance and overdominance?

Response: There's unfortunately not a lot of data. It's very hard to detect epistasis within a single population. It's just statistically very hard to find, so it's very hard to say how much there really is. My impression is that epistasis becomes of increasing importance when you have inbred populations and wide crosses. The other thing that I think you need to realize is that in many of the examples, you really can't tell dominance from epistasis. So it becomes a difficult question to answer unless you have lots of lines to compare.

Response (M. Cooper): In the experimental work that we are doing with wheat, we have realized some of the difficulties with the power in some of the experiments in getting reliable estimates. We've gone to a lot of trouble and invested some resources in trying to find out how important epistasis is for wheat. working in a common inbred population, so I can't compare it to dominance, but we find that epistasis is quite large relative to the estimates of the additive components of variance. Because of the care we took in our experiments, those estimates are quite a bit larger than their standard errors. The other thing that we are finding is that we encounter a lot of epistasis by environment interaction, and I think that that has a fair bit to do with some of the stepwise genetic improvements that we actually see over time historically through the breeding work that has been undertaken in wheat, certainly in our region.

Response (J.F. Crow): I think one reason why in selection experiments there can be epistasis and not have much of an effect is what Kimura called a quasilinkage equilibrium or quasilinkage disequilibrium, whichever you choose. But he found that, in long continued selection with loose linkage, the linkage disequilibrium parameter and the epistatic variance are opposite in sign and very nearly equal, so that actually you will make better progress by ignoring epistasis if there's been a long period of selection than by taking it into account. I don't know whether that's ever been tested practically or not.

Response: Can I just comment on that? One of the things that we find quite interesting is that if we work with crosses (and we are making an inference from a small number of experiments here where many of the crosses that we've looked at have gone through pedigree breeding programs) we don't find much epistasis. But if we take material unrelated to the germplasm that we're working with, and then cross that with material that's adapted, we see quite a lot of epistasis and yield improvement.

Ayusha, Indian Agricultural Research Institute, India, and Sun Qi Xin, Beijing Agricultural University, China: How much importance can you give to epistasis in the expression of panmictic heterosis compared with dominance and overdominance?

Response: I think that's an open question. I really don't, I really can't answer that. We know that dominance is important, but I don't think that we know that epistasis isn't. So deciding which is the most important of the two is hard.

Response (K.R. Lamkey): I think that with the way baseline heterosis is modeled, epistasis may not be important because it's dealing with the population average and not individual lines. If you look at individual hybrids then epistasis could become important and can either increase or decrease the baseline.

Response: That's a good point, because one of the things that I've been working on is essentially the difference between what I'm doing and a diallel cross. I'm asking people to look at the performance of individuals, and, in fact, I think that epistasis may become far more important when we take the performance of individuals rather than the averages across lines or across lines which are not completely inbred.

R. Phillips, University of Minnesota, USA: You indicated that additive genetic variance increases with inbreeding. Do you think that non-classical explanations such as increased mutability, methylation, etc. could be involved?

Response: There have been several studies that have shown an increase in heritabilities following population bottlenecks. I think it's interesting that these started showing up when people realized that heritabilities could increase. So I think they may have occurred before, but people never thought to report them. But the thing is that we see the increase in heritability immediately following the bottleneck, often within one or two generations. I can't imagine that a higher level of mutation would affect that, and I also think that it's unlikely that methylation could take affect that quickly, nor can I think of any physiological mechanism that would cause the increase.

P. Peterson, Iowa State University, USA: This is addressed to the epistatic people. For thirty years, we have pretty well described a number of epistatic interactions at the molecular level. Is there any connection between molecularly described, well known interactions at the molecular level, with epistatic interactions at the population level?

Response: I think the answer is basically no. What you're describing is what we are calling physiological epistasis. You have to have physiological epistasis in order for the statistical epistasis to be present. The converse is not true. That is you can show very strong epistatic interactions at the molecular level, and they may show up as additive effects at the population level. It just depends on the gene frequencies. And it is a well known feature of populations that there's no very good relationship between molecular variation and quantitative genetic variation.

P. Peterson, Iowa State University, USA: I disagree. It depends on what you are measuring. Say you're measuring anthocyanin in maize. The yield that anthocyanin. That's all. That's a quantitative measure. You could cross two lines. You could have an inbred with almost zero. Another inbred, also almost zero. And you could maximize it in the F_1. Isn't that an interaction at the molecular level that we know about that could be applied to yield performance?

Response: Of course you can. I guess I misunderstood your point. I am coming from an evolutionary biology perspective thinking in natural populations. In natural populations there's no real relationship between quantitative genetic variation and molecular variation. Certainly it's true that if you have identified alleles and you can combine them through hybrids, it will work fine.

QUESTIONS FOR J.B. MIRANDA FILHO

J. Suárez, Argentina: When you state that inbreeding is an effect existing within a population and that heterosis is an effect between populations, do you mean that a different set of loci act in each situation?

Response: No. Maybe I was not very clear in my statement. I would say that inbreeding depression and heterosis can not be considered the same phenomenon in opposite directions. Heterosis results from crossing and inbreeding is an intra-population effect.

M. Sanchez and B. Dhillon, Punjab Agricultural University, India: How does linkage influence the rate of inbreeding depression.

Response: I'm sorry but I don't have enough experience in this area to answer this question. I pass to Dr. Lamkey.

Response (K.R. Lamkey): Linked dominant genes are probably important in heterosis. I don't think there is any question about this. The person who has written about this best in agronomic crops is probably Ted Bingham. He had an article in *Crop Science* recently on linkats and his tetraploid and diploid alfalfa. He also wrote a really nice paper for a heterosis symposium sponsored by the Crop Science Society of America and the American Society of Horticultural Science last year that explained that well, so hopefully you'll be able to read about that.

QUESTIONS FOR M. COOPER

P. Sun, Dairyland Research, Int., USA: On the subject of genetic gain, are you talking about exploiting additive genetic variance, what is the most effective generation to achieve genetic gain? I'm just wondering what aspect of genetic gain you were talking about. Is it exploiting additive genetic variance?

Response: In the studies that we have conducted, we find that we are making as good a gain as we are going to make with S_1 selection. We've looked at going to S_2, for example, and we find that there are tradeoffs that don't really allow us extra gain in the whole system. So we are sticking with S_1 for the moment.

Response (K.R. Lamkey): In the experiments we have been doing, the results have been in terms of genetic gain and recurrent selection. We have a paper coming out in four or five months showing that S_2 recurrent selection was the best performing method in comparison to five other methods that we looked at in a common base population. However, we have been unable to get S_2 selection to work in other populations. So it's population specific, unfortunately, as is everything we have been talking about here today.

Comment (Sharnum, Indian Sorghum Program, India): I want to make two comments on the discussion that has just taken place. One of the queries raised here is whether an inbred line can be equal to the F_1 hybrid? The answer is in the Indian program. We have developed certain varieties over a period of time, which are as high yielding as the hybrids, which were released in the past. And I think if we look the theory of the heterosis, which is being presented, we have to assume that the heterosis is always productive. Heterosis is not necessarily productive. It may be productive, it may be zero, and it may be negative. I would just like to comment on the Dr. Cooper's paper also. He has been a very good collaborator, and the implications of his studies are evident in the cooperative trials conducted over a large number of locations. His studies and the presentation that he made to the Indian Program show that we can substantially reduce the number of locations and still have very effective selection.

Chapter 10

Genetic Diversity and Heterosis

A. E. Melchinger

INTRODUCTION

Genetic diversity among and within genera, species, subspecies, populations, and elite breeding materials is equally of interest in plant genetics and breeding. While taxonomists and germplasm banks are primarily interested in the higher levels of this hierarchy, plant breeders are mainly concerned with the diversity among and within breeding populations and elite germplasm, because it largely determines the future prospects of success in breeding programs.

Before 1970, established methods for measuring genetic diversity between taxonomic units have relied on pedigree analysis and morphological, physiological or cytological markers as well as biometric analysis of quantitative and qualitative traits, heterosis or segregation variance in crosses. In the following two decades, isozymes have been successfully used in numerous taxonomic and evolutionary studies (Hamrick & Godt, 1997); however, they often failed in the classification of elite breeding materials due to the limited number of marker loci available and the low level of polymorphism.

The development of molecular markers such as RFLPs, RAPDs, AFLPs, and SSRs in recent years removed most of the limitations associated with isozymes. Meanwhile, dense marker linkage maps have been developed in all major crop species. Because the new marker systems reveal differences at the level of DNA, they provide an extremely powerful tool for assessment of genetic diversity in cultivated and wild plant species. The technical details and principles behind the various molecular markers as well as their general uses in diversity studies and plant breeding programs have been described in a recent review (Westmann & Kresovich, 1997).

In this chapter, we summarize recent results on the use of different types of DNA markers for the assessment of genetic diversity at the intra-specific level, especially for germplasm grouping and pedigree analysis. Second, we regard the relationship between genetic diversity and heterosis and focus particularly on the advantages of heterotic groups. Third, we consider the possibilities to predict heterosis and hybrid performance from molecular markers.

USE OF DNA MARKERS FOR GROUPING OF GERMPLASM AND PEDIGREE ANALYSIS

Measures of Genetic Diversity Based on DNA Markers

The final result of marker assays with all types of DNA markers listed above is a characteristic banding pattern (also referred to as DNA fingerprint) for each individual. If the mode of inheritance is known and each band can be assigned an allele of a marker locus, estimates of genetic distance (GD) between two taxonomic units can be calculated from differences in the allele frequencies at multiple marker loci. In evolutionary or selection studies, the GD measure proposed by Nei (1978) is usually preferred, whereas in germplasm surveys mostly the Rogers' distance (RD) (Rogers, 1972) and the modified Rogers' distance (MRD) (Goodman & Stuber, 1983) are employed. The RD is distinguished by desirable genetic properties, which makes it attractive in pedigree analysis of related genotypes, while MRD is appealing in studies of heterosis for reasons described in Eq. [3] below. With unknown relationships among bands and alleles of marker loci, it is common practice to determine the proportion of mismatches in the bands between two individuals either according to the Nei and Li (1979) distance (NLD) or the measure originally proposed by Jaccard (1908). Link et al. (1995) recommended the latter for marker systems such as RAPDs and AFLPs, but for codominant markers such as RFLPs and SSRs both measures lead to identical ranking of GD estimates among inbreds.

Given the high costs of DNA marker assays, knowledge of the sampling variance associated with GD estimates is needed to determine the minimum number of markers required for a given level of precision in the GD estimates. The latter depends on the number of markers assayed, their coverage of the genome and the degree of polymorphism in the materials investigated. If N markers are randomly sampled over the genome, the standard error (SE) of the RD between homozygous inbreds can be calculated as SE = RD(1-RD)/N (Dubreuil et al., 1996), which is identical to the jackknife estimator of SE used in previous studies (Melchinger et al., 1991). Based on this calculation of SE, Barbosa-Neto et al. (1997) derived expressions of confidence intervals of GD estimates and the minimum number of markers required to achieve a desired precision. Alternatively, the SE of GD estimates can be determined by the bootstrap procedure (Tivang et al., 1994). When marker loci have been chosen to optimize the genome coverage, both estimators of SE overestimate the actual variance of GD estimates (Dubreuil et al., 1996).

For comparison of the level of genetic diversity among different populations for a given marker set or among different types of markers for a given population of taxonomic units, various genetic parameters can be used. These include (i) the percentage of polymorphic loci, (ii) the average number of alleles per polymorphic locus, and (iii) the diversity index of Nei (1973), also known as the polymorphic index content (PIC) $H = 1 - \sum x_i^2$, where x_i is the frequency of the ith allele, averaged across loci. The latter corresponds to the expected average heterozygosity under Hardy-Weinberg equilibrium. With several populations, the proportion of the genetic diversity residing among populations (G_{ST}) can be determined as (Nei, 1973): $G_{ST} = 1 - \overline{H}_S / H_T$, where $\overline{H}_S$ is the mean genetic diversity within populations and H_T is the total genetic diversity across all populations.

Comparison of Different Types of DNA Markers

With the development of new DNA marker technologies in recent years, the question raised which of them are most suitable for various applications in plant genetics and breeding. Comparison of RFLPs and RAPDs in diversity studies with maize (*Zea mays* L.) (Hahn et al., 1995), barley (*Hordeum vulgare* L.) (Russell et al., 1997), and sorghum [*Sorghum bicolor* (L.) Moench] (Yang et al., 1996) revealed only moderate agreement in GD estimates, which was largely attributed to the lower reproducibility of RAPDs. AFLPs generally showed a lower level of polymorphism per band than RFLPs, but due to the large number of bands simultaneously analyzed per primer combination, they usually had the highest marker index among all available marker systems (for review see Russell et al., 1997). The latter provides an overall measure of marker efficiency to discriminate among genotypes. GD estimates based on RFLPs and AFLPs generally showed high correlations in studies with barley (Russell et al. 1997) and maize (Melchinger et al., 1998) and also good agreement with pedigree data. Comparisons of RFLPs with SSRs in maize also revealed high correlations ($r = 0.95$) in GD estimates and close agreement ($r \geq 0.80$) with pedigree data (Smith et al., 1997). While PIC values were similar for both types of markers, SSRs showed fewer deviations from a Mendelian mode of inheritance than RFLPs. Altogether, the authors concluded that SSR technology offers advantages over RFLPs in terms of reliability, reproducibility, discrimination, standardization, and cost effectiveness. In studies with autogamous crops having a narrow genetic base such as barley, rice (*Oryza sativa* L.), and soybean (*Glycine max Merr.*), SSRs always revealed the highest level of polymorphism (Russell et al., 1997). Moreover, SSRs and RFLPs both possess the desirable property that banding patterns can usually be interpreted in terms of alleles at mapped loci (Smith et al., 1997). In conclusion, SSR and AFLP markers show great promise to complement or substitute RFLPs for many applications in plant breeding and variety protection.

Grouping of Germplasm with the Aid of DNA Markers

Information about the relationships among breeding materials and the genetic diversity in the available germplasm is important for the choice of parents in plant breeding programs. This applies particularly to hybrid breeding, where recognition and exploitation of heterotic patterns between different sources of germplasm are important for success (see arguments in a later section). Two relevant questions in this context are: (i) How divergent are lines from the same and different germplasm pools? (ii) Which criteria and biometric methods allow a faithful grouping of germplasm?

Comprehensive studies of genetic diversity based on RFLPs, AFLPs, and SSRs have been reported in several crop species including rice (*Oryza sativa* L.) maize and barley (Table 10–1). Irrespective of the type of DNA markers employed and the materials investigated, combinations of genotypes from different germplasm groups had on average significantly greater mean GD than combinations of lines from the same germplasm group. The relative increase in the mean GD values (ΔGD%) observed for line combinations between as opposed to within germplasm groups was only moderate in those maize studies including only pairs of unrelated lines but it was much higher, if related pairs of lines were not excluded. Interestingly, ΔGD% was comparatively high in studies with barley and rice, reflecting the low level of genetic diversity within germplasm pools as compared with between germplasm pools in autogamous crops. In general, GD estimates for individual pairs of lines showed a wide range within and between germplasm groups and usually overlapped.

Despite this fact, principal coordinate analysis (PCoA) based on GD estimates determined by RFLPs provided a distinct separation of lines from different germplasm groups in maize (Melchinger et al., 1991; Messmer et al., 1992; Dubreuil et al., 1996) and barley (Melchinger et al., 1994). Likewise, cluster analysis (CA) of 148 U.S. maize inbreds (Mumm and Dudley, 1994) and 116 European and U.S. maize inbreds (Dubreuil et al., 1996) as well as B- and R-lines in sorghum (Ahnert et al., 1996) and sunflower (Hongtrakul et al., 1997) partitioned the lines in accordance with their origin from different breeding groups and pedigree information. As an example, Fig. 10–1 shows the results of a cluster analysis of CIMMYT maize inbreds. Lines from the same germplasm group (Antigua, ETO, Tuxpeño, P47/PL32) form distinct subgroups. Based on our experiences, PCoA is suitable for faithful portrayals of the relationships between larger groups of lines but distances between similar lines can be distorted. In contrast, CA is reliable for depicting close relationships between lines because these clusters are merged at the beginning in the hierarchical process. Mumm et al. (1994) validated the results of CA by various methods and found that classifications depicted in the phenogram represented the true associations among maize inbreds reasonably well. In addition, they recommended the MRD as the most suitable GD measure for classification purposes. Since PCoA and CA complement each other to a certain degree, Messmer et al. (1992) recommended to perform both types of analysis in parallel in order to extract the maximum amount of information from molecular data.

Table 10–1. Mean genetic distance ($\overline{GD}$) among lines within and between germplasm groups in maize, barley, and rice determined with different DNA markers.

Materials			Mean genetic distance ($\overline{GD}$)				DNA markers		Reference
Group A	Group B		A × A	B × B	A × B	ΔGD (%)†	Type	Number	
Maize inbreds									
Flint	Dent	U‡	0.54	0.54	0.59	9	RFLP	203 CEC§	Messmer et al., 1992
RYD§	LSC§	U	0.54	0.57	0.60	9	RFLP	83 CEC	Melchinger et al., 1991
RYD	LSC		0.36	0.37	0.51	40	RFLP	135 CEC	Ajmone Marsan et al., 1998
(Same materials)			0.35	0.31	0.52	58	AFLP	6 PC§	
RYD	LSC		0.46	0.52	0.63	28	RFLP	63 CEC	Dubreuil et al., 1996
Flint	RYD		0.50	0.46	0.66	38	RFLP	63 CEC	
Flint	LSC		0.50	0.52	0.68	34	RFLP	63 CEC	
Barley cultivars									
Winter	Spring	U	0.15	0.16	0.24	61	RFLP	136 CEC	Melchinger et al. 1994
Rice cultivars									
SE Asia	SE China		0.24	0.48	0.51	40	RFLP + SSR	117 CEC, 10 PC	Zhang et al. 1994

† ΔGD (%) = [2 $\overline{GD}$(A × B) / ($\overline{GD}$(A × A) + $\overline{GD}$(B × B))] × 100 - 100.
‡ U : A × A and B × B involved only unrelated pairs ($f < 0.1$) of lines.
§ CEC, clone–enzyme combinations; PC, primer combinations; RYD, Reid Yellow Dent; LSC, Lancaster Sure Crop.

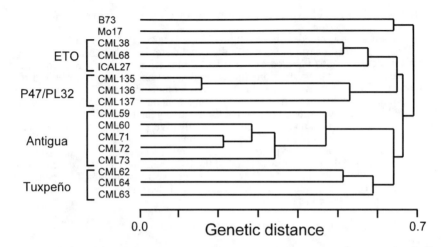

Fig. 10–1. Cluster analyses of CIMMYT maize lines from four different germplasm groups (ETO, Population 47/Pool 32, Antigua, Tuxpeño) based on genetic distances calculated from RFLP data of 256 clone–enzyme combinations (original data kindly provided by D. Hoisington et al., 1993, CIMMYT). Inbreds B73 and Mo17 from the U.S. Cornbelt were included for comparison.

Given the wide range in GD values for line combinations within germplasm pools and the relatively small differences in mean GD among groups (see Table 10–1), Melchinger et al. (1991) arrived at the following conclusions: (i) reliable classification of a line of unknown origin to established breeding pools requires determination of its GD to a large number of representative lines from each germplasm group and (ii) GD estimates must be determined with sufficient accuracy. With regard to the latter, AFLPs and multiplexing of SSRs are particularly appealing, because a larger number of marker bands can be evaluated in a single assay. With regard to the former, Messmer et al. (1992) calculated for each line an average genetic distance (AGD) to a core group of lines from each germplasm group. For all pure flint and dent lines, the AGD to other lines from the same heterotic group was significantly smaller than the AGD to lines from the other heterotic groups. Moreover, for lines of mixed origin the AGD provided a good indicator for the proportion of ancestors from each germplasm group.

In conclusion, numerous experimental studies clearly demonstrated that with a sufficient number of RFLPs, SSRs or AFLPs, it is possible to reveal genetic relationships among breeding materials and genetic resources with any degree of precision required. In hybrid breeding, this information is useful for a clearer description of existing heterotic groups and identification of new heterotic groups in a systematic manner (see later section). Moreover, inbreds of unknown genetic origin could be assigned to established heterotic groups. In all breeding categories, DNA markers could assist the breeder in the choice of genetically diverse parents for breeding programs and a systematic introgression of new germplasm to established breeding pools. Last but not least, molecular markers can help to monitor the level of genetic diversity in (i) breeding materials and cultivars over time (see e.g., Labate et al. and Smith et al., 1999, this publication) and (ii) germplasm collections for a meaningful formation of core collections.

Analysis of Pedigree Relationships with DNA Markers

Traditionally, breeders have employed the coancestry f devised by Malécot (1948) as an indirect measure to quantify the genetic similarity between related individuals. Calculation of f is based on the rules of Mendelian inheritance, probability theory, and simplifying assumptions (absence of selection, mutations, genetic drift, and unrelatedness of individuals without known common ancestors). Although these assumptions are rather unrealistic in a breeding context, the coancestry has been employed in several crops to (i) determine the relative genetic contribution of ancestors to released cultivars and (ii) estimate the effective population size of a given germplasm pool and monitor its changes over time.

Based on theory, one expects a linear relationship between the coancestry f of related inbreds and the expectation $\mathcal{E}(GD)$ of the GD determined as RD from molecular data (Melchinger et al., 1991). Accordingly, for two inbreds related by pedigree with coancestry f and unknown marker genotypes of their ancestors, an estimate of their GD can be obtained as

$$\hat{GD} = \overline{GD}\ (1-f) \tag{1}$$

where $\overline{GD}$ refers to the mean GD of unrelated homozygous lines from the respective germplasm group with regard to the investigated set of marker loci. Conversely, if the GD of two individuals is known, their coancestry can be estimated as

$$\hat{f} = 1 - GD/\overline{GD}. \tag{2}$$

Better GD estimates can be obtained, if the marker genotypes of a complete set of progenitors are known (Melchinger, 1993).

Experimental results on the correlation between f and GD estimates based on DNA markers in different crops are summarized in Table 10–2. In studies with maize inbreds of various origins, f was tightly correlated with GD and explained more than 60% of the variation in GD between related lines. In contrast, in studies with barley cultivars as well as sorghum and wheat (*Triticum aestivum* L.) lines correlations of f with GD estimates based on RFLP or AFLP markers were moderate to low ($|r| < 0.68$).

As discussed in detail by Melchinger (1993), f and GD estimates represent fundamentally different concepts for measuring genetic similarity of related individuals and are affected by different sources of error. The tight correlations between both measures observed for maize suggests that these errors are of minor importance. Hence, in so far as pedigree data are correct, they provide reliable descriptors of genetic similarity in maize. Conversely, GD estimates can be used to check the validity of pedigree data and detect possible errors (Messmer et al., 1992). By the same token it has been proposed to employ GD estimates for identification of essentially derived varieties (EDVs) in plant variety protection according to the revised UPOV rules (Lange, 1996). A critical issue in this context is the definition of an appropriate threshold for the minimum GD of EDVs, which depends on the germplasm in question as well as the specific set of markers employed. The poor correlations of f with GD in barley, sorghum, and wheat compared with the high correlations in maize requires further research. One possible explanation might be that pedigrees in autogamous and partially allogamous species are more complex and relationships between the original parental germplasm pools may exist but are generally unknown (Graner et al., 1994).

In addition to measuring GD by averaging marker information across the entire genome, classical pedigree analysis can be extended by mapped DNA markers with codominant inheritance to follow the inheritance of specific chromosome regions through multiple generations (Lorenzen et al., 1995). With appropriate computer software, it is possible to draw pedigrees for individual linkage groups and to trace the ancestral sources of markers in a given line (Boutin et al., 1995). This type of analysis can provide a deeper insight into the genetic contribution of main ancestors as well as

Table 10–2. Correlations (r) of genetic distance (GD) estimates based on DNA markers with coancestry (f) of related genotypes in various crop species.

Materials–origin	f†	$r(GD, f)$	Type	Number‡	Reference
			DNA markers		
Maize lines					
U.S. Cornbelt	-	- 0.80	RFLP	80	Smith et al., 1997
(Same materials)	-	- 0.81	SSR	131	
European flint	0.14	- 0.84**	RFLP	188 CEC	Messmer et al., 1993
European dent	0.23	- 0.91**	RFLP	188 CEC	
Miscellaneous	-	- 0.77**	RFLP	63 CEC	Dubreuil et al., 1996
Barley cultivars					
Winter	0.26	- 0.26**	RFLP	136 CEC	Graner et al., 1994
Spring	0.33	- 0.32*	RFLP	136 CEC	
Spring	0.13	- 0.39*	AFLP	8 PC	Schut et al., 1997
Sorghum lines					
R-lines	-	- 0.46**	RFLP	104 CEC	Ahnert et al., 1996
B-lines	-	- 0.43**	RFLP	104 CEC	
Wheat lines	0.35	- 0.68**	STS	63 PC	Martin et al., 1995
	-	- 0.55**	SSR	23 PC	Plaschke et al., 1995

*,** Significant at the 0.05 and 0.01 probability levels, respectively.
† Mean of f values for all pairs considered.
‡ CEC, clone–enzyme combinations; PC, primer combinations.

the role of selection and drift (Lorenzen et al., 1995; Bernardo et al., 1997; Labate et al., 1997). It also may help to detect putative linkages between molecular markers and mono- or oligogenic traits.

RELATIONSHIP BETWEEN GENETIC DISTANCE AND HETEROSIS

This part is essentially an appraisal of heterotic groups. I start with a well-known result on the relationship between genetic distance and heterosis and summarize the advantages of inter-group over intra-group crosses in various breeding categories. We then deal with the question: How should germplasm be organized for optimum exploitation of heterosis? Finally, we provide criteria and review approaches for identification of heterotic groups and patterns.

Quantitative Genetic Expectations and Experimental Results

Consider two populations $\pi1$ and $\pi2$ in Hardy-Weinberg equilibrium and their hybrid population $\pi1 \times \pi2$. Assuming two alleles per locus and no epistasis, we obtain the following relationship between the mean of these populations (Falconer & Mackay, 1996):

$$\Delta H(\pi1 \times \pi2) = \mu_{\pi1 \times \pi2} - (\mu_{\pi1} + \mu_{\pi2})/2 = \sum_i y_i^2 d_i = \sum_i MRD_i^2 (\pi1, \pi2) d_i, \qquad [3]$$

where ΔH is the panmictic-midparent heterosis (PMPH) in the terminology of Lamkey and Edwards (1998), d_i is the dominance effect at locus i, y_i is the difference in gene frequencies and MRD_i^2 ($\pi1$, $\pi2$) the square of the modified Rogers' distance at locus i between $\pi1$ and $\pi2$. Consequently, with directional dominance ($d_i > 0$), the hybrid population is expected to outperform the mean of its two parent populations and ΔH is expected to increase with increasing genetic distance between $\pi1$ and $\pi2$.

This expectation was first confirmed by Moll et al. (1962) in an experiment with maize. In a second study, Moll et al. (1965) concluded that there is an optimum level of GD, after which PMPH and hybrid performance decline due to the lack of adaptation of one or both parent populations or fertility problems; however, in both studies the geographic distance of the origin of populations was used instead of direct measures of GD.

Using the definitions and result of Lamkey and Edwards (1999, this publication), we have

PMPH = inbred–midparent heterosis - baseline heterosis,   [4]

where the first and second term on the right hand side corresponds to the mean of midparent heterosis (MPH) in inter- and intra-group hybrids, respectively. Hence, the association between PMPH and parental genetic distance can be studied by a comparison of these two groups.

Comparison of Intra- and Inter-Group Hybrids

Table 10–3 summarizes the increase Δ in the means of GD, MPH, and hybrid yield (F1P) for balanced sets of inter- over intra-group hybrids for various crops. In maize, inter-group hybrids betweeen lines from Reid Yellow Dent (RYD) and Lancaster Sure Crop (LSC), which are the main heterotic groups in the U.S. Cornbelt, had a 21% greater GD, 42% greater MPH, and 21% greater F1P compared to intra-group hybrids. Similar results were observed in crosses between European flint and dent lines and also in crosses between Petkus and Carsten, the two heterotic groups in hybrid rye (*Secale cereale* L.). By comparison, the increase in GD, MPH, and F1P of inter- over intra-group hybrids were somewhat lower between the two major heterotic groups of hybrid rice in China and potential heterotic groups in faba bean (*Vicia faba* L.). In conclusion, these experimental findings are in accordance with the above-mentioned theoretical expectation.

In addition to a better exploitation of heterosis, inter-group hybrids also have an advantage with regard to a more favorable ratio of general (GCA) to specific combining ability (SCA) variances. This is demonstrated by data from a diallel experiment with six European flint and six dent lines originally published by Dhillon et al. (1990). In the ordinary diallel analysis of all 66 crosses (including intra- and inter-group hybrids), the ratio of $\sigma_{SCA}^2 : \sigma_{GCA}^2$ amounted to 3.9 for forage yield and 12.9 for ear yield (Table 10–4); however, when the analysis was restricted to the subset of 36 flint × dent inter-group hybrids, this ratio was only 0.50 and 0.30, respectively, due to a dramatic decrease in σ_{SCA}^2. With predominance of σ_{GCA}^2 over σ_{SCA}^2, early testing becomes more effective and promising hybrids can be identified and selected mainly based on their prediction from GCA effects, which makes hybrid breeding more efficient.

Management of germplasm in genetically diverse heterotic groups offers a further advantage for hybrid breeding in that crosses between related lines are automatically avoided, when we aim at inter-group hybrids for commercial use. This implies that inbreeding may affect the means of intra-group hybrids but not of inter-group hybrids. Consequently, a high selection intensity within each heterotic group is not harmful for the hybrid mean of inter-group hybrids, at least for a short term. Finally, with inter-group hybrids it is possible to exploit the contribution of multiplication effects to MPH, if the two parental heterotic groups display substantial differences in their yield components such as the Minor and Major germplasm in faba bean (Link et al., 1995).

Table 10–3. Increase in the mean genetic distance (GD), mid-parent heterosis (MPH), and hybrid performance (F1P) for yield of inter-group hybrids over intra-group hybrids in various crops. $\Delta = [100 * \pi1 \times \pi2 / (\pi1 \times \pi1 + \pi2 \times \pi2)/2] - 100$, where $\pi1$ and $\pi2$ refer to the two heterotic groups.

Crop	Heterotic group		Inter- vs. intra-group hybrid			Reference
	$\pi1$	$\pi2$	ΔGD	ΔH	ΔF1P	
			----------- % -----------			
Maize	RYD†	LSC†	21	42	21	Dudley et al., 1991
	Flint	Dent	19	33	16	Dhillon et al., 1993
Rye	Petkus	Carsten	-	16	21	Hepting, 1978
Rice	SE China	SE Asia	12	9	16	Zhang et al., 1995
Faba bean	Minor	Major	7	15	5	Link et al., 1996

† RYD, Reid Yellow Dent; LSC, Lancaster Sure Crop.

Table 10–4. Estimates of general (GCA) and specific combining ability (SCA) variances in diallel crosses and factorial (F × D) crosses of six European flint and six dent inbred lines for forage and ear yield (original data taken from Dhillon et al., 1990).

Trait	σ^2_{GCA}		σ^2_{SCA}		$\sigma^2_{SCA} : \sigma^2_{GCA}$	
	Diallel	F × D†	Diallel	F × D	Diallel	F × D
Forage yield	49.9	77.7	194.0	38.9	3.9	0.50
Ear yield	5.24	17.5	67.9	5.3	12.9	0.30
Mean ratio	1 : 2.5		9 : 1		25 : 1	

† Refers to the average GCA variance of flint and dent lines.

Usefulness of Heterotic Groups in Other Breeding Categories

Hitherto the concept of heterotic groups has only been used in hybrid breeding; however, it has a much wider scope. This is obvious by considering a clone as an asexually propagated hybrid. Thus, heterosis could be exploited best in exactly the same fashion as in sexually propagated hybrids (Schnell, 1978). This suggests that clones bred by crossing parents from genetically diverse germplasm groups should be superior in yield over those developed from crosses between parents of the same germplasm group. To our knowledge, this approach has not been put into practice so far.

With respect to open-pollinated varieties (OPVs) and synthetics, quantitative genetic theory indicates that in the population $(\pi 1 \cup \pi 2)^r$ obtained by random mating of the two parent populations $\pi 1$ and $\pi 2$, we can exploit one-half of ΔH:

$$\mu_{(\pi 1 \cup \pi 2)^r} = (\mu_{\pi 1} + \mu_{\pi 2})/2 + 0.5 \ \Delta H(\pi 1 \times \pi 2). \qquad [5]$$

With larger values of ΔH, the performance of $(\pi 1 \cup \pi 2)^r$ often outyields the higher performing parent population. Thus, in the case of OPVs or synthetics too, it can be rewarding to have separate heterotic groups that are intercrossed to establish composites released to the farmer. This approach is especially attractive in combination with hybrid breeding and/or reciprocal recurrent selection, when the breeder has already established distinct heterotic groups, and was advocated in a comprehensive breeding system (Eberhart et al., 1967) and in breeding of synthetics (Gallais, 1991).

Grouping of Germplasm for Optimum Exploitation of Heterosis

Provided that an effective system for large-scale production of hybrid seed is available, there are currently numerous attempts to start hybrid breeding in allogamous or partially-allogamous crops such as rice, wheat, rapeseed (*Brassica napus* L.), and faba bean (Melchinger & Gumber, 1998). A basic question in this context is: How should germplasm be organized in heterotic groups for optimum exploitation of heterosis? A theoretical answer is presented here, assuming positive heterosis and absence of epistasis.

Suppose we have a set Ω of homozygous inbreds and subdivide these lines into two disjoint subsets A and B of equal size. Then, we obtain the following relationship:

$$\overline{MRD^2}(A,B) = \overline{MRD^2}(\Omega) + 0.5 \ \overline{MRD^2}(\pi A, \pi B), \qquad [6]$$

where the bar indicates averaging over all pairs of lines in A and B and in Ω, respectively, and πA and πB refer to the population of inbreds in set A and B, respectively. Combining Eq. 3 and 6 yields

$$\overline{MPH}(A \times B) = \overline{MPH}(\Omega) + 0.5\Delta H(\pi A \times \pi B). \qquad [7]$$

The first term on the right hand side of Eq. 6 and 7 depends only on Ω and, hence, is fix irrespective of the subdivision of Ω into A and B. Consequently, the mean heterosis and mean hybrid performance depend only on the second term $\Delta H(\pi A \times \pi B)$. Based on Eq. 3, this term becomes maximum, if $\overline{MRD^2}(\pi A, \pi B)$ is maximum. In conclusion, that partition of Ω which maximizes the difference in gene frequencies between the two subgroups will maximize the mean heterosis and hybrid performance in intergroup crosses between them.

Criteria and Approaches for Identification of Promising Heterotic Groups and Patterns

In a recent review, Melchinger and Gumber (1998) recommended the following criteria for choice of heterotic groups and patterns in hybrid breeding:
1. high mean performance and large genetic variance in the hybrid population;
2. high *per se* performance and good adaptation of the parent populations to the target region(s);
3. low inbreeding depression, if hybrids are produced from inbreds.
If hybrid seed production is based on a CMS system, the choice of parent populations also must consider the existence of maintainers and effective restorers. In practice, the choice of heterotic patterns has been mainly based on the performance of the corresponding hybrid population because other criteria such as the genetic variance can be determined only with great expenditures.

Basically, the approaches for identification of promising heterotic patterns depend largely on the source materials. With a smaller number of populations it is common practice to evaluate diallel or factorial crosses among them. This approach has been taken to identify suitable heterotic patterns in rye (Hepting, 1978) as well as tropical and subtropical maize in several studies conducted at CIMMYT (Beck et al., 1990, 1991; Crossa et al., 1990; Vasal et al., 1992a,b,c). It also has been employed to identify alternative heterotic patterns for maize in the U.S. Cornbelt (Kauffmann et al., 1982; Mungoma & Pollak, 1988) and western Europe (Moreno-Gonzalez et al., 1997; Ordas, 1991). While most of these studies considered GCA and SCA effects of the populations as well as ΔH on a percentage basis, we recommend to emphasize hybrid performance of the population crosses as main selection criterion.

If a large number of germplasm is available and established heterotic groups or proven testers exist, the performance of testcrosses between them is usually taken as main criterion for the choice and grouping of materials. For example, assignment of 92 CIMMYT maize inbreds to Tropical Heterotic Groups was based on their testcross performance for yield in combination with two dent and two flint tester lines (Vasal et al., 1992d,e). Lines showing negative SCA with the former but positive SCA with the latter were assigned to group "A" and vice versa, lines having negative SCA with the dent and positive SCA with the flint tester were assigned to group "B". This approach also has been applied to broaden the genetic basis of established heterotic groups with germplasm of similar heterotic response (Gutierrez-Gaitan et al., 1986). A way to improve the efficiency of this approach might be to first assess the marker-based GD of the new germplasm to the existing heterotic groups and, based on this information, evaluate only promising testcrosses with representative testers.

When a large number of germplasm exists but no established heterotic groups are available, it is important to first identify groups of genetically similar germplasm. As outlined in the first part of this review, this can be accomplished most accurately and reliably by GD estimates based on DNA markers. In a second step, one can then produce and evaluate diallel or factorial crosses among representative genotypes from each group or, alternatively, among composites produced from genotypes in each group. Finally, promising groups can be selected as heterotic groups or patterns based on mean hybrid performance and other criteria mentioned above.

As a final remark, we would like to stress that the choice of heterotic groups and patterns is of fundamental importance in hybrid breeding, because it usually predetermines the base materials for generations of breeders. Once heterotic groups have been established and improved by numerous cycles of selection, it is difficult to develop competitive ones. Hence, considerable expenditures are justified at the beginning of a hybrid breeding program to arrive at an optimal decision. Looking at the history of successful heterotic patterns in various crops, there is evidence suggesting that adapted populations isolated by time and space are the most promising candidates for heterotic patterns (Melchinger & Gumber, 1998).

CORRELATIONS OF GENETIC DISTANCES BASED ON DNA MARKERS WITH HETEROSIS AND HYBRID PERFORMANCE

Predicting hybrid performance has always been a primary objective in all hybrid crop breeding programs. The development of molecular marker techniques has provided new tools for heterosis prediction and DNA markers have been used extensively in investigating correlations between parental GD and F1P or MPH, especially in maize. Previous studies with isozymes had demonstrated that GD estimates were in most cases positively associated with F1P for grain yield (for review see Stuber, 1994); however, it remained unclear whether the low correlations between both measures observed in most cases were due to poor coverage of the genome or whether other causes were involved.

In view of the breeding situation in different crops and based on quantitative genetic considerations, Melchinger (1993) proposed to distinguish three cases: (i) crosses between related lines, (ii) crosses between unrelated lines from the same germplasm group, and (iii) crosses between (unrelated) lines from different heterotic groups. Taking this aspect into account, Table 10–5 gives an overview of investigations with RFLPs, AFLPs, and SSRs reported in the literature. In studies involving crosses between related lines as well as intra- and inter-group crosses, the correlations of parental GD with F1P and MPH were moderate to high in maize (Smith et al., 1990; Boppenmaier, 1994; Ajmone Marsan et al., 1998), rapeseed (Diers et al.,1996), and rice (Zhang et al., 1995). The marker system (AFLPs vs. RFLPs) seemed to have little influence on the correlations (Ajmone Marsan et al., 1998), while they were generally higher with a wider range in the f values of crosses between related lines. When crosses among related lines ($f > 0$) were excluded from the calculations, the correlations were smaller and only of moderate size except for a study in rice (Saghai Maroof et al., 1997). Correlations for intra-group crosses were moderate to high in maize (Dhillon et al., 1993; Ajmone Marsan et al., 1998) but not different from zero in rice (Saghai Maroof et al., 1997). Likewise, correlations of GD with F1P and MPH were always close to zero for inter-group crosses in maize (Boppenmaier et al., 1992; Dhillon et al., 1993). Investigations with intra-group crosses including those among related lines found no significant associations of GD with F1P and MPH in soybean (Cerna et al., 1997) and wheat (Martin et al., 1995). This might be due to the low level of heterosis in these two crops (<15%) as compared with maize, rapeseed, and rice.

Based on results from various studies in maize, Melchinger (1993) summarized the relationship between parental GD and MPH in a schematic representation shown in Fig. 10–2. Experimental evidence from other studies listed in Table 10–3 as well as quantitative genetic expectations support this conclusion. Accordingly, for crosses among related lines, there exists a tight association between GD and MPH for yield characters because both measures are a linear function of f and, hence, decrease with increasing f. For intra-group crosses, the correlation r(GD, MPH) is generally positive too. This can be explained by (i) hidden relatedness between some parents considered as being unrelated based on their pedigree records and (ii) presence of the same linkage phase between QTL and marker loci in the maternal and paternal gametic arrays of intra-group hybrids, which results in a positive covariance between GD and MPH (Charcosset et al., 1991). In contrast, no significant association between both measures exists for inter-group hybrids. In this case, the maternal and paternal gametic arrays may differ in the linkage phase for many QTL–marker pairs; as a consequence, positive and negative terms cancel each other in their net contribution to the covariance (GD, MPH), resulting in a low or zero correlation (Charcosset & Essioux, 1994). Fig. 10–2 also illustrates that high estimates of r(GD, MPH) can be expected if correlations are calculated across different types of crosses due to group effects. This is because both GD and MPH are expected to increase from crosses among related lines to intra-group crosses and further to inter-group crosses.

It has been common practice in most studies to determine GD estimates from a set of DNA markers chosen for good coverage of the entire genome but not for linkage to QTL influencing heterosis of the target trait. Theoretical investigations (Charcosset et al., 1991) and computer modeling (Bernardo, 1992) demonstrated that with intra- and inter-group crosses the correlation r(GD, MPH) is expected to decrease if (i) QTL influencing heterosis are not closely linked to markers used for calculation of GD estimates and vice versa (ii) markers employed for calculation of GDs are not linked to QTL. Hence, increasing the marker density alone will not necessarily improve the ability to predict MPH by GD estimates; rather, markers must additionally be selected for tight linkage to QTL affecting heterosis of the target trait in the germplasm under study. This is corroborated by the comparison of results obtained with 209 AFLPs vs. 135 RFLPs (Ajmone Marsan et al., 1998) and a study by Dudley et al. (1991); however, attempts to identify QTL–marker associations by use of multiple regression techniques, in which observed phenotypic values for a fixed set of hybrids are regressed on their coded marker genotypes, must be regarded with great caution, because the number of variables from which the regressors are selected, is usually as large or even greater than the number of phenotypic observations. Moreover, comparison of QTL mapping results from different populations in maize (Stuber, 1995;

Table 10–5. Correlations (*r*) of genetic distance (GD) between parental lines based on DNA markers with F_1 performance (F1P) and mid-parent heterosis (MPH) for yield of their hybrids in various crops.

Type of hybrids				$r(GD,\cdot)$		DNA	
f >0	intra	inter	N†	F1P	MPH	markers	Reference

Maize

f >0	intra	inter	N	F1P	MPH	DNA markers	Reference
+	+	+	123	0.93**	0.87**	RFLP	Smith et al., 1990
+	+	+	64	0.61**	0.49**	RFLP	Boppenmaier, 1994
+	+		62	0.48**	0.45**	RFLP	Dhillon et al., 1993
	+		14	0.70**	0.58*	RFLP	Dhillon et al., 1993
		+	36	- 0.23	- 0.10	RFLP	Dhillon et al., 1993
		+	66	- 0.06	- 0.04	RFLP	Boppenmaier et al., 1992‡
+	+	+	78	0.51**	(0.36**)§	AFLP	Ajmone Marsan et al., 1998
+	+		53	0.23	(- 0.08)	RFLP	Ajmone Marsan et al., 1998
	+		15	0.47	(0.31)	RFLP	Ajmone Marsan et al., 1998

Rapeseed

| + | + | + | 21 | 0.49* | 0.58* | RFLP | Diers et al., 1996 |

Rice

+	+	+	28	0.31	0.53	RFLP + SSR	Zhang et al., 1995
	+	+	28	0.79**	0.47*	RFLP + SSR	Saghai Maroof et al., 1997
	+		21	0.22	- 0.15	RFLP + SSR	Saghai Maroof et al., 1997

Soybean

| + | + | | 48 | - 0.29 | 0.08 | RFLP | Cerna et al., 1997 |

Wheat

| + | + | | 21 | 0.27 | 0.33 | RFLP | Martin et al., 1995 |

*, ** Significant at the 0.05 and 0.01 probability levels, respectively.
† N, Number of crosses examined.
‡ Yield, forage yield in this study.
§ Values in parentheses refer to correlations based on RFLP data.

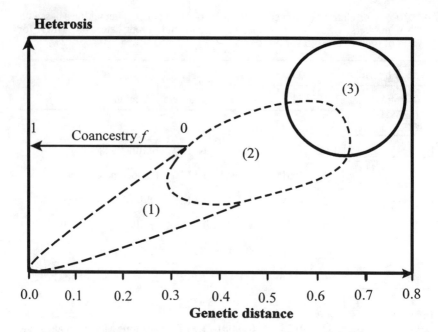

Fig. 10–2. Schematic representation between mid-parent heterosis for yield and
parental genetic distance based on unselected DNA markers covering the entire
genome. (1) Crosses between related lines; (2) intra-group crosses between unrelated lines; (3) inter-group crosses.

Lübberstedt et al., 1998) suggests that QTL regions affecting a given trait are not
consistent across different germplasms.

In conclusion, the potential application of DNA markers in hybrid breeding
depends very much upon whether divergent heterotic groups have been established or
not. If well-established heterotic groups are not available, marker-based GD estimates
can be used to avoid producing and testing of crosses between related lines. Furthermore, crosses with inferior MPH could be discarded prior to field testing based on
prediction. Another potential application exists, if new lines of unknown heterotic
pattern or inbreds developed from crosses between parents from different heterotic
groups (e.g,. commercial hybrids) are to be evaluated for testcross performance. In this
case, GD estimates could assist the breeder in the choice of appropriate testers for
evaluating the combining ability of lines; however, with regard to the typical situation
of hybrid breeding, in which crosses are produced between lines from genetically
divergent heterotic groups, GD estimates based on an unselected set of DNA markers
alone are not promising for predicting hybrid performance or any of its components.
This must not be confused with the proposal of Bernardo (1994) to use coancestry
coefficients determined by DNA markers in combination with the BLUP (best linear
unbiased prediction) approach for prediciton of related hybrids, which holds greater
promise (Charcosset et al., 1998).

CONCLUSIONS

In summary, the results accumulated in recent years with numerous crops
indicate that DNA markers such as RFLPs, AFLPs, and SSRs are well suited for

- assessment of the genetic diversity in germplasm and monitoring of its development over time,
- grouping and classification of cultivars, breeding materials, and genetic resources,
- pedigree analysis at the level of the entire genome and specific chromosome regions, and
- choice of parents for establishing base populations.

These applications are of direct use in the management of germplasm banks as well as in breeding programs and plant variety protection.

Theoretical and experimental results demonstrate that organisation of germplasm in genetically divergent germplasm pools is beneficial for optimum exploitation of heterosis. This applies not only to hybrid breeding but also to breeding of clones and open-pollinated or synthetic varieties. DNA markers provide a powerful tool for identifying groups of genetically similar germplasm; however, heterotic response between these groups cannot be predicted from genetic distances based on DNA markers but must be evaluated in field trials. Likewise, prediction of heterosis and hybrid performance from parental genetic distances can be applied in hybrid breeding only under special circumstances but not to inter-group crosses between parents from genetically diverse heterotic groups.

ACKNOWLEDGMENTS

The author is indebted to A. Gallais, H.C. Becker, and A. Charcosset for helpful comments and suggestions on the manuscript. Gratitude is also expressed to R. Meyer and B. Devezi for their assistance in typing the manuscript and preparing the figures.

REFERENCES

Ahnert, D., M. Lee, D.F. Austin, C. Livini, W.L. Woodman, S.J. Openshaw, J.S.C. Smith, K. Porter, and G. Dalton. 1996. Genetic diversity among elite sorghum inbred lines assessed with DNA markers and pedigree information. Crop Sci. 36:1385–1392.

Ajmone Marsan, P., P. Castiglioni, F. Fusari, M. Kuiper, and M. Motto. 1998. Genetic diversity and its relationship to hybrid performance in maize as revealed by RFLP and AFLP markers. Theor. Appl. Genet. 96:219–227.

Barbosa-Neto, J.F., C.M. Hernández, L.S. O'Donoughue, and M.E. Sorrells. 1997. Precision of genetic relationship estimates based on molecular markers. Euphytica 98:59–67.

Beck, D.L., S.K. Vasal, and J. Crossa. 1991. Heterosis and combining ability among subtropical and temperate intermediate-maturity maize germplasm. Crop Sci. 31:68–73.

Beck, D.L., S.K. Vasal, and J. Crossa. 1990. Heterosis and combining ability of CIMMYT's tropical early and intermediate maturity maize (*Zea mays* L.) germplasm. Maydica 35:279–285.

Bernardo, R. 1992. Relationship between single-cross performance and molecular marker heterozygosity. Theor. Appl. Genet. 83:628–634.

Bernardo, R. 1994. Prediction of maize single-cross performance using RFLPs and information from related hybrids. Crop Sci. 34:20–25.

Bernardo, R., A. Murigneux, J.P. Maisonneuve, C. Johnsson, and Z. Karaman. 1997. RFLP-based estimates of parental contribution to F_2- and BC_1-derived maize inbreds. Theor. Appl. Genet. 94:652–656.

Boppenmaier, J. 1994. Genetische Diversität für RFLPs bei europäischen Maisinzuchtlinien und ihre Beziehung zur Hybridleistung von Kreuzungen innerhalb und zwischen Formenkreisen. Dissertation Universität Hohenheim, Stuttgart.

Boppenmaier, J., A.E. Melchinger, E. Brunklaus-Jung, H.H. Geiger, and R.G. Herrmann. 1992. Genetic diversity for RFLPs in European maize inbreds: I. Relation to performance of flint × dent crosses for forage traits. Crop Sci. 32:895-902.

Boutin, S.R., N.D. Young, L.L. Lorenzen, and R.C. Shoemaker. 1995. Marker-based pedigrees and graphical genotypes generated by Supergene software. Crop Sci. 35:1703–1707.

Cerna, F.J., S.R. Cianzio, A. Rafalski, S. Tingey, and D. Dyer. 1997. Relationship between seed yield heterosis and molecular marker heterozygosity in soybean. Theor. Appl. Genet. 95:460–467.

Charcosset, A., B. Bonnisseau, O. Touchebeuf, J. Burstin, P. Dubreuil, Y. Barrière, A. Gallais, and J.-B. Denis. 1998. Prediction of maize hybrid silage performance using marker data: Comparison of several models for specific combining ability. Crop Sci. 38:38-44.

Charcosset, A., and L. Essioux. 1994. The effect of population structure on the relationship between heterosis and heterozygosity at marker loci. Theor. Appl. Genet. 89:336-343.

Charcosset, A., M. Lefort-Buson, and A. Gallais. 1991. Relationship between heterosis and heterozygosity at marker loci: a theoretical computation. Theor. Appl. Genet. 81:571-575.

Crossa, J., S.K. Vasal, and D.L. Beck. 1990. Combining ability estimates of CIMMYT's tropical late yellow maize germplasm. Maydica 35:273–278.

Dhillon, B.S., J. Boppenmaier, W.G. Pollmer, R.G. Herrmann, and A.E. Melchinger. 1993. Relationship of restriction fragment length polymorphisms among European maize inbreds with ear dry matter yield of their hybrids. Maydica 38:245–248.

Dhillon, B.S., P.A. Gurrath, E. Zimmer, M. Wermke, W.G. Pollmer, and D. Klein. 1990. Analysis of diallel crosses of maize for variation and covariation in agronomic traits at silage and grain harvests. Maydica 35:297–302.

Diers, B.W., P.B.E. McVetty, and T.C. Osborn. 1996. Relationship between heterosis and genetic distance based on restriction fragment length polymorphism markers in oilseed rape (*Brassica napus* L.). Crop Sci. 36:79–83.

Dubreuil, P., P. Dufour, E. Krejci, M. Causse, D. de Vienne, A. Gallais, and A. Charcosset. 1996. Organization of RFLP diversity among inbred lines of maize representing the most significant heterotic groups. Crop Sci. 36:790–799.

Dudley, J.W., M.A. Saghai Maroof, and G.K. Rufener. 1991. Molecular markers and grouping of parents in maize breeding programs. Crop Sci. 31:718–723.

Eberhart, S.A., M.N. Harrison, and F. Ogada. 1967. A comprehensive breeding system. Züchter/Genet.Breed.Res. 37:169–174.

Falconer, D.S., and T.F. Mackay. 1996. Introduction to quantitative genetics. 4th ed. Longman Group Ltd., London.

Gallais, A. 1991. Three-way and four-way recurrent selection in plant breeding. Plant Breeding 107:265–274.

Goodman, M.M., and C.W. Stuber. 1983. Races of maize. VI. Isozyme variation among races of maize in Bolivia. Maydica 28:169–187.

Graner, A., W.F. Ludwig, and A.E. Melchinger. 1994. Relationships among European barley germplasm: II. Comparison of pedigree and RFLP data. Crop Sci. 34:1199-1204.

Gutierrez-Gaitan, M.A., H. Cortez-Mendoza, E.N. Wathika, C.O. Gardner, M. Oyervides-Garcia, A.R. Hallauer, and L.L. Darrah. 1986. Testcross evaluation of Mexican maize populations. Crop Sci. 26:99–104.

Hahn, V., K. Blankenhorn, M. Schwall, and A.E. Melchinger. 1995. Relationships among early European maize inbreds: III. Genetic diversity revealed with RAPD markers and comparison with RFLP and pedigree data. Maydica 40:299–310.

Hamrick, J.L., and M.J.W. Godt. 1997. Allozyme diversity in cultivated crops. Crop Sci. 37:26–30.

Hepting, L. 1978. Analyse eines 7 x 7-Sortendiallels zur Ermittlung geeigneten Ausgangsmaterials für die Hybridzüchtung bei Roggen. Z. Pflanzenzüchtg. 80:188-197.

Hongtrakul, V., G.M. Huestis, and S.J. Knapp. 1997. Amplified fragment length polymorphisms as a tool for DNA fingerprinting sunflower germplasm: genetic diversity among oilseed inbred lines. Theor. Appl. Genet. 95:400–407.

Jaccard, P. 1908. Nouvelles recherches sur la distribution florale. Bull. Soc. Vaud Sci. Nat. 44:223–270.

Kauffmann, K.D., C.W. Crum, and M.F. Lindsey. 1982. Exotic germplasm in a corn breeding program. III. Corn Breeder's School 18:6–39.

Labate, J.A., K.R. Lamkey, M. Lee, and W.L. Woodman. 1997. Molecular genetic diversity after reciprocal recurrent selection in BSSS and BSCB1 maize populations. Crop Sci. 37:416–423.

Lange, P. 1996. Plant breeders' rights and patents: new developments and concepts for cooperative solutions. Vortr. Pflanzenzüchtg. 33:57–69.

Link, W., C. Dixkens, M. Singh, and M.M.A.E. Schwall. 1995. Genetic diversity in European and Mediterranean faba bean germplasm revealed by RAPD markers. Theor. Appl. Genet. 90:27–32.

Link, W., B. Schill, A.C. Barbera, J.I. Cubero, A. Filippetti, L. Stringi, E. von Kittlitz, and A.E. Melchinger. 1996. Comparison of intra- and inter-pool crosses in faba beans (*Vicia faba* L.). I. Hybrid performance and heterosis in Mediterranean and German environments. Plant Breeding 115:352–360.

Lorenzen, L.L., S. Boutin, N. Young, J.E. Specht, and R.C. Shoemaker. 1995. Soybean pedigree analysis using map-based molecular markers: I. Tracking RFLP markers in cultivars. Crop Sci. 35:1326–1336.

Lübberstedt, T., A.E. Melchinger, S. Fähr, D. Klein, A. Dally, and P. Westhoff. 1998. QTL mapping in testcrosses of flint lines of maize: III. Comparison across populations for forage traits. Crop Sci. 38:1278-1289.

Malécot, G. 1948. Les mathématiques de l'hérédite. Masson et Cie, Paris.

Martin, J.M., L.E. Talbert, S.P. Lanning, and N.K. Blake. 1995. Hybrid performance in wheat as related to parental diversity. Crop Sci. 35:104–108.

Melchinger, A.E. 1993. Use of RFLP markers for analyses of genetic relationships among breeding materials and prediction of hybrid performance. p. 621–628. *In* D.R. Buxton et al. (ed.) International Crop Science I. CSSA, Madison, WI.

Melchinger, A.E., A. Graner, M. Singh, and M.M. Messmer. 1994. Relationships among European barley germplasm: I. Genetic diversity among winter and spring cultivars revealed by RFLPs. Crop Sci. 34:1191–1198.

Melchinger, A.E., and R.K. Gumber. 1998. Overview of heterosis and heterotic groups in agronomic crops. p. 29–44. *In* K.R. Lamkey and J.E. Staub (ed.) Concepts and breeding of heterosis in crop plants. CSSA, Madison, WI.

Melchinger, A.E., R.K. Gumber, R.B. Leipert, M. Vuylsteke, and M. Kuiper. 1998. Prediction of testcross means and variances among F_3 progenies of F_1 crosses from testcross means and genetic distances of their parents in maize. Theor. Appl. Genet. 96:503-512.

Melchinger, A.E., M.M. Messmer, M. Lee, W.L. Woodman, and K.R. Lamkey. 1991. Diversity and relationships among U.S. maize inbreds revealed by restriction fragment length polymorphisms. Crop Sci. 31:669–678.

Messmer, M.M., A.E. Melchinger, J. Boppenmaier, E. Brunklaus-Jung, and R.G. Herrmann. 1992. Relationships among early European maize (*Zea mays* L.) inbred lines: I. Genetic diversity among flint and dent lines revealed by RFLPs. Crop. Sci. 32:1301–1309.

Messmer, M.M., A.E. Melchinger, J. Boppenmaier, and R.G. Herrmann. 1993. Relationships among early European maize (*Zea mays* L.) inbred lines: II. Comparison of pedigree and RFLP data. Crop Sci. 33:944–950.

Moll, R.H., J.H. Longquist, J.V. Fortuna, and E.C. Johnson. 1965. The relation of heterosis and genetic divergence in maize. Genetics 52:139–144.

Moll, R.H., W.S. Salhuana, and H.F. Robinson. 1962. Heterosis and genetic diversity in variety crosses of maize. Crop Sci. 2:197–198.

Moreno-Gonzalez, J., F. Ramos-Gourcy, and E. Losada. 1997. Breeding potential of European flint and earliness-selected U.S. Corn Belt dent maize populations. Crop Sci. 37:1475–1481.

Mumm, R.H., and J.W. Dudley. 1994. A classification of 148 U.S. maize inbreds: I. Cluster analysis based on RFLPs. Crop Sci. 34:842–851.

Mumm, R.H., L.J. Hubert, and J.W. Dudley. 1994. A classification of 148 U.S. maize inbreds: II. Validation of cluster analysis based on RFLPs. Crop Sci. 34:852–865.

Mungoma, C., and L.M. Pollak. 1988. Heterotic patterns among ten Corn Belt and exotic maize populations. Crop Sci. 28:500–504.

Nei, M. 1973. Analysis of gene diversity in subdivided populations. Proc. Natl. Acad. Sci. USA 70:3321–3323.

Nei, M. 1978. Estimation of average heterozygosity and genetic distance from a small number of individuals. Genetics 89:583–590.

Nei, M., and W.-H. Li. 1979. Mathematical model for studying genetic variation in terms of restriction endonucleases. Proc. Natl. Acad. Sci. USA. 76:5269–5273.

Ordas, A. 1991. Heterosis in crosses between American and Spanish populations of maize. Crop Sci. 31:931–935.

Plaschke, J., M.W. Ganal, and M.S. Röder. 1995. Detection of genetic diversity in closely related bread wheat using microsatellite markers. Theor. Appl. Genet. 91:1001–1007.

Rogers, J.S. 1972. Measures of genetic similarity and genetic distance. Studies in genetics VII. Univ. of Tex. Publ. 7213:145–153.

Russell, J.R., J.D. Fuller, M. Macaulay, B.G. Hatz, A. Jahoor, W. Powell, and R. Waugh. 1997. Direct comparison of levels of genetic variation among barley accessions detected by RFLPs, AFLPs, SSRs and RAPDs. Theor. Appl. Genet. 95:714–722.

Saghai Maroof, M.A., G.P. Yang, Q. Zhang, and K.A. Gravois. 1997. Correlation between molecular marker distance and hybrid performance in U.S. southern long grain rice. Crop Sci. 37:145–150.

Schnell, F.W. 1978. Progress and problems in utilizing quantitative variability in plant breeding. Plant Res. and Dev. 7:32–43.

Schut, J.W., X. Qi, and P. Stam. 1997. Association between relationship measures based on AFLP markers, pedigree data and morphological traits in barley. Theor. Appl. Genet. 95:1161–1168.

Smith, J.S.C., E.C.L. Chin, H. Shu, O.S. Smith, S.J. Wall, M.L. Senior, S.E. Mitchell, S. Kresovich, and J. Ziegle. 1997. An evaluation of the utility of SSR loci as molecular markers in maize (Zea mays L.): Comparisons with data from RFLPs and pedigree. Theor. Appl. Genet. 95:163–173.

Smith, O.S., J.S.C. Smith, S.L. Bowen, R.A. Tenborg, and S.J. Wall. 1990. Similarities among a group of elite maize inbreds as measured by pedigree, F1 grain yield, grain yield heterosis, and RFLPs. Theor. Appl. Genet. 80:833–840.

Stuber, C.S. 1994. Breeding multigenic traits. p. 97–115. In Phillips, R.L. and I.K. Vasil (ed.). DNA-based markers in plants. Kluwer Academic. Boston.

Stuber, C.W. 1995. Mapping and manipulating quantitative traits in maize. TIG 11:477–481.

Tivang, J.G., J. Nienhuis, and O.S. Smith. 1994. Estimation of sampling variance of molecular marker data using the bootstrap procedure. Theor. Appl. Genet. 89:259–264.

Vasal, S.K., G. Srinivasan, D.L. Beck, J. Crossa, S. Pandey, and C. De Leon. 1992a. Heterosis and combining ability of CIMMYT's tropical late white maize germplasm. Maydica 37:217–223.

Vasal, S.K., G. Srinivasan, J. Crossa, and D.L. Beck. 1992b. Heterosis and combining ability of CIMMYT's subtropical and temperate early-maturity maize germplasm. Crop Sci. 32:884–890.

Vasal, S.K., G. Srinivasan, F. Gonzalez C., G.C. Han, S. Pandey, D.L. Beck, and J. Crossa. 1992c. Heterosis and combining ability of CIMMYT's tropical x subtropical maize germplasm. Crop Sci. 32:1483–1489.

Vasal, S.K., G. Srinivasan, G.C. Han, and F. Gonzalez C. 1992d. Heterotic patterns of eighty-eight white subtropical CIMMYT maize lines. Maydica 37:319–327.

Vasal, S.K., G. Srinivasan, S. Pandey, H.S. Cordova, G.C. Han, and F. Gonzalez C. 1992e. Heterotic patterns of ninety-two white tropical CIMMYT maize lines. Maydica 37:259–270.

Westmann, A.C., and S. Kresovich. 1997. Use of molecular marker techniques for description of plant genetic variation. p. 9–48. In J.A. Callow et al. (ed.) Biotechnology and plant genetic resources. CAB Int., Oxford, England.

Yang, W., A.C. de Oliveira, I. Godwin, K. Schertz, and J.L. Bennetzen. 1996. Comparison of DNA marker technologies in characterizing plant genome diversity: Variability in Chinese sorghums. Crop Sci. 36:1669–1676.

Zhang, Q., Y.J. Gao, M.A. Saghai Maroof, S.H. Yang, and J.X. Li. 1995. Molecular divergence and hybrid performance in rice. Molec. Breed. 1:133–142.

Zhang, Q., B.Z. Shen, X.K. Dai, M.H. Mei, M.A. Saghai Maroof, and Z.B. Li. 1994. Using bulked extremes and recessive class to map genes for photoperiod-sensitive genic male sterility in rice. Proc. Natl. Acad. Sci. USA. 91:8675–8679.

Chapter 11

Effect of Hybrid Breeding on Genetic Diversity in Maize

J. S. C. Smith, D. N. Duvick, O. S. Smith, A. Grunst, and S. J. Wall

CHANGES IN GENETIC DIVERSITY WITHIN CORN BELT MAIZE

The extent and quality of maize (*Zea mays* L.) germplasm diversity that is now widely cultivated in the USA reflects usage by farmers and breeders in this and previous centuries. "What this superior germ plasm is and how it is used constitute a story of surpassing importance to the modern world," (Henry A. Wallace as cited in Jenkins, 1936). This statement rings true even more today, for population growth and environmental needs demand that productivity be increased still further.

Sturtevant (1899) documented 69 different open-pollinated varieties of flint maize and 323 varieties of dent maize. But farmers did not use this diversity equally. Twenty-five varieties were most widely grown at the beginning of the 20th century (Bowman & Crossley, 1908) and chief among these was Reid Yellow Dent. It is from this germplasm base that most of U.S. hybrid maize has been developed. Open-pollinated varieties dominated U.S. maize acreage until the mid-1930s. By 1941, hybrid acreage in major U.S. Corn Belt states had reached from 75 to 97% (Duvick, 1996). Hybrids displaced open-pollinated varieties because, "Many open-pollinated plants were semi-barren and it seemed almost all lodged badly. (They were) ... weak-rooted and stalk-rot susceptible" (Baker, 1984). The hybrid breeding and production method is conceptually simple and highly effective (Duvick, 1996) and hybrids provide a natural form of intellectual property protection that early on helped to attract privately funded commercial investments.

Plant breeders use and release genetic diversity in time. The objectives of this study, therefore, are to examine changes in diversity among a set of inbreds that formed parents of hybrids that were widely used in the U.S. Corn Belt during each decade from 1930 to the present. Comparisons will be made to publicly bred inbred lines that were widely used in the U.S. Corn Belt from the 1950s through the 1980s. Therefore, we examined genetic diversity using pedigree and isozymic data for 71 inbreds that formed parents of 44 important widely used U.S. Corn Belt hybrids released by Pioneer Hi-Bred International, Inc. since 1930 (Table 11–1). The numbers of inbreds that were used in each decade are given in Table 11–2. The mean percentage contribution of each pedigree component was calculated for each decade under the assumption that each inbred was used in equal proportion within each decade. For purposes of comparisons with germplasm that was bred in the public domain we similarly calculated mean percentage contribution by pedigree for public inbred lines adapted to the central U.S. Corn Belt and that were reported as widely used by Zuber and Darrah (1980) from the 1950s through the

Table 11–1. Maize hybrids examined in this survey.

Decade of usage	Hybrid†	Parents‡ F × M	Decade of usage	Hybrid	Parents F × M
1930	307	SX × SX	1970	3301A	IN × IN
	317	SX × SX		3366	IN × IN
	322	SX × SX		3388	SX × IN
	330	SX × SX		3517	SX × IN
	351	IN × SX		3529	SX × bc
				3541	SX × bc
1940	336	SX × SX			
	340	SX × SX	1980	3362	IN × IN
	339	SX × SX		3377	IN × IN
	344	SX × SX		3378	IN × IN
	352	SX × SX		3379	IN × IN
	350B	SX × SX		3475	IN × IN
1950	301B	SX × SX	1990	3260	IN × IN
	328	SX × SX		3279	IN × IN
	329	SX × SX		3335	IN × IN
	347	SX × SX		3375	IN × IN
	354	SX × SX		3394	IN × IN
	354A	SX × SX		3417	IN × IN
1960	3206	SX × SX		3489	IN × IN
	3306	IN × IN		3496	IN × IN
	3334	IN × IN		3563	IN × IN
	3376	IN × IN			
	3390	SX × IN			
	3571	SX × IN			
	3618	SX × SX			

† All Pioneer® brand proprietary hybrids.
‡ F, Female; M, Male; IN, Inbred; SX, Single Cross; bc, 7/8 Backcross.

1980s. These public lines were B14, C103, and Oh43 (1950s), B14, B37, C103, and Oh43 (1960s), A619, A632, B14A, B37, B73, H84, Mo17, N28, and Va26, (1970s) and A619, A632, A634, A635, B37, B64, B68, B73, and Mo17 (1980s). Pedigree distances were calculated as (1 − s) where s = Malecot's Coefficient of Relatedness or Coefficient of Parentage. Isozyme data were collected using 23 loci. Distances
were calculated using a modified Rogers' Distance. Polymorphic Index Content (PIC), a genetic diversity measure was calculated from:

$$PIC = 1 - \sum_{i=1}^{n} f_i^2$$

where f_i^2 is the frequency of the ith allele. PIC values were calculated across all 23 isozyme loci.

Table 11–2. Diversity among inbreds and among hybrids used within each decade for the hybrids that were studied.

Decade	No. of inbreds	Inbred pedigree	Diversity isozymes	No. of hybrids	Hybrid pedigree	Diversity isozymes	PIC†
1930	10	0.929	0.572	5	0.967	0.297	0.31
1940	11	0.950	0.555	6	0.940	0.211	0.29
1950	14	0.942	0.562	6	0.832	0.209	0.31
1960	15	0.927	0.551	7	0.911	0.324	0.31
1970	12	0.861	0.552	6	0.861	0.353	0.29
1980	8	0.891	0.528	5	0.832	0.310	0.25
1990	14	0.830	0.470	9	0.793	0.274	0.22

† PIC, Polymorphic Index Content.

PEDIGREE AND ISOZYMIC DATA

Pedigree and isozymic distances among Pioneer® brand hybrids and among inbred lines that were used within each decade are presented in Table 11–2. Fifty-four alleles were present at 23 isozyme loci giving 2.35 alleles per locus. Mean pedigree distances between inbreds increased through to the 1940s (i.e., similarities according to pedigrees decreased). Pedigree distances between inbreds then fell for each decade until the 1980s. During the 1990s, pedigree distances among inbreds were the lowest of any decade; i.e., pedigree similarities between inbreds were the highest of any decade. Isozymic distances between inbreds fell in the 1940s, recovered in the period from 1950 through the 1970s and then fell to a historical low in the 1990s. Pedigree distances among Pioneer® brand hybrids used within each decade fell from the 1930s through the 1950s. Pedigree distances between hybrids increased in the 1960s but then fell during each subsequent decade. By the 1990s, pedigree distances between hybrids were the lowest of any decade under study. Isozymic distances among hybrids declined to a historic low during the 1950s but then increased to a peak in the 1970s. Isozymic distances between hybrids declined during the last two decades but remained higher than during the 1940s or 1950s. Mean PIC values across all hybrids used within each decade remained at or close to 0.31 from the 1930s through to the 1970s but then declined to a seven decade low in the 1990s. Genetic diversity within individual hybrids varied according to the type of cross. Double-cross hybrids had a mean of 56.6% heterozygous isozymic loci and mean PIC of 0.26. Modified single-cross hybrids had a mean of 46.4% heterozygous isozymic loci and a mean PIC of 0.21. Single-cross hybrids had a mean of 33.1% heterozygous isozymic loci and a mean PIC of 0.16.

Isozymic and pedigree data indicate changes in germplasm diversity across decades. Five isozymic loci showed fresh diversity in hybrids after the 1930s. Acp1-3 appeared in the 1940s, Acp4-3 appeared in the 1980s, Est4-5 appeared in the 1980s, Got1-6 appeared in the 1940s; Phi1-5 appeared in the 1950s and reappeared in the 1990s. Fifty-seven pedigree components describe the Pioneer and the publicly bred germplasm reported upon in this study. Twenty-three components contributed ≥1% of pedigree during at least one decade (Table 11–3). For germplasm used in Pioneer® brand hybrids, Reid contributed proportionately most during the 1940s and then declined, but still remained predominant. By the 1970s, most Reid germplasm came through Iowa Stiff Stalk Synthetic parentage. Lancaster Sure Crop achieved a peak of usage during the 1940s and despite an upturn in the 1970s, then declined to historically low levels. Several sources had low to moderate levels of usage during middle period decades (AB8, BH212, Boone

Table 11–3. Mean pedigree constitution during each decade for inbreds that were used as parents for hybrids where a constituent contributed ≥ 1% during at least one decade.

Pedigree	Decade						
	1930	1940	1950	1960	1970	1980	1990
Pioneer proprietary germplasm:							
A111	0	0	0	3.0	4.2	0	0
AB8	0	0	0	3.0	4.0	0	0
BH212	0	0	0	0	1.0	0	0
Boone County White	0	0	4.0	0.9	1.0	0	0
Clarage (via BSSS)	0	0	0	0.75	2.71	2.33	2.39
Dockendorf	0	0	0	0	0	6.0	1.0
FC	0	0	0	0	0	6.0	2.7
HY	10.0	0	4.0	6.76	6.71	2.33	2.39
K140	0	5.0	4.0	0	0	0	0
Lancaster Sure Crop	0	15.9	10.0	5.0	8.6	3.0	2.9
Leaming	30.0	0	0	1.5	9.77	12.63	7.05
LLE	0	0	4.0	6.0	0	0	2.8
M3204	0	0	0	0	2.1	0	0
Maiz Amargo	0	0	0	0	0	3.3	5.5
Midland	0	0.5	7.0	10.0	0.9	5.0	2.0
Minnesota 13	0	0	6.0	4.0	2.2	6.8	3.1
Prolific Composite	0	0	0	0	0	0	1.8
Reid-non BSSS	60.0	74.0	56.0	39.8	22.9	24.0	13.9
Reid-BSSS	0	0	0	8.8	28.0	26.4	27.4
Reid-total	60.0	74.0	56.0	48.6	50.9	50.4	41.3
SMPRS5	0	0	0	0	0	0	2.7
SRS 303	0	0	0	0	0	0.8	1.1
Public inbreds:							
Clarage (via BSSS)	--	--	2.1	3.1	3.8	4.5	--
GE440	--	--	0	0	5.6	0	--
HY	--	--	2.1	3.1	3.8	4.5	--
Lancaster Sure Crop	--	--	50.0	37.5	9.7	9.7	--
Leaming	--	--	4.2	6.3	7.6	8.9	--
Maiz Amargo	--	--	0	0	2.8	0	--
Minnesota 13	--	--	8.3	6.3	4.2	8.0	--
Pride of Saline	--	--	0	0	5.6	0	--
Reid-non BSSS	--	--	8.3	6.3	9.0	7.6	--
Reid-BSSS	--	--	25.0	37.5	45.3	53.6	--
Reid-total	--	--	33.3	43.8	54.3	61.2	--
White composite	--	--	0	0	2.8	2.8	--

County, White, A111, M3204) but then disappeared. Midland and Minnesota 13 first appeared in widely used hybrids in the 1940s and 1950s and remained through subsequent decades. Dockendorf, FC, Maiz Amargo, Prolific Composite, SMPRS5 and SRS303 appeared recently. For public germplasm, major pedigree components were Reid and Lancaster. During the four decades from 1950, the proportion of Reid increased from 33% to 61%. Lancaster began this period at 50%, falling to 37.5% in the 1960s with a precipitous drop to 9.7% during the next two decades. Three pedigree components (GE440, Pride of Saline, White Composite) were present only in the public germplasm while 13 components (A111, AB8, BH212, Boone County White, Dockendorf, FC, K140, LLE, M3204, Midland, Prolific Composite, SMPR55 and SR303) were only present in the germplasm released by Pioneer. Two of these components (Prolific Composite and SMPR55) appeared in Pioneer germplasm in the 1990s, a decade that was not studied for public lines. Seven components were present in public and proprietary Pioneer germplasm. These were Clarage, HY, Lancaster, Sure Crop, Leaming, Maiz Amargo, Minnesota 13 and Reid.

TRENDS AND FUTURE DIRECTIONS

Self-pollination to produce highly homozygous inbred lines that can be stably reproduced and which, when crossed in appropriate combinations can make high yielding hybrids, represent biological phenomena that are ideally suited to the effective exploration and utilization of genetic diversity in production agriculture. The ability to maintain inbred lines as trade secrets and the need for annual purchase of seed by farmers led to significant participation in research and product development by private investors from the very beginnings of hybrid maize. Farmers chose to annually purchase hybrids after it had been demonstrated that hybrid seed was an investment opportunity for the grower because of performance advantages over the previously relied upon landrace open-pollinated varieties.

Self-pollination and detailed record keeping allow portions of the maize germplasm pool, originally represented by open-pollinated farmer–landrace varieties, to be reconstituted and reclassified in terms of families of inbred lines or pedigree backgrounds that can be characterized by agronomic performance, per se and in hybrid combinations. Inbreeding depression created difficulties for seed producers which they overcame by making F1 hybrid parents to create double-cross commercial varieties. During the 1950s, breeders began to organize germplasm into heterogeneous pools such as Stiff Stalk and non-Stiff Stalk that were generally unrelated by pedigree and maintained in that fashion to allow efficient exploitation of heterosis. The per se performance attributes of inbred lines improved, presumably because inbreeding allowed for selection against deleterious recessive alleles, so that by the 1970s all of the commercial hybrids reported upon here were single cross.

There were relatively high pedigree and isozymic distances between inbred lines up until the 1960s. The figures for the 1930s and 1940s reflect the origin of first generation lines from a relatively broad base of germplasm. In contrast, diversities between hybrids declined from the 1930s through the 1950s as pedigree connections between inbred lines increased and as double cross hybrids, each made with four lines, had common parental components. Diversities between inbreds fell after the 1960s as pedigree connections increased and in spite of introductions of germplasm that was originally exotic to the U.S. Corn Belt. Genetic diversity among hybrids showed a different trend in the 1960s and 1970s. Diversity among hybrids increased in these decades as different single cross hybrids had fewer of their parental inbreds in common than had been the case with double cross hybrids and because inbred parents from different heterotic pools were unrelated by pedigree. The transition from double-cross to single-cross hybrids increased the proportion of genetic diversity apportioned between hybrids compared to that present within individual hybrids.

Coefficients of parentage (COP) have continued to increase due to a persistent reliance upon Reid Yellow Dent and because germplasm that is exotic to the U.S. Corn Belt usually undergoes 10 to 30 years of breeding with inbred lines that are well adapted to that region before it appears in the genetic background of widely used hybrids. COP among maize hybrids ranged from 0.03 (in the 1930s) to a high of 0.21 (in the 1990s) and are numerically equivalent or higher than other major U.S. grain crops (Souza & Sorrels, 1989; Martin et al., 1991; Mercado et al., 1996; Kim & Ward, 1997). But numerical comparisons of COP made across genera with significant differences in their breadths of founder germplasm are probably meaningless in terms of revealing relative levels of genetic diversity. In addition, there is greater diversity of cultivar usage in maize than for small grain cereals. For example, 79 to 99% of U.S. wheat (*Triticum aestivum* L.) acreage is planted to primary cultivars (Cox et al., 1986), but only 24% of U.S. maize acreage meets that level of usage (Smith et al., 1992).

Fluctuations of isozymic allele frequencies indicate qualitative changes of germplasm in hybrids. Fourteen alleles that had become extinct in hybrids reappeared in one or more subsequent decades. Widely used hybrids of the 1940s and 1950s had very different germplasm from those of the 1930s (see usage for HY, K140, Lancaster Sure Crop, Leaming, Krug, Osterland, Wilson). During the 1950s, 1960s and 1970s, new germplasm appeared through AB8, BH212, Boone County, White, Clarage, LLE, A111, M3204, Midland and Minnesota 13. Other new sources appeared in widely used hybrids during the 1980s and 1990s (Dockendorf, FC, a revival of Leaming, Maize Amargo, Prolific Composite, SMPRS5 and SRS303). While Reid Yellow Dent has been the major contributor to widely used U.S. maize germplasm during this century, usage has declined from its peak.

Comparisons of germplasm in public and proprietary Pioneer inbred lines are only possible during the four decades for which usage data of public lines were available. During this period, public inbred lines were comprised of 10 pedigree backgrounds that each contributed ≥1% during at least one decade; the Pioneer germplasm originated from 17 backgrounds. None of the pedigree components that were partitioned to either public or Pioneer lines had a usage of ≥10% in any decade. Of the pedigree components that were commonly used, only Leaming, Lancaster Sure Crop and Reid made contributions ≥10% per decade; however, even these similarities are superficial. There were very clear differences in usage of Lancaster and Reid germplasm between public and this proprietary germplasm. During the 1950s and 1960s, usage of Lancaster was up to 10-fold greater for the public germplasm. Usage of Reid germplasm that did not originate from the Iowa Stiff Stalk Synthetic was up to five-fold greater for this proprietary germplasm. Conversely, usage of Reid originating via the Iowa Stiff Stalk Synthetic was greater in the public germplasm. But usage of publicly bred inbreds is now less relevant to establishing maize germplasm usage in the USA. Darrah and Zuber (1986) found that 92% of maize production came from hybrids made with at least one proprietary line. Frey (1996) reported that 94% of U.S. field corn breeders were employed by the private sector. Pedigrees of privately produced hybrids are proprietary information. Consequently, germplasm usage can no longer be gauged by traditional surveys of public line usage and so molecular marker data increasingly provide the means to compare germplasm. Isozymic and zein profiles indicated continued usage of B73, A632, Oh43, and Mo17 or closely related derivatives in many hybrids (Smith, 1988). More discriminative restriction fragment length polymorphism (RFLP) data showed that while most hybrids were dissimilar, there were notable exceptions where hybrids had very similar profiles (Smith et al., 1992). Many hybrids that were widely used in 1992 had RFLP profiles showing similarities to B73 or related lines (Smith, 1997).

Goodman (1985) could identify only 19 U.S. maize hybrids with exotic germplasm and these represented only 0.9% of the acreage. Where will new useful diversity come from? Are U.S. maize breeders limiting progress that could otherwise by made in advancing maize productivity by not making more effective use of exotic germplasm? Rasmussen and Phillips (1997) argue that diversity may be created *de novo*. Exotic germplasm could be another source of new and useful diversity. But effective sourcing of exotic and unadapted germplasm requires classification into U.S. heterotic groups, adaptation, and prebreeding that can take 25 to 30 years to accomplish. Activities that require 20 to 40 year time frames before germplasm will appear in farmers' fields are too lengthy and risky for individual privately funded breeders to engage in alone. Yet a review of phylogenies shows that all U.S. elite maize germplasm is dependent upon a basic foundation of adaptation through population improvement and genetic enrichment. U.S. hybrid maize was initially successful due to investments by both the public and private sectors. A coordinated public–private sector program of genetic enrichment is needed to help provide new germplasm for the future. New information and technologies from genome sequencing, physiological and genetic studies can provide new tools to help breeders more effectively utilize a broader base of genetic resources. Researchers, conservators and breeders must remain at the forefront in melding biology and technologies into the biotechnology of agriculture by making the most productive use of genetic resource diversity so that the growing food and industrial demands of consumers can be met in an environmentally sound way.

REFERENCES

Baker, R. 1984. Some of the open-pollinated varieties that contributed the most to modern hybrid corn. p. 1–19 *In* Proc. of the 20th Annual Illinois Corn Breeders School. Univ. of Illinois, Urbana-Champaign.

Bowman, M.L., and B.W. Crossley. 1908. Corn growing, judging, breeding, feeding and marketing. Published by the authors, Des Moines, IA.

Cox, T.S., J.P. Murphy, and D.M. Rogers. 1986. Changes in the red winter wheat regions of the United States. Proceedings of the National Academy Science USA 83:5583–5586.

Darrah, L.L., and M.S. Zuber. 1986. 1985 United States farm maize germplasm base and commercial breeding strategies. Crop Sci. 26:1109–1113.

Duvick, D.N. 1996. Plant breeding, an evolutionary concept. Crop Sci. 36:539–549.

Frey, K.J. 1996. National plant breeding study: I. Human and financial resources devoted to plant breeding research and development in the United States in 1996. Spec. Rep. 98, Iowa Agric. and Home Econ. Exp. Stn., Ames.

Goodman, M.M. 1985. Exotic maize germplasm: Status, prospects, and remedies. Iowa State J. Res. 59:497–527.

Jenkins, M.T. 1936. Corn improvement. USDA Yearbook of Agriculture. U.S. Gov. Print. Office, Washington, DC.

Kim, H.S., and R.W. Ward. 1997. Genetic diversity in eastern US soft winter wheat (*Triticum aestivum* L. em. Thell.) based on RFLPs and coefficients of parentage. Theor. Appl. Genet. 94:472–479.

Martin, J.M., T.K. Blake, and E.A. Hockett. 1991. Diversity among North American Spring barley cultivars based on coefficients of parentage. Crop Sci. 31:1131–1137.

Mercado, L.A., E. Souza, and K.D. Kephart. 1996. Origin and diversity of North American hard spring wheats. Theor. Appl. Genet. 93:593–599.

Rasmussen, D.C., and R.L. Phillips. 1997. Plant breeding progress and genetic diversity from *de novo* variation and elevated epistasis. Crop Sci. 37:303–310.

Smith, J.S.C. 1988. Diversity of United States hybrid maize germplasm; isozymic and chromatographic evidence. Crop Sci. 28:63–69.

Smith, J.S.C., O.S. Smith, S. Wright, S.J. Wall, and M. Walton. 1992. Diversity of US hybrid maize germplasm as revealed by restriction fragment length polymorphisms. Crop Sci. 32:598–604.

Smith, S. 1997. Cultivar identification and varietal protection. p. 283–299. *In* G. Caetano-Anolles, and P.M. Gresshoff (ed.) DNA markers, protocols, applications, and overviews. Wiley-VCH, New York.

Souza, E., and M.E. Sorrells. 1989. Pedigree analysis of North American oat cultivars released from 1951 to 1985. Crop Sci. 29:595–601.

Sturtevant, E.L. 1899. Varieties of corn. Bull. 57. USDA, Washington, DC.

Zuber, M.S., and L.L. Darrah. 1980. 1979 U.S. corn germplasm base. p. 234–249. *In* H.D. Loden and D.W. Wilkinson. (ed.) Proc. of the 35th Annual Corn and Sorghum Industry Res. Conf. Chicago, IL. 9–11 Dec. Am. Seed Trade Assoc., Washington, DC.

Chapter 12

Population Genetics of Increased Hybrid Performance between Two Maize Populations under Reciprocal Recurrent Selection

J. A. Labate, K. R. Lamkey, M. Lee, and W. L. Woodman

INTRODUCTION

Heterosis, the superiority in one or more characteristics of crossbred organisms relative to their inbred parents, is the basis of the modern cultivars utilized in maize (*Zea mays* L.). Heterosis is of interest in nondomesticated species due to its relevance to the question "how much polymorphism is maintained in natural populations due to selection?" (Berger, 1976). For maize and certain other domesticated species that employ inbred lines to produce commercial hybrids, knowledge of the mechanisms of gene action producing heterosis could contribute to advances in breeding techniques.

One method used to evaluate the existence of heterosis involves measuring multilocus heterozygosity levels in individuals sampled from a population with molecular genetic markers and correlating heterozygosity with a trait believed to reflect fitness, e.g., fecundity, viability, growth rate, or developmental stability. A vast number of these studies, many involving natural populations, have been published during the previous three decades. A general consensus is that a significant positive correlation between multilocus heterozygosity and fitness surrogates has been documented for a systematically wide range of organisms, although it is not a universal phenomenon (recently reviewed by Britten, 1996; Mitton, 1994; Zouros & Foltz, 1987).

The two genetic mechanisms most commonly invoked to explain heterosis are dominance and overdominance. The dominance hypothesis explains heterozygote superiority as a result of the masking of deleterious recessive alleles in an individual, whereas the hypothesis of overdominance postulates an advantage of heterozygosity per se, e.g., through differences in biochemical properties of homozygote vs. heterozygote encoded single-locus products (Berger, 1976). Dominance cannot be distinguished in practice from pseudooverdominance, associative overdominance, or dominance-correlation heterosis. These are all synonyms of heterosis due to the joint action of genes associated in negative gametic phase disequilibrium. Some causes of gametic disequilibrium are directional selection, recombination suppression, inbreeding, or small effective population size (see Houle, 1989 for references).

Heterosis is relevant to the study of several subdisciplines within biology (e.g., plant and animal breeding, mating system evolution, developmental genetics); many good reviews differing slightly in their emphases have been published recently. Sedcole (1981) reviewed examples from plant breeding from approximately 1930 to 1980. Tsaftaris (1995) provided a review of recent molecular techniques used to study heterosis in plants, e.g., looking at RNA amount poly-morphism (RAP), protein amount polymorphism (PAP), or DNA methylation lev-els. A review of heterosis as it relates to plant inbreeding depression can be found in Ritland (1996).

Berger (1976) reviewed theoretical mechanisms for the superiority of het-erozygosity per se for protein polymorphisms; many of these have intuitive appeal. Presently, there are only a few well-documented instances of overdominance as a mechanistic explanation of heterosis in natural populations (see Mitton, 1994). The same can be said for domesticated species. One maize example is often cited: Schwartz (1973) found that active and stable heterodimers of alcohol dehydroge-nase (Adh) are made up of two monomers, one of which is inactive (but stable) and the other of which is labile (but relatively active). The paucity of examples of sin-gle-locus heterosis may not be due to its infrequency, but may be because it is dif-ficult to study, and overdominance is infrequently the sole supporting hypothesis.

Crow (1993, p. 15) recently reviewed genetic evidence that has led to the disfavor of the overdominance hypothesis in lieu of simple dominance as an expla-nation of heterosis in maize. Most importantly, researchers have found positive evidence for pseudooverdominance. This came from experiments in which hybrid maize populations were advanced several generations and recombination broke up linkages between favorable dominant and deleterious recessive alleles (e.g., Gardner and Lonnquist, 1959). Additional reasons for accepting the dominance hypothesis, according to Crow (1993), are a larger deleterious mutation load than originally generally believed (measured in a few species) and successful selection for relatively high yielding maize inbred lines (compared to early hybrids). The mutation load can explain the observed 15 to 20% grain yield increases observed in maize hybrids over their panmictic base populations, and high yielding inbred lines would not be possible if overdominance was the mechanism underlying high yield.

QUANTITATIVE GENETIC EVIDENCE FROM THE BSSS X BSCB1 RECIPROCAL RECURRENT SELECTION PROGRAM

In spite of the general acceptance of dominance as the explanation for het-erosis in maize today, this was not true 50 years ago. Comstock et al. (1949) pro-posed a breeding method for maize that they termed recurrent reciprocal selection (now known as reciprocal recurrent selection, RRS). Their motivation for devel-oping the method was, as they stated, to discover a selection method that would be effective regardless of the level of dominant gene action. They proposed that RRS would be beneficial for instances in which overdominance, or situations analogous to overdominance (repulsion phase linkages), existed or when interactions of non-allelic genes (epistasis) were important; it would also exploit additive genetic ef-fects. In theory, RRS is intended to improve the performance of an interpopulation cross of two genetically divergent populations. One cycle of RRS involves devel-opment of genetic units within populations (e.g., S_1 lines, first-generation progenies from self-fertilized individuals), reciprocal crosses of genetic units between popu-lations, phenotypic evaluation of these testcrosses, and selection of progenies based on testcross results. Selected progenies are then mated within each population. The next cycle of selection is initiated from these. RRS is designed to allow for genetic recombination within populations to maintain quantitative genetic varia-tion, while minimizing inbreeding. The maintenance of two separate gene pools allows a different allele to be fixed within each population. For loci where this is achieved, interpopulation hybrids are assured to be heterozygous.

Two maize populations, Iowa Stiff Stalk Synthetic (BSSS) and Iowa Corn Borer Synthetic #1 (BSCB1), are currently in their 14th cycle of RRS in the Coop-erative Federal-State maize breeding program at Iowa State University. Increased grain yield of the interpopulation cross has been the primary target of selection, with reduced grain moisture at harvest and increased resistance to root and stalk lodging as secondarily selected traits. Selection has been highly successful; mean grain yield of the interpopulation cross improved 77% by Cycle 11, relative to Cycle 0, with concurrent favorable responses in the other traits (Keeratinijakal & Lamkey, 1993a).

Midparent heterosis for BSSS(R) and BSCB1(R) was estimated as the difference between the mean of the interpopulation cross and the mean of the two parental populations. Inbreeding depression (the reduction in the mean value of a character produced by inbreeding) was measured for the interpopulation cross by selfing their F_1. Steady increases in heterosis and inbreeding depression for grain yield over 11 cycles were found (Keeratinijakal & Lamkey, 1993a). These were interpreted as resulting from an increase, over time, in heterozygosity of the interpopulation cross. Using Smith's (1983) model Keeratinijakal & Lamkey (1993b) partitioned the genetic response to selection of BSSS(R) and BSCB1(R) into components due to additive and dominance effects and looked for evidence of overdominance. They found (partial to complete) dominance effects to be more important than additive effects in the interpopulation cross, with no evidence for overdominance. Diversity analysis (Moll & Hanson, 1984) of the two populations supported this interpretation. Directional dominance for grain yield and a difference in the frequencies of alleles affecting grain yield between the original populations were also inferred.

Similar results have been reported for other maize RRS programs. Eyherabide and Hallauer (1991a,b) reported on reciprocal full-sib recurrent selection in the BS10 and BS11 populations. They found significant increases in midparent heterosis and inbreeding depression for grain yield in the interpopulation cross over eight cycles of selection. They also detected directional dominance and different frequencies of alleles with dominance effects for grain yield between the Cycle 0 populations. They suggested that selection had caused changes in frequencies of alleles with dominant effects in a different set of loci for each population or that different isoalleles with dominant effects had been selected in each population. Hanson and Moll (1986) also concluded that overdominant gene effects were not evident in the Jarvis and Indian Chief populations after 10 cycles of RRS; alleles having additive or dominant effects were selected.

MOLECULAR MARKER EVIDENCE FROM THE BSSS X BSCB1 RECIPROCAL RECURRENT SELECTION PROGRAM

We have genotyped samples from three populations within BSSS(R) and BSCB1(R), representing three different stages in their selective history (see Labate et al., 1997 for complete details). BSSS and BSCB1 synthetic populations trace back to 16 and 12 inbred lines, respectively. These collections of inbred lines are herein referred to as progenitor (P) populations. Cycle 0 populations were formed by several generations of random-mating bulked seed obtained from a series of crosses between progenitor inbred lines. These BSSS(R) and BSCB1(R) Cycle 0 populations were the starting material for RRS. Finally, we have genotypes from samples from both populations after twelve cycles of RRS (Cycle 12).

The molecular markers used were 82 nuclear genomic restriction fragment length polymorphism (RFLP) loci randomly distributed across all 20 chromosomal arms. The markers were assumed to be selectively neutral, i.e., the alleles at a locus would not differ measurably in their effects on the selected traits. The probes were chosen for their high levels of polymorphism and extensive coverage of the genome. One-hundred individuals from each Cycle 0 and Cycle 12 population were chosen at random for genotyping, as well as single individuals from each of 28 progenitor inbred lines (two of the BSSS progenitor inbred lines had been lost; however, the two parental lines of one of these were included). Each of the 82 RFLP probes was considered to be a single locus, and variants at each locus were assumed to be allelic.

Genetic Diversity

We found that mean gene diversity, expected heterozygosity under random-mating, was initially quite high within BSSS(R) and BSCB1(R). This also can be thought of as the probability of obtaining a heterozygote when two alleles are sampled at random from the population. This probability was around 60% in both progenitor populations. After 12 cycles of RRS, mean gene diversity had decreased to near 30% in each. Coinciding with this, the mean number of alleles per locus in BSSS(R) and BSCB1(R) dropped from about four to less than three. A further question was of interest. Looking at the *total* gene pool of BSSS(R) and BSCB1(R), what happened to genetic diversity over 12 cycles of RRS? If two alleles were sampled at random, one from each population, what would be the probability of obtaining a heterozygote? The increases in heterosis and inbreeding depression of the interpopulation cross seen in the quantitative genetic analyses suggested that the interpopulation cross was becoming more heterozygous. The pooled mean genetic diversity for the progenitor populations was estimated to be 63% and for the Cycle 12 populations, approximately 66%. The two estimates were not significantly different based on their standard errors.

Because of the assumption of selective neutrality of the RFLP markers, the lack of increase in interpopulation gene diversity was not completely unexpected. In fact, the estimated loss of mean genetic diversity *within* each population conformed to theoretical expectations (Nei, 1987, Eq. 13.12) of genetic drift of neutral alleles (i.e., random changes in allele frequency caused by gametic sampling each generation). We could see that, in the face of substantial loss of diversity within each population, the between population genetic diversity had remained high. Genetic diversity is a function of the numbers of alleles at a locus and allelic frequencies. This implied that, in general, different alleles had reached high frequencies in BSSS(R)C12 and BSCB1(R)C12.

Results from a principal components analysis (PCA) (Rohlf, 1994) of the 428 individuals sampled from BSSS(R) and BSCB1(R) populations are shown in Fig. 12–1. Each point represents an individual separated in a three-dimensional space based on the presence/absence of 391 alleles (genotypes for 82 loci). The progenitor lines do not form two discrete clusters according to which population they formed. BSSS(R) and BSCB1(R) were initially nearly genetically identical. By Cycle 0, BSSS(R) and BSCB1(R) seem to be distinct from each other. In the absence of genetic drift and selection, the Cycle 0 populations should have remained clustered with the progenitors. We have inferred that maintenance for several decades of BSSS(R)C0 and BSCB1(R)C0 has altered their genetic constitutions. This was especially evident in BSSS(R), for which it seemed that many rare alleles present in P were not sampled in the modern representatives of Cycle 0 (Labate et al. 1997). By Cycle 12, BSSS(R) and BSCB1(R) were substantially diverged. The separation between the Cycle 0 and Cycle 12 populations include a component due to genetic drift, because a limited number of lines (10 to 20) were selected and recombined each cycle, and a component due to selection, because the recombined lines were not chosen at random.

So far, the results presented have focused on *mean* diversity, and genetic changes across *all* loci. By examining the data, it was clear that some of the loci had experienced extreme changes in allele frequencies over the course of selection. The pertinent question became, "Have any of the loci experienced allele frequency changes that were too large to be explained by genetic drift?" Even though the markers fit a neutral model based on mean levels of gene diversity, this did not preclude that some of the allele frequency changes had been influenced by selection. This could have come about directly through selection or, more probably, through genetic hitchhiking. The hitchhiking effect is seen when selection at a locus changes the frequencies of neutral alleles at closely-linked loci and is conditioned on initial linkage disequilibrium between the loci.

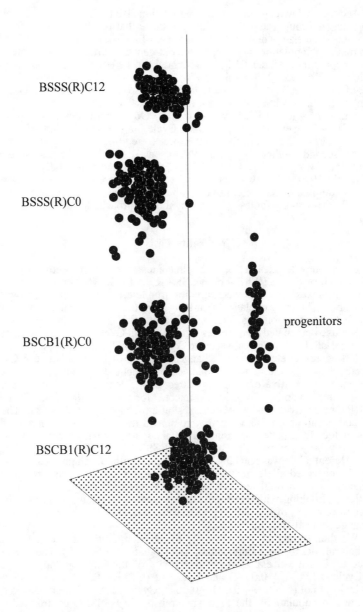

BSSS(R)C12

BSSS(R)C0

BSCB1(R)C0

progenitors

BSCB1(R)C12

Fig. 12–1. Principal components analysis of Iowa Stiff Stalk Synthetic (BSSS) and Iowa Corn Borer Synthetic #1 (BSCB1) based on genotypes of sampled individuals at 82 RFLP loci. The six sampled populations include progenitor inbred lines, populations before RRS (C0 populations), and populations after 12 cycles of RRS (C12 populations). Progenitor populations do not form two distinct groups.

Effective Population Size

Accurate knowledge of effective population size (N_e) is a key to discerning genetic changes brought about by drift from those that result from selection. Effective population size is defined as the number of individuals in an idealized (i.e., random mating) population that would undergo genetic drift at the same rate as the observed population. Under RRS, N_e is thought to be equal to the number of selected lines each cycle (Vencovsky, 1978). If all parents leave exactly the same number of offspring, N_e is expected to equal $2N - 1$ (Kimura, 1983, p. 41). When the number of selected lines has varied, N_e can be calculated as the harmonic mean of the number of selected lines over all cycles. Our empirical estimates based on the loss of mean genetic diversity between Cycle 0 and Cycle 12 supported an N_e equal to the harmonic mean of the number of selected lines, $N_e = 12$ (Labate et al., 1997). A second method (Waples, 1989a), based on allele frequency changes across all loci, was used to estimate N_e for BSSS(R) and BSCB1(R) populations (Labate et al., 1999). The two methods agreed; N_e is approximately the harmonic mean of the number of selected lines over all cycles. The 95% confidence intervals obtained for N_e using Waples' (1989a) method approached, but did not overlap with, $(2N - 1)$.

Neutrality Tests

Given our estimates of N_e, we applied a test of selective neutrality (Waples, 1989b) to each of the 82 RFLP loci in BSSS(R) and BSCB1(R) populations. The null hypothesis was: the observed variation in allele frequency between two time points can be sufficiently explained as arising from the sampling of a population, of size N_e, that has undergone t generations of genetic drift.

We used estimated frequencies at time points Cycle 0 and Cycle 12 as initial and final allele frequencies, assumed $N_e = 12$ (or 23), and $t = 12$ generations (cycles). Because allele frequency changes at many of the loci between Cycle 0 and Cycle 12 were too large to be explained by genetic drift alone, we interpreted these changes as positive evidence for directional selection and/or genetic hitch-hiking. The null hypothesis of drift was rejected for 11 and 17 loci in BSSS(R) and BSCB1(R), respectively, using Waples' test at a probability level of 5%. The loci were found on all chromosomes and were spread throughout the genome. These nonneutral loci fit a pattern of complementary genetic changes between the two populations. Only one was shared between BSSS(R) and BSCB1(R), and at that locus a different allele was reaching high frequency within each population.

The observed allele frequencies at the 27 loci are illustrated in Fig. 12–2. Frequencies of nonneutral alleles are shown at Cycle 0 and Cycle 12 for both populations. Looking within a population at nonneutral alleles identified for that population, rejection of the null hypothesis was associated with an approximately 60% change in an allele's frequency.

We then estimated gene diversity of the interpopulation cross, comparing the 55 neutral loci to the 27 nonneutral loci (Labate et al., 1999). The nonneutral loci increased in mean expected heterozygosity of the interpopulation cross between Cycle 0 (0.664 ± 0.0352) and Cycle 12 (0.776 ± 0.0537) whereas the 55 neutral loci did not (Cycle 0 = 0.603 ± 0.0243, Cycle 12 = 0.595 ± 0.0384). Comparing the two populations, the 11 nonneutral loci in BSSS(R) contributed to the increase in interpopulation heterozygosity more than the 17 nonneutral loci in BSCB1(R). A partial explanation for this can be found by studying Fig. 12–2, parts c and d. Many of the 17 nonneutral BSCB1(R) alleles were at high frequencies in BSSS(R) at Cycle 0 and remained high in BSSS(R) at Cycle 12 (e.g., bnl835, bnl749, umc155). These loci underwent marked *decreases* in interpopulation expected heterozygosity.

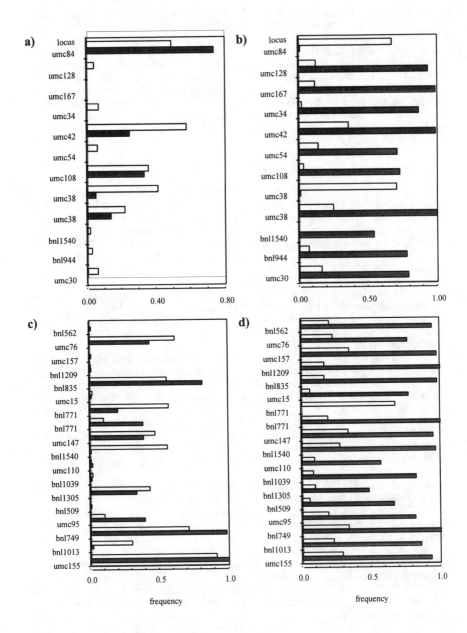

Fig. 12–2. Allele frequencies at Cycle 0 (white bars) and Cycle 12 (filled bars) for 27 nonneutral loci identified in the BSSS(R) and BSCB1(R) populations. a) frequencies in BSCB1(R) for 11 nonneutral loci in BSSS(R), b) frequencies in BSSS(R) for 11 nonneutral loci in BSSS(R), c) frequencies in BSSS(R) for 17 nonneutral loci in BSCB1(R), d) frequencies in BSCB1(R) for 17 nonneutral loci in BSCB1(R).

One prediction under RRS is that if a favorable allele exists in both populations, selection will be more effective for that allele in the population within which it is more common (Cress, 1967). At about one-half of the nonneutral loci, the favored allele was at an initial frequency of less than 10% in the reciprocal population and remained low. The other loci didn't conform to this predicted pattern (Fig. 12–2). Possible reasons for this are (i) at most loci, there were more than two alleles in BSSS(R) and BSCB1(R), so the dynamics of selection were not predicted by this simple model; (ii) intralocus, complete dominance was not the genetic mechanism for increasing the selected allele; or (iii) in the instance of genetic hitchhiking, interlocus correlation (two-locus disequilibrium) patterns were different within BSSS(R) and BSCB1(R).

CONCLUSIONS

Heterosis for grain yield in the interpopulation cross has increased in the BSSS(R) and BSCB1(R) RRS program, and the two populations have become quite genetically diverged from each other. The use of molecular markers has provided some insight into the roles of selection and genetic drift in BSSS(R) and BSCB1(R). Theoretical studies (Li, 1978) have shown that the absolute value of the selection coefficient for an allele must be greater than $1/N_e$ for selection to overcome genetic drift. This assumes a Wright-Fisher model of random genetic drift of neutral alleles (see Hartl & Clark, 1989, p. 351). The selection coefficient is the relative gametic contribution of a particular genotype compared with the most favored genotype in the population (Falconer & Mackay, 1996, p. 26). Our findings imply that a large fraction of loci in the maize genome, about 33% of those surveyed, were affected by selection. If $N_e = 12$ as estimated, then selection coefficients were at least 8%.

Although yield has not been the only agronomic trait selected, it has been emphasized. If yield is affected by many loci that are densely distributed throughout the genome and that carry large phenotypic effects, it is easy to understand why fixation of the most favored genotype in an inbred line derived from an improved population is difficult. Other population genetic studies where molecular markers were used also found that a large fraction of scored loci affected yield (Stuber et al., 1980, 1992), although some studies (e.g., Brown & Allard, 1971; Kahler, 1983) have found that genetic drift could explain observed allele frequency changes. The earlier studies used allozyme loci; DNA-based markers are much more informative in maize.

Stuber et al. (1992), using 67 RFLP loci and nine isozyme loci, genotyped sets of lines descending from a cross originating between two maize inbred lines. When they regressed mean trait value on percent heterozygous marker loci, they found a high correlation between grain yield and proportion of heterozygous markers. A large fraction of the genome was found to affect yield (markers significantly associated with yield were found on all 10 chromosomes), even though this experimental design was limited to detecting regions polymorphic between the two original inbreds.

Reciprocally selected populations should continue to provide a suitable experimental system within which to study relationships between multilocus heterozygosity and phenotype. In this genetic system recombination is prohibited at the interpopulation level, allowing fixation of balanced intralocus or interlocus gene action in the interpopulation cross. Testing theories of gene action requires estimation of parameters such as mutation rates, selection pressure, recombination distances, and inbreeding coefficients (Zouros & Foltz, 1987). It should be possible to obtain more accurate measures of these parameters in maize selection programs than in natural populations.

ACKNOWLEDGMENTS

This study was supported in part by a grant from Pioneer Hi-Bred International, Inc. to K. R. Lamkey and M. Lee. J. A. Labate is funded by the USDA-ARS Postdoctoral Research Associate Program. This work is a joint contribution of the Corn Pest and Crop Genetics Research Unit, USDA-ARS, Dep. of Agronomy, Iowa State Univ. and Journal Paper J-17614 of the Iowa Agriculture and Home Economics Experiment Station, Ames, Iowa, 50011-1010. Project No. 3082 and 3134, and supported by Hatch Act and State of Iowa Funds.

REFERENCES

Berger, E. 1976. Heterosis and the maintenance of enzyme polymorphism. Am. Nat. 110:832-839.

Britten, H.B. 1996. Meta-analysis of the association between multilocus heterozygosity and fitness. Evolution 50:2158-2164.

Brown, A.H.D., and R.W. Allard. 1971. Effect of reciprocal recurrent selection for yield on isozyme polymorphisms in maize (*Zea mays* L.). Crop Sci. 11:888-893.

Comstock, R.E., H.F. Robinson, and P.H. Harvey. 1949. A breeding procedure designed to make maximum use of both general and specific combining ability. J. Am. Soc. Agron. 41:360-367.

Cress, C.E. 1967. Reciprocal recurrent selection and modifications in simulated populations. Crop Sci. 7:561-567.

Crow, J.F. 1993. Mutation, mean fitness, and genetic load. Oxford Surv. Evol. Biol. 9:3–42.

Eyherabide, G.H., and A.R. Hallauer. 1991a. Reciprocal full-sib recurrent selection in maize: I. Direct and indirect responses. Crop Sci. 31:952–959.

Eyherabide, G.H., and A.R. Hallauer. 1991b. Reciprocal full-sib recurrent selection in maize: II. Contributions of additive, dominance, and genetic drift effects. Crop Sci. 31:1442–1448.

Falconer, D.S., and T.F.C. Mackay. 1996. Introduction to quantitative genetics. 4th ed. Longman. Essex, England.

Gardner, C.O., and J.H. Lonnquist. 1959. Linkage and the degree of dominance of genes controlling quantitative characters in maize. Agron. J. 51:524–528.

Hanson, W.D., and R.H. Moll. 1986. An analysis of changes in dominance-associated gene effects under intrapopulation and interpopulation selection in maize. Crop Sci. 26:268–273.

Hartl, D.L. and A.G. Clark. 1989. Principles of Population Genetics. 2nd ed. Sinauer Associates, Inc., Sunderland, MA.

Houle, D. 1989. Allozyme-associated heterosis in *Drosophila melanogaster*. Genetics 123:789–801.

Kahler, A.L. 1983. Effect of half-sib and S1 recurrent selection for increased grain yield on allozyme polymorphisms in maize. Crop Sci. 23:572–576.

Keeratinijakal, V., and K.R. Lamkey. 1993a. Responses to reciprocal recurrent selection in BSSS and BSCB1 maize populations. Crop Sci. 33:73–77.

Keeratinijakal, V., and K.R. Lamkey. 1993b. Genetic effects associated with reciprocal recurrent selection in BSSS and BSCB1 maize populations. Crop Sci. 33:78–82.

Kimura, M. 1983. The neutral theory of molecular evolution. Cambridge Univ. Press, Cambridge.

Labate, J.A., K.R. Lamkey, M. Lee, and W.L. Woodman. 1997. Molecular genetic diversity after reciprocal recurrent selection in BSSS and BSCB1 maize populations. Crop Sci. 37:416–423.

Labate, J.A., K.R. Lamkey, M. Lee, and W.L. Woodman. 1999. Temporal changes in allele frequency in two directionally selected maize populations. Theor. Appl. Genet. (accepted).

Li, W.-H. 1978. Maintenance of genetic variability under the joint effect of mutation, selection and random drift. Genetics 90:349–382.

Mitton, J.B. 1994. Molecular approaches to population biology. Annu. Rev. Ecol. Syst. 25:45–69.

Moll, R.H., and W.D. Hanson. 1984. Comparisons of effects of intrapopulation vs. interpopulation selection in maize. Crop Sci. 24:1047–1052.

Nei, M. 1987. Molecular evolutionary genetics. Columbia Univ. Press, New York.

Ritland, K. 1996. Inferring the genetic basis of inbreeding depression in plants. Genome 39:1–8.

Rohlf, F.J. 1994. NTSYS-pc: Numerical taxonomy and multivariate analysis system. Ver. 1.80. Exeter Software, Setauket, NY.

Schwartz, D. 1973. Single gene heterosis for alcohol dehydrogenase in maize: The nature of the subunit interaction. Theor. Appl. Genet. 43:117–120.

Sedcole, J.R. 1981. A review of the theories of heterosis. Egypt. J. Genet. Cytol. 10:117–146.

Smith, O.S. 1983. Evaluation of recurrent selection in BSSS, BSCB1, and BS13 maize populations. Crop Sci. 23:35–40.

Stuber, C.W., S.E. Lincoln, D.W. Wolff, T. Helentjaris, and E.S. Lander. 1992. Identification of genetic factors contributing to heterosis in a hybrid from two elite maize inbred lines using molecular markers. Genetics 132:823–839.

Stuber, C.W., R.H. Moll, M.M. Goodman, H.E. Schaffer, and B.S. Weir. 1980. Allozyme frequency changes associated with selection for increased grain yield in maize (*Zea mays* L.). Genetics 95:225–236.

Tsaftaris, S.A. 1995. Molecular aspects of heterosis in plants. Physiol. Plant. 94:362–370.

Vencovsky, R. 1978. Effective size of monoecious populations submitted to artificial selection. Brazil J. Genetics 1:181–191.

Waples, R.S. 1989a. Temporal variation in allele frequencies: Testing the right hypothesis. Evolution 43:1236–1251.

Waples, R.S. 1989b. A generalized approach for estimating effective population size from temporal changes in allele frequency. Genetics 121:379–391.

Zouros, E., and D.W. Foltz. 1987. The use of allelic isozyme variation for the study of heterosis. Isozymes 13:1–59.

Chapter 13

Broadening the Genetic Diversity in Maize Breeding by Use of Exotic Germplasm

M. M. Goodman

INTRODUCTION

Maize (*Zea mays* L.) improvement throughout the world has been accompanied by a narrowing germplasm base, as newer lines and varieties have been derived from intercrosses of existing elite materials. Most other crops have followed the same pattern. For maize, there is no indication that improvement rates have been adversely affected by this narrowing germplasm base (Duvick, 1990), but there is concern that bottlenecks may restrict breeding flexibility and slow response to new opportunities, pests, pathogens, and agronomic practices in the future.

Use of temperate maize germplasm in tropical breeding programs is poorly documented (for an exception see Kim et al., 1987), but clearly increasing. Many newer tropical hybrids contain at least some temperate germplasm, often as a part of only one side of a hybrid pedigree; however, the use of exotic maize germplasm in temperate areas is better documented. Surveys of exotic germplasm in U.S. hybrids suggest that the amount of exotic germplasm used is generally small, but rising (Goodman, 1985; 1998).

Basically, there are three geographic types of maize germplasm exotic to temperate areas. The most widely-used exotic sources are those from other temperate areas, such as Argentine, European, and South African germplasm used in the USA (Table 13–1). Conversely U.S. germplasm is used in other temperate regions. The second widely-used source of exotic germplasm is that from the lowland tropics, representing races or varieties such as Cuban Flint, Suwan, Tusón, and Tuxpeño. Thus far these have been used rather sparingly in the USA, mostly as sources of pest- or disease-resistance. The third potential source of exotic germplasm for temperate areas is from the highland tropics (races like Chalqueño, Cuzco, Sabanero, or San Geronimo); these have had less use in the USA due to their low tolerance for heat and other stresses.

Use of exotic germplasm in U.S. maize hybrids has increased almost three-fold during the past 12 years, with total exotic-germplasm use increasing from about 1% in 1984 (Goodman, 1985) to almost 3% in 1996 (Table 13–1). The largest part of this increase has come from the use of *temperate* exotic germplasm. Use of *tropical* exotic germplasm in U.S. hybrids is much smaller (only about 0.1% in 1984; about 0.3% in 1996). Most widely-sold U.S. hybrids that contain exotic germplasm have small percentages (2 to 6% is typical) of temperate, exotic germplasm from the insect-resistant, Argentine cultivar, Maíz Amargo (mostly via B68). A few have higher percentages, typically 12 to 25%, most often from the related French lines F2 or F7. U.S. hybrids containing tropical germplasm show the same pattern, but at much lower levels of exotic germplasm: widely sold hybrids with tropical maize germplasm

Table 13–1. Use of exotic germplasm in U.S. hybrids in 1996 (adapted from Goodman, 1998b).

Area	Average % exotic germplasm	Area	Average % exotic germplasm
Argentina	1.99		
Australia	0.03	Caribbean	0.23
Europe	0.46	Mexico	0.07
South Africa	0.08		
Temperate total	2.56	Tropical total	0.30

usually contain from 1 to 5% tropical germplasm, but a few less-popular hybrids contain 25 to 50% tropical germplasm.

Maize breeding throughout the temperate parts of the world is rapidly becoming well-integrated, with phytosanitary restrictions and intellectual property rights being the most important barriers to germplasm, line, and hybrid exchange between organizations. Thus, the focus of this chapter is the use of tropical maize germplasm in temperate areas. Five questions are considered:

1. What types of maize germplasm sources are most promising?
2. How can choices be made among the many possibilities within a specific type?
3. How difficult is the problem of photoperiod response?
4. Can tropical maize germplasm lead to yield improvement, in addition to its current role as a last-resort source of disease- and insect-resistance?
5. How can field-based (or molecular-marker-assisted) selection be conducted so as to maximize any positive yield-potential available within a specific, variable, exotic accession or hybrid?

SOURCES OF PROMISING GERMPLASM

While there is little published information comparing selection success with major types of germplasm (from crosses of inbred lines, hybrids, elite families, synthetics, or germplasm accessions), the overwhelming opinion among maize breeders with exotic experience is that inbred lines or hybrids are more promising sources than are populations with no history of inbreeding. Table 13–2 suggests about a 100-fold advantage for improved sources with a history of inbreeding over elite synthetics improved by familial selection. Since initial evaluation of tropical sources must usually be based on yield-trial data from the tropics, germplasm-accession sampling has, until recently, been especially discouraging. The LAMP project (Salhuana, 1995; Salhuana & Sevilla, 1995; Salhuana et al., 1997) has alleviated much of that problem recently. Still, many potentially-available lines, hybrids, and accessions exist for which no performance data are available or accessible.

SELECTING AMONG AVAILABLE POSSIBILITIES

Let us choose tropical hybrids as the category of germplasm for use as base material for a temperate-targeted breeding program. To choose among the available tropical hybrids, it would be helpful to know their pedigrees (often unavailable), and more importantly their agronomic performance, preferably from multiple tropical locations. Once a candidate set of hybrids is identified from such data, they can be grown in short-day winter nurseries alongside elite, domestic breeding stocks to judge their maturities (independent of daylength effects) and agronomic features in comparison to standard, domestic stocks. Late maturity, even under short days, and poor agronomic performance eliminate many candidates either at the initial winter-nursery

Table 13–2. Summary statistics for first-cycle, largely tropical, line development at North Carolina State University (adapted from Goodman, 1992).

Category	Tropical hybrid derivatives†	Tropical synthetic derivatives†
	------------- no. -------------	
Total Number of Nursery Plots Used for Line Development:	306	7395
Numbers of Plots for Families with No Testable Lines:	42	247
Numbers of Plots for Families with All Topcrosses Yielding Less than the Check Median:	97	4060
Numbers of Lines where Topcross Yield Exceeded Check Median:	71	17

† Tropical hybrid derivatives were 100% tropical; the tropical synthetic derivatives ranged from 25 to 75% tropical, the remainder being elite U.S. germplasm.

screening or in subsequent, segregating, breeding nurseries. The better candidates can be crossed with elite lines representing important heterotic groups, tested and subjected to inbreeding (for example, at southern locations in the USA).

PHOTOPERIOD

Photoperiod or daylength response is mostly a cosmetic problem. It can be eliminated, even from descendants of 100%-tropical crosses, and is mostly an inconvenience to the uninitiated who try to grow tropical materials for crossing in summer, rather than short-day, winter nurseries. Experimental results at Iowa State University suggest that photoperiod response can readily be eliminated or greatly diminished by mass selection within populations as diverse as ETO, Suwan, Tusón, and Tuxpeño (Hallauer, personal communication). Similar results were reported for derivatives of tropical hybrids handled by pedigree selection at North Carolina State University (Holley & Goodman, 1988).

TROPICAL GERMPLASM FOR YIELD IMPROVEMENT

Tropical germplasm has traditionally been used in the USA as a last-resort source of disease- and insect-resistance. Holley and Goodman (1988) reported that topcrosses of derivatives of tropical hybrids were sometimes competitive with commercial hybrids; similar results were reported by Goodman et al. (1990) and by Uhr and Goodman (1995). Cargill and Northrup King provided topcross seed of all recently-released North Carolina (NC) lines crossed with LH132, a short-statured, Stiff-Stalk-Synthetic line, and LH150, a tall, southern, non-Stiff-Stalk-Synthetic line. At the time the yield trials were conducted, all NC lines from NC250 to NC300, representing all NC line releases from 1980 to the early 1990s, were tested in such single-crosses. Three of the 32 NC lines tested were of 100%-tropical origin: NC296, NC298, and NC300. Only better-performing crosses are listed in Table 13–3, which demonstrates that tropical lines can compete with domestic lines in some tropical × domestic combinations. (LH132 × NC258 is sold by several companies and

Table 13–3. Highest yielding North Carolina (NC) line topcrosses from trials conducted 1991 to 1994 at Clayton, Lewiston, and Plymouth, NC.

Pedigree†	Grain Yield	Moisture	EP‡	Ear height	Days to tassel	
	t ha⁻¹	%	%	cm	no.	
NC258 × LH132	8.7	18.9	95	96	74	*NC lines*
NC268 × LH150	9.3	16.6	95	116	76	*<10%*
NC280 × LH150	8.6	17.0	95	116	75	*exotic*
NC292 × LH150	8.7	16.7	98	117	74	
NC294 × LH150	8.7	17.3	94	116	76	
NC296 × LH132	9.6	17.1	96	107	75	*NC lines*
NC298 × LH132	8.6	18.0	96	106	76	*100%*
NC300 × LH132	8.4	18.1	97	100	75	*exotic*
Mo17*Ht* × B73*Ht*	7.6	15.2	94	104	72	
Pioneer 3165	8.8	19.0	94	109	77	*Checks*
Dekalb 689	8.9	17.1	97	111	76	
LSD (0.05)§	0.4	0.4	4	4	1	
CV%	6	3	5	5	1	

† NC lines of largely Lancaster (NC258), B73 (NC268, NC280, NC292, NC294) or tropical (NC296, NC298, NC300) origin were crossed to Holden tester lines (LH132 is a Stiff-Stalk line; LH150 is a non-Stiff-Stalk line).
‡ EP, percentage of erect plants at harvest.
§ LSD and CV based on entry x environment interaction term.

Table 13–4. Comparisons of NC296A topcrosses and commercial hybrids for 1992 to 1994 trials at Clayton, Lewiston (not 1993), and Plymouth, NC.

Entry	Commercial hybrid Grain Yield	Mois.‡	EP‡	GLS§	NC296A cross with hybrid† Grain Yield	Mois.‡	EP‡	GLS§
	t ha⁻¹	%	%		t ha⁻¹	%	%	
B73*Ht* × Mo17*Ht*	7.5	16.1	92	3.0	9.5	18.0	83	6.9
Dekalb 689	8.8	18.0	91	4.9	8.2	19.3	82	7.3
LH132 × LH82	7.1	16.8	97	5.1	9.1	18.0	89	6.7
NK N8727	9.1	18.7	95	4.0	9.8	19.8	90	6.4
Pioneer 3140	8.8	18.0	97	4.7	8.4	19.8	89	6.7
Pioneer 3162	8.8	18.8	97	3.3	9.2	20.0	89	6.7
Pioneer 3165	8.9	19.8	89	5.5	9.1	20.8	83	6.5
Pioneer 3379	7.8	15.8	97	3.5	9.2	18.0	89	6.8
Pioneer 3394	8.6	16.4	98	2.7	9.7	17.8	90	6.9
Average	*8.2*	*17.5*	*95*	*3.9*	*9.2*	*19.1*	*87*	*6.8*
LSD (0.05)¶	0.7	0.7	8	1.5	0.7	0.7	8	0.7
CV%	8	4	9	10	8	4	9	8

† NC296A is a temperate-adapted line that was derived from a cross of two tropical hybrids, Pioneer X105A from Jamaica and H5 from CENTA (Centro Nacional de Tecnologia Agricola), El Salvador. Three of the lines in H5 developed by Jesus Merino, CENTA; fourth was from the Rockefeller program in Central America.
‡ Mois., percentage of moisture; EP, percentage of erect plants at harvest.
§ Gray leaf spot rated only in 1992 at two locations, scored on a 9 = no disease, 1 = dead basis.
¶ LSD and CV based on entry x environment interaction term.

NC296 is used in at least a few commercial hybrids in the USA and Mexico). Data indicate that NC296A, another all-tropical inbred, is a source for potentially new factors for yield for U.S. breeding programs; it also has promise for gray leaf spot- and southern rust-resistance (Table 13–4, rust data not shown). All NC296A crosses were essentially immune to southern rust in 1992 (all were rated 9 on a 1 to 9 scale; the most-resistant commercial hybrid, Pioneer 3140 scored 7.5, the others ranged from 2.5 to 5.0). This experiment was conducted as a demonstration to attempt to persuade private breeders that a tropical source could serve to increase yields, even of elite domestic hybrids, which were chosen to represent a range of high-response and stress-resistant, commercial hybrids. The experiment included nine hybrids and the same nine hybrids crossed to NC296A. The NC296A crosses out-yielded the hybrids themselves by an average of about 1 t ha^{-1} at a cost of 1.6% grain moisture and 8% lower standability. A similar experiment was conducted with NC296 and 366-7 (two sister lines of NC296A), using a different set of seven hybrids (Table 13–5). The results were similar, although the differences were smaller: a yield gain of 0.7 t ha^{-1} at a cost of 0.85% moisture for the 50%-tropical topcrosses. There was no difference in mean lodging between the commercial hybrids and their topcrosses with these two temperate-adapted, tropical inbreds, despite the effects of a tropical storm (Bertha) and a major hurricane (Fran) in 1996.

The use of tropical germplasm usually introduces more lodging, higher moisture at harvest, taller plants, and susceptibility to smut. However, higher yield and more GLS- and southern rust-resistance suggest that NC296A and its sister lines should not be ignored, despite less-than-perfect appearance and weak roots. Problems related to maturity, standability, and smut-susceptibility are readily improved; NC296 was a first cycle, 100%-tropical, temperate-adapted inbred derived from first-cycle tropical hybrids based upon lines that themselves were not fully inbred. We are now testing third cycle inbreds from all-tropical germplasm, and have

Table 13–5. NC296-type topcrosses† vs. commerical hybrids. Year 1996: Clayton, NC; Lewiston, NC;Year 1995: Clayton, Lewiston, Plymouth, NC.

Hybrid	Commercial Grain			× NC296 Grain			× 366-7 Grain		
	Yield	Mois.‡	EP‡	Yield	Mois.‡	EP‡	Yield	Mois.‡	EP‡
	t ha^{-1}	%	%	t ha^{-1}	%	%	t ha^{-1}	%	%
B73*Ht*×Mo17*Ht*	6.8	16.6	68	8.2	18.4	81	8.2	18.1	80
DeKalb 743	7.9	19.5	78	8.3	19.6	80	8.5	19.4	83
NK N8727	8.4	19.1	86	8.8	19.4	87	8.4	19.5	84
Pioneer 3245	8.6	17.8	84	9.3	18.8	86	9.1	18.9	83
Pioneer 3394	8.1	16.5	92	8.5	17.7	82	8.8	18.2	84
Pioneer 3283W	7.3	17.5	86	7.9	18.1	84	8.0	18.5	84
Pioneer 3287W	7.1	18.3	81	8.0	18.8	77	7.6	18.7	81
Average:	*7.7*	*17.9*	*82*	*8.4*	*18.7*	*82*	*8.4*	*18.8*	*83*
DeKalb 689	7.6	18.1	78	*Various*					
LH132 × LH51	7.7	17.3	84	*Check*					
Pioneer 3085	8.1	19.7	66	*Hybrids*					
Pioneer 3165	8.1	19.9	74						
LSD (0.05)§	0.8	0.7	11						
CV%	8	3	11						

† NC296 and 366-7 are sister lines both derived from the cross of the same two tropical hybrids as NC296A (see Table 13–4).
‡ Mois., percentage of moisture; EP, percentage of erect plants at harvest.
§ LSD and CV based on entry x environment interaction term.

concluded that two factors restrict our progress with all-tropical derivatives much more than the problems mentioned above:

1. Poor germination and seedling vigor under adverse (cold, wet, cloudy) spring growing conditions, and
2. A lack of available high-yielding, early maturing, lowland-tropical inbreds or hybrids. Most all-tropical hybrids are full-season hybrids that are almost as well suited to fence building as grain production, even when grown under short days.

We are attempting to remedy the late-maturity, high moisture problem by selection for earlier dry-down and seem to have had reasonable success (Hawbaker et al., 1997). More cold, wet springs similar to 1997 will probably solve our spring-vigor problem.

SELECTION WITHIN A VARIABLE EXOTIC SOURCE

The Germplasm Enhancement of Maize (GEM) project, a follow-up project to LAMP, is a multi-institutional, public-private, cooperative endeavor to quickly inject elite exotic germplasm into public and private breeding programs (Salhuana et al., 1994). The breeding populations used in GEM include (i) elite germplasm accessions identified by LAMP crossed to elite, domestic, private lines; (ii) tropical hybrids crossed to elite, domestic, private lines; and (iii) the breeding populations in (i) and (ii) crossed to second elite, domestic, private lines from the same heterotic group but from a second company. Nearly all large domestic companies participate, as does one Argentine company, Morgan. Almost every public maize breeding program in the USA is included in the project. The companies contribute germplasm, through crosses which they make using their own lines, nursery space and labor for selfing, and yield-trial space. Public programs are involved with disease- and insect-resistance, value-added trait characterization, and breeding work. The effort is led by Linda Pollak of Iowa State University and Martin Carson from North Carolina State University, both USDA-ARS researchers. Wilfredo Salhuana chairs the steering committee, which represents both the private and public sectors.

One immediate question arose concerning which generation of selfing should be used for topcross testing of the (germplasm accession × private line) breeding crosses. These 50%-exotic populations start as variable F_1s, where the F_1 variation represents one-half the additive genetic variation in the variable germplasm accession. However, once they have been selfed twice, the variation among families is dominated by the newly-generated variance of the racial cross arising as a result of segregation between the elite, domestic Corn Belt inbred and the exotic accession (Table 13–6). Thus, one must test individual F_1 plants or F_1S_1 (F_2) families if efficient selection is to be done within the accession, and many plant breeders familiar with germplasm accessions of maize would want to select within such accessions. This same principle applies to QTL studies with variable exotic populations. If it is important to select *within* such populations, then selection must be done early, before newly generated exotic × elite variation is expressed, or very large sample sizes (not hundreds, but thousands) will be needed to assure that the specific QTL of interest within the variable exotic will be identified (see Beavis, 1994).

RATIONALE AND CONCLUSIONS

The maize breeding program at North Carolina State devotes much of its effort to the development of largely tropical lines adapted to temperate environments. To the best of our knowledge, we are the only organization in the world that emphasizes line development from 50 to 100% tropical germplasm for use in temperate areas. There is a strong possibility that such daylength-neutral, largely-tropical germplasm may also be useful in the tropics and subtropics. Yield trials in Argentina and Brazil (for example, see Goodman et al., 1990), and breeding and production use in India and Mexico support that concept. It is among our all-tropical lines that we have found

Table 13–6. Distribution of readily transmissible genetic variation of a line × accession cross and of subsequent selfed lines (from Goodman, 2000).

Generation	Average genetic variation†	
	Within	Among
(Accession × Line) F_1	0	σ_A^2
(Accession × Line) F_1S_1	σ_B^2	σ_A^2
(Accession × Line) F_1S_2	$\sigma_B^2/2$	$\sigma_A^2 + \sigma_B^2/2$
(Accession × Line) F_1S_3	$\sigma_B^2/4$	$\sigma_A^2 + 3\sigma_B^2/4$
(Accession × Line) F_1S_4	$\sigma_B^2/8$	$\sigma_A^2 + 7\sigma_B^2/8$
⋮	⋮	⋮

† σ_A^2 is the readily heritable variation transmitted from the accession (generally one-half the additive genetic variance within the accession itself plus minor epistatic and usually unknowable dominance effects). σ_B^2 is the variance arising in the F_2 (F_1S_1) generation from elite × exotic segregation; for simplicity, it is treated here as additive. F_1S_i is the ith selfed generation derived from the variable F_1 population.

our highest testcross yields, greatest southern leaf blight resistance, best southern rust resistance, and reliable gray leaf spot resistance, equivalent to that of our best all-temperate line, NC258. We have yet to determine whether we can successfully mine germplasm accessions directly for useful lines, but the initial data look promising (Stuber, 1978; Castillo-Gonzalez & Goodman, 1988; Eberhart et al., 1995; Holland & Goodman, 1995).

Most of the line development work with largely tropical materials that has reached yield-trial stage at North Carolina State has been from either tropical synthetics or tropical hybrids. Work is in progress using lines from CIMMYT, IITA, and Thailand (the latter restricted to Ki4 and Ki11 as Thai inbreds and hybrids are difficult to obtain in the USA due to phytosanitary restrictions).

Preliminary data from the GEM project are quite encouraging, but only a single-year's data are complete for individual families, and those are only from a single population, Chis 775. Basically, the topcross of one F_1 family (of the variable accession Chis 775 crossed to a non-Stiff-Stalk inbred) outperformed the checks, while several others were close to the checks. Given the range of variation expected in the F_2 and in subsequent generations, it should be possible to extract superior, 50%-Chis 775 families. Line development will depend upon the prevalence of deleterious recessives. If they are not pervasive, then direct development of 50%-tropical lines should be possible. We will also cross these families with our existing, temperate-adapted, tropical lines and proceed with line development. At Iowa State University, and at cooperating private companies, identified elite families will be crossed to elite Midwestern lines for the development of 25%-tropical, 75%-domestic lines. By the summer of 2001, there should be extensive proprietary and public testing of topcrosses of partially inbred lines tracing to the GEM project. By the year 2005, some of this material could begin to reach farmers' fields.

It is very likely that the rate of use of exotic germplasm will continue to increase as data become available to identify the most promising sources of exotic germplasm and as these sources are converted to more readily-used lines and populations, free from daylength restrictions and agronomic flaws. These will be especially useful for QTL (quantitative trait loci) studies such as those of Ragot et al. (1995), where extensive field experimentation is required to accurately estimate genetic effects (and for which pre-adapted materials or several generations of backcrossing are usually needed).

There is little doubt that virtually all breeding stocks, germplasm accessions, and wild or weedy crop relatives, no matter how unpromising they may appear phenotypically, contain untapped alleles or allelic-combinations that could be used for plant breeding, with adequate investment in conventional or marker-assisted selection (Tanksley & Nelson, 1996). The more distant the relationship to current elite lines, the greater the probability that there is something unique to discover. Unfortunately, due to the usual amount of backcrossing necessary for most agronomic evaluations of quantitative traits (as opposed to single-gene, qualitative traits) from unadapted sources, there is a low likelihood to find exactly what is sought. For example, to evaluate an unadapted maize accession agronomically, the equivalent of two backcrosses is needed in North Carolina (for Iowa three are needed). To do the same thing with teosinte would require *at least* an additional backcross. With each backcross, the possibility of loss of the allele of interest increases, and this allele probably is not fixed in the original cross-pollinated accession. Thus, most breeding and applied molecular genetics programs are apt to remain concentrated on elite, improved germplasm as long as genetic advances resulting in improved lines continue at the current rate of 1 to 2% per year.

It is certainly possible to dissect the genetic structure of a second, third, or fourth backcross population of an exotic maize or teosinte accession using a saturating array of molecular markers combined with standard factorial-regression anaysis, bulked-segregant anaysis, or other analytical techniques (Tanksley & Nelson, 1996). The number of distinct families [somewhere between hundreds and thousands (Beavis, 1994)] required to detect differences of the order of 5 to 10% for quantitative traits, such as yield or standability, is particularly daunting in light of the existence of over 250 highly variable races of maize and several species and races of teosinte that are at least equally as geneticly variable as the races of maize. None of the teosintes have been seriously screened for favorable characteristics, and some individual teosinte accessions appear to have genetic variation levels comparable to that found within entire species of some self-pollinated crops (J. Jesus Sanchez, INIFAP, unpublished data). Despite LAMP, lack of preliminary evaluation also is the case for over one-half of the Latin American accessions—and some entire races—of maize. Under these sorts of circumstances, the use of pre-adapted, elite inbred lines of exotic maize, such as B103, NC296, NC298, or NC300, would be the first choice for molecular-marker work with maize exotics. Elite, proven inbreds from tropical breeding programs, even though they lack U.S. adaptation, would be the second choice because (i) they have passed severe selection for numerous favorable attributes and (ii) being commercial-quality inbreds, they have been purged of many, if not most, deleterious recessive alleles that are carried by highly heterogeneous and heterozygous populations of open-pollinated maize and teosinte. A critical advantage of pre-adapted materials is that they can be evaluated as F_2S_1 families, rather than BC_2 (or even BC_4) families. With each backcross, half of the additive variation is lost. Thus, differences of 20% among F_2S_1 families are reduced to only 5% among BC_2 families, and differences of <5% are often difficult to detect in many breeding programs.

Clearly, use of exotic germplasm will be needed in the USA to provide insurance against certain diseases or pests that are prevalent elsewhere in the world but currently absent in the USA In some cases, the USA lacks elite lines with resistance. Examples include streak virus from Africa, Rio Quarto virus from Argentina, and African gray leaf spot, a much more aggressive form than the relatively late-developing strains found in the USA It would be prudent to spend some of the millions of dollars currently being invested in genetic engineering projects for herbicide resistance on more critical projects to develop adequate safeguards against such readily-identifiable and potent threats to agricultural security as these diseases. Adequate response will require both the use of exotic germplasm (that is where the resistance is), the use of molecular-marker technology (to efficiently transfer the resistance), and international collaboration (because the current disease sites—fortunately for the USA—lie outside U.S. borders). At least some of these diseases will reach the

USA; failure to have an adequate defense could be devastating, since few, if any, elite lines have adequate resistance at present.

While we work on temperate-adapted, tropical line development at North Carolina State and colleagues such as Dudley (1984) and Crossa (1989) develop theory for selecting optimal populations and ideal percentages of exotic germplasm, natural selection itself is apt to provide the ultimate force to encourage the widespread use of genetic resources.

REFERENCES

Beavis, W.D. 1994. The power and deceit of QTL experiments: Lessons from comparative QTL studies. Annu. Corn Sorghum Research Conf. Proc. 49:250–266.

Castillo-Gonzalez, F., and M.M. Goodman. 1988. Agronomic evaluation of Latin American maize accessions. Crop Sci. 29:853–861.

Crossa, J. 1989. Theoretical considerations for the introgression of exotic germplasm into adapted maize populations. Maydica 34:53–62.

Dudley, J.W. 1984. Theory for the identification and use of exotic germplasm in maize breeding programs. Maydica 29:391–407.

Duvick, D.N. 1990. The romance of plant breeding and other myths. p. 39–54. *In* J.P. Gustafson (ed.) Gene manipulation in plant improvement, Plenum Press, New York.

Eberhart, S.A., W. Salhuana, R. Sevilla, and S. Taba. 1995. Principles for tropical maize breeding. Maydica 40:339–355.

Goodman, M.M. 1985. Exotic maize germplasm: Status, prospects and remedies. Iowa State J. Res. 59:497–527.

Goodman, M.M. 1992. Choosing and using tropical corn germplasm. Annu. Corn Sorghum Res. Conf. Proc. 47:47–64.

Goodman, M.M. 1998. Reseaarch policies thwart potential payoff of exotic germplasm. Diversity. 14:30-35.

Goodman, M.M. 2000. Incorporation of exotic germplasm into elite maize lines: Maximizing favorable effects of the exotic source. Theor. Pop. Biol. 57:(In press.)

Goodman, M.M., F. Castillo-Gonzalez, and J. Moreno. 1990. Choosing and using exotic maize germplasm. Illinois Corn Breeders School Proc. 26:148–171.

Hawbaker, M.S., W. H. Hill, and M.M. Goodman. 1997. Application of recurrent selection for low moisture content in tropical maize (*Zea mays* L.): I. Testcross yield trials. Crop Sci. 37:1650–1655.

Holland, J.B., and M.M. Goodman. 1995. Combining ability of tropical maize accessions with U.S. germplasm. Crop Sci. 35:767–773.

Holley, R.N., and M.M. Goodman. 1988. Yield potential of tropical hybrid corn derivatives. Crop Sci. 28:213–217.

Kim, S.K., Y. Efron, F. Khadr, J. Fajemisin, and M.H. Lee. 1987. Registration of 16 maize-streak resistant tropical maize parental inbred lines. Crop Sci. 27:824–825.

Ragot, M., P.H. Sisco, D.A. Hoisington, and C.W. Stuber. 1995. Molecular-marker-mediated characterization of favorable exotic alleles at quantitative trait loci in maize. Crop Sci. 35:1306–1315.

Salhuana, W. 1995. Latin American maize project leaves untapped legacy of agricultural riches. Diversity 11(4):6.

Salhuana, W., L. Pollak, and D. Tiffany. 1994. Public/private collaboration proposed to strengthen quality and production of USA corn through germplasm enhancement. Diversity 10(1):77–78.

Salhuana, W., and R. Sevilla. 1995. Latin American Maize Project (LAMP) Stage 4 Results from Homologous Areas 1 and 5. Natl. Seed Storage Lab., Fort Collins, CO.

Salhuana, W., R. Sevilla, and S.A. Eberhart. 1997. Final Report, Latin American Maize Project. Pioneer Hi-Bred Int., Johnston, IA

Stuber, C.W. 1978. Exotic sources for broadening genetic diversity in corn breeding programs. Ann. Corn and Sorghum Res. Conf. Proc. 33:34–47.

Tanksley, S.W., and J.C. Nelson. 1996. Advanced backcross QTL analysis: a method for the simultaneous discovery and transfer of valuable QTLs from unadapted germplasm into elite lines. Theor. Appl. Genet. 92:191–203.

Uhr, D.V., and M.M. Goodman. 1995. Temperate maize inbreds from tropical germplasm. I. Testcross yield trials. Crop Sci. 35:779–784.

Chapter 14

Mechanisms Contributing to Genetic Diversity in Maize Populations

P. A. Peterson

INTRODUCTION

Maize Genome is Volatile

The maize (*Zea mays* L.) genome has more than two billion pairs of bases (approximately 2400 megabases). Much of this maize genome is repetitive DNA (60 to 80%) (Flavell et al., 1974). The maize genome is five times greater than the rice genome and one-eighth as large as the wheat (*Triticum aestivum* L.) genome. The maize genome is composed of a large number of transposing elements. The predominant type in maize is the retroelement, which, in contrast to the transposable elements replicates via reverse transcription through an RNA intermediate. Many of these elements are repeated in flanking parts of genes, and these repeat pairs are likely to be long terminal repeats (LTR) with several families of retrotransposons collectively accounting for >25% of the maize genome (San Miguel et al., 1996). In general, these retroelements are found not in coding sequences of genes but in flanking sequences. Especially interesting, they are not known to interfere with gene function. Thus, the maize genome carries much DNA, the function of which is not known.

The demonstration of an ever-changing maize genome is self-evident in the plethora of lines rescued from the cycling of the BSSS population (Sprague, 1946). The genetic gain that has been demonstrated between the maize lines of the 1930s to those of the 1980s and beyond is a clear demonstration of the ever-available variability for selection (Russell, 1974, 1985, 1991; Duvick, 1992). Most studies assign a value of 50 to 70% for the genetic component of the realized gain. Realistically, this is an underestimated value because the genetic contribution has accommodated the improved husbandry. Though one could ascribe recombination of favorable alleles as a basis for these advances, other studies provide additional causes for this advance.

In an effort to investigate the basis of this genetic gain, Sprague et al. (1960) initiated a study with doubled haploids. The rationale in this investigation was that these doubled haploids represented in their origin completely homozygous material. In their evaluation of S_3 through S_6 progenies, Sprague et al. (1960) measured nine quantitative traits. Significant differences between generations were ascribed to mutational changes. These authors concluded that there were 4.5 mutations per attribute per 100 gametes tested.

Questioning whether the process of developing these monoploid lines through haploidization was a factor in this generally high mutation rate, Russell et al. (1963) undertook a similar study of six long time inbreds maintained by ear-to-row selection. These authors measured nine traits for five generations. They estimated the rate of mutation as 2.8 mutations per attribute per 100 gametes. Though this rate of mutation is approximately one-half the rate of the monoploid study, it is nevertheless substantial. These two studies substantiate the current findings of excessive polymorphism in maize and support the contention of the volatility of the maize genome.

Another example of polymorphism that exists between inbred lines was studied by David Grant, while working at Pioneer Hi-Bred International, (personal communication). He measured the DNA sequence variation at RFLP loci between six unrelated lines (<9% relatedness by pedigree, 1.9% average similarity). The initial report included data for 72 loci distributed at random in the genome, and PCR amplified the regions from each inbred. Single-pass sequencing runs were made from each end of the amplified product, resulting in 250 to 900 bases of sequence with an average of 500 for each locus. Polymorphisms were identified after alignment of the sequences.

The number of polymorphisms observed per locus ranged from 0 to 35. These changes included single-base changes and both large and small insertion–deletion events (including SSRs). Because the amount of sequence obtained at each locus was variable, the number of polymorphisms per locus was normalized to 1 kilobase (kb) of sequence. As expected, the most frequent number of single-base changes at a locus was 0 (21% of the loci). Surprisingly, only a single locus had one change, while 15 (21% of the loci) had either 2 or 3 changes. The remainder of the observed polymorphisms was also distributed unequally —there were peaks at 6 to 7 (10%), 10 to 11 (10%), and 13 to 14 (11%) changes per kb. The distribution of insertion–deletion events showed a similar peak at 2 to 3 kb^{-1} (10%), a second more pronounced one at 5 to 7 (20%) and a third at 10 to 11 (7%) events kb^{-1}. These results imply that (i) the individual polymorphisms may not have arisen independently, (ii) the high frequency of three-base changes may be a result of a meiotically related mutagenic event, and (iii) the mechanism for their generation may act over a distance of at least hundreds of bases.

Molecular diversity among developing populations of maize also was investigated by Labate et al. (1997). Starting with progenitor lines of two longtime populations (BSSSR and BSCB1) and evaluating the genetic diversity after 12 cycles of selection, they came to the following conclusions: (i) no single reciprocal recurrent progenitor made excessive genetic contribution to either C0 or C12; (ii) the two populations diverged after 12 cycles; (iii) polymorphism level and gene diversity level decreased; and (iv) the mean number of alleles per locus decreased from four to about less than three. Of the 391 alleles at 82 loci, 25% were unique (limited to a single progenitor inbred line of the BSSS population). Questions arise as to the nature of the uniqueness and why some ended in extinction by C12. Yet 20% of the unique alleles were maintained by C12. This study focused on an interesting question: What happens to individual alleles in a longtime population maintenance program (progenitor to C12)? What is needed is a refined study (DNA sequence) of individual alleles, especially unique ones that survived the population maintenance cycles.

Native Variability in Maize Breeding Populations

When plant breeding populations are considered, a question arises as to the wealth of variability found in these populations. Maize breeding populations (Lamkey, 1992) show an infinite amount of variability that can be selected in the population improvement process (Hallauer et al., 1983). This variability has been evident in the long-term Illinois oil and protein lines (Dudley, 1977) and displayed in the molecular expression of isozyme and diversity frequency (Goodman & Stuber, 1983) and with restriction fragment length polymorphisms (RFLP) analysis (Helentjaris et al., 1985; Sughroue & Rocheford, 1994).

Because of the well-known advances in maize population improvement process (Hallauer, 1985), there is the problem of the origin and source of this infinitely available variability in these populations (Keeratinijakal & Lamkey, 1993). For example, the BSSS(C0) and BSSS(R) populations have been closed populations since the original development of these populations (Sprague, 1946). The genomic content of this population has not had additional introduced germplasm, and all the selection during the cyclical experimentation has come from selection *within* the population during the screening process (Lamkey et al., 1991). At least three options for the identification of the source of this variability are possible. First, most prominent among the forces generating changes is the *reassembly of genetic materials* originating from *intergenic recombination;* better combinations are assembled by unloading unfavorable genes closely linked to favorable genes and by reassembling these favorable units, thus better combinations are combined in a composite. Such an assembly was developed by the Stuber group in finding via RFLP analysis a favorable segment in a quantitative trait locus (QTL) search study and inserting this acknowledged segment into an already-developed elite inbred (Stuber, 1997). Such a process is accomplished in a recurrent selection breeding protocol (Hallauer, 1985; Hallauer & Miranda Fo, 1988). A second option would include the rearrangement or a *recombination of the genomic materials via intraallelic and intragenic* recombination as a contributing force to this variability. Developing new gene sequences via homologous exchanges (Xu et al., 1995) could lead to better-performing gene sequence arrays. This process would include codon swapping that would be translated into new protein sequences. This rate of change is slow, and the optimal contribution has not been demonstrated. The third option is that *altered gene nucleotide sequence and base substitution* changes originate during the population maintenance program. These genetic changes could come from arrays of nucleotide sequences that would include altered, deleted, or additional nucleotides that make up both the coding and regulatory components of genes. The originator of these changes would be from known or-as-yet undiscovered transposons or an as-yet unknown mutation process. The plant improvement program had been in process before the discovery of transposons and their possible role in the mutation process. Transposons are known to be involved in the origination of novel sequences via a mutagenic process.

TRANSPOSONS

Genetic and Molecular Aspects of Transposons

Transposable elements have been a major study in maize genetics for one-half century (McClintock, 1948, 1951; Peterson, 1994). The genetic study in the premolecular era established (i) the different systems of transposons (at least 10), (ii) the functional and deficient elements and the accompanying pattern differences, (iii) the diversity of elements, (iv) transposons' effect on gene expression, and (v) transposons' distribution in native populations. With the development of the molecular biology of transposons, many features of transposon genetics were clarified including the molecular aspects of structure, size, the terminal inverted repeats (TIR), the target site duplication (TSD) of each of the specific systems that add nucleotides to the genome (Fedoroff et al., 1983; Pereira et al., 1985), gene nucleotide alterations accompanying excision of an element (Sachs et al., 1983), and subsequent change in gene products (Schwarz-Sommer et al., 1985; Wessler et al., 1986).

Transposons in Native Populations

The altered nucleotide gene sequences associated with most transposon excisions from genes has generated discussion on the role of transposons as a force in evolution (Schwarz-Sommer et al., 1985). The discovery of functional (genetically active) transposons in numerous maize breeding populations (Peterson & Friedemann, 1983; Peterson & Salamini, 1986; Peterson, 1986; Cormack et al., 1988; Lamkey et al., 1991; Seo & Peterson, 1995) and the distribution of the varied inserts in most genes examined in a number of plant species (Okagaki & Wessler, 1988; Bureau & Wessler, 1992; Bureau et al., 1996; Wessler, 1996) has piqued the interest of investigators on the role of transposons in native populations.

Though most actively studied in maize, transposons are found in most plant species studied (Peterson, 1987, 1994) and have been intensively investigated in *Drosophila* populations (Engels, 1983; Kidwell, 1985). Their role in the native populations has generated diverse experimentation (Mackay, 1986) in demonstrating their effect on these populations.

Case for Transposons

What, then, makes the case for transposons relative to the variability-inducing process in organisms? The first clues about transposon effects on gene nucleotide sequence came from a genetic analysis of *a1-m1* and the excision products *al-m(papu)* allele (Peterson, 1956, 1961). These non-responding [*a1m(nr)*] alleles are nonfunctional for the *A1* product, are no longer responding to *En,* and are essentially null alleles. Gene alterations detected with genetic methods were finally confirmed with the molecular isolation of the *adh1* allele harboring a *Ds* insert (Sachs et al., 1983). Quite importantly and equally significant were the unexpected target-site duplications. With this process, nucleotides are *added* to the genome. Here, it was quite evident from an examination of the altered nucleotide sequences that the excision of the *Ds* element was not perfect; the repair process (Saedler & Nevers, 1985) of the target site (TSD) leaves a new sequence and, consequently, a new coding sequence yielding an altered or nonfunctional protein. Is this a process for nature to provide a process for developing new coding sequences? This TSD change has been amply confirmed with subsequent studies at all loci and with all transposons that have been molecularly analyzed (*Ds* at *bz1,* Dooner et al., 1985; *Mu1* at *a1,* Brown & Sundaresan, 1991; Civardi et al., 1994; *Mu1* at *kn1,* Lowe et al., 1992; Athma & Peterson, 1991; review, Timmermans et al., 1996; *Ds1* at *adh1,* Dawe et al., 1993; *I/dSpm*-induced *wx* changes, Wessler et al., 1986; Schwarz-Sommer et al., 1985; *Mu3*-induced promoter-scrambling, Kloeckener-Gruissem & Freeling, 1995). These changes have been confirmed in experimental heterologous species harboring transposons (Wisman et al., 1997).

Transposon-induced changes in gene sequences have been found in natural populations. These changes were uncovered serendipitously in the process of analyzing gene sequences. One such case includes the regulatory gene, *C1,* and two of its derivatives. One allele, *C1-I,* a longtime occupant in native South American maize populations, is a dominant suppressor of color; the suppressive effect on coloration is caused by a truncated protein arising from an abortive transcript (stop codon) that binds and dominates promoter sites of structural genes. Because the protein lacks an acidic domain, it cannot activate transcription on the anthocyanin structural genes, and, in this way, color is not expressed (suppressed). What is unique about the *C1-I* promoter is the inclusion of a Box II region (Scheffler et al., 1994) that makes the product super active. With this promoter, more *C1-1* transcripts are generated and thus, dominate the structural gene promoter sites, and consequently excluding "normal" proteins. The sequence in the neighborhood of this Box II has been associated with transposon-induced footprints. Further, the *SC1* (super *C*) allele has the same promoter (Box II) and similarly expresses a strong allelic expression. (This expression is based on the highly competitive nature of this allele when combined with *C1-I(C1-I/^SC1)* in modifying *C1-I*'s suppressive capacity.

Transposon changes do show positive improvement and include the transposon-induced changes at the promoter of the *nivea* locus of *Antirrhinum* (Sommer et al., 1988), and the *Ds* excision-induced revertants at the *sh2* locus that led to striking yield increase (Giroux et al., 1996). Others examples will be uncovered as other alleles are analyzed.

With this array of varied results, why does the natural process include deleterious effects with the occasional positive effect? But in plant domestication, the process of neoteny is important in fine-tuning a plant's performance (Lester, 1989). Yet, in summary, transposons are pervasive in maize populations and with the added TSD nucleotides and subsequent imperfect excisions do provide a milieu for nucleotide alteration.

GENE CHANGES

The intragenic exchange considered is the exchange between chromosome strands that yield a change that occurs within the transcribed portion of the gene. The exchange results in the formation of chimeric gene fusion. This exchange process would need heteroalleles that differ in a coding sequence which, by the exchange, would yield a new product.

Intercistronic

A^b Allele

This type of exchange among supposed alleles was uncovered at the preDNA-discovery era. When two definable phenotypes ascribed to the same locus (defined by inheritance analysis) yield a different third phenotype, via altered flanking markers, then an intralocus exchange is identified. Such was the case in the A^b study (Laughnan, 1952). Here, two distinct phenotypes identified with two different alleles, A^d and A^b, at the *A1* locus (alleles in South American indigenous populations) yielded a third phenotype via displaced exchange of homologous segments that were in a tandem array and consequently yielded the new phenotypes (Fig. 14–1). In this case, A^b included alpha (α) and βeta (β) homologous segments in tandem array. When synapsis occurred between $a1^{dt}$ (a) and alpha (α) of the

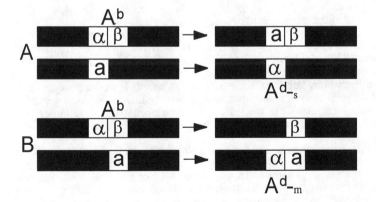

Fig. 14–1. The exchanges at the *A1* locus in maize (Laughnan, 1952).

alpha- beta array, an A^d-s (α alone) stable was generated. When synapsis was be-tween a^{dt} and beta, aα (A^d-mutable) was generated. Here, new complexes of gene arrays were recovered (Fig. 14–1) at approximately 7.73×10^{-4}.

R Locus Complex

The *R* locus represents another case of tandem repeats with opportunities for displaced synapsis that result in new arrays and consequently new alleles. The *R* locus (Dellaporta et al., 1988) has an abundance of alleles with distinguishable phenotypes. These alleles include *R-r* (colored plant and seed), *R-g* (green plant, colored seed), *r-r* (colored plant, colorless seed), and *r-g* (colorless plant and seed). Each of these alleles encodes a distinct, recognizable phenotype and mutate inde-pendently (Stadler, 1948) and include exchange between components (Stadler and Emmerling, 1956). The *R-r* complex is organized as a tandem duplication and is approximately 0.14 to 0.16 map units. The complex includes three tandem compo-nents P, Q, and S with some, but not total, commonality. For example, the Q com-ponent is a truncated P component while the P and S share a 3' homology, and the S itself is in tandem array of two components. This partial homology generates displaced synapsis because the P and S components have conserved regions at the 3' region where exchanges take place (Robbins et al., 1991; Walker et al., 1997). From this displaced synapsis, new tandem arrays are generated. For example, *r-r* derivatives are generated by exchanges at the homologous 3' regions of S and P yielding a P and s duplication that is phenotypically a colored plant (P) and color-less seed (s). This displaced intragenic recombination frequency is 3.5×10^{-4}. This frequency is compared to 10.9×10^{-4} at *wx* (Nelson, 1968) and 13.0×10^{-4} at *bz1* (Dooner, 1986). Obviously, this process also leads to codon swapping.

Intragenic

Within Coding Regions of Genes

In most studies of intragenic recombination in maize, crosses with hybrid heteroalleles recover the wild type. In this scenario, the mutant alleles were non-functional either by point mutations, deletions, or insertions usually resulting in nonexpression of the alleles that did not overlap with the defective parts of the other allele of the heterocomplex. The exchange yielded a crossover without the defective portions of each of the alleles and thus was functional. This type of in-tragenic study was first analyzed at the waxy locus (Nelson, 1968). By recovering wild type [a pollen assay that distinguishes *wx* (red) from *Wx* (purple)], defective positions of each of the alleles could be mapped within the gene.

Intragenic recombination was targeted at several other genes: at *a1* (Brown & Sundaresan, 1991; Civardi et al., 1994); *b* (Patterson et al., 1995), *gl1* (Salamini & Lorenzoni, 1970), *r* (Dooner & Kermicle, 1986; Büschges et al., 1997). The rates of recombination at these loci ranged from 0.9 to 1.3×10^{-3}. Because of this higher rate of recombination than that estimated for the whole maize genome, tran-scribed regions are more prone to recombination activity than is prevalent generally in the genome. Note that these studies are based on phenotypic selection of wild type recovery of known heteroalleles and, therefore, obligate recovery of the wild type sequence.

So what is the actual amount of recombination in the maize genome that is not biased to recovering wild type and thus demands a transcribing region that is only 20% of the maize genome? To overcome this experimental bias, Timmer-mans et al. (1996) based their recombination study on restriction fragment poly-morphic loci (RFLP) and consequently with no selection criteria as to the recovery of wild type. With an F_2 population of a cross of two distinct maize inbreds (A188 × W64A) with seven RFLP loci covering an 83 cM region of chromosome 7L, they identified two recombinant events. In a detailed study of one of the recombinants, several features were apparent. The event did occur in a 534 bp region of perfect homology between parent alleles "embedded within a 2773-bp unique sequence" though no transcript was identified. Further, the area identified was not methylated.

What can be concluded from the Timmermans et al. (1996, 1997) study although this was based on only one recombinant allele that was thoroughly analyzed? These authors indicated that the crossover occurred within a region of perfect homology (200-250 bp) that was not methylated. Further, there were no new polymorphisms in this recombinant indicating that the crossover was conservative, there was fidelity in the crossover and no additional bp were added. This supports the contention that no new combinations were generated though codon swapping could have occurred. That is, there would be a new codon such as **AGACCG** changing to **AAACCG** leading to a protein sequence change from arginine-proline to lysine-proline. Whether the substitution of the **A** nucleotide for the **G** in the arginine-to-lysine change in the protein would have the effect seen in the transposon-effected change in the *Sh2* revertant (Giroux et al., 1996) has not been established, but the possibility is there. Not surprisingly, two insertions (10 kb and 860 bp) were found by Timmermans et al. (1996) in the A188 allele within the probe but more than two kb from the crossover site. This was not surprising because insertions are often found in maize genes as was first recognized in the *wx-m8* allele (Schwarz-Sommer et al., 1985). Of course, some exchanges may not reveal themselves via non cross-over of flanking marker exchanges (Büschges et al., 1997).

There are situations that enhance the frequency of recombination. In several analyses, recombination is increased many fold in the vicinity of inserts responding to active transposons (Greenblatt, 1981). Certainly the repair process in the vicinity of transposase-induced double-strand breaks (Saedler & Nevers, 1985) would generate recombination events (Fig. 14–2). The proteins involved in editing and proofing the fusion of single-strand breaks following the excision would provide a suitable milieu for the exchange to occur.

But what of the recombination process? Note than an exchange resulting in a linkage breakup involves the operation of a number of proteins that facilitate the reunion and fusion of the newly formed chromosome strands. The double-strand break is shepherded by editing and proofing proteins. In bacteria studies where the analysis of the recombination proteins is most complete, a number of genes have been identified that are involved along with the consequences following mutations in the genes governing this process (Fig. 14–2; Cox, 1997).

CONCLUSIONS ON MECHANISMS CONTRIBUTING TO DIVERSITY

There is ample evidence for recurring change in maize breeding materials accompanying the selection process. Though the recombining of existing diversity

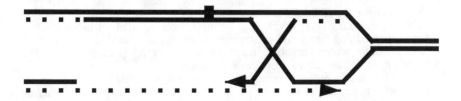

Fig. 14–2. The repair of gaps and the Holiday Junction in recombination.

cannot be adequately quantified, the generation of new changes in the development of maize breeding materials is possible and is proposed as a source of the genetic gain visualized over the decades in maize breeding programs.

Transposons

Transposon activity, for which there is ample evidence in maize lines, has been proposed as a contributor to this diversity. By their very nature, transposons add nucleotides to the maize genome. The number varies with the specific transposon from a minimum of 3 with the *En/Spm* system to 8 for *Ac/Uq* and more with *Mu*. And, following excision which is often imperfect and leaving a changed sequence as a residue of the target-site duplication, new codons and consequently a changed protein is generated. Yet this activity might not be adequate to account for the estimated rate of change of 2.8 to 4.8 per 100 gametes per trait (Sprague et al., 1960; Russell et al., 1963). Yet, the transposition process is prevalent, and the frequency of transposition with the excision process and its consequences cannot be ignored.

Recombination Process

Recombination does take place in transcribed regions leading to novel recombinant strands. The novel part of this process includes codon swapping but not the inclusion of additional nucleotides as in the case of transposon activity. If the Timmermans et al. (1996) study could be generalized, the pairing process preceding exchange requires perfect homology of at least 200 to 250 bp to facilitate the exchange. In view of the numerous inserts that pervade the maize genome, this required homology of a stretch of DNA sequence is a further limitation of exchange events. In the Timmermans et al. (1996) study of random, unselected exchange events, there were two exchanges in a 83 cM region among 96 chromosomes. Because the exchange occurred with the chromosome strands in perfect homology in a relatively long stretch of chromosome (83 cM), it is unlikely that this activity would equal the rate of change proposed in the Sprague et al. (1960) and Russell et al. (1963) study.

Consensus

In order to accommodate the facts presented, a multiple set of processes is needed. Likely, intergenic, intragenic, and transposon activity combined are contributing to the *mutagenicity mechanism* generating diversity. Without basic information on the molecular changes found in successive cycles of plant breeding materials such as the BSSS population, it is not possible to identify the major player in this genetic gain scenario. If the meiotic recombination process is mutagenic, then it is likely that this characteristic could lead to an adequate amount of change. The mutagenicity would be promoted by inadequate fidelity of the proteins involved in the Holiday Junction assembly as these ever-recurring new DNA stands are originating (Fig. 14–2). The multiple proteins involved in the process could lead to editing errors and subsequent deletions (Saveson & Lovett, 1997). Coupled with the proteins involved in transposon activity, adequate diversity in the form of substituted and added nucleotides is possible.

One case of a process in nature providing an opportunity for change can be seen in an examination of the error process in bacterial replication. A nonerror mutant was isolated and subjected to a (chemostat) survival test (Drake, 1992). Bacterial populations without the mutant (the wild type) were a more viable entity suggesting that the natural process sustains error-maintaining mechanisms in the replication process. This result further suggests that the natural process covets a variability-inducing mechanism.

A Proposal

Sequences of cloned genes: The most direct approach to the variability problem is the comparison of sequencing of genes of a closed BSSS population (Labate et al., 1997) that includes a number of cycles. This approach seems to be a mountainous and costly task. However, by the judicious choice of the units to be cloned and sequenced, the task can be considered reasonable.

ACKNOWLEDGMENTS

Contribution from the Department of Agronomy, Iowa State Univ. and Journal Paper J–17577 of the Iowa Agric. and Home Econ. Exp. Stn. Project 3176, and supported by Hatch Act and State of Iowa funds.

REFERENCES

Athma, P., and T. Peterson. 1991. Ac induces homologous recombination at the maize P locus. Genetics 128:163–173.

Brown, J., and V. Sundaresan. 1991. A recombination hotspot in the maize A1 intragenic region. Theòr. Appl. Genet. 81:185–188.

Bureau, T.E., and S.R. Wessler. 1992. Tourist: A large family of small inverted repeat elements frequently associated with maize genes. The Plant Cell 4:1283–1294.

Bureau, T.E., P.C. Ronald, and S.R. Wessler. 1996. A computer-based systematic survey reveals the predominance of small inverted-repeat elements in wild-type rice genes. Proc. Natl. Acad. Sci. USA 93:8524–8529.

Büschges, R., K. Hollricher, R. Panstruga, G. Simons, M. Walter, A. Frljters, R. van Daelen, T. van der Lee, P. Diergaarde, J. Groenendijk, S. Topsch, P. Vos, F. Salamini, and P. Schulze-Lefert. 1997. A novel control element of plant pathogen resistance. Cell 88:695–705.

Civardi, L., Y. Xia, K.J. Edwards, P.S. Schnable, and B.J. Nikolau. 1994. The relationship between genetic and physical distances in the cloned *a1-sh2* interval of the *Zea mays* L. genome. Proc. Natl. Acad. Sci. USA 91:8268–8272.

Cormack, J.B., D.F. Cox, and P.A. Peterson. 1988. Presence of the transposable element *Uq* in maize breeding material. Crop Sci. 28:941–944.

Cox, E.C. 1997. *mutS*, proofreading, and cancer. Genetics 146:443–446.

Dawe, R.K., A.R. Lachmansingh, and M. Freeling. 1993. Transposon-mediated mutations in the untranslated leader of maize adh1 that increase and decrease pollen-specific gene expression. Plant Cell 5:311–319.

Dellaporta, S.L., I.M. Greenblatt, J.L. Kermicle, J.B. Hicks, and S.R. Wessler. 1988. Molecular cloning of the *R-nj* gene by transposon tagging with *Ac*. p. 263–281. *In* J.P. Gustafson and R. Appels (ed.) Chromosome and function: impact of new concepts. Plenum Press, New York.

Dooner, H.K. 1986. Genetic fine structure of the bronze locus in maize. Genetics 113:1021–1036.

Dooner, H.K., and J.L. Kermicle. 1976. Displaced and tandem duplication in the long arm of chromosome 10 in maize. Genetics 82:309–322.

Dooner, H.K., E. Weck, S. Adams, E. Ralston, M. Favreau, and J. English. 1985. A molecular genetic analysis of insertion in the bronze locus in maize. Molec. Gen. Genet. 200:240–246.

Drake, J.W. 1992. Mutation rates. BioEssays 14:137–140.

Dudley, J.W. 1977. 76 generations of selection for oil and protein in maize. p. 459–473. In E. Pollak et al. (ed.) Proc. Int. Congr. Quant. Genet. Iowa State Univ. Press, Ames.

Duvick, D.N. 1992. Genetic contributions to advances in yield of U.S. maize. Maydica XXXVII:69–79.

Engels, W.R. 1983. The P family of transposable elements in drosophila. Ann. Rev. Genet. 17:315–344.

Fedoroff, N., S. Wessler, and M. Shure. 1983. Isolation of the transposable maize controlling elements Ac and Ds. Cell 35:235–242.

Flavell, R.B., M.D. Bennett, J.B. Smith, and D.B. Smith. 1974. Genome size and the proportion of reiterated nucleotide sequence DNA in plants. Biochem. Genet. 12:257–269.

Giroux, M.J., J. Shaw, G. Barry, B. Greg Cobb, T. Greene, T. Okita, and L.C. Hannah. 1996. A single gene mutation that increases maize seed weight. Proc. Natl. Acad. Sci. USA 93:5824–5829.

Goodman, M.M., and C.W. Stuber. 1983. Races of maize. VI. Isozyme variation among races of maize in Bolivia. Maydica XXVIII:169–187.

Greenblatt, I.M. 1981. Enhancement of crossing-over by the transposable element, modulator, in maize. Maydica XXVI:133–140.

Hallauer, A.R. 1985. Compendium of recurrent selection methods and their application. Crit. Rev. Plant Sci. 3:1–33.

Hallauer, A.R., and J.B. Miranda, Fo. 1988. Quantitative genetics in maize Breeding. Revised Ed. Iowa State Univ. Press, Ames.

Hallauer, A.R., W.A. Russell, and O.S. Smith. 1983. Quantitative analysis of Iowa Stiff Stalk Synthetic. Stadler Symp. 15:83–104.

Helentjaris, T., G. King, M. Slocum, C. Siedenstrang, and S. Wegman. 1985. Restriction fragment polymorphisms as probes for plant diversity and their development as tools for applied plant breeding. Plant Molec. Biol. 5:109–118.

Keeratinijakal, V., and K.R. Lamkey. 1993. Genetic effects associated with reciprocal recurrent selection in BSSS and BSCB1 maize populations. Crop Sci. 33:78–82.

Kidwell, M.G. 1985. Hybrid dys-genesis in Drosophila melanogaster: nature and inheritance of P element regulation. Genetics 111:337–350.

Kloeckener-Gruissem, B., and M. Freeling. 1995. Transposon-induced promoter scrambling: a mechanism for the evolution of new alleles. Proc. Natl. Acad. Sci., USA 92:1836–1840.

Labate, J.A., K.R. Lamkey, M. Lee, and W.L. Woodman. 1997. Molecular genetic diversity after reciprocal recurrent selection in BSSS and BSCB1 populations. Crop Sci. 37:416–423.

Lamkey, K.R. 1992. Fifty years of recurrent selection in the Iowa Stiff Stalk Synthetic maize population. Maydica XXXVII:19–28.

Lamkey, K.R., P.A. Peterson, and A.R. Hallauer. 1991. Frequency of the transposable element *Uq* in Iowa Stiff Stalk Synthetic maize populations. Genet. Res. Camb. 57:1–9.

Laughnan, J.R. 1952. The action of allelic forms of the gene *A* in maize. IV. On the compound nature of A^b and the occurrence and action of its Ad derivatives. Genetics 37:375–395.

Lester, R.N. 1989. Evolution under domestication involving disturbance of genic balance. Euphytica 44:125–132.

Lowe, B., J. Mathern, and S. Hake. 1992. Active *Mutator* elements suppress the knotted phenotype and increase recombination at the *Kn1-O* tandem duplication. Genetics 132:813–822.

Mackay, T.F. 1986. Transposable element-induced fitness mutations in *Drosophila melanogaster*. Genet. Res. Cambridge. 48:77–87.

McClintock, B. 1948. Mutable loci in maize. Carnegie Inst. Wash. Year Book 47:155–169.

McClintock, B. 1951. Chromosome organization and genic expression. Cold Spring Harbor Symp. Quant. Biol. 16:13–47.

Nelson, E. 1968. The waxy locus in maize. I. The location of the controlling element alleles. Genetics 60:507–524.

Okagaki, R.J., and S.R. Wessler. 1988. Comparison of non-mutant and mutant waxy genes in rice and maize. Genetics 120:1137–1143.

Patterson, G.I., K.M. Kubo, T. Shroyer, and V.L. Chandler. 1995. Sequences for paramutation of the maize *b* gene map to a region containing the promoter and upstream sequences. Genetics 140:1389–1406.

Pereira, A., Zs. Schwarz-Sommer, A. Gierl, I. Bertram. P.A. Peterson, and H. Saedler. 1985. Genetic and molecular analysis of the enhancer (*En*) transposable element system of *Zea mays*. EMBO J. 4:17–23.

Peterson, P.A. 1956. An *a1* mutable arising in pg^m stocks. Maize Genet. Coop. Newsl. 30:82.

Peterson, P.A. 1961. Mutable *a1* of the *En* system in maize. Genetics 46:759–771.

Peterson, P.A. 1986. Mobile elements in maize: A force in evolutionary and plant breeding processes. Stadler Genet. Symp. 17:47–78.

Peterson, P.A. 1987. Mobile elements in plants. CRC Crit. Rev. Plant Sci. 6:105–208.

Peterson, P.A. 1994. Transposable elements in plants. p. 363–373. *In* C.J. Arntzen (ed.) Encyclopedia of Agricultural Science. Academic Press, New York.

Peterson, P.A., and P.D. Friedemann. 1983. The Ubiquitous controlling-element system and its distribution in assorted maize testers. Maydica XXVIII:213–249.

Peterson, P.A. and F. Salamini. 1986. Distribution of active transposable elements among important corn breeding populations. Maydica XXXI:163–172.

Robbins, T.P., E.L. Walker, J.L. Kermicle, M. Alleman, and S.L. Dellaporta. 1991. Meiotic instability of the *R-r* complex arising from displaced intragenic exchange and intrachromosomal rearrangement. Genetics 129:271–283.

Russell, W.A. 1974. Comparative performance for maize hybrids representing different eras of maize breeding. p. 81–101. *In* D. Wilkinson (ed.) Proc. 29th Annu. Corn Sorghum Res. Conf. Am. Seed Trade Assoc., Washington, DC.

Russell, W.A. 1985. Comparisons of the hybrid performance of maize inbred lines developed from the original and improved cycles of BSSS. Maydica XXIX:375–390.

Russell, W.A. 1991. Genetic improvement of maize yields. Adv. Agron. 46:245–298.

Russell, W.A., G.F. Sprague, and L.H. Penny. 1963. Mutations affecting quantitative characters in long-time inbred lines of maize. Crop Sci. 3:175–178.

Sachs, M.M., W.J. Peacock, E.S. Dennis, and W.L. Gerlach. 1983. Maize Ac/Ds controlling elements: A molecular viewpoint. Maydica XXVIII:289–302.

Saedler, H., and P. Nevers. 1985. Transposition in plants: A molecular model. EMBO J. 4:585–590.

Salamini, F. and C. Lorenzoni. 1970. Genetical analysis of glossy mutants of maize. III. Intracistron recombination and high negative interference at the gl_1 locus. Mol. Gen. Genetics 108:225–232.

San Miguel, P., A. Tikhonov, Y-K. Jin, N. Motchoulskaia, D. Zakharov, A. Melake-Berhan, P.S. Springer, K.J. Edwards, M. Lee, Z. Avramova, and J.L. Bennetzen. 1996. Nested retrotransposons in the intergenic regions of the maize genome. Science(Washington, DC) 274:765–768.

Saveson, C.J., and S.T. Lovett. 1997. Enhanced deletion formation by aberrant DNA replication in *Escherichia coli*. Genetics 146:457–470.

Scheffler, B., P. Franken, E. Schutt, A. Schrell, H. Saedler, and U. Wienand. 1994. Molecular analysis of *C1* alleles in *Zea mays* defines regions involved in the expression of this regulatory gene. Molec. Gen. Genet. 247:40–48.

Schwarz-Sommer, Zs., A. Gierl, H. Cuypers, P.A. Peterson, and H. Saedler. 1985. Plant transposable elements generate the DNA sequence diversity needed in evolution. EMBO J. 4:591–597.

Seo, B-S., and P.A. Peterson. 1995. A transposable element in diverse corn lines, *Ubiquitous* (*Uq*): Allelism test. Theor. Appl. Genet. 90:1188–1197.

Sommer, H., U. Bonas, and H. Saedler. 1988. Transposon-induced alterations in the promoter region affect transcription of the chalcone synthase gene of Antirrhinum majus. Molec. Gen. Genet. 211:49–55.

Sprague, G.F. 1946. Early testing of inbred lines of corn. J. Am. Soc. Agron. 38:108–117.

Sprague, G.F., W.A. Russell, and L.H. Penny. 1960. Mutations affecting quantitative traits in the selfed progeny of doubled monoploid maize stocks. Genetics 45:855–866.

Stadler, L.J. 1948. Spontaneous mutation at the *R* locus in maize. I. Race differences in mutation race. Am. Nat. 82:289–314.

Stadler, L.J., and M. Emmerling. 1956. Relation of unequal crossing over to the interdependece of the *R-r* elements (P) and (S). Genetics 41:124–137.

Stuber, C.W. 1997. 25 years of searching and manipulating QTLs in maize in genetics, biotechnology and breeding of maize and sorghum. *In* A. S. Tsaftaris (ed.) Royal Society of Chemistry Information Services. Cambridge, England.

Sughroue, J.R., and T.R. Rocheford. 1994. Restriction fragment length polymorphism differences among Illinois long-term selection oil strains. Theor. Appl. Genet. 87:916–924.

Timmermans, M.C.P., O.P. Das, and J. Messing. 1996. Characterization of a meiotic crossover in maize identified by a restriction fragment length polymorphism-based method. Genetics 143:1771–1783.

Timmermans, M.C.P., O.P. Das, J.M. Bradeen, and J. Messing. 1997. Region-specific *cis-* and *trans*-acting factors contribute to genetic variability in meiotic recombination in maize. Genetics 146:1101–1113,

Walker, E.L., W.B. Eggleston, D. Demopulos, J. Kermicle, and S.L. Dellaporta. 1997. Insertions of a novel class of transposable elements with a strong target site preference at the *r* locus of maize. Genetics 146:681–693.

Wessler, S.R. 1996. Plant retrotransposons: Turned on by stress-induced. Current Biol. 6:1–5.

Wessler, S.R., G. Baran, M. Varagona, and S. Dellaporta. 1986. Excision of *Ds* produces *waxy* proteins with a range of enzymatic activities. EMBO J. 5:2427–2432.

Wisman, E., G.H. Cardon, P. Fransz, and H. Saedler. 1998. The behaviour of the autonomous maize transposable element *En/Spm* in *Arabidopsis* allows efficient mutagenesis. Plant Molec. Biol. 37:989–999.

Xu, X., A-P. Hsai, L. Zhang, B.J. Nikolau, and P.S. Schnable. 1995. Meiotic recombination break points resolve at high rates at the 5' end of a maize coding sequence. The Plant Cell 7:2151–2161.

Chapter 15

Genetic Diversity and Heterosis

Discussion Session

QUESTIONS FOR A.E. MELCHINGER

T. Hohls, North Carolina State University, USA: Do you have an explanation for why the correlation between genetic distance (GD) measures and hybrid performance is so low with inter-group crosses? Could it be that the GD measures proposed by Nei and Li (1978) and Rogers (1972) are inaccurate?

Response: Looking at studies in maize on the relationship between mid-parent heterosis (MPH) with GD based on isozymes, RFLPs, or AFLPs, there is no indication that increasing the number of polymorphic markers used for calculation GD measures improves the correlation. This suggests that inaccuracy in the GD measures is not a major cause for the low correlations with MPH in inter-group hybrids, and increasing marker density does not necessarily improve predictive power. This is confirmed by quantitative-genetic theory (see Charcosset & Essioux, 1994, cited in my chapter) showing that the covariance between specific genetic distance and SCA is a sum of positive terms for intra-group hybrids. However, with inter-group hybrids there can be positive and negative terms, which may cancel each other, resulting in a low correlation r(GD, MPH). Therefore, chances to predict MPH in inter-group hybrids could only be improved if markers tightly linked to QTL affecting heterosis of a trait are known.

A. Sierra, ICAMEX, Mexico: Would you explain why you do not use high parent heterosis (HPH)?

Response: According to quantitative-genetic theory, HPH is a function of both additive and dominance effects, whereas mid-parent heterosis (MPH) simply is a linear function of dominance effects. For this reason, I see no rational basis for predicting HPH from GD measures, while in the case of MPH, Eq. [3] provides such a basis.

A. Charcosset, INRA, France: Markers provide an efficient tool to discard crosses between related inbreds. Given this situation, are heterotic groups still necessary?

Response: I agree that DNA markers are a powerful tool to avoid crosses between related inbreds, which are automatically avoided with inter-group crosses; however, as mentioned in my lecture, inter-group hybrids offer several additional advantages over intra-group hybrids that are not offset with DNA markers. This makes me believe that heterotic groups will still have a firm place in hybrid breeding in the future.

Sun Qi Xin, Beijing Agricultural University, China, and B. Cukadar-Olmedo, CIMMYT, Mexico: For crops with heterotic patterns unavailable, such as wheat, how reliable is it to use molecular markers to establish heterotic patterns?

Response: When starting a new hybrid breeding program in a crop such as wheat, genetic relationships among breeding lines that might be taken as base materials are often unknown. In this situation, DNA markers can be of great help to recognize groups of genetically similar and dissimilar materials; however, the heterotic response among these groups cannot usually be predicted from GD measures, but rather they must be determined by producing and evaluating testcrosses among groups in field trials. Based on the heterotic response and genetic similarities determined by DNA markers, some of these groups could subsequently be merged to form heterotic groups for future use in hybrid breeding. With regard to wheat, I would be optimistic that DNA markers such as AFLPs, SSRs or sequence-tagged sites (STS) markers allow a highly reliable grouping of the existing germplasm that might be useful for hybrid breeding.

E. Redoña, Philippine Rice Research Institute, Philippines: What type of genetic distance is best for each type of molecular marker? Are all equally effective?

Response: For marker systems such as isozymes, RFLPs or SSRs, where each band usually corresponds to an allele of a marker locus inherited in a Mendelian fashion, one can choose among several distance measures, most of which require that allele frequencies be determined at individual marker loci. In this ideal case, I would recommend choosing the GD measure based on (i) desirable mathematical and genetic properties (see Melchinger, 1993) and (ii) the intended use of the GD measure. For example for predicting MPH, we prefer in our lab to use MRD^2 due to the expected linear relationship with MPH (see Eq. [3]). For AFLPs and RAPDs, with a large number of bands whose allelic relationships are unknown, one can basically only choose between GD measures that count the number of matching and mismatching bands. Use of these distance measures (Nei & Li, Jaccard) can be problematic with heterozygous individuals, large differences in the number of bands per individual, or missing marker bands.

D. Hess, CIMMYT, Mexico: What is the maximum number of heterotic groups that a maize breeder can use efficiently?

Response: My recommendation would be to work with a very small number of heterotic groups for a given target region, ideally two or three. By focusing resources on breeding efforts in few heterotic groups, it is possible to achieve a high selection intensity and guarantee a high rate of progress in these heterotic groups. With a greater number of heterotic groups, the question arises, which pairs could/should be used as complementary heterotic patterns. It might be of interest to remember that 20 years ago, E.J. Wellhausen (1978) recommended that tropical maize breeders should focus their efforts on two broad-based heterotic groups: (i) a dent composite and (ii) a flint composite. Within these two main groups, three subpopulations of early, medium and late maturity should be formed, but obviously there could be a flow of germplasm among these subpopulations within each heterotic group. Based on my experience, I fully agree with this proposal.

F. Sanchez, Chapingo University, Mexico: If there is a well-established heterotic pattern, wouldn't one parent be the best tester of the other parent and vice versa?

Response: I agree that with a well-established heterotic pattern, lines from one heterotic group should be tested for combining ability in combination with tester(s) (population, single crosses, elite inbred(s)) from the other heterotic group and vice versa. The choice of the most appropriate tester becomes relevant, if no established heterotic pattern is available or lines are of mixed origin (e.g., developed from commercial hybrids). In these cases, DNA markers could help to choose testers with a large genetic distance that are most likely not related or least related to the line to be tested.

David Beck, CIMMYT, Mexico: Can you expand on your discussion of comparisons between the ratio of $\sigma^2_{SCA} : \sigma^2_{GCA}$ by Dhillon et al. (1990) and your subsequent analysis. Did you expect to see less σ^2_{SCA} in flint × dent crosses? Why was this so?

Response: My review of the literature (see Table 10–3) shows that intra-group crosses have generally lower hybrid performance and mid-parent heterosis than inter-group crosses. Therefore, in a complete diallel of flint and dent lines as analyzed by Dhillon et al. (1990), flint × flint and dent × dent crosses have mostly large negative SCA effects, whereas flint × dent crosses generally have large positive SCA effects, i.e., we have an inflation of σ^2_{SCA} due to group effects. In contrast, when we restrict the analysis to the factorial of flint × dent crosses, some of them have positive and some negative SCA effects, because their sum is equal to zero. Consequently, the SCA effects and σ^2_{SCA} are much smaller in magnitude than in the diallel case. This experimental result is consistent with quantitative-genetic theory showing that under an additive-dominance model, σ^2_{SCA} is always smaller in the hybrid population $\pi1 \times \pi2$ than in $(\pi1 \cup \pi2)^r$.

B.S. Dhillon, Punjab Agricultural University, India: How can the concept of heterotic pools be useful in breeding of OPVs? Crossing two very diverse types will need a lot of resources and time for the development of reasonably uniform OPVs.

Response: My point is that heterotic groups are advantageous in hybrid breeding, but when OPVs are desired in addition to hybrids (as applies to many developing countries), the two populations could be combined and the resulting progeny further improved through recurrent selection for development of OPVs. The latter proposal (suggested by E. J.Wellhausen, cited above, to CIMMYT maize breeders) exploits half of the panmictic midparent heterosis present in hybrids. Uniformity could be a problem for OPVs, but this depends on the morphological diversity of the two heterotic groups and the level of heterogeneity accepted by farmers and authorities for release of varieties.

QUESTIONS FOR J.S.C. SMITH

J. Janick, Purdue University, USA: The pie charts of genetic contributions by pedigree to hybrids show increasing diversity with time, but the decreasing genetic distances between inbreds and between hybrids show decreasing diversity in time. Please explain.

Response: The pie charts show new diversity entering into the U.S. Corn Belt mostly from other regions in the USA but which are exotic to the U.S. Corn Belt. Maiz Amargo is one example of new diversity coming recently from outside the USA. Other factors are at play that affect overall diversity between inbreds and hybrids each decade. First, there is an appreciable component (approximately 50%) of the germplasm that is Reid Yellow Dent and so has now established a 60 year period of continued sourcing and breeding of related lines. Second, germplasm that is exotic to the US Corn Belt itself requires at least two to three decades of pedigree breeding and enhancement with adapted U.S. Corn Belt germplasm before it can make a useful contribution to the most widely grown U.S. Corn Belt hybrids. So pedigree connections must be made with this germplasm before it can be useful new diversity to the U.S. Corn Belt.

S. Taba, CIMMYT, Mexico: You concluded your talk with mention of the importance of maize genetic resource conservation. What is the strategic work that needs to be done regarding conservation for the future?

Response: A broad and representative range of genetic diversity needs to be properly conserved, evaluated, and documented. Genetic resources need to be made accessible, enhanced, and re-evaluated. Technological advances are needed to help us make more effective use of a broad array of genetic diversity. These technological developments are now beginning to be available. Breeders who fail to make use of new technologies and knowledge to breed new improved varieties using a broader base of genetic diversity will be failing to recoup much of these investments they have made into those technologies. We need to use success stories in the use of plant genetic resource diversity to help advocate for the continuing conservation and evaluation of these resources. I will briefly outline a seven point plan for the improved conservation and use of plant genetic resources:

1. Show success such as the results provided by Major Goodman and by the LAMP and GEM projects.
2. Need advocates and visionaries such as Bill Brown, Tim Reeves, Norman Borlaug and Ken Frey who is proposing a germplasm enhancement program for the USA.
3. Need to educate Chief Executive Officers about the importance of conserving and making more accessible plant genetic resource diversity.
4. Need to educate the public: a public that in industrially developed countries has increasingly benefited from improved use of plant genetic resources but which is increasingly intellectually and geographically isolated from the land.
5. Need to educate children who in the very near future will be tax payers, stock holders, and voters so that they are appreciative of the need to invest for the long term, in basic research and conservation of plant genetic resources.
6. Need to raise awareness that exotic plant genetic resource diversity can help improve agricultural yields but that long term work is required to evaluate and enhance the germplasm and this requires significant amounts of public and of private investments. We need to build a basis of support for investing in future generations.

And, if we are not successful in achieving these ends the last item of the plan will exert itself:

7. Despoliation of the environment, famine, malnutrition and rising food prices. These events will certainly raise the investments made into research and conservation but only after a failure to garner huge returns that do accrue from investing in agriculture, unnecessary suffering especially of the most disadvantaged in society, the irreparable loss of genetic resources and the spectre of international strife fueled by poverty and by actual and perceived inequities.

QUESTIONS FOR J.A. LABATE

H. Gevers, Quality Seed Co., South Africa: Were the parents at each cycle always selected strictly at random? If not, would you expect a different response?

Response: No, I don't believe selection of the tested lines is completely at random. There is some phenotypic selection for favorable-appearing plants at that stage. Ideally, we would like to sample the entire range of alleles available within a population for testing in the interpopulation cross, and selection at this stage may be prohibiting this. It could be limiting the response, but we really don't know.

F. Márquez-Sánchez, Chapingo University, Mexico: What would have happened if instead of doing RRS, the populations BSSS and BSCB1 had been improved separately and then crossed at cycle 12?

Response: I would predict that the interpopulation cross would not have improved as much as it did because the selection pressure would have been different. There's no way of knowing what would really happen without doing the experiment.

D. Borchardt, KWS, Germany: Could you link the map position of the nonneutral loci to any QTL for the respective traits under selection (as assessed in other experiments)?

Response: It's not easy to do this yet because we don't have enough resolution in our study to infer that delimited chromosomal regions have been selected. Also, several traits have been selected simultaneously, and this further limits our ability to associate genotypic regions with phenotype.

C. Aekatasanawan, Kasetsart University, Thailand: Does the population cross of cycle 12 × cycle 12 give the maximum yield and maximum heterosis?

Response: From the data I showed, it looked like there might have been a leveling off of heterosis by cycle 10. We don't know the answer to this question because the subsequent cycles haven't been evaluated.

M. Carena, Iowa State University, USA: Can you give your opinion about genetic diversity? Do you believe that different geographic origins of populations will predict heterosis?

Response: In my opinion, genetic variation due to different geographic origins will not imply that the different populations will act well as complementary heterotic groups.

A. Gallais, INRA, France: What type of RRS have you used?

Response: Initially half-sib, and then full-sib starting at cycle 10.

A. Cardinal, Iowa State University, USA: You showed in the cluster analysis that cycle 0 differed from the progenitors. What is your explanation? What implications does this have for population conservation or germplasm management?

Response: The cycle 0 populations are genetically different from the progenitors, a lot of this difference can be explained by the loss of rare alleles in the cycle 0 populations after being maintained for several decades. Possibly the population sizes during maintenance haven't been large enough.

P. Peterson, Iowa State University, USA: What happened to the locus bnl771 in BSCB1(R) (an allele at high frequency in cycle 0 that went to low frequency in cycle 12)?

Response: This type of change occurred at several loci. It implies that another allele at the locus reached high frequency, but sometimes not enough for Waples' test to be significant. Locus umc84 in BSSS(R) is another example.

J. Brewbaker, University of Hawaii, USA: How would you use molecular markers to reduce the amount of work required for RRS?

Response: I don't think molecular markers are going to provide shortcuts in the selection program right now in terms of marker-assisted selection. They have been useful, e.g., for estimating effective population size and measuring loss of rare alleles. Eventually they may help in decisions made about optimal effective population size in RRS.

QUESTIONS FOR M. M. GOODMAN

R. Bertram, USAID, USA: Can you comment a bit more on the actual or potential value of temperate-adapted materials in maize breeding for the tropics and subtropics?

Response: The temperate-adapted, tropical lines from North Carolina are finding use in Mexico, Central America, India, and China. At least a part of the reason for their wide adaptation is greatly reduced photoperiod response.

W. Trevisan, Cargill International Seeds, Brazil: If you could restart a GEM program today, would you start with CIMMYT accessions or tropical/subtropical elite hybrids or lines?

Response: My choice would be to begin with elite, tropical hybrids or lines. It is generally easier to start with hybrids (i) because they are available, and (ii) data are available. The subtropical materials I have tried to use at North Carolina State University (except for the Florida Synthetic) have been rather disappointing.

S. Vasal, CIMMYT, Thailand: Your comment that hybrid data from CIMMYT is not available is not true. Since 1994 CIMMYT has started international hybrid trials and this type of data is available on request.

Response: A great deal of data is available, but very little of it that I have been able to locate is directly connected to CML release lines, with the exception of two *Maydica* papers from several years ago. This isn't of much consequence to those in the tropics, who can and should generate their own topcrosses and yield trials, but it is a major roadblock for use in temperate regions, where such trials are not feasible and the results from working with a specific tropical line will not be available until an investment of about 15 years of breeding work is made. If such data exist, I, and many others who also have had difficulty obtaining such data, would very much like to see them.

M. Relende, Agroceres, Brazil: Does a greater harvest index come along with selection for earliness in tropical materials or does the harvest index remain unchanged when you adapt tropical materials?

Response: Generally speaking, there is a large increase in harvest index with selection for earliness. Perhaps the prime example of that is Elmer Johnson's Tuxpanito.

N. Singh, Directorate of Maize Research, India: Introgression of tropical materials has been done very systematically to improve disease and pest resistance of temperate germplasm, but vice-versa has not been done to improve tropical germplasm. How can field-based selection be conducted to maximize any positive effect available within a specific, variable, exotic accession or hybrid?

Response: Early topcross testing of F_1S_1, F_2S_1, or BC_1S_1 families appears to be the best way to identify promising exotic and temperate combinations. S.K. Kim at IITA certainly demonstrated that this could be done, and private programs, such as Cargill, DeKalb, and Pioneer now have successful hybrids that are based in part on tropical lines introgressed with temperate germplasm.

G. Edmeades, CIMMYT, Mexico: You referred to photoperiod sensitivity as largely cosmetic and easily dealt with. What time-efficient method of eliminating photoperiod sensitivity do you recommend?

Response: The answer depends upon the goals at hand. If the purpose is to use a tropical line's germplasm as a portion of a breeding project, then F_2, F_3, and F_4 segregation from crosses with elite, temperate lines will quickly eliminate photoperiod-sensitive segregates, when these are grown under long days. We regularly do this at North Carolina State University, most often using all-tropical lines like NC296, NC298, or NC300 as the insensitive, temperate source. Thus, we currently have a large series of F_5 and F_6 lines from crosses of the earlier releases of CML lines with NC tropical lines. These are now reasonably adapted to Raleigh, NC, and after two to three additional generations for screening and selfing (usually until the F_6 we do plant-to-plant mating within unadapted crosses), they will go into topcrosses for yield-trial evaluations.

If one wishes to convert a specific population (a line, hybrid, accession, synthetic or family), then a backcross scheme using selected F_2 plants for backcrossing is needed, like the well-known sorghum conversion project. While one could do a QTL project to localize the loci responsible for photoperiod and use laboratory-assisted selection for this, that approach seems a bit academic —something that should be done, but not something that would routinely be used. Conversion takes time, under even ideal circumstances, and working with crosses or backcrosses instead provides the opportunity to not only reduce photoperiod sensitivity but also to make breeding progress, while conversion pretty much retains the status quo. To justify photoperiod conversion of a line, the line would need to be truly superior to justify the loss of breeding progress over the several-year conversion process. The same caveat applies to such genetic engineering wonders as *Bt*, glyphosphate-resistance, glufosinate-resistance, and virus resistance.

A. Ismail, Agriculture Research Center, Egypt: Temperate materials, including Mo17 and B73, do not perform well in Egypt. Combinations of inbreds of subtropical origin from CIMMYT with inbreds isolated from temperate material are the best.

Response: I certainly would not recommend direct use of typical U.S. lines, whether they be B73, FR1064, or LH195, for production use in the tropics or semi-tropics, but Mo17 crosses with ETO materials, for example, do often result in useful, well-adapted lines containing 50% (by pedigree) Mo17, and these can be used in tropical/semitropical single-crosses, three-way, or four-way crosses. Narrow-based synthetics, such as a synthetic of B73-derived lines, are often a better choice for initiating such work, rather than placing all bets on a single B73-type line.

Valdes, Universidad Autonimo de Nuevo Leon, Mexico: (i) What are the perspectives for highland germplasm in future hybrid breeding in USA? (ii) How high is heterosis for yield in tropical $\times$ U.S. crosses compared with U.S. $\times$ U.S. crosses; are there gains for traits other than disease- and insect-resistance?

Response: (i) Highland tropical germplasm has thus far been more useful in the extreme north of the USA and in Canada than in the U.S. Corn Belt, and that is likely to continue to be the case. The utility is largely for good germination, early seedling vigor, and common rust resistance. (ii) The specific combining abilities (for yield) of tropical $\times$ temperate crosses tend to be quite high compared with that of temperate $\times$ temperate crosses.

B. Bowen, Pioneer Hi-Bred, International, USA: What potential does Steve Tanksley's Advanced Backcrossing Strategy have for introgressing advantageous alleles from tropical germplasm into U.S. Corn Belt material?

Response: There is little doubt that advantageous alleles exist in even the most unpromising sources, and these can be identified by relatively unsophisticated, but currently fairly costly, means. I suspect some of this will be done, if only for academic reasons. Still, if biotechnological tools are to be most wisely used, the most agronomically promising exotics will be probed first, very much in the manner of conventional searching for disease resistance: first check elite lines; second check older and not-so-elite lines; third check foreign elite lines, then follow the same sequence for synthetics. If necessary, go to germplasm accessions, and, if all else fails, try related, cross-compatible species.

What is not clear to me is when molecular techniques are an adequate substitute for certain field-selection procedures, particularly in the case of multiple-trait selection or traits where epistasis is important. Single-locus, non-epistatic traits are, in themselves, rather novel to most practicing maize breeders, but are, thus far, the strongest domain for molecular genetics.

S. Pandey, CIMMYT, Mexico: Since there are many participants here from the tropics, please comment on the possibilities and usefulness of using temperate germplasm in the tropics.

Response: The potential value of temperate material in the tropics is well illustrated by private breeding programs and by the program initiated by S. K. Kim at IITA. Very few new, privately-developed, tropical hybrids from the multi-nationals, at least, lack some temperate germplasm. Kim was the first to demonstrate that this can be done very quickly in a public program, but Jim Brewbaker in Hawaii has been active in this area for years. The major temperate contributions are usually root strength, yield potential, and additional heterosis.

QUESTIONS FOR P.A. PETERSON

E. Redona, Philippines Rice Research Institute: Are transposons also involved in the generation of novel variation in other crops such as rice, or is the variation generated more through mutation and recombination, etc.?

Response: First, it is not definitely proven that transposons are the basis of variation in maize. Yes, changes are caused by transposons, but whether the significant variation is needs more research. Yes, transposon related MITS such as *Ditto-Os2* in rice appear to relate to a TATA box (Bureau et al., 1996).

Qi Xin Sun, Beijing Agricultural University, China: Is the shut-down of the gene mechanism related to heterosis?

Response: The reduction of the tassel size is an example. To save the starch from the excess pollen of the large tassel, a smaller tassel could shift resources to greater yield. But this does not explain the heterotic effect.

F. Troyer, USA: What is the selection advantage of repeated DNA? And why does it exist?

Response: Basically, in nature's processes it is "Use it or lose it". It is very abundant, and even though it is not coding, it is pervasive in the genome. With the current unraveling of gene sequences, some repeated DNA may be found to be related to critical gene components (see Bureau, previously cited).

B. Bowen, Pioneer Hi-Bred, International, USA: Why would you restrict your sequencing proposal to genes involved in secondary metabolism?

Response: If my resources were limited, I would start there. Further, secondary metabolism would provide more options. See for instance, the five to six genes leading to DIMBOA.

M. Lee, Iowa State University, USA: Are there active transposons in rice? I remember observations at the rice-blast disease resistance gene Xa21 with *Ac* sequences in rice.

Response: Yes, there are transposon sequences in rice, and they are numerous (cited previously, Bureau et al.). But they are not in coding sequences, and further, none have been found that are active. *Ac* like sequences have been found in a number of crops.

J. Brewbaker, University of Hawaii, USA: If an inbred-based synthetic like BSSS has only two transposons, do you expect much higher numbers in open-pollinated populations (like Tuxpeno, etc.)?

Response: The two transposons in BSSS are active. There are silent copies of a number of others. Remember that BSSS has a narrow base (16 inbreds), and the two active transposons entered the population via the introduced inbreds. Yes, when the resources are available, I will test those other populations for active transposons. And you are correct, populations such as corn have active transposons.

Chapter 16

Biochemistry, Molecular Biology, and Physiology of Heterosis

C. W. Stuber

INTRODUCTION

Heterosis (or hybrid vigor) is manifested in hundreds of millions of hectares of field and vegetable crops throughout the world. This phenomenon is one of the primary reasons for the success of the commercial hybrid maize (*Zea mays L.*) industry as well as the success of plant breeding endeavors in many other crop and horticultural plants; however, the causal factors at the physiological, biochemical, and molecular levels are today almost as obscure as they were at the time of the conference on heterosis held in 1952 (Stuber, 1994).

Although the effects of the heterosis phenomenon have been quantified in a wide variety of plant studies, the underlying genetic basis has not been satisfactorily explained despite many attempts to do so. The genetic theories for heterosis (which have changed little since 1952) include: (i) dominance, including linked dominant favorable factors; (ii) true overdominance, which is nearly impossible to distinguish from pseudo-overdominance (i.e., closely linked loci at which alleles having dominant or partially dominant advantageous effects are in repulsion phase linkage); and (iii) certain types of epistasis. Elucidation of these theories remains a major challenge. Biometrical approaches can only evaluate average (or net) genetic effects on heterosis; however, studies that have used biochemical, physiological, or molecular approaches may provide some limited insights into a better understanding of this phenomenon. Even though the biochemical and/or physiological responses reported may only represent manifestations of heterosis at a level other than at the overall mature plant stage (as measured in the field), these responses should be closer to the gene level and may ultimately help to elucidate the genetic basis of heterosis. A better understanding of the mechanisms underlying this phenomenon should enhance the ability of plant breeders to generate new genotypes that may be used directly as hybrids or may form the basis for future selection programs.

BIOCHEMICAL, MOLECULAR, AND PHYSIOLOGICAL EVIDENCE

DNA Methylation

Hepburn et al. (1987) studied DNA methylation in plants and reported several examples in the literature that provided evidence for a relationship between methylation and suppression of gene activity. Extrapolations of these observations (A. Hepburn, personal communication) might implicate the degree of DNA methylation as a molecular regulator of heterosis. This hypothesis provides for a gradual accumulation of methylation during selfing, which is then released and/or repatterned when the selfed lines are crossed to generate hybrids. R.L. Phillips presented a model for heterosis (that remains to be tested) invoking a relationship between gene expression and methylation in which methylation might turn on genes previously inactive, thus leading to creation of recessive and dominant genes, respectively (presented at the Workshop on Molecular Basis of Heterosis in Plants, March 19–20, 1991, University of Minnesota).

Tsaftaris et al. (1997) also implicated cytosine methylation in genomic DNA as a mechanism for controlling gene regulation in maize. They reported that DNA methylation was found to be genotype, tissue, and developmental stage specific. They also found that hybrids were less methylated than inbreds, which is consistent with the hypothesis proposed by Hepburn (see previous paragraph). In addition, their data showed that inbreds varied in their methylation status and improved lines were less methylated than older low yielding lines. Tsaftaris et al. (1997) also reported that stressful environments (such as high density planting) increased the level of methylation, which, in turn, suppressed genetic expression throughout the genome. F_1 hybrids, however, were found to be more resistant to such a density induced increase in methylation than their parental inbreds. They suggested that an epigenetic DNA modification such as this (which may alter gene expression) could be heritable because such a change can occur in a plant cell that may become a gamete.

Molecular Heterosis

Although rare, examples of heterosis associated with a single genetic locus have been reported. For example, Schwartz and Laughner (1969) have demonstrated single gene heterosis for alcohol dehydrogenase activity in maize. For maize glutamic dehydrogenase, Pryor (1974) found a single molecular form in F_1 hybrids. More recently, Hall and Wills (1987) reported an example of conditional overdominance at an alcohol dehydrogenase locus in yeast. They suggested that their finding could be considered equivalent to that of sickle cell hemoglobin in man and shows promise as a tool for investigating the physiological basis for overdominance.

These studies demonstrate a hybrid response at the individual enzyme level; however, the overall plant response involves the interactions of many enzymes associated with numerous genetic loci. For example, Hageman et al. (1967) made a thorough study of the role of nitrate reductase in the heterotic response of maize hybrids and found considerable variation in the activity of this enzyme among the inbreds tested. In spite of the wide range in activity, they were unable to causally associate nitrate reductase activity with heterosis. Results such as these indicate the difficulty in the dissection of the individual components of heterosis, which, undoubtedly, form a complex interactive system.

Srivastava (1981) strongly suggests that heterotic hybrids are endowed with a more balanced metabolism than the inbred parents. He also indicates that the metabolic advantage of the gene products at the cellular or organelle level could be viewed as being produced by nonallelic complementation rather than by intergenic complementation pertaining to only one gene locus. Demonstrations of multimeric *hybrid enzymes* or *isozyme spectrum* due to nonallelic or intergenomic interaction provides evidence for a buffering mechanism at the subcellular level in the hybrids. As stated by deVienne and Rodolphe (1985), protein complementation (or interallelic complementation) represents an extreme expression of molecular interactions since only the heteropolymeric form is active; however, the complementation is not always positive. The presence in an oligomer of both mutant and wild (normal) protomers can result in complete inactivation of the enzyme. Molecular interactions between subunits that are all active have been reported in the literature. For example, Scandalios et al. (1972) studied maize catalase, which is a tetrameric enzyme, and showed that different combinations of alleles from two loci resulted in isozymes which differed in biochemical properties, such as specific activity, temperature sensitivity, photosensitivity, and inhibitor sensitivity. In most instances, heterotetramers generated by either intragenic or intergenic hybridization exhibit clear interaction phenomena, sometimes including molecular heterosis. It is of interest to note, however, that electrophoretic mobility of isoenzymes generally shows additivity, that is, a heterodimer usually migrates exactly between the two homodimeric forms. Tsaftaris (1995) points out that molecular interactions among subunits of polymeric proteins occurs not only for enzymatic proteins, but also (and more importantly) it occurs for transcription factor proteins and regulatory proteins, in general.

Leonardi et al. (1991) correlated protein-amount polymorphisms with field performances of ten single-cross maize hybrids. They concluded that genes controlling protein amounts, particularly those with multiple effects, directly affected the expression of hybrid vigor. In their discussion, they suggested that the greater two parental genotypes differ in their profile of gene expression during development, the greater would be the possibilities of interactions at the molecular level. These interactions would be expressed as nonadditive, and heterosis could be considered as a measure of these interactions. In addition, it was suggested that protein-amount polymorphisms, as well as the frequency of nonadditivity, might allow for the definition of predictors of heterosis based on molecular indicators.

Metabolic Balance and Physiological Bottlenecks

Some researchers have approached the biochemical–physiological explanations of heterosis from a more complex view. Hageman et al. (1967) proposed the metabolic balance concept as a determinant of the hybrid response and suggested that traits such as plant growth and grain yield reflect the end results of a series of biochemical reactions, each of which is governed by one or more specific enzymes. According to Hageman and Lambert (1988) and Schrader (1985), the metabolic concept of heterosis is dependent upon the coordination of all reactions and systems for efficient growth under a given environment. The elements of this concept are: (i) inbred lines in a crop such as maize tend to have unbalanced metabolic systems with some enzymes controlled at high levels, some at medium levels, and some genetically limited to low or ineffective levels of activity, (ii) highly homozygous lines of maize may have some important enzyme reactions that are severely limiting metabolism, and (iii) the specific limitations probably differ among individual lines, which may be overcome in a heterozygote by the appropriate choice of complementary inbred parents, resulting in a better-balanced metabolic system in the resulting hybrid.

The metabolic balance concept is analogous to the concept of limiting factors or physiological bottlenecks (as outlined by Mangelsdorf, 1952), in which the bottlenecks may reside at different loci (and hence in different metabolic pathways) in different inbreds. Schrader (1985) outlined a simplified model system for two inbreds (I and II) that differ for units of enzyme activity for two of three metabolic pathways: A, B, and C (see the three columns under "Relative enzyme units" in Table 16–1). With Schrader's model, the hybrid will exhibit intermediate enzyme activities and will have a more balanced metabolism than either of the inbred parents. A limitation or bottleneck in pathway A in inbred I, and a limitation or bottleneck in pathway B in inbred II are overcome in the F_1 hybrid (Rhodes et al., 1992). I have added three columns under *Relative effect* to Schrader's model (Table 16–1) showing hypothetical results when the combined effects of these three enzyme pathways are either: (i) constrained by the pathway with the lowest *Relative enzyme units* under column headed *Lowest*, (ii) functioning in an additive fashion, column headed *Additive*, or (iii) functioning in a multiplicative fashion, column headed *Multiplic.*

It is of interest to note that in the Heterosis Conference held in 1952, Mangelsdorf (1952) suggested several postulates that are relevant to the metabolic concept of heterosis and genotype by environment interactions as follows:

1. At each moment throughout its life, the physiological processes of even the most vigorous organism are limited to their prevailing rates by bottlenecks or limiting factors.

Table 16–1. Hypothetical model for two inbreds and a heterotic F_1.

	Relative enzyme units†			Relative effect		
Genotype	Pathway A	Pathway B	Pathway C	Lowest	Additive	Multiplic
Inbred I	1	3	2	1	6	6
Inbred II	3	1	2	1	6	6
F_1 hybrid	2	2	2	2	6	8

† After Schrader, 1985.

2. The physiological bottleneck at any given moment results from the interaction of a particular locus (referred to as the bottleneck locus) with the remainder of the genotype and the environment at that moment.
3. The physiological bottleneck may be ameliorated, or even removed, by correcting the particular feature of the environment contributing to the bottleneck.
4. The bottleneck may be ameliorated by substituting a more effective allele at the bottleneck locus, provided that such an allele is available.
5. The amelioration, or removal of a bottleneck, by improving the environment or by substituting a better allele at the bottleneck locus, will permit an increase in the rate of the essential physiological process. This increase may be either small or large, depending upon the point at which the next ensuing bottleneck becomes limiting. The substitution of a more efficient allele at a bottleneck locus in a certain genotype under a particular environment may result in a large gain. The substitution of the same allele in a different genotype or under another environment may result in little or no gain. Difficulties would be encountered in analyzing the inheritance of genes affecting yield and other quantitative characters, which are subjected to the influence of a varied and fluctuating array of genetic–environmental bottlenecks.
6. The differences between the weakest inbred and the most vigorous hybrid are merely those of degree. Each represents an integration of the many genetic–environmental bottleneck effects under which each is subjected. The weak inbred has been throttled down by one or more bottlenecks to a low level. The superior hybrid is able to grow much greater, even attaining what might be termed extreme vigor. However, both the weak inbred and the vigorous hybrid have throughout their lives been restricted to their respective levels by their genetic–environmental bottlenecks.
7. Success in developing higher-yielding genotypes depends largely upon the ability of the breeder to substitute more effective alleles at the bottleneck loci and to accomplish this without establishing new and equally serious bottlenecks at other loci.

Rhodes et al. (1992) state that an interesting feature of this concept postulated by Mangelsdorf (1952) is that it predicts that for each inbred there is likely to be at least one major bottleneck locus and potentially a series of less serious (minor) bottleneck loci, the latter of which will manifest themselves only when the bottleneck at the major locus is ameliorated. For a hybrid to exhibit heterosis, it is likely that the two inbred parents must have bottlenecks at different loci, either in different metabolic pathways or at different steps in the same metabolic pathway. Thus analyses of segregating populations generated from a vigorous hybrid would be expected to reveal at least two major independently segregating loci determining productivity, with perhaps a series of minor loci contributing relatively smaller effects. This concept is not inconsistent with the results reported by Stuber et al. (1992) in the cross of B73 x Mo17. If the challenge to the breeder is to identify, map, and then manipulate these potential bottleneck loci in each inbred (Mangeldsorf, 1952), then DNA-based marker-facilitated technology should provide the tool to accomplish this objective. Obviously, map location does not necessarily define function. The more difficult challenge to the plant physiologist or molecular biologist is to define the precise imperfections in metabolism that these loci may determine, that is, to equate function to location. The use of markers derived from candidate genes with known functions for quantitative trait locus (QTL) mapping may be helpful in this endeavor. As Rhodes et al. (1992) state, "Only when both location and function have been elucidated will the question of dominance versus overdominance likely be resolved for each locus contributing to heterosis."

The concept of metabolic balance resembles the dominance hypothesis, however, the heterozygote does not necessarily have to exhibit a character (enzyme level) that is equal to (dominance), approaches (partial dominance), or which is greater than (overdominance) either of the parents, in order to explain heterosis. Schrader (1985) reported that the heterozygote usually exhibits intermediate enzyme levels with respect to the inbred parents. Among maize inbreds, two- to three-fold differences in the levels of enzymes, which catalyze important biochemical reactions, are not uncommon. Generally, the observed mean enzyme level of an F_1 maize hybrid closely approximates the mean of the two parents (Schrader, 1985). Thus, the character enzyme level usually shows additivity.

If additivity is the rule rather than the exception for the character enzyme level, does this imply that genetic variability for this character is for the most part irrelevant to heterosis? Kacser and Burns (1981) attempt to resolve this apparent dilemma by considering the relationship between the character enzyme level and the character metabolic flux. They describe flux as a function of environmental parameters, which are assumed to be constant, and a composite of genetically determined enzyme parameters that are proportional to enzyme concentration and are modifiable by mutation. It is so defined that if any one activity is reduced to zero, the flux will be reduced to zero, resulting in a metabolic block. With only a single enzyme contributing to the character metabolic flux, a 50% reduction in activity will reduce the flux by one-half, and the character will show additivity; however, as the number of enzymes of a pathway increases, a 50% reduction in any one enzyme will have a progressively smaller effect on flux. Therefore, the character metabolic flux will show increasing levels of dominance depending upon the number of enzymes in the pathway. This concept reveals that alleles that affect the character enzyme level in a fundamentally additive fashion can effect the character metabolic flux in an additive, partially dominant, or dominant fashion depending on the number of enzymes in the pathway. Although the effects of genes governing traits such as productivity may be additive at the enzyme level, the consequence of nonlinearity between enzyme level and metabolic flux in complex metabolic pathways may, therefore, appear as a dominant effect (Kacser & Burns, 1981).

An obvious implication of these concepts for plant physiologists is that physiological selection criteria (such as metabolic flux measurements) may not necessarily provide a more sensitive indicator of heterosis than net productivity per se because the latter integrates flux over time. Plant breeders have been successful in evaluating and exploiting heterosis by comparing the relative productivity of heterozygotes with homozygotes. This success arises from the fact that productivity is an exponential and time-dependent function of the flux via the prevailing bottleneck in metabolism, wherever this may be located. The poor progress in defining the causal factors of heterosis at the biochemical level stems from the fact that bottlenecks can potentially reside at any one of a great many loci essential for growth, and that the immediate effects on physiological processes of allele substitution at these loci is very difficult to evaluate experimentally (Rhodes et al., 1992).

Complementary Effects of Parents

In a review of heterosis by Sinha and Khanna (1975), the authors indicate that processes such as germination, respiration, and photosynthesis can be split into components that show Mendelian dominance in F_1 hybrids. In heterotic hybrids the parents usually bring together contrasting, but complementary characters that could have multiplicative effects. This then leads to the types of gene interactions that result in complementation of associated physiological and biochemical processes. For example, during germination complementation could occur with regard to water absorption, amylase activity, phytase activity, respiratory activity, and the number of leaf primordia. In photosysthesis, complementation could involve carboxylases and cyclic and noncyclic photophosphorylation. Sinha and Khanna (1975) further suggest that interaction of genes at the processes level, such as those involved in the components of photosynthesis, is complementary and provides only a limited advantage over the parents; however, the end products of complementary effects interact multiplicatively. It is this interaction that eventually makes the F_1 hybrids outstandingly superior to their parents.

In a QTL fine-mapping study using materials generated from the cross of maize inbreds B73 and Mo17, Graham et al. (1997) showed that a major QTL on chromosome 5, that appeared to act in an overdominant fashion, could be partitioned into at least two smaller QTLs. Dominant gene action was observed at each of these two QTLs, with the favorable dominant allele from B73 in one QTL and the favorable dominant allele from Mo17 in the other QTL. Thus, what had appeared to be a single overdominant locus was partitioned into two loci linked in repulsion phase with dominant gene action at each. Further interpretations of the data indicated that the longer ear of Mo17 with more kernels per row positively complemented the shorter B73 ear.

An investigation of a sorghum [Sorghum bicolor (L.) Moensch] cross by Gibson and Schertz (1977) provided an example of a hybrid that displayed the complementary effects of its parents. The hybrid has a high crop growth rate during grain filling (characteristic of its female parent) and a highly effective conversion of dry matter to grain (as did its male parent). In a study of cotton (*Gossypium hirsutum* L.), Wells and Meredith (1986) attempted to better understand physiological alterations related to heterosis by monitoring growth and partitioning of dry matter in four upland cotton cultivars and their F_1 progenies. The heterotic growth pattern observed in this study, although related to larger leaf-area production, did not appear to be the result of greater proportional leaf-area partitioning. Instead, the final increase of leaf area in the hybrids (when compared with the parents) appeared to be the result of early growth responses. In a study of three cultivars (different from the above study) and their F_1 progenies, Wells et al. (1988) examined the relationship between leaf area and canopy photosynthesis. Their data showed that hybrids produced bigger plants that intercepted more light than their respective parents, and hence have increased photo synthetic rates on a per plant basis. They found that during early growth stage apparent photosynthesis (based on CO_2 exchange rates) was associated significant with leaf area per plant; however, later in growth, this association weakened as mutu. shading of the leaves occurred. The conclusions were similar to those above in tha much of the heterotic response appeared to be related to early seedling growth and early leaf area development.

Leaf areas of four maize inbreds (CM7, CM49, CL3, and CG8) and four hybrids (CM49 × CL3, CM49 × CG8, CM7 × CM49, and CM7 × CG8) were analyzed by Pavlikova and Rood (1987) in terms of numbers of leaves and lengths and widths of individual leaf blades. The lengths and widths were further analyzed in terms of numbers, widths, and lengths of long abaxial epidermal cells. Their results showed incomplete dominance for increasing leaf number and overdominance for increasing individual leaf blade area. They concluded that the overdominance for increasing individual leaf area resulted from the complementation of incomplete dominance for increasing cell length and width and overdominance for cell number in length and width. They concluded further that both complementation of dominance effects and overdominance for increased mitotic activity were involved in heterosis for leaf area in the maize hybrids studied. Pavlikova and Rood (1987) also suggested that this type of leaf area analysis may be useful to breeders attempting to increase the leaf area of maize because it reduces the complex, polygenic trait of leaf area into easily measured components that display less complex inheritance patterns.

Complementation at the Organelle Level

Srivastava (1981) emphasized that heterosis in plants often results in enhanced conservation of energy. If this phenomenon is observed in mitochondria and chloroplasts, it could have importance for increasing crop yield through the manipulation of organelle genomes.   The role of mitochondrial heterosis and complementation as an explanation for the basic mechanism of heterosis was first reported by McDaniel and Sarkissian (1966). They reported that mitochondria isolated from seedlings of maize hybrids and their inbred parents showed differential efficiencies of oxidative phosphorylation in the synthesis of ATP. Mitochondria from a nonheterotic hybrid showed the same phosphorylative efficiency as those of its inbred parents. This was interpreted as the classical expression of heterosis, at the organelle level, judged by the efficiency of enzyme catalysis. McDaniel (1986), in a review article, discussed this phenomenon extensively. Attempts to corroborate the mitochondrial complementation approach in other laboratories and in other plant species or to use this phenomenon in plant breeding have met with minimal success.

Chloroplast heterosis and complementation associated with several parameters have been observed in several economically important crops (Srivastava, 1981). The term *chloroplast complementation* is generally used to indicate that greater activity is observed from a 1:1 mixture of isolated parental chloroplasts when compared with the midparental value. Higher photosynthetic rates in isolated chloroplasts of hybrids of several crop species have been observed in the seedling stages. Srivastava (1981) further states that "It is likely that heterotic hybrids are endowed with superior systems of cholorplasts and mitochondria, and such superiority is provided by genomic and intergenomic interactions."

Phytohormones and Heterosis

The role of the phytohormone, gibberellin (GA), has been investigated in several crop plants and in trees with respect to the regulation of hybrid vigor. In a study of four commercially important maize inbreds and their 12 F_1 hybrids, Rood et al. (1988b, 1990) found that the hybrids contained higher concentrations of endogenous GAs than their parental inbreds. Accelerated shoot growth of the inbreds generated by the addition of exogenous GA_3 indicated that a deficiency of endogenous GA limited the growth of the inbreds, which could be a factor in inbreeding depression. Conversely, they suggested that the increased endogenous concentration of GA in the hybrids, relative to the inbreds, could provide a phytohormonal basis for heterosis of shoot growth in maize.

Rood et al. (1992) also studied the association of GA with shoot growth in four inbred lines and two F_1 hybrids of sorghum. The results were similar to what they had found in maize. At some stages, the sorghum inbreds appeared to be limited in growth due to a partial deficiency of endogenous GA. Positive responses to exogenous GA_3 application suggested that the GA deficiency was a limitation to their growth. Thus, Rood et al. (1992) concluded that GAs are involved in heterosis for shoot growth in sorghum. In another investigation, Rood (1995) found that the rate of GA metabolism was positively correlated with growth in sorghum, being faster in a rapid-growing hybrid and slower in slower-growing parental inbreds. Rood (1995) concluded that in both maize and sorghum, and more broadly in various field crop plants and trees, high

GA content and rapid GA metabolism are associated with, and probably at least partially responsible for, rapid shoot growth and particularly, rapid height growth. He also concluded that hybrids are particularly fast-growing and the correlation between GA content, GA metabolism, and growth rate in inbreds versus hybrids further suggests that heterosis for rapid growth rate is partially mediated by GA.

Effects of Environmental Stress on Heterosis

Environmental variability may affect the relationships of specific physiological components of heterosis; however, in a study of the influence of temperature on heterosis for several maize seedling growth traits, Rood et al. (1988a) found that the level of heterosis for these traits could not be explained simply by the ability of a hybrid to better tolerate cool temperatures. They concluded that hybrids derived from a group of four elite inbred lines displayed heterosis similarly under either favorable or cool temperature conditions.

In an effort to determine how QTLs might behave under several environmental stress variables, we conducted a major study in which QTLs affecting heterosis for grain yield and several other traits were mapped under eight combinations of stress and non-stress related variables (Stuber, 1997; LeDeaux & Stuber, 1997). The variables were low (drought) and optimum moisture, low (deficient) and optimum nitrogen, and low (about 36 000 plants ha^{-1}) and high (about 72 000 plants ha^{-1}) plant density. For the two traits, grain yield and ear height, most of the marker loci were found to be linked to QTLs that affect at least one of these two traits; however, only a few were found to show significant interaction with environmental stresses. Even though the yield level of the least stressful combination of variables was 10 times greater that that of the most stressful combination, the QTL mapping pattern differed very little under the two regimes. Veldboom and Lee (1996) conducted a maize mapping study in two different environments and also found very few differences in the QTL mapping patterns. However, they did not impose the same severity of stress that we did in the studies conducted in North Carolina. Results from these studies indicate that lines and hybrids bred for superior performance in non-stress environments should also perform well in stress environments, as least for the stresses used in these investigations.

Additional conclusions from the study reported above (LeDeaux & Stuber, 1997) included: (i) most QTLs for grain yield (in the materials from the B73 × Mo17 derived populations) act dominantly or (pseudo)overdominantly, but most QTLs for ear height act additively or show partial dominance, and (ii) in a separate analysis that focused on locus × locus interaction effects, several pairs of unlinked QTLs were found to significantly affect yield in both of the backcross populations evaluated, indicating that epistasis is a factor contributing to heterosis for grain yield in maize.

Blum et al. (1990) studied two grain sorghum hybrids and their parental lines in the greenhouse under a gradient of ambient temperatures and two water regimes (well-irrigated and drought up to heading). Significant heterosis was found for biomass, grain yield per plant, and grain number per panicle. No heterosis occurred for harvest index, indicating that heterosis in grain yield was due to heterosis in biomass. Neither growth duration nor leaf area could explain heterosis in biomass. When carbon exchange rate (CER) data were subjected to a stability analysis, the two hybrids had greater CER than their respective parents, especially under conditions favoring high CER. When extreme stress conditions developed, the hybrid's performance depended on its genetic background more than on heterosis.

In some earlier studies Blum et al. (1990) reported that sorghum hybrids produced more dry matter in proportion to their increase in leaf area. This is supported by the finding that sorghum hybrids had the same ratio of dry matter to leaf area as their parents. Heterosis must come from a greater or a more efficient source of assimilates in the hybrid in order to account for a larger plant with a greater leaf area. The heterotic yield component in this case was grain number per panicle, and not kernel weight. The greater grain yield in the hybrids was achieved with about the same harvest index as the parents, meaning that hybrids did not excel in relative dry matter partitioning to grain. Irrespective of the moisture regime, hybrids produced more grain in proportion to their greater biomass.

CONCLUSIONS

A general conclusion based on the above data and hypotheses is that selection criteria based on molecular, biochemical, or physiological measurements will not necessarily provide a more sensitive indicator of heterosis than net productivity per se (whether it be for grain yield or some other quantitatively inherited trait) because net productivity integrates such things as metabolic flux over time. Plant breeders will continue to capitalize on the heterosis phenomenon, in spite of the relative lack of information regarding the biochemical, physiological, molecular, and genetic factors involved in the expression of the phenomenon. As new technology at all levels (molecular, cellular, tissue, organismal) unfolds, exciting new methodologies should become available to make plant breeding more efficient and more precise.

REFERENCES

Blum, A., S. Ramaiah, E.T. Kanemasu, and G.M. Paulsen. 1990. The physiology of heterosis in sorghum with respect to environmental stress. Ann. Bot. 65:149–158.

de Vienne, D., and F. Rodolphe. 1985. Biochemical and genetic properties of oligomeric structures: a general approach. L. Theor. Biol. 116:527–568.

Gibson, P.T., and K.F. Schertz. 1977. Growth analysis of a sorghum hybrid and its parents. Crop Sci. 17:387–391.

Graham, G.I., D.W. Wolff, and C.W. Stuber. 1997. Characterization of a yield quantitative trait locus on chromosome five of maize by fine mapping. Crop Sci. 37:1601–1610.

Hageman, R.H., and R.J. Lambert. 1988. The use of physiological traits for corn improvement. p. 431–461. In G.F. Sprague and J.W. Dudley (ed.) Corn and corn improvement. 3rd ed. Agro. Monog. No. 18. ASA, CSSA, SSSA. Madison, WI.

Hageman, R. H., E. R. Leng, and J. W. Dudley. 1967. A biochemical approach to corn breeding. Advan. Agron. 19:45–86.

Hall, J.G., and C. Wills. 1987. Conditional overdominance at an alcohol dehydrogenase locus in yeast. Genetics 117:421–427.

Hepburn, A.G., F.C. Belanger, and J.R. Mattheis. 1987. DNA methylation in plants. Developmental Genetics. 8:475–493.

Kacser, H., and J.A. Burns. 1981. The molecular basis of dominance. Genetics 97:639–666.

LeDeaux, J.R., and C.W. Stuber. 1997. Mapping heterosis QTLs in maize grown under various stress conditions. p. 40–41. Book of abstracts: Symposium on the Genetics and Exploitation of Heterosis in Crops, 17–22 Aug. CIMMYT, Mexico City.

Leonardi, A., C. Damerval, Y. Herbert, A. Gallais, and D. de Vienne. 1991. Association of protein amount polymorphism (PAP) among maize lines with performances of their hybrids. Theor. Appl. Genet. 82:552–560.

Mangelsdorf, A.J. 1952. Gene interaction in heterosis. p. 321–329. *In* J.W. Gowen (ed.), Heterosis. Iowa State College Press, Ames.

McDaniel, R.G. 1986. Biochemical and physiological basis of heterosis. CRC Crit. Rev. in Plant Sci. 4:227–246.

McDaniel, R.G., and I.V. Sarkissian. 1966. Heterosis: Complementation by mitochondria. Science (Washington, DC) 152:1640–1642.

Pavlikova, E., and S. B. Rood. 1987. Cellular basis of heterosis for leaf area in maize. Can. J. Plant Sci. 67:99–104.

Pryor, A.J. 1974. Allelic glutamic dehydrogenase isozymes in maize —a single hybrid isozyme in heterozygotes. Heredity 32:397–401.

Rhodes, D., G.C. Ju, W-J. Yang, and Y. Samaras. 1992. Plant metabolism and heterosis. Plant Breed. Rev. 10:53–91.

Richey, F.D. 1942. Mock-dominance and hybrid vigor. Science (Washington, DC) 96:280–281.

Rood, S.B. 1995. Heterosis and the metabolism of gibberellin A_{20} in sorghum. Plant Growth Regulation 16:271–278.

Rood, S.B., R.I. Buzzell, and M.D. MacDonald. 1988a. Influence of temperature on heterosis for maize seedling growth. Crop Sci. 28:283–286.

Rood, S. B., R. I. Buzzell, D. J. Major, and R. P. Pharis. 1990. Gibberellins and heterosis in maize: Quantitative relationships. Crop Sci. 30:281–286.

Rood, S. B., R. I. Buzzell, L. N. Mander, D. Pearce, and R. P. Pharis. 1988b. Gibberellins: A phytohormonal basis for heterosis in maize. Science 241:1216–1218.

Rood, S. B., J. E. T. Witbeck, D. J. Major, and F. R. Miller. 1992. Gibberellins and heterosis in sorghum. Crop Sci. 32:713–718.

Scandalios, J.G., E.H. Liu, and M.A. Campeau. 1972. The effects of intragenic and intergenic complementation on catalase structure and function in maize: a molecular approach to heterosis. Arch. Biochem. Biophys. 153:695–705.

Schrader, L.E. 1985. Selection for metabolic balance in maize. p. 79–89. *In* J.E. Harper et al. (ed.) Exploitation of physiological and genetic variability to enhance crop productivity. Waverly Press, Baltimore, MD.

Schwartz, D., and W.J. Laughner. 1969. A molecular basis for heterosis. Science (Washington, DC) 166:626–627.

Sinha, H.K., and R. Khanna. 1975. Physiological, biochemical, and genetic basis of heterosis. Adv. Agron. 27:123–174

Srivastava, H.K. 1981. Intergenomic interaction, heterosis, and improvement of crop yield. Adv. Agron. 34:117–195.

Stuber, C.W. 1994. Heterosis in plant breeding. Plant Breed. Rev. 12:227–251.

Stuber, C.W. 1997. Case history in crop improvement: Yield heterosis in maize. p. 197–206. *In* A. H. Paterson (ed.) Molecular analysis of complex traits. CRC Press, Boca Raton, FL.

Stuber, C.W., S.E. Lincoln, D.W. Wolff, T. Helentjaris, and E.S. Lander. 1992. Identification of genetic factors contributing to heterosis in a hybrid from two elite maize inbred lines using molecular markers. Genetics 132:823–839.

Tsaftaris, A.S. 1995. Molecular aspects of heterosis in plants. Physiol. Plant. 94:362–370.

Tsaftaris, A.S., M. Kafka, and A. Polidoros. 1997. Epigenetic changes in maize DNA and heterosis. p. 125–130. *In* A.S. Tsaftaris (ed.) Genetics, biotechnology and breeding of maize and sorghum. The Royal Society of Chemistry, Cambridge, England.

Veldboom, L.R., and M. Lee. 1996. Genetic mapping of quantitative trait loci in maize in stress and nonstress environments: I. Grain yield and yield components. Crop. Sci. 36:1310–1319.

Wells, R., and W.R. Meredith, Jr. 1986. Heterosis in upland cotton. I. growth and leaf area partitioning. Crop Sci. 26:1119–1123.

Wells, R., W.R. Meredith, Jr., and J.R. Williford. 1988. Heterosis in upland cotton: II. Relationship of leaf area to plant photosynthesis. Crop Sci. 28:522–525.

Chapter 17

Towards Understanding and Manipulating Heterosis in Crop Plants — Can Molecular Genetics and Genome Projects Help?

M. Lee

INTRODUCTION

Hybrid cultivars are bred and grown for a variety of reasons which include the favorable vigor of the hybrid progeny and systematic exploitation of heterosis. In their entirety, schemes for breeding and production of hybrid cultivars have steps at which new information or technology could make significant improvements to some of the steps, the overall system, and possibly, the product. Before some of the improvements are realized, however, it will be necessary to achieve a higher level of understanding of basic plant biology in the context of nature and agriculture.

This may be an immense task and, depending on the objectives, it may be an unrealistic one; however, given the rate of change and new developments in the technical and information infrastructures for basic biological sciences, and the novel approaches they permit, it is becoming more difficult to delineate the possible from the impossible and the practical from the impractical. Thus, there is some reason to believe that new elements of basic biology, including molecular genetics, will enhance our understanding of heterosis and ultimately, our ability to exploit it further.

How will molecular genetics help? At this time, identifying the paths to understanding a topic as complex as heterosis requires clairvoyance at many junctions because we often lack the basic information for making decisions, devising strategies and conducting tests. This is changing. There have been a few demonstrations that techniques and information derived from molecular genetics would benefit several aspects of the breeding and production of hybrid cultivars. Also, we are in the early stage of a flood of raw data from plant genome projects, major efforts destined to affect many areas of plant biology, biotechnology, and agriculture. As the data are distilled into information and assembled into technologies, some paths will emerge, and so will some obstacles. The purpose of this chapter is to assess some recent, current and future developments in molecular genetics and biology for their potential to improve our understanding of heterosis and hybrid breeding systems.

To provide some context for the assessments, a few assumptions were made about the future: (i) crop productivity will become more dependent on crop genetics; (ii) basic knowledge of nature (plant biology and the supporting environment) will be central to crop-based agriculture; (iii) yield remains the primary goal (i.e., produce and protect plant biomass); and (iv) hybrid cultivars are for the general good of humanity. These assessments are based on recently established and evolving techniques (the recent past), current developments (the present) and their subsequent phases (the future).

THE RECENT PAST: DNA MARKERS AND GENETIC ENGINEERING

Two recent developments that could become important components of hybrid production schemes are DNA markers (e.g., RFLPs, SSRs, AFLPs) and genetically engineered male sterility systems (GEMSS; Mariani et al., 1992). These early derivatives of molecular genetics each provide different utilities, some of which could have profound effects on the efficiency and productivity of hybrid breeding schemes.

DNA markers have been used to explore several issues related to heterosis such as genetic mapping of QTLs, genetic diversity and heterotic groups, and prediction methods. These subjects will not be reviewed at length herein because they are discussed separately by several speakers at this conference and in previous reviews (Lee, 1995). At this point, it is clear that the utility of DNA markers is still rather limited and under development. First, the marker systems are still too slow or expensive for the scale and speed of many breeding programs and objectives. But, faster and less expensive marker systems are being developed and breeders are optimistic about their utility (Lee, 1995). Second, it has become clear that the mostly anonymous markers must have reliable and meaningful information associated with them (e.g., breeding values) in order to have utility in a breeding program. Attaching information to the markers will take time, special efforts, and additional human and physical resources; however, a recent U.S. patent (Johnson, 1996) suggests that DNA fingerprints of maize (*Zea mays* L.) inbreds and new approaches to data analysis can be combined to predict the performance of hybrid progeny for grain yield. Such predictive power would allow breeding programs to focus testing resources on the most promising candidate hybrids and become more efficient in terms of resource allocation.

Genetically engineered male sterility systems (GEMSS) have the potential to dramatically affect the production and productivity of hybrid cultivars. Admittedly, the first generations of these systems have several features which limit them to a few species, but there are ways to obviate those limitations and it is reasonable to assume that better systems will be created. For example, genetic transformation is hindered by some obstacles such as genotypic limitations, transgene silencing, and the number of genes carried on a construct; however, there are good reasons to believe that those obstacles will be circumvented or overcome through advancements that allow transfer of large inserts (>100 Kb) of DNA, pollen transformation and monocot transformation with *Agrobacterium tumefaciens*. The previously impractical or impossible objectives may soon become routine accomplishments. Eventually, GEMSS could be very important for several crops by improving or enabling the production of hybrid cultivars and their value could well surpass the immediate benefits related to seed production.

The production of hybrid cultivars depends on some system of controlled pollination and each system entails some compromise and sacrifice. For example, cytoplasmic male sterility (CMS) requires the development and maintenance of separate male and female (seed) gene pools. Progeny of the female gene pool must have a very reliable and stable male-sterile phenotype when producing the hybrid seed. Typically, only a subset of the female gene pool has the genetic constitution that confers the desired phenotype in a reliable manner. The consequence of this requirement is that the female gene pools are often shallower, or less diverse than the male gene pools, e.g., sorghum [Sorghum bicolor (L.) Moench] (Ahnert et al., 1996). As a result, genetic gain of the hybrid cultivars is largely dependent on variation in the male gene pool. This is a major constraint placed on genetic gain. Also, such systems create risks associated with genetic uniformity of the organellar genomes of the female gene pools.

How could GEMSS improve these situations? Besides enabling large-scale production of hybrid cultivars for some crops for the first time, good GEMSS might rely on inducible control of the male-sterile phenotype (e.g., induce with a chemical spray) such that the common components of CMS, the sterile, maintainer and restorer lines and related breeding activities are not needed. Potentially, such systems could enable breeders to use and recombine germplasm with fewer restrictions. Of course, this assumes that the inducible gene(s) must be introduced through transformation, at least once, and that the limitations of contemporary transformation systems continue to diminish.

THE PRESENT: GENOMICS

The advent of the genomics era and attendant genome projects has introduced the biological sciences to the information age, or the raw data age when it is now feasible and reasonable to suggest that entire eukaryotic genomes be sequenced and analyzed. What becomes of that information and how it is used are largely unknown. What does one do with the complete DNA sequence of most of the 30 000 to 80 000 genes of their target organism? How do we relate these genes to pathways, responses, and traits? How will we identify donor species as sources of truly novel genes and phenotypes for transfer to the target species? How will we incorporate knowledge of gene expression into our assessments of genetic variation and diversity?

In many ways, the early phases of the genome projects are extensions of descriptive biology and analogous to the construction of libraries or gene banks: nobody really knows what is contained collectively therein but many are convinced the raw data and material are or will be enlightening and valuable. After the raw material and observations (e.g., DNA sequence and functional analysis) are collected and distilled, it becomes reasonable to take steps toward further understanding and rational manipulation of the target genomes.

At this time, there are a few reasonably comprehensive plant genome projects. The effort in *Arabidopsis* is relatively advanced but those of crop species are in their early stages, particularly in the public sector. The sizes of the crop genomes, and the relatively diffuse and poorly funded efforts make the tasks of a genome project especially challenging. Rice (*Oryza sativa* L.), with the smallest genome of the major grain crops, has a genome size three times that of *Arabidopsis* and has been estimated to contain 10 000 more genes (30 000 vs. 20 000 genes; S. McCouch cited in Cohen, 1997). The genomes of other major crops are usually much larger and may encode 80 000 to 100 000 expressed genes. How do we prepare to investigate and query such genomes and their interactions? The experiences and evolution of the Human Genome Project(s) indicate that such endeavors benefit from some centralized planning, coordination and concentration of resources. Crop plant research often lacks such organization. So, successful development and implementation of crop plant genome initiatives will require public and private institutions to learn some lessons and adapt some principles from previous efforts.

What are genomics and genome projects? The definitions and components vary but they usually comprise an integrated assembly of approaches with the goal being a comprehensive assessment of the content and expression of an entire genome, starting with the DNA and functional analysis of its products and ending with the phenotypes. Obviously, genome projects and the term *genomics* overlap or duplicate extant areas of biological research. The key differences are the intended depth and breadth of inquiry associated with determining the DNA sequence of all the genes or the entire genome of an organism. Typically, the components of genome projects may be placed into at least three broad areas: enabling infrastructure, functional genomics, and bioinformatics. Bioinformatics (using the rules and principles of the biological, physical, and chemical sciences to extract useful information from raw molecular data) will not be discussed herein but its role is obvious and essential because it underlies and permeates all other components (e.g., Botstein & Cherry, 1997).

The enabling infrastructure provides the raw material and data and the methods for collating and verifying those items within the appropriate biological context. The infrastructure typically includes cDNA and genomic sequence data, maps (genetic maps of mutants, DNA markers, candidate genes and quantitative trait loci; physical maps based on chromosome breakpoints and libraries of large inserts of DNA such as bacterial artificial chromosomes and radiation hybrids), and efficient transformation systems for complementation tests and other functional analyses.

Assuming we obtain the complete DNA sequence of all the maize genes or the entire rice genome, how do we learn about the functions of the genes, features of the genome, their interactions and their roles in mediating phenotypes? The first step typically involves a branch of bioinformatics, sequence analysis and comparison coupled with DNA sequence data bases from other organisms (and the hope that someone else has found a putative function for a DNA sequence related to your new DNA sequences). This step will not be treated herein as a brief and recent summary has discussed its real and potential merits (Botstein & Cherry, 1997).

Functional genomics provides opportunities for directly linking molecules to their biological role(s) and phenotype(s) as well as providing clues about an unexplored dimension of genetic variation, the various aspects of gene expression. Data bases of DNA sequences and related analyses may provide matches (i.e., strong hints of identity and function) for 40 to 70% of the genes emerging from any genome project. On that basis, one may begin to generate hypotheses about the *putative* function(s) of newly sequenced genes. This alone may be quite useful but the hypotheses need to be tested in the target species. Also, the other genes (30 to 60%) are not strongly matched with any putative function or homologue. For crop genomes such as maize, there may be 30 000 to 40 000 unmatched and unknown genes at the completion of a cDNA (EST) sequencing project. Also, such projects will miss many genes represented by rare transcripts or by those difficult to clone for other reasons. So, there is clearly a need to complement and extend the information provided by the sequencing component of the project.

Large-scale functional analyses usually include hybridization analyses of high density arrays of cDNA clones (e.g., Shena et al., 1995) and modified insertional mutagenesis with transposable elements to generate knock-out mutations or to identify regulatory elements (i.e., gene machines or traps; Sundaresen, 1996). Insertional mutagenesis strategies, while limited by the inherent one-gene-at-a-time process, are valuable because they provide direct links between a phenotype (e.g., mutant), a gene interrupted by a known and cloneable transposon insertion, and the intermediary molecules (mRNA and protein). Also, these strategies may isolate genes that are not readily cloned or identified by sequencing projects.

The high density hybridization arrays and similar methods provide the means for assessing patterns of gene expression, investigating gene function(s) and developing diagnostic assays for various applications. If you have identified tens of thousands of genes to some degree, how do you learn how the organism deploys them, alone or in concert, to maintain or mediate a phenotype in response to various treatments? Which genes are expressed or suppressed in the organ, tissue and cell of interest? It is now possible to make simultaneous and rough assessments on tens of thousands of genes and their mRNA levels in various tissues, environmental conditions and genetic backgrounds. While these types of assessments, and their enabling machines are relatively nascent, their development is worth pursuing because they permit the exploration of a plant's biological circuitry and totally unknown components of genetic variation; gene expression and interactions. I think it is safe to say that an understanding of heterosis will be highly dependent on knowledge of those two components (Leonardi et al., 1991; Guo & Birchler, 1994).

GENOMICS: THE GOOD NEWS

At this early stage there are several positive developments related to the genomics era. First, the genomics era (coupled with a very favorable era in the financial markets and advancements in computing power) has provided the motivation and capital for rapid technological advancements in DNA sequencing, cloning, expression analysis and other areas. These advancements have made it possible to organize and conduct several genome projects, and sometimes, complete some phases of the projects well ahead of schedule and at lower cost. The initial result has been a flood of raw data and dreams of sequencing the entire genomes of important crop species. Much of the data are in a form such that anyone with a reasonably powerful personal computer, with the right connections (electronic), can access, query and analyze the data. Ultimately, these raw data will be converted into information and insight, and a better understanding of many aspects of crop plant biology as indicated by the comparisons of the first plant disease resistance genes (Staskawicz et al., 1995). The advancements have permitted more comprehensive assessments and comparisons of crop plants' genetic and physiological repertoires across taxa (Paterson et al., 1995) such that it may be possible to identify universal or unique features of phenomena like heterosis among groups of crop species. Such comparisons will facilitate gene isolation and the ability to endow crops with new and valuable phenotypes. Soon, it should be possible to explore new dimensions of genetic variation, the various stages of gene expression, and to assess their roles in determining and mediating phenotypes. Ultimately, these developments will strengthen the links among areas of basic plant biology and then connect them with plant breeding and crop improvement. Certainly, elements of a crop's heterotic response are interwoven with many aspects of plant metabolism and development. As our awareness and sophistication grows with the new information, it should be possible to make more rational and information-driven assessments and strategies.

In some ways, the plant genomics era is at the same stage of development where crop-based agriculture was during the periods of crop domestication of a few thousand years ago (or soybean domestication of a few decades ago!). At that time, there was a lot of raw germplasm (or information). Probably few of us could have imagined or would have predicted how the germplasm has been distilled and molded to meet the demands of today's production environments. The first generation of data from the genome projects has many of the same features of the early crop germplasm and will have to undergo a similar process of data domestication and enhancement (and hopefully, distribution).

GENOMICS: THE BAD NEWS

Transitions are rarely easy and the arrival of genomics is a major transition for some in plant science and technology. Like any new venture, genome projects must be sold and bought. During this process, the benefits are often speculative in nature and the immediate costs are very real. Herein lies some bad news about genome projects: they only provide some of the pieces of the puzzle. Complete cDNA or genomic sequencing only identifies a portion of the genes because some genes are not readily cloned (e.g., rare transcripts) or they are not easily discerned from noncoding regions of the genome (e.g., pseudogenes, remnants of transposons and subtleties of gene structure and codon usage in plants). Genes identified on the basis of sequence comparisons with data bases assume only a putative function. Often, the putative function must be confirmed and its role defined in the context of the target organism, and perhaps in the context of certain environmental parameters. Genome projects may be generating a wealth of context dependent information. Therefore, much testing and assembling remains after the sequencing and first assessments of gene function.

Of course, genome projects have a cost. By the standards of funding for basic and applied plant science research, they are very expensive. If they are conducted at the expense of the extant and pathetically underfunded plant science research efforts, they are a disaster. Viewed in another light, plant genome projects are a bargain and a wise allocation of resources. For the price of two F-16 fighter jets ($36 million each) each year we would be making an investment in the food security of 5.6 billion people, the economies of most countries and many components of the global environment.

Another major concern of the genomics era and genome projects is related to intellectual property rights in an increasingly capitalistic world (Swaminathan & Jana, 1992; Lee, 1998). The reductionist nature of genome projects will provide many opportunities to partition and sequester biological systems into intellectual property. While such activity may serve a useful purpose for some, and sometimes benefit the majority, it can also be divisive, counterproductive, unjust, and ignorant. Perhaps we are fortunate that Darwin, Beal, East and Schull made their discoveries with hybrid maize; or, that crops were domesticated before the current era of intellectual property rights in plant science.

Perhaps the worst news associated with plant genome projects is from the plants themselves. From what little we know, they are amazingly complex and interactive biological systems. These systems will take some time and effort to decipher, and when that happens, we may learn that their components may resist or exceed our meager powers of manipulation.

COMPLEXITY OF PLANT LIFE: SOME PERSPECTIVES

In 1982 the complete genomic sequence of the virus, bacteriophage lambda, was reported (Sanger et al., 1982). Encompassing slightly more than 48.5 kilobase pairs of DNA and encoding dozens of genes, the achievement has been hailed as a landmark event in modern biology (Old & Primrose, 1994). Yet, a review of the genetic controls of lambda's life cycle (Ptashne, 1986) rarely refers to the sequencing effort such that one gets the impression that the complete DNA sequence contributed very little to our understanding of such a simple virus. To what extent did knowledge of the DNA sequence elucidate features of lambda's lifestyle? In the case of lambda, much biological research preceded the sequencing effort by virtue of the temporal availability of technology. Also, much biological research proceeded the arrival of the sequence data. Similar situations may be observed with other more complex, model systems such as *E. coli* and yeast. Perhaps the first big lesson from the genome projects for those species was that, despite decades of relatively focused research, we still had a lot to learn about those simple organisms. What will be the experience with plants? No doubt, the challenges of understanding the multifaceted aspects of a plant's life cycle in a dynamic environment are considerably more formidable. Given that plant biological research still lags far behind those model systems, the plant genome projects could have a greater influence on our understanding of plants and our abilities to modify them.

The presently available information suggests that plants, like other organisms, depend on complex and interactive networks encoded in their genomes (Strohman, 1994). For example, transcription of one structural gene requires the binding and interactions of products from several other genes, the transcription factors. What are the transcription factors and what are their signals in plants? Transcription also is influenced by the chemical and physical disposition of the surrounding chromatin as well as the orientation and positions of regulatory regions of the gene. The significance of these variables is well established for some eukaryotic genes but it has not been determined for many plant genes.

One of the first revelations of plant molecular genetics was the size and redundancy of plant nuclear genomes. Obviously, we were well aware of the extent of polyploidy and there were some very good early hints of segmental duplications; however the degree of duplication of the coding regions of plant genomes, the prevalence of multigene families and localized sequence duplication has been surprising and frustrating to molecular biologists (Strohman, 1994; Pickett & Meeks-Wagner, 1995) and probably very advantageous to plants. This feature of plant nuclear genomes will hinder attempts to understand and manipulate it.

The duplicity of the genome is reflected by what little we know about plant metabolism (Varner, 1995). Plants typically encode duplicate versions of enzymes (i.e., isozymes) that differ in levels of expression and tissue specificity. There are also enzymes with overlapping functions by virtue of common domains. In addition, plant metabolism, including primary metabolism, contains alternate routes to the same compounds (e.g., sucrose breakdown; ap Rees, 1995). Secondary metabolism in plants becomes more complex and we should anticipate encountering similar themes on a regular basis. Such life strategies and capacities are good and bad news for those wishing to manipulate plant metabolism. The variability among species indicates some degree of flexibility such that it should be possible to add or modify enzymes at some junctions in order to add or delete a metabolic or developmental option. Of course, the trick will be to identify such aspects of metabolism because some components may be more or less amenable to change (Rhodes et al., 1992; ap Rees, 1995). Such context-dependent knowledge may be acquired primarily through trial and error.

With such genomes and an immobile existence, plants obviously need to perceive, discern and process many internal and external signals and manage many stresses. This information is then somehow integrated to trigger and drive developmental events. It is likely that a large percentage of the genes in plants are somehow involved in perceiving, transferring, and receiving information from numerous cues. We are only beginning to identify components of some signal transduction pathways in crop plants.

At all levels, plants are highly interactive and dynamic organisms. The opportunities for interaction certainly reflect important aspects of plants' life strategies developed over the course of evolution. The interactive and redundant nature of plant biology has probably been one of the essential features underlying crop plants' capacity to adapt to the challenges of agriculture and will be one of the more frustrating barriers to understanding it for the purpose of further manipulation. To understand and manipulate a phenomena such as heterosis, we may not need to understand all aspects of crop biology but we will certainly need to know much more about some of the components. Identifying those components within the metabolic milieu will be the first challenge.

THE FUTURE: DATA DOMESTICATION AND TESTING

Before we are able to make rational judgements about the utility of molecular genetics for understanding and manipulating heterosis, it will be necessary to experience the future of plant biology in the wake of the genome projects. The immediate future will include a process that collects relatively huge volumes of raw data about plants and converts it into meaningful information.

Once we have the information, or some of it, then it becomes reasonable to make more objective assessments of goals and limits, devise and test strategies, and make progress in a systematic manner. It also becomes possible to ask better questions and to have a chance of obtaining definitive answers about crop plant biology. This phase will only be possible if we adopt an attitude of exploration and discovery because at this time we know so little. We are forced to make too many guesses because we lack fundamental information.

AN ASSESSMENT: HOW WILL WE IDENTIFY GENES INVOLVED IN HETEROSIS?

Among other purposes, this conference was organized to summarize our knowledge and understanding of heterosis in crop plants. One conclusion from the conference may be that our knowledge of the genetic aspects of heterosis has advanced slightly since the 1950s. Another conclusion may be that our knowledge of the biological basis of heterosis (e.g., the genes, their functions, and their phenotypes) has not advanced; consequently, our understanding of heterosis would follow a similar trajectory. How will we make progress in these areas?

Recent developments of techniques and materiel have made it possible to explore the genetic and biological architecture of heterosis in crop plants. This era was initiated in maize through the use of DNA markers and genetic mapping with a population derived from the widely grown single-cross hybrid, B73 × Mo17 (Stuber et al., 1992). Recently, a set of 350 recombinant inbred lines (RILs) has been derived from that population after four generations of intermating (Covarrubias-Prieto et al., 1989; Lee, 1997). Intermating increased the frequency of recombinant gametes such that the genetic map expanded nearly fourfold (Beavis et al., 1992). The increased genetic resolution of this population should greatly increase the precision of QTL mapping. By backcrossing the RILs to both parents, it should be possible to map QTLs for heterosis. The availability of high density arrays of many maize cDNA clones (known and unknown) should also permit exploration of the status of one level of gene expression in relation to heterosis and the interactions between the genomes of the inbred parents. The use of sequenced cDNA clones will provide opportunities to discover which genes are associated with aspects of heterosis and to compare them to those related to heterosis in other crop species such as rice (Yu et al., 1997). Also, it may be possible to study changes in patterns of gene expression in populations subjected to reciprocal recurrent selection for hybrid performance (Labate et al., 1997).

SUMMARY, CONCLUSIONS, AND QUESTIONS

Can molecular genetics improve our understanding and manipulation of heterosis? Given our ignorance of the subjects and the history of prior attempts, some skepticism may be appropriate. Also, the complexity of the plants and their numerous interactions with the environment will make it difficult or impossible for molecular genetics to identify universal aspects of heterosis suitable for alteration. We may not find a biologically-based unifying mechanism or pathway of heterosis. Certainly, molecular genetics alone will not accomplish this, especially if there are not adequate human resources to analyze, explore, and integrate new sources of information while maintaining the highly effective infrastructure of plant breeding.

There are also reasons, however, to believe that molecular genetics will contribute to progress in exploiting heterosis in crop plants. Until very recently, most of basic biology has had little or no impact on genetic improvement of crops. We are beginning to witness contributions of molecular biology in crop improvement for pest resistance, weed control, and grain quality, but the recent advances are being achieved with very limited basic information about the entire system, the plant. As impressive and challenging as they may be, I equate those achievements to putting accessories or ornaments on an automobile. It may be possible to exploit that paradigm for quite some time but more meaningful improvements may require changes to fundamental aspects of the plant. Fortunately, the pieces of the systems are being defined and I believe our abilities to assemble and understand them will improve. There will be many false positives, but if nothing is ventured, then nothing is gained. Given the essential purpose of the crops, the venture is easily justified.

ACKNOWLEDGMENTS.

Journal Paper J-17687 of the Iowa Agriculture and Home Economics Experiment Station, Ames, Project 3134, supported by the Hatch Act, State of Iowa Funds, USDA-NRI Plant Genome Award 97-35300-4939, and the Iowa Corn Promotion Board. The author thanks C.W. Stuber and two anonymous reviewers for their helpful suggestions.

REFERENCES

Ahnert, D., M. Lee, D.F. Austin, C. Livini, W.L. Woodman, S.J. Openshaw, J.S.C. Smith, K. Porter, and G. Dalton. 1996. Genetic diversity among elite sorghum lines assessed with DNA markers and pedigree information. Crop Sci. 36:1385-92.

ap Rees, T. 1995. Prospects of manipulating plant metabolism. TIB 13:375–379.

Beavis, W.D., M. Lee, D. Grant, A.R. Hallauer, T. Owens, M. Katt, and D. Blair. 1992. The influence of random mating on recombination among RFLP loci. Maize Genet. Coop. Newsl. 66:52–53.

Botstein, D., and J.M. Cherry. 1997. Molecular linguistics: extracting information from gene and protein sequences. Proc. Natl. Acad. Sci. USA. 94:5506–5507.

Cohen, J. 1997. Corn genome pops out of the pack. Science (Washington, DC) 276:1960–1962.

Covarrubias-Prieto, J., A. R. Hallauer, and K. R. Lamkey. 1989. Intermating F2 populations of maize. Genetika. 21:111–126.

Guo, M., and J. A. Birchler. 1994. Trans-acting dosage effects on the expression of model gene systems in maize aneuploids. Science 266:1999–2002.

Johnson, R. 1996. Process for predicting the phenotypic trait of yield in maize. U.S. patent 5,492,547.

Labate, J.A., K.R. Lamkey, M. Lee, and W.L. Woodman. 1997. Molecular genetic diversity after reciprocal recurrent selection in BSSS and BSCB1 maize populations. Crop Sci. 37:416–423.

Lee, M. 1995. DNA markers and plant breeding programs. Adv. Agron. 55:265–344.

Lee, M. 1997. A new mapping population for high resolution genetic mapping in maize. Plant and Animal Genome V, Jan. 12–16. San Diego, CA.

Lee, M. 1998. Genome projects and gene pools: New germplasm for plant breeding? Proc. Natl. Acad. Sci. USA 95:2001–2004.

Leonardi, A., C. Damerval, Y. Hebert, A. Gallais, and D. de Vienne. 1991. Association of protein amount polymorphisms among maize lines with performance of their hybrids. Theor. Appl. Genet. 82:552–560.

Mariani, C., V. Gossele, M. de Beuckleer, M. de Block, R.B. Goldberg, W. de Greef, and J. Leemans. 1992. A chimeric ribonuclease-inhibitor restores fertility to male sterile plants. Nature (London) 347:737–741.

Old, R.W., and S.B. Primrose. 1994. Principles of Gene Manipulation: an Introduction to Genetic Engineering. Blackwell Science, London.

Paterson, A.H., Y.-r. Lin, Z. Li, K.F. Schertz, J.F. Doebley, S.R.M. Pinson, S.-C. Liu, J.W. Stansel, and J.E. Irvine. 1995. Convergent domestication of cereal crops by independent mutations of corresponding genetic loci. Science (Washington, DC) 269:1714–1718.

Pickett, F.B., and D.R. Meeks-Wagner. 1995. Seeing double: Appreciating genetic redundancy. The Plant Cell 7:1347–1356.

Ptashne, M. 1986. A genetic switch: Gene control and phage lambda. Cell Press & Blackwell Scientific Publ., Cambridge, MA.

Rhodes, D., G.C. Ju, W.-J. Yang, and Y. Samaras. 1992. Plant metabolism and heterosis. Plant Breed. Rev. 10:53–91.

Sanger, F., A.R. Coulson, G.-F. Hong, D.F. Hill, and G.B. Petersen. 1982. Nucleotide sequence of bacteriophage lambda DNA. J. Molec. Biol. 162:729–73.

Shena, M., D. Shalon, R.W. Davis, and P.O. Brown. 1995. Quantitative monitoring of gene expression with a complementary DNA microarray. Science (Washington, DC) 270:467–470.

Staskawicz, B.J., F.M. Ausubel, B.J. Baker, J.G. Ellis, and J.D.G. Jones. 1995. Molecular genetics of plant disease resistance. Science (Washington, DC) 268:661–667.

Strohman, R. 1994. Epigenesis: The missing beat in biotechnology? Biotechnology 12:156–164.

Stuber, C.W., S.E. Lincoln, D.W. Wolff, T. Helentjaris, E.S. Lander. 1992. Identification of genetic factors contributing to heterosis in a hybrid from two elite maize inbred lines using molecular markers. Genetics 132:823–839.

Sundaresan, V. 1996. Horizontal spread of transposon mutagenesis: new uses for old elements. Trends Plant Sci. 1:184–190.

Swaminathan, M. S., and S. Jana. 1992. The Impact of Plant Variety Protection on Genetic Conservation. p. 257–263. *In* M.S. Swaminathan and S. Jana (ed.) Biodiversity, implications for global food security. Macmillan India, Madras, India.

Varner, J. E. 1995. Foreward: 101 reasons to learn more plant biochemistry. The Plant Cell. 7:795–796.

Yu, S.B., J.X. Li, C.G. Xu, Y.F. Tan, Y.J. Gao, X.H. Li, Qifa Zhang, and M. Saghai Maroof. 1997. Importance of epistasis as the genetic basis of heterosis in an elite rice hybrid. Proc. Natl. Acad. Sci. USA 94:9226–9231.

Chapter 18

Epigenetic Changes in Maize DNA and Heterosis

A. S. Tsaftaris, M. Kafka, A. Polidoros, and E. Tani

INTRODUCTION

Although the biological basis of heterosis remains unknown, plant breeders have made wide use of this phenomenon. Among other reasons, farmers prefer F_1 hybrids not only for higher yields but also for stable yields in different locations and in different years. F_1 hybrids, by being able to successfully encounter varying kinds of stresses imposed in different locations and/or years, exceed homozygous lines in their stability of performance. The use of isozymes and molecular markers, for assessing parental inbred line genetic distance and its possible relation to F_1 hybrid performance, has received impetus during the 1980s and 1990s. A major conclusion from these efforts has been that, although the parental genetic distance was highly correlated with F_1 performance when lines with similar pedigrees were crossed, nonsignificant correlations were found when lines with dissimilar pedigrees were used (Frei et al., 1986; Tsaftaris, 1987, 1990, 1995; Lee et al., 1989; Melchinger et al., 1990; Smith et al., 1990, Stuber et al., 1992).

Information concerning the quantitative expression of genes in parental inbreds and in hybrids and its possible relation to heterosis and hybrid vigor resulted from the so called RAP analysis (RNA Amount Polymorphism) and PAP analysis (Protein Amount Polymorphism) quantifying gene expression at the RNA or the protein level, respectively. Early studies were carried out concerning the amount of total RNA synthesis in heterotic hybrids at early developmental stages (Cherry et al., 1961; Nebiolo et al., 1983). Later Romagnoli et al. (1990) studied the gene expression of primary root tips of a heterotic F_1 and its parental inbreds by means of the relative abundance of tissue specific mRNAs as well as of the in vitro translated products of these mRNAs. The results indicated that increased expression of certain loci is responsible for vigor manifestation. Similarly, Tsaftaris and Polidoros (1993) analyzed the level of mRNA expression of 35 loci in maize (*Zea mays* L.) parental lines and their high and low heterotic hybrids and it was found that the high heterotic hybrid exceeded quantitatively the low heterotic one and all parental lines in mRNA production. PAP analysis reported by de Vienne et al. (1994, 1996) provided similar information at the protein level. These results suggest the significance of both regulatory proteins (and their encoding genes) and the mechanisms of regulation of gene activity in manifestation of complicated phenomena such as heterosis (Tsaftaris, 1990; Leonardi et al., 1991; Tsaftaris & Polidoros, 1993; de Vienne et al., 1996; Tsaftaris & Kafka, 1998).

One such mechanism involved in regulating the amount of expression output of genes is methylation of cytosine residues in their DNA (Cedar, 1984). The extent and distribution of genomic DNA methylation was found to be significantly correlated with the rate of expression of many genes examined not only in plants but in all higher organisms as well. Methylation of certain residues (frequently cytosine within a dinucleotide sequence CpG or CpNpG, as typically found in plant genomes, where N is equal to any nucleotide except G) is an important post replication modification of DNA (Gruenbaum et al., 1981). Cytosine methylation has been implicated in several functions of molecular significance including modulation of gene expression, position effect, variegation, transgene inactivation, quelling, transvection, and others (Hollick et al., 1997). Finally, differential methylation may constitute an important, although little studied, source of DNA polymorphism (Silva & White, 1988; Messeguer et al., 1991). Although methylation has been implicated in all of the above processes, many questions as to function still remain. It is known that DNA methylation requires the involvement of enzymes such as methyltransferases, but how methyltransferases recognize the sequences to be methylated and the numbers and types of methyltransferases involved are mostly unknown (Reich et al., 1992; Richards, 1997).

We report here the results of our studies on the extent of cytosine methylation in maize genomic DNA, its variation among different genotypes (parental inbred lines and hybrids) and developmental stages, as well as its possible involvement in manifesting heterosis. The role of maize growth conditions in the field on the extent of genomic DNA methylation under stressful or more favorable conditions was also examined.

The plant genetic materials analyzed in this work consisted of:

1. Ten S_2 lines selected from cycle 6 (C6) of a population that originated from the F_2 of the single-cross F_1 hybrid (FS68 × NE2), the F_1 hybrid itself, and the progenitor parental lines, FS68 and NE2.
2. Seven old inbred lines of different pedigrees commonly used commercially as parents in hybrid seed production (H95, Grl41341, Mo17, H99, H108, Pa402, and W64A)
3. Five S_4 lines (102A, 22M, 30B, 60M, and 4D) selected from the F_2 of the commercial hybrid LORENA after five years of mass and pedigree honeycomb selection (Fasoulas and Fasoula, 1995) and two of their hybrids (60M × 30B and 22M × 102A).

The plant materials used were grown at the Experimental Farm of the Aristotelian University of Thessaloniki using the honeycomb experimental design, at a 1.5 × 1.5 m distance. In order to study the impact of the environment, some genotypes were grown with two different plant densities: 1.5 m (spaced) and 25 cm (dense) distance between individual plants with a density of 0.513 plants m^{-2} and 18.5 plants m^{-2}, respectively.

DNA METHYLATION IN MAIZE

Maize DNA was found to be highly methylated and methylation was found to be genotype specific as well as to be affected by the environmental conditions. Maize genomic DNA methylation was detected after treating genomic DNA with isoschizomeric restriction enzymes sensitive or insensitive to cytosine methylation such as Hpa II-MspI and Ecor II-BstNI (data not shown; it was impossible to acquire accurate quantitative data). When HPLC was used for quantifying percentage of cytosine methylation, methylation was found to vary from 32.8% in 66 day-old leaves of the inbred 60M at the dense planting to 21.8% in 44 day-old leaves of the 50B inbred at the spaced planting. Average methylation for all samples (excluding the samples taken from plants at the dense planting) was 27.6%, a value very close to the 27.2% estimated by Amasino et al. (1990). Our earlier studies (Tsaftaris et al., 1997) indicated that methylation of maize DNA also is tissue and developmental stage specific.

METHYLATION IN INBREDS AND HYBRIDS

For HPLC analysis of cytosine methylation of maize DNA, 5 µg of the hydrolyzed DNAs were dissolved in 100 µL 0.01 N HCL and a 20 µL aliquot was injected into a chromatogram. The molar concentrations of cytosine and 5-methylcytosine (5-mC) were determined using the HPLC method of Patel and Gopinathan (1986), with some modifications, by comparing the heights of the peaks in the chromatogram with those of the standards made with pure bases.

All five bases (A,T,G,C,5-mC) were separated according to their different retention times. The results were expressed as percent of methylcytosine (5-mC) to the total (5-mC+C) cytosines of DNA. Repetitive chromatograms of the samples tested in this study never showed deviations higher than 0.05% for the same sample. DNA methylation of the 10 S_2 second cycle lines was examined in conjunction with DNA methylation of the original progenitor parental lines and their F_1 hybrid as shown in Table 18–1. The FS68 parental line (a very low yielding inbred) showed 31.4% methylation, one of the highest values observed. The second parental line, NE2, showed 28.3% methylation and the F_1 hybrid showed 27.4%. Thus, the hybrid had a lower percentage of methylation, even lower than the lower parent (in this case NE2) with 28.3% methylation. The ten S_2 very high yielding lines had an average methylation of 24.9% with variance of 0.54. Their methylations varied from 25.8% (Line 4A) to 23.8% (Line 2B). These values are significantly lower, not only from percentage methylation of their original progenitor parental lines, FS68 and NE2, but also from the percentage methylation of the F_1 indicating that breeding for higher yield in the original F_2 population for five continuous generations using the mass honeycomb selection design was accompanied by a concomitant lowering of the methylation level in the derived S_2 lines.

Table 18–2 shows the percentage of DNA methylation for seven commercial lines of different pedigrees. Methylation varied from 32.1% (line W64) to 27.7% (line Mo17) and all lines had a higher methylation level compared with the S_2 selected lines.

Table 18–1. 5-mC content measured by HPLC analysis in two very old low yielding parental lines, their F_1 hybrid, and 10 new second cycle S_2 selected maize inbred lines of similar pedigree derived from the hybrid.

Parental lines		F_1 hybrid		Selected S_2 lines			
FS68	31.4	FS68		2B	23.8	2D	24.3
NE2	28.3	×	27.4	2F	25.4	3A	24.7
		NE2		3E	25.4	4A	25.8
				4E	24.1	5D	24.3
				5E	25.6	6AT	25.6

Table 18–2. 5-mC content measured in seven old commercial maize inbred lines of different pedigrees commonly used as parents in hybrid seed production.

Genotypes	W64A	GRL	H95	H99	H108	Pa402	Mo17
% Methylation	32.1	30.6	29.9	29.2	29.2	28.7	27.7

EFFECTS OF GROWTH CONDITIONS ON METHYLATION

Table 18–3 reports the results from measurements of methylation in another set of inbreds and two of their hybrids. There were five S_4 lines selected from the F_2 of the commercial hybrid LORENA. Two more lines, the old line NE2 and an S_2 line (see Table 18–1), also were included in the experiment for comparisons. In spaced plantings, inbreds showed an average 27.18 and 27.54% methylation in the two stages, respectively, which is comparable but relatively lower than for the older commercially used inbreds (Table 18–2), but significantly higher than the selected high yielding selected S_2 lines (see Table 18–1).

What is important though, considering the role of density induced stress, is that in the early stage when the density induced stress is negligible, inbreds showed no difference in their percentage of methylation. On the contrary, in the later stage (Stage B) when the effect of density became apparent, the percentage of methylation in inbreds was significantly higher in the dense conditions. The situation is different in the hybrids. In the late stage in dense conditions (Stage B), hybrids showed exactly the same methylation in dense and in space planting. This indicates that, differently from inbreds, the hybrids are more resistant to density induced stress, at least when characterized by their total genome methylation.

DIFFERENCE IN THE METHYLATION DETECTED WITH THE CRED-RA TECHNIQUE

The effect of density stress on the methylation of genomic DNA of the genetic material was verified using the Coupled Restriction Enzyme Digestion and Random Amplification (CRED-RA) procedure. The CRED-RA technique used in this study is based on the following hypothesis: a DNA fragment cannot be amplified if it contains a specific restriction site in the region between two primer binding sites and that site is cut by RE digestion prior to PCR. If DNA methylation of the restriction

Table 18–3. 5-mC content measured in two stages (A = 40 and B = 60 days after pollination) under two different conditions of growth in the field in five S_4 lines selected from the F_2 of the commercial hybrid LORENA and two of their hybrids. Two more lines, the old NE2 and an S_2 line (see Table 18–1) also were included for comparisons.

| Genotypes | Stage A | | Stage B | |
	Spaced	Dense	Spaced	Dense
NE2	28.20	28.80	29.30	30.60
5C	30.20	30.30	29.80	30.10
102A	23.10	23.07	24.20	28.71
22M	30.60	27.24	26.22	22.40
30B	21.80	27.70	25.91	27.30
60M	28.30	30.50	27.90	32.87
4D	28.10	28.20	29.47	32.39
Mean	27.18	27.97	27.54	29.19
60M × 30B	31.25	27.32	26.10	26.50
22M × 102A	28.62	25.85	26.20	26.40
Mean	29.93	26.58	26.15	26.45

site prevents digestion at the site within the genomic fragment, the fragment can be amplified. However the amplified product will be susceptible to cleavage because the restriction site will not be methylated during DNA amplification. Thus, DNA methylation can be identified by comparing the banding patterns of template DNA amplified without restriction and template DNA amplified after restriction. CRED-RA was used to identify methylation and to map methylation polymorphism in citrus (Cai et al., 1996) and maize (Tsaftaris et al., 1997).

For this analysis, DNA samples were digested with the methylation sensitive Hpa II restriction enzyme for 4 hours at 37° C prior to running PCR. Nucleic acids then were purified using ethanol precipitation and redissolved in 10 ul sterile water. PCR was performed using the universal primer M13, and other 10-mer random primers with the cycles profile as described by Welsh et al. (1991) .

In Fig. 18–1 one such example of CRED-RA analysis in DNA of an S_4 line grown in the dense or spaced conditions is shown after using the M13 universal primer (13 nucleotides long). As shown in this figure a sequence of 850bp and 600bp were methylated in dense planting but unmethylated in widely spaced planting. Similar results were obtained in other S_4 lines and their hybrids using the same or other decamers as primers (data non shown).

These results can be summarized as follows:

1. F_1 hybrids were found to have less methylated sites than their parents.
2. There were more methylated sites under dense planting (more stressful) conditions in relation to spaced planting.
3. F_1 hybrids were found more resistant to this density induced methylation in comparison to their parents.

These results are in agreement with our previous data where we have studied the role of growth conditions in the methylation of another individual sequence, namely the Ac element of maize. When we quantified the frequency of demethylation (thus activation) of a methylated Ac element, in dense or spaced conditions, we found that for three consecutive years, demethylation of the methylated element was significantly more frequent in spaced plants than in plants grown under more stressful, dense conditions (Kafka, 1996; Tsaftaris & Kafka, 1998).

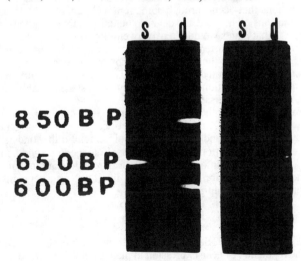

Fig. 18–1. CRED-RA analysis of DNA sequences of DNA isolated from dense (d) and space (s) conditions from an S_4 line using the M13 universal primer for the amplification reaction.

CONCLUSIONS

The results presented in this work show that there is extensive variation in the methylation status of maize depending on genotype, developmental stage and the conditions of plant growth. Although the differences in the level of methylation occur (as shown by the study of the total methylation) in a very narrow range, these differences could be very significant if they implicate that critical cytosine residues (e.g., in promoter areas of genes) undergo demethylation or de novo methylation.

The decrease in DNA methylation in maize inbreds bred for higher yield indicates that selection could change the level of methylation. The uniform low methylation of all S_2 high yielding second cycle lines tested in this study suggests that improvement is related to a decrease of methylation after five continuous generations of selection in wide spacing. On the other hand, percentage of methylation of the 10 S_2 lines of similar pedigree was smaller and varied less than that of the seven commercial inbreds of different pedigrees used today for hybrid seed production.

Environmental conditions of plant growth tend to affect genomic DNA methylation. Less favorable stressful conditions result in more methylated DNA (less expressed) and, in general, heterotic hybrids are more resistant to such density induced genomic DNA methylation. This resistance of the hybrid genome to genome methylation and, consequently, suppression of many genes under different stresses imposed through higher density could be at the core of F_1 hybrid high yield and, possibly more importantly, F_1 hybrid stable yield. Advancements made in the breeding of superior performing F_1 hybrids have been associated with planting at higher densities. In fact, recent hybrids tend to surpass older hybrids in performance only when grown under the high planting densities now used in the fields for most farmers.

Plant development has some unique aspects that are radically different than animal development. Most structures in the mature plant develop after embryogenesis by the meristematic actions. By delaying the bulk of growth and differentiation until after embryogenesis, plants can fashion their form to the environment. In this phenotypic plasticity, in response to different environmental conditions, F_1 hybrids show superior performance over their parental homozygous lines.

The relatively late divergence of somatic and reproductive lineages also allows hereditary alterations that occur through somatic development to be transmitted to subsequent generations. The need for developmental plasticity and environmental interaction suggests that plants would intensively employ epigenetic regulatory strategies that can give heritable, often reversible, changes in their genetic information without immediate altering of their primary nucleotide sequence. These epigenetic strategies might also be elaborated in plants, because of the opportunity to select and to transmit to the next generation metastable epigenetic stages established through development and the somatic tissue.

DNA methylation is in the core of these epigenetic changes. It is a cause of new phenotypic variation. DNA methylation redefines the concept of genetic variation and the way the genotype and the environment interact. Further understanding of its role and mechanism of function should reveal a whole, genome wide, mechanism of regulation of the genetic information and further understanding of important phenomena such as transvection, paramutation, quelling, transgene inactivation, inbreeding depression and its antipode heterosis.

REFERENCES

Amasino, R.M., M.C John, M. Klaas, and D.N. Crowell. 1990. Role of DNA methylation in the regulation of gene expression in plants, p.187–198. In G. A. Clawson et al. (ed.) Nucleic acid methylation. Alan R. Liss, New York.

Cai, Q., C.H. Guy, and G.A. Moore. 1996. Detection of cytosine methylation and mapping of a gene influencing cytosine methylation in the genome of Citrus. Genome 39:235–242.

Cedar, H. 1984. DNA methylation and gene expression. p.147–164. *In* A. Razin et al. (ed.) DNA methylation. Biochemistry and biological significance. Springer Verlag, New York.

Cherry, J.H., R.H. Hagerman, J.N. Rutger, and B.J. Jones. 1961. Acid-soluble nucleotides and ribonucleic acid of different corn inbreds and single-cross hybrids. Crop Sci. 1:133–137.

De Vienne, D., J. Burstin, S. Gerber, A. Leonardi, M. Le Guilloux, A. Murigneux, M. Beckert, N. Bahrman, C. Damerval, and M. Zivy. 1996. Two-dimensional electrophoresis of proteins as a source of monogenic and codominant markers for population genetics and mapping the expressed genome. Heredity 76:166–177 .
De Vienne, D., A. Maurice, A. Leonardi, J. M. Josse, and C. Damerval. 1994. Mapping factors controlling protein expression. Cell Molec. Biol. (Noisy-le-Grand) 40:29–39.

Fasoulas, A.C., and V.A. Fasoula. 1995. Honeycomb selection designs. Plant Breed. Rev. 13:87–139.

Frei, O.M., C.W. Stuber, and M.M. Goodman. 1986. Use of allozymes as genetic markers for predicting performance in maize single cross hybrids. Crop Sci. 26:37–42.

Gruenbaum, Y., T. Naveh-Many, H. Cedar, and A. Razin. 1981. Sequence specificity of methylation in higher plant DNA. Nature (London) 292:860–862.

Hollick, J.B., J.E. Dorweiler, and V.L. Chandler. 1997. Paramutation and related allelic interactions. Trends Genet. 13:302–308.

Kafka, M. 1996. Regulation of gene expression in inbred lines and in hybrids of maize (*Zea Mays* L). Ph. D. Diss., Aristotelian Univ. of Thessaloniki, Thessaloniki, Greece.

Lee, M., E.B. Goldshalk, K.R. Lamkey, and W.L. Woodman. 1989. Association of restriction fragment lengh polymorphisms among maize inbreds with agronomic performance of their crosses. Crop Sci. 29:1067–1071.

Leonardi, A., C. Damerval, Y. Hebert, A. Gallais, and D. De Vienne. 1991. Association of protein amount polymorphism (PAP) among maize lines with performances of their hybrids. Theor. Appl. Genet. 82 :552–560.

Melchinger, A.E., M. Lee, K.R . Lamkey, A.R. Hallauer, and W.L. Woodman. 1990. Genetic diversity for restriction fragment length polymorphisms and heterosis for two diallel sets of maize inbreds. Theor. Appl. Genet. 80:488–496.

Messeguer, R., M. Ganal, J.C. Steffens, and S.D. Tanskley. 1991. Characterization of the level, target sites and inheritance of cytosine methylation in tomato nuclear DNA. Plant Molec. Biol. 16:753–770.

Nebiolo, C.M., W.J. Kaczmazczyk, and V. Ulrich. 1983. Manifestation of hybrid vigor in RNA synthesis parameters by corn seedling protoplasts in the presence and absence of gibberellic acid. Plant Sci. Lett. 28:195–206.

Patel, C.V., and K.P. Gopinathan. 1986. Determination of trace amounts of 5-methylocytosine in DNA by reversed-phase high-performance liquid chromatography. Anal. Biochem. 164:164–169.

Reich, N.O., C. Olsen, F. Osti, and J. Murphy. 1992. In vitro specificity of EcoRI DNA methyltransferase. J. Biol. Chem. 267:15802–15807.

Richards, E.J. 1997. DNA methylation and plant development. Trends Genet.13:319–323.

Romagnoli, S., M. Maddaloni, C. Livini, and M. Motto. 1990. Relationship between gene expression and hybrid vigor in primary root tips of young maize (*Zea mays* L.) plantlets. Theor. Appl. Genet. 80:767–775.

Silva, A.J., and R. White. 1988. Inheritance of allelic blue-prints for methylation patterns. Cell 54:145–152 .

Smith, O.S., J.S.C. Smith, S.L. Bowen, R.A. Tenborg, and S.J. Wall. 1990. Similarities among a group of elite maize inbreds as measured by pedigree, F_1 grain yield, grain yield heterosis and RFLPs. Theor. Appl. Genet. 80:833–840.

Stuber, C.W., S.E. Lincoln, D.W. Wolff, T. Helentjaris, and E.S. Lander. 1992. Identification of genetic factors contributing to heterosis in a hybrid from two elite maize lines using molecular markers. Genetics 132:823–839.

Tsaftaris, A.S. 1987. Isozymes in plant breeding. p. 103–119. *In* M.C. Rattazzi et al. (ed.) Isozymes: Current topics in biological and medical research. Alan R. Liss, New York.

Tsaftaris, A.S. 1990. Biochemical analysis of inbreds and their heterotic hybrids in maize. p. 639–664. *In* A. Ogita and C. Markert (ed.) Isozymes: Structure, function and use in biology and medicine. Wiley-Liss, New York.

Tsaftaris, A.S. 1995. Molecular aspects of heterosis in plants. Physiol. Plant. 94:362–370.

Tsaftaris, A.S., and M. Kafka. 1996. The role of epigenetic changes of DNA in plant breeding. p. 5–10. *In* S. Zotiset al. (ed.) Proc. of XI Panhellenic Conf., Plant Breeding and Problems in Modern Agriculture. Florina, Greece.

Tsaftaris, A.S., and M. Kafka. 1998. Mechanisms of heterosis in crop plants. J. Crop Prod. 1:95–111.

Tsaftaris, A.S., M. Kafka, and A. Polidoros. 1997. Epigenetic modifications of total genomic maize DNA: The role of growth conditions. p.125–130. *In* A. S. Tsaftaris (ed.) Genetics, biotechnology and breeding of maize and sorghum. The Royal Soc. of Chem. Press, Cambridge, England.

Tsaftaris, A.S., and A.N. Polidoros. 1993. Studying the expression of genes in maize parental inbreds and heterotic and non-heterotic hybrids. p. 283–292. *In* A. Bianci et al. (ed.) Proc. XII Eucarpia Maize and Sorghum Conf., Bergamo, Italy.

Welsh, J., R.J. Honeycutt, M. Mc Clelland, and B.W.S. Sobral. 1991. Parentage determination in maize hybrids using the arbitrarily primed polymerase chain reaction (AP-PCR). Theor. Appl. Genet. 82:473–476.

Chapter 19

Potential Heterosis Associated with Developmental and Metabolic Processes in Sorghum and Maize

J. D. Eastin, C . L. Petersen, F. Zavala-Garcia, A. Dhopte, P. K. Verma,
V. B. Ogunlea, M. W. Witt, V. Gonzalez Hernandez, M. Livera Munoz, T. J. Gerik,
G. I. Gandoul, M. R. A. Hovney, and L. Mendoza Onofre

INTRODUCTION

Organizers of this symposium recognized the enormous value of heterosis or hybrid vigor in increasing world wide production of sorghum [*Sorghum bicolor* (L.) Moench], maize (*Zea mays* L.), and other crops along with the need for further increasing the rate of production improvement in view of our rapidly expanding population. They further recognize that our knowledge about the genetics, physiology, biochemistry or molecular bases of hybrid vigor is meager. Expanding such knowledge should enhance our ability to better use hybrid vigor if we are able to successfully discover the approximate limiting order of yield dependent traits for which useful heterotic responses can be found. This may be especially true in applying newer biotechnology techniques if we can characterize the limiting functional traits in question in sufficient biochemical detail and if the corresponding working genes can be cloned.

Definitions of heterosis vary somewhat but usually have general commonality with the definition given by Brewbaker (1964) "Heterosis is a genetic expression of the beneficial effects of hybridization." Again, increased use of heterosis to achieve yield gains will likely depend considerably on advancing our understanding of heterosis developmentally and metabolically in a sizable proportion of the individual yield limiting processes that contribute to Brewbaker's total "beneficial effects of hybridization". How do we go about developing this understanding to better use heterosis?

Development of grain is complex involving many essential processes over a long period of time. One cannot hope to investigate all processes and must, therefore, narrow the research focus first by determining critical growth stages and then estimating the yield limiting order of major contributing developmental process and their controlling metabolic (biochemical) processes within stages. Determining critical developmental stages is difficult but has been achieved to reasonable degrees for a number of crops; however, characterization of controlling biochemical processes is sparingly addressed. Discerning any biochemically limiting factors requires finding or often developing contrasting genotypes or germplasm pools for the character(s) in question. Developing appropriate germplasm is often a productive practical exercise in itself. Even without knowing controling biochemical process details, excellent screening progress can be made on things like abiotic stresses simply by knowing the sensitive developmental stages and pressuring the germplasm with the abiotic stresses desired. Useful commercial germplasm can emerge from screenings set up to study biochemical contrasts.

PHYSIOLOGICAL OVERVIEW AND RESULTS

Developmental Processes and Yield Components

Most physiological–developmental processes research on yield limitations has been conducted primarily to explain yield differences rather than for the sake of explaining heterosis mechanistically. Historically, in the late 1950s and early 196's, crop physiology was a relatively new discipline and its probable role in crop improvement was quite diffuse and uncertain. The application of infrared analyzers to field photosynthesis measurements, use of radioactive carbon in translocation research, and other such technologies were stylish, and information on essential physiological processes was accumulating. In addition, the traditional agronomic investigations on dates and rates of seeding, fertility trials, etc. remained in full swing. Scientists studied biochemistry, traditional agronomy and photosynthesis but basically there was little interaction between the biochemical level and the traditional agronomy field experiment level. Unfortunately, developmental physiology was hardly mentioned at that time which left a large integration void between field experiments and molecular biology. While in its infancy in the 1960s, field definition of whole plant physiological processes was continuing, and a good information base was being built especially on maize. The information was not, however, particularly useful to plant breeders in improving yields. Evans' (1993) remarkable review and analysis of crop adaptation and yield outlines the extraordinary complexity of assessing yield limitations in a given environment and physiologically assisting breeders in improving crops. For example, we initiated sorghum research at Nebraska on photosynthesis, stomatal physiology, light interception, translocation, nitrogen nutrition, and stress responses in the mid 1960s, but after three or four years, we didn't know a great deal more about yield limitations that would help breeders than at the onset of our program. Again, sound scientific information was accumulating but a useful physiologically based breeding focus evaded us. Useable explanations of yield limitations for breeders were scant. Reflections on this took us back to the field to analyze developmental patterns of sorghum genotypes known to yield differently due to height and other differences. One major approach we used involved determining why parents and hybrids yield differently. To do this Eastin and Sullivan (1974) developed simple developmental stage terminology (later expanded by Eastin) appropriate to yield and yield component analyses as follows:

Growth Stage 1 (GS1)--planting to panicle initiation (PI)--the vegetative stage. 1-1---planting, 1-2---germination, 1-2-1---epicotyl emergence, 1-2-2---hypocotyl emergence, 1-3---seedling emergence, and 1-4---leaf numbering.

Growth Stage 2 (GS2)--PI through floret differentiation (FD) to bloom--floral developmental stage---setting seed number potential. 2-1---panicle initiation, 2-2---floret differentiation, 2-3---microsporogenesis, and 2-4---megasporogenesis. (Lee et al., 1974; Dhopte, 1984).

Growth Stage 3 (GS3)--bloom to physiological maturity (BL)(Eastin et al., 1963)---grain fill when seed weight is determined within genetic limits. 3-1---anthesis, 3-2---endosperm cell walls formed, 3-3---milk stage, 3-4---soft dough, 3-5---hard dough, 3-6---dark layer, and 3-7---drydown.

Being open ended, this system can be altered to accommodate most existing or new management–research problems. Table 19–1 illustrates the utility of this system from early 1970s research using vegetative, floral and grain filling developmental terminology. All the hybrids tested yielded higher than their parent means except for Redlan × TX414. Hybrid RS671 had an 18.8% higher grain yield than the parent mean, largely by virtue of a 16.4% faster grain fill rate because grain fill duration was only 2.5% longer (nonsignificant) than the parent mean. By contrast, the other heterotic hybrids had from 10 to 15% faster grain fill rates and filled grain 10 to 12% longer. Both GS3 duration and grain fill rate (production efficiency) contributed to

POTENTIAL HETEROSIS

Table 19–1. Grain yield data for sorghum hybrids and respective parents at Mead, NE.

Genotype	Days to maturity	Grain yield	Hybrid % over parent $\bar{x}$	Production rate Grain	Hybrid % over parent	GS3†	Grain filling period and production rate Hybrid % over parent	Grain	Hybrid % over parent ($\bar{x}$)
	d	kg ha^{-1}	%	kg ha^{-1} d^{-1}	%	d	%	kg ha^{-1} d^{-1}	%
RS 671‡	108	7461§	18.8	69.1		39.8		187.5	(16.4)#
Redland	112	6747¶		60.2	14.7	38.8	2.5	173.9	7.8
TX415	108	5817		53.9	28.2	38.8	2.5	149.9	25.0
RS626	105	7816§	25.9	74.4		42.6		183.5	(14.3)
CK60	106	6277		59.2	25.6	39.8	7.0	157.7	16.3
TX414	103	6184		60.0	24.0	37.8	12.6	163.3	12.3
RS625	105	8053§	21.9	76.7		42.7		188.6	(10.2)
Martin	103	7288¶		70.8	8.3	38.2	11.7	191.8	-1.6
TX414	103	5925		57.5	33.3	37.8	12.9	156.7	20.3
RS610	106	8250§	27.1	77.8		44.0		187.5	(14.9)
CK60	106	6464		61.0	27.5	39.8	11.0	162.4	15.4
7078	104	6521		62.7	24.0	39.8	11.0	163.8	14.4
RT	108	6982	6.0	64.6		40.9		170.7	(-0.6)
Redlan	112	6891		61.5	5.0	38.8	5.4	177.6	-4.0
TX414	103	6280		61.0	5.9	37.8	8.2	166.1	2.8

† GS3 grain fill.
‡ The first genotype of a set is the hybrid followed by the female and then a pollinator parent.
§ Significant difference between hybrid and parent mean.
¶ Significant difference between parents.
# Mean of hybrid over parents in parentheses.

improve yield in these hybrids while in RS671 only higher production rate was a major heterotic contributor to yield.

The next question to be addressed for sorghum improvement was developmental stage sensitivity particularly during GS2 because that is when seed number potential is set within genetic limits. Also, it usually is the period of greatest heat and water stress in the U.S. Great Plains. Ogunlela (1979) elevated night temperature 5°C above ambient in the field and noted the greatest yield reduction (28%) at approximately the late microsporogenesis stage (FD1–FD7, see Table 19–2). Manjarrez et al. (1989) later confirmed that microporogenesis was the most sensitive stage to drought stress followed by early grain filling (about ten days after anthesis) (see also Dickinson, 1976). Eastin et al. (1988) partially reviewed stage sensitivities used and outlined development of a preanthesis field stress screening method. Basically stress was induced near microsporogenesis by overplanting on a sandy soil under 350 mm of rainfall and evaluating yields compared with less stressed plots.

Given the complexity of stress resistance in sorghum, a population approach was chosen to assure high numbers of recombinations in developing germplasm to test this.screening model. Derivation of lines for this research was done by first randomly mating appropriate populations in a favorable environment at Mead, NE, where 150 to 200 fertile S_1 plants with high yield potential were selected; S_2 plants were then grown in a winter nursery where three to five plants were selected per row to generate 500 to 1000 S_3 families that were subjected to stress screening as outlined above (Eastin et al., 1988).

Table 19–2. Influence of night temperature on yield and other characteristics of RS671 grain sorghum at Lincoln, NE. Night temperatures in the field were regulated at 5°C above ambient. Values in parenthesis are percentage reduction from the control except for weight per 1000 seeds, which is change from control. (Ogunlela, 1979).

Treatment time	Grain yield	Seed number	Seed weight	Grain dry matter rate
	g plant^{-1}	no. plant^{-1}	g 1000 seed^{-1}	g grain GS3 d^{-1} plant^{-1}
Control	66.9	2659	26.6	2.09
PI_1 to PI_7†	59.3 (11%)	2333 (12%)	27.2 (2%)	1.85 (11%)
PI_8 to FD_1‡	53.4 (20%)	2174 (18%)	27.8 (5%)	1.71 (18%)
FD_1 to FD_7	48.0 (28%)	1855 (30%)	29.7 (12%)	1.49 (29%)
FD_8 to BL_1§	52.7 (21%)	2176 (18%)	27.8 (5%)	1.66 (21%)
BL_1 to BL_7	55.9 (16%)	2223 (16%)	25.5 (-4%)	1.80 (14%)

† PI, panicle initiation (subscripts are days).
‡ FD, floret differentiation (stamen and pistil primordia).
§ BL, bloom.

Results from this physiologically based screening approach appear useful, judging from our tests as well as independent U.S. Regional Research tests. For example, pollinators of the top six hybrids plus the 10th hybrid in D.T. Rosenow's 1989 U.S. Regional Research test at Lubbock, TX (Texas A&M University) came from our screening program (top yield of 2700 kg ha^{-1}). Commercial checks DK46 and DK41Y ranked 16th and 17th, Wheatland × TX430 was 23rd and RS671 was 31st out of 42 total entries. Hybrids 5 (46038A × N91) and 6 (46031A × N91) in the Lubbock test were also put in a high yield environment test at Mead, NE (J.F. Pedersen, U.S. Regional Research test) in 1989 and ranked 1 and 2 with a top yield of 9103 kg ha^{-1}. These results suggest that this screening approach allows us to select for both stress resistance and high yield potential to fit sorghum's variable environments where yields range from crop failure to the equivalent of good maize production conditions even at the same location in different years. A number of other Lubbock Regional Research Tests have given similar results about 80% of the time. Success on the high yield end probably comes from making the original S_1 selections for high yield in a favorable environment and then stress testing later. The one visible difference in the N91 pollinator hybrids in the above regional tests is that they have larger than normal (bold) seeds that led us to look closer at the seed weight component of yield counter to conventional wisdom.

Table 19–3 shows yield and yield component data along with grain fill lengths for Wheatland and TX3042 females pollinated by N91 and sister line 17473. Because N91 and 17473 are similar, the respective Wheatland (W) and TX3042 hybrid means will be discussed. The W hybrid mean of 7367 kg ha^{-1} exceeds the TX3042 hybrid

Table 19–3. Associations between sorghum grain yield, length of grain fill (GS$_3$), rate of grain fill, seed m^{-2} and seed weight.

Genotype	Grain yield and components			Days to		Rate of grain fill
	Grain yield	Seed number	Seed weight	Bloom	GS3	
	kg ha^{-1}	seeds m^{-2} (1000)	g 100 seeds^{-1}	d	d	kg ha^{-1} GS3d^{-1}
W† × 17473	7884	22.4	3.49	62.8	39.8	198
W × 91	6849	19.3	3.56	62.6	40.4	169
Mean 1	7367	20.9	3.53	62.7	40.1	184
TX3042 × 17473	6611	23.2	2.87	57.3	29.6	223
TX3042 × N91	6341	23.2	2.74	57.3	30.1	208
Mean 2	6476	23.2	2.81	57.3	30.1	216
Mean 1/Mean 2	1.14		1.26	1.09	1.33	
Mean 2/Mean 1		1.11				1.17

† W, Wheatland

mean of 6 476 kg ha^{-1} by 14%, even though the TX3042 hybrids have 23 200 seeds m^{-2} compared with 19 300 seeds m^{-2} for the W hybrids (11% superiority for TX3042 hybrids). Wheatland hybrid yield is superior because the seeds are 26% larger (3.53 vs. 2.81 g 100 seeds^{-1}). The larger seed weight appears attributable to a 33% longer grain fill period for the W hybrids (40.1 vs. 30.1 days). With pollinators being common and females being different, the seed size is clearly a female effect in this case. However, other observations indicates that the sister R (pollinator) lines N91 and 17473 tend to give high progeny seed weights as does TX430 (pollinator in W crosses) which seems generally not to have been recognized in the literature even though W × TX430 was popular in the mid 1960s. Therefore, experiments were set up to compare male and female effects on sorghum hybrid yields and yield components. Selected seed size and seed number references were reviewed by Eastin et al. (1990).

Table 19–4 contains yield and yield component data on the parent lines used. Note, for A lines, that Wheatland (3.05 g 100 seeds^{-1}) and 46038 (2.28 g 100 seeds^{-1}) represent the extremes in seed weight with TX3042 being near the middle; all have similar numbers of seeds m^{-2}. The 17473 (NE stress resistant) and TX2737 pollinators have the same seed weight with 17667 being smaller. Both the Wheatland (W) and 46038 × 17473 hybrids had about 25% heterosis for yield and from 15.8 to 19.4 heterosis for g 100 seeds^{-1} while maintaining or slightly increasing seeds m^{-2} (6.4 and 4.5 %, respectively) above the parent means, which is contrary to conventional thinking regarding how to improve yield. Many sorghum breeders feel there is a strong negative correlation between seed number and seed weight and, therefore, ignore seed weight. While that negative correlation occurs often, exceptions like the above appear noteworthy. One might have expected the W × TX2737 hybrid to show heterosis but it didn't in this case. However, the 46038 × TX2737 hybrid yielded 35% more than the parent mean which was accounted for by 27% heterosis in seeds m^{-2} plus a 7% trend toward increased seed weight. The combination of the smallest seed weight female (46038) × the smallest male (17667) gave about 20% heterosis for both seed number and seed weight. Using the W female with 17667 gave 12% grain yield heterosis, which is accounted for solely by 11.7% heterosis for seeds m^{-2}. Summarily then, seeds m^{-2} traditionally correlates positively with grain yield while seed weight usually does not correlate positively. The general opinion that there is an inverse correlation between seeds m^{-2} and g 100 seeds^{-1} has led people to generally ignore seed weight as an important yield component probably because yield components are rarely measured in sorghum tests, thereby obscuring the observations noted above. The notable exceptions in these data suggest that g 100 seeds^{-1} merits greater attention. The seed weight trait, however, is difficult to work with because of apparent linkages between seed size control genes and plant type control genes.

Tables 19–3 and 19–4 represent sorghum responses in temperate environments which are rather different from responses in Marin, Mexico, a more tropical setting. Table 19–5 illustrates grain fill days of 25 to 30 rather than up to 40 in Nebraska. The heat units involved at the two locations may be less divergent than the days. Wheatland × 17473 is still a top hybrid but heterosis for seed weight is only 5% compared with 15.9% at Lincoln, Nebraska. At least no decline in seed weight occurred in Mexico while heterosis for seeds m^{-2} is a startling 93% to account for most of the yield heterosis of 104%. The other clear example of heterosis for seeds m^{-2} is 46038 × 17473. The line 17473 is a product of the Nebraska–Kansas preanthesis stress screening program mentioned earlier. Significantly, all of the hybrids show heterosis or trends toward heterosis for seed weight, which apparently is a noteworthy stress recovery and/or stress resistance characteristic in that hot Mexican environment. This is less obvious in temperate environments where seeds m^{-2} will often be on the order of 15 000 to 25 000 rather than the 5 000 to 12 000 seeds m^{-2} in the Mexico test under lower populations. Longer term observations are needed on this.

The r^2 value for regressing kg ha^{-1} on seeds m^{-2} is 0.838 and supports the conventional idea that higher seeds m^{-2} satisfactorily accounts for higher yields; however, the r^2 for yield vs. kg ha^{-1} GS3 d^{-1} is 0.870 implicating metabolism during

Table 19–4. Grain yield and yield components (seeds m⁻² and seed weight) of sorghum hybrids compared with their parental line means at Lincoln, NE.

Hybrids	Grain			Seed number			Seed weight (size)		
	Yield	Parent $\bar{x}$	% hybrid deviation from parent $\bar{x}$	Number	Parent $\bar{x}$	% hybrid deviation from parent $\bar{x}$	Weight	Parent $\bar{x}$	% hybrid deviation from parent $\bar{x}$
	kg ha⁻¹	kg ha⁻¹	%	seeds m⁻²	seeds m⁻²	%	g 100 seeds⁻¹	g 100 seeds⁻¹	%
Females									
Wheatland	7777			25843			3.05		
46038	6021			26504			2.28		
TX3042	7401			25662			2.89		
Pollinators									
17473	9002			33510			2.78		
TX2737	8593			30772			2.80		
17667	5645			25374			2.25		
Wheatland X									
17473	1066	8530	25.0	31603	29677	6.4	3.38	2.92	15.8
TX2737	8342	8154	2.3	28404	28308	0.3	2.95	2.93	1.0
17667	9526	6711	12.1	28603	25609	11.7	2.64	2.65	0.0
46038 X									
17473	9471	7652	23.8	31358	30007	4.5	3.02	2.53	19.4
TX2737	9847	7276	35.3	36294	28638	26.7	2.72	2.54	7.0
17667	8404	5833	44.1	31025	25939	19.6	2.72	2.27	19.8

Table 19-5. Sorghum hybrids and parent line yield, yield components, and heterosis responses in Marin, Mexico, 1997.

Genotype	Grain yield	Seed weight	Seed no.	% heterosis using parent $\bar{x}$			Grainfill	
				Grain yield	Seed weight	Seed no.	Period	Rate
	kg ha⁻¹	g 100 seeds⁻¹	seeds m⁻²	%	%	%	d	kg ha⁻¹ d⁻¹
Wheatland/17473	3579	2.93	12202	104	5	93	28.6 BCDE†	125
P8310‡	3066	2.87	10693				27.1 DEFG	113
TX430	2411	2.60	9273				25.8 FG	93
W/TX430	2360	3.40	6941	7	17	-10	25.8 A	91
46038/17473	2074	3.00	6914	58	41	13	28.5 ABCD	73
46038/TX430	2068	2.80	7385	17	24	-1.6	27.7 CDEF	75
Wheatland (W)	1984	3.23	6135				28.4 BCDE	70
W/TX2737	1883	3.17	5945	11	15	-3	29.9 AB	63
TX3042/17473	1870	3.40	5501	33	44	-8	29.3 ABCD	64
46038/TX2737	1838	2.87	6412	47	35	8	27.8 BCDEF	66
TX3042/TX2737	1753	3.00	5842	30	28	2	29.2 ABCD	60
17473	1521	2.33	6526				25.4 G	60
TX2737	1399	2.30	6081				29.3 ABC	48
TX3042	1297	2.40	5405				28.7 BCDE	45
46038A	1110	1.93	5742				26.5 EFG	42

† Duncan's multiple range test (0.05).

‡ Pioneer Hi-Bred International hybrid.

grain fill as a potentially important factor perhaps apart from or in addition to the influence of sink size (seeds m^{-2}) in regulating respiration by virtue of influencing the plant's demand for respiratory products (Beevers, 1970). While the low r^2 for kg ha^{-1} vs. g 100 seeds^{-1} of only 0.216 reinforces the value of a high seeds m^{-2}, inspection of Fig.19-1 illustrates that considerable variability for yield (from 1 750 to 3 500 kg ha^{-1}) is noted at about 2.9 g 100 seeds^{-1} suggesting that selection for seed size without sacrificing seed numbers is a realistic option. Note that the top hybrid has the 17473 pollinator, which was selected under temperate stress conditions (Western Kansas) for a high, stable seed number. The comparatively high seed number seems to carry over in the Marin test. Again, the notion that one can select for higher seed weight without sacrificing or even increasing seed number is viable. Possibly heterotic response selection for greater seed size (and longer grain fill) is needed in temperate zones and a higher, more stable seed number in the tropics is needed without sacrificing the alternative yield component in both cases. It appears that such goals are feasible if yield components are considered and monitored in breeding programs. For example, we are attempting to improve yields by putting heterosis for larger seed size into our NP39R preanthesis stress resistant tan plant population by maintaining the 17473 type of heterotic seed size response (W × 17473 or W × TX430 type responses). Selection consists of recombining about equal numbers of seeds from high seed number and large seed size heads for recombination in each cycle. Hopefully this will maintain the heterotic seed weight response, either not very common in the industry or not commonly recognized, and still retain high seed numbers. Five to eight recombination cycles will likely be required to overcome the apparent deleterious seed size control linkages referred to earlier before the proper seed number–seed weight combinations for a range of environments can be expressed and assessed. Obviously, the optimum combination for a particular environment is an adaptive value, the potential flexibility for which we likely will not understand until some of the apparent linkages involving seed size control genes are broken up and a serious, commercial program on manipulating seed weight in conjunction with seed number is attempted.

Yield Gains

Ziegler (1990) considered that the increase in yield for many crops over the preceding 50 years was due to changes in allocation of assimilates rather than changes in total production. Similarly, Evans (1993) concluded his review analysis of yield improvement stating "We have seen, for example, that the predominant improvements so far have not been in the efficiency of the major metabolic and assimilatory processes, but in the patterns of partitioning and the timing of development." The obvious is that improvements in harvest index (HI) for grain are indeed a function of changes in patterns of partitioning or preferably termed changes in differentiation and growth patterns; however, these partitioning allocation pattern differences must eventually be traceable to metabolic reactions. While the efficiency of major metabolic and assimilatory processes may not have been altered much by plant breeders (Evans, 1993), it seems likely that the manner in which they interact collectively and complement each other probably has been manipulated subconsciously and perhaps substantially by plant breeders and must influence overall production efficiency. Knowing where to begin in searching for and exposing limiting metabolic functions–processes is difficult but often may best come from evaluating limiting developmental stage times and then evaluating the relative limitations of major essential processes (systems) including photosynthesis and respiration at limiting stages. Lopez-Peralta et al. (1991) reported that sucrose level influenced the growth and development of young sorghum panicles excised during GS2 and cultured in vitro. Panicle growth dependency on sucrose level implicates either photosynthesis level available to supply adequate substrate or capacity of respiration and coupled syntheses to elaborate stored energy into growth or both. In this connection, the work of Ogunlela (Table 19–2) illustrated: (i) that night time temperature effects were

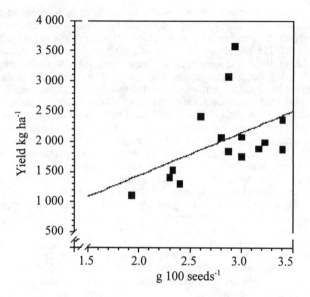

Fig. 19–1. Sorghum grain yield vs. g 100 seeds^{-1} at Marin, Mexico, 1997.

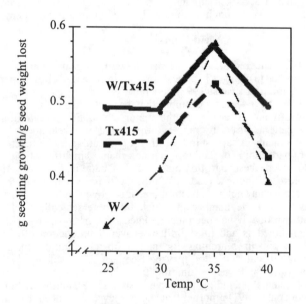

Fig. 19–2. Gram of etiolated sorghum seedling growth per gram of seed weight lost (production efficiency) at different temperatures.

substantial causing up to 28% yield reduction and (ii) that the night treatments did not influence photosynthesis materially (Ogunlela & Eastin, 1984, 1985). Therefore, night metabolism effects appear to be more limiting than day effects during floret development.

Other evidence in sorghum which supports the relevance of investigating dark reaction systems (respiration associated events) as a generally limiting factor was reported by Rice and Eastin (1986). They measured root dry matter accumulation in nutrient solutions from panicle initiation through hard dough and also root respiration was measured with an oxygen electrode on RS671 (stress sensitive) and DeKalb C46 (stress resistant). Stress resistant C46 had a generally lower respiration rate and produced twice as much root dry matter per g of oxygen consumed. No effort was made to evaluate the efficiency difference in terms of growth or maintenance respiration; however, such observations render the judgement of Ogunlela and Eastin (1984,1985) and Eastin et al (1983) on the relative importance of dark vs. light metabolic reactions more tenable. The grain fill rate differences in Tables 19–1, 19–2, 19–3, and 19–5 add evidence supporting metabolic differences as contributing to yield differences (perhaps sometimes through effects of heterosis). Some will argue, however, that the production rate–metabolic differences in Tables 19–2 and 19–5 might be a direct consequence of relative differences in seed number (sink size). That is hard to establish unequivocally but again, even if it is true, the seed number differences established during GS2 still must be traceable to metabolic differences controlling differentiation and growth at the time from panicle initiation and development up to anthesis (GS2). These seed number differences subsequently no doubt influence the grain fill metabolic or production rate differences noted in the tables cited above. Nonetheless, there almost certainly are metabolic efficiency differences involved also (Rice & Eastin, 1986).

One final line of evidence led to pursuing more detailed dark respiration work. After Ogunlela's work we developed a simple test to expose dark production efficiency differences between genotypes and within a genotype at different temperatures. It is based on weighing sorghum seeds, germinating them in a rolled paper towel to the point of radicle appearance, and then placing different lots in the dark at 25, 30, 35 and 40°C. Harvests are made on linear parts of the respective temperature growth curves, seeds are weighed separately from new growth and the grams of new growth g^{-1} of seed weight lost (efficiency estimates) are plotted against temperature. Figure 19–2 illustrates one common result where there is often little difference between lines (W, Wheatland) and hybrids (W × TX415) at the optimal temperature but, as temperature diverges from near optimum in either direction, the hybrid is often superior. Results seem to offer a parallel example of where heterosis is temperature dependant as noted by McWilliam and Griffing (1964) in maize and by McWilliam et al. (1969) in *Phalaris*.

Based on the developmental physiology research of Ogunlela (1979; and Ogunlela and Eastin, 1985) above and subsequent research of Dhopte (1984) on night reaction limitations, it seems logical to look at dark respiration because it is the primary dark reaction process whereby energy from photoassimilate breakdown is coupled directly or indirectly to a host of essential plant maintenance and growth processes.

Metabolic Research

Respiration (R) rate, as a yield related function, is not well understood. Lambers (1985) reviewed this complex subject and cited references plus his own work suggesting that slow respiration rates sometimes correlate positively with yield but the relative merits of slow versus fast respiration are not clear. In our laboratory Chris Petersen is evaluating whole etiolated sorghum seedling CO_2 evolution and also mitochondria extracted from these seedlings. Genotypes are Redlan (heat stress sensitive line–SS), 17473 (stress resistant line–SR) and Dekalb DK46 (SR hybrid).

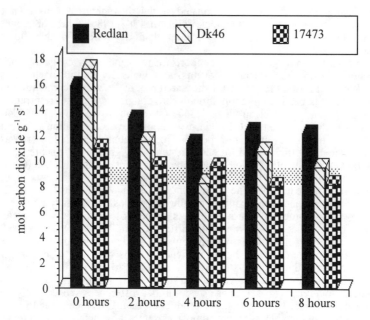

Fig. 19–3. Etiolated sorghum seedling CO_2 evolution response to heat shock (42°C exposure) duration. Respiration rates of non heat shocked seedlings at 28°C fall within the shaded band area for all three genotypes. Plotted values are respiration rates at 42°C after 0, 2, 4, 6, and 8 hours of continuing heat shock at 42°C.

Fig. 19–3 shows a shaded area between 7 and 9 moles CO_2 evolution/g^{-sec} which encompasses the R rates of all three genotypes at 28°C. Plotted values are R rates at 42°C after 0, 2, 4, 6, and 8 hours of exposure to 42°C (heat shock). The SS line almost doubles R at 42°C (compared to R at 28°C) and declines some during heat shock treatments but does not return to near the original 28°C R levels after 4 hours of heat shock as do DK46 and 17473. Both Redlan (high value) and 17473 (low value) may be damaged to some degree relative to DK46 after from 4 to 6 hours of 42°C exposure that may or may not relate to mitochondrial function.

On the other hand, Cruz-Garcia et al. (1995) claim that respiration rate cannot be regarded as a limiting factor in the early germination events of artificially deteriorated maize seeds, because they did not find reductions in the seed respiration rate (measured as a whole seed) during the first 24 hours of germination, whereas that deterioration induced a strong loss of seed germinability and of seedling vigor; however, more recent work by Santiago and Gonzales (unpublished data) showed a detectable and early reduction in respiration of the scutellum isolated from deteriorated maize seeds as compared with those from undeteriorated seeds. This reduction in respiration preceded the loss in seed germinability.

In depth interpretations of the Petersen sorghum respiration system and the Cruz-Garcia and Santiago-Gonzalez maize respiration system as limiting processes are difficult because there are no measures of coupling between substrate oxidation (CO_2 evolution or O_2 consumption) and released energy use for germination, differentiation, growth, etc. Mitochondrial analyses can be useful in this respect.

Farrar (1985) lists mitochondrial function as three fold: "metabolism of carbon compounds, electron transport, and recycling of the phosphorylated and reduced products ATP, NADH, and NADPH." Carbon compound byproducts furnish carbon skeletons for essential syntheses while the citric acid cycle and electron transport provide for energy demands of various essential syntheses via production of reducing equivalents and ATP. To investigate all suspected respiration coupled essential syntheses is impractical so mitochondrial function is evaluated as a broad indicator of dark metabolism efficiency. A measure of mitochondrial function and integrity is discerned by the organelle's ability to efficiently couple useful energy production (ATP formation) tightly to oxygen reduction (ADP:O); an ADP:O ratio of 2 is theoretically ideal when using exogenous NADH as the substrate.

Verma, in our laboratory, associated stress resistance with production of low molecular weight heat shock proteins (hsps) in sorghum in 1993. Also, in our laboratory, Petersen subsequently demonstrated that heat shocking sorghum from 35 to 42°C induces production of 22 kD hsps in mitochondria. He has demonstrated that the 22 kD hsp is similar but not identical to the 22 kD hsp isolated in maize (Lund et al., 1998). Also, induction of 22 kD hsps improved mitochondrial integrity (higher ADP:0 ratio) in all sorghum genotypes tested. However, the temperatures inducing synthesis of mitochondrial 22 kd hsps varied among sorghum genotypes as did the hsp quantities produced. Some stress resistant (SR) sorghums induce hsp production at 35°C while most induce at 38°C and higher but damage may have occurred at 38°C+. Following two hours of seedling heat shock at 42°C, mitochondria were extracted and tested at 42°C. Respective ADP:0 ratio for 17473 (SR), DK46 (SR) and Redlan (stress susceptible) were 1.6, 1.4, and 1.1 clearly suggesting a 31% superiority in energy conservation–utilization under stress for 17473 over Redlan. After extended heat shock (four hours or more) DK46 had a superior ADP:0 ratio over the others which was coupled with a lower O_2 uptake rate. Electrophoresis patterns appear to show higher 22 kD protein levels in DK46 and 17473 than Redlan after four and six hours of 42°C heat shock, however, prolonged heat shock may damage Redlan and 17473 to some degree.

Mitochondrial results generally complement gas exchange results (Petersen above) in that DK46 mitochondria are more stable than Redlan and 17473 mitochondria, which may relate to Redlan's higher, less stable etiolated seedling respiration rate and 17473s low rates at 42°C (Fig. 19–3) which are likely not well coupled in whole seedlings judging from the mitochondrial work. Expansion of this type of research to line–hybrid evaluations will likely provide some good information eventually on heterosis in metabolic events.

The emphasis placed here on metabolism is not new but research attention has been slight. Notions about physiological blocks and heterosis were considered at the last international heterosis conference (Mangelsdorf, 1952). Hageman et al. (1967) stressed the fact that grain growth and yield are the end results of a series of biochemical reactions and recognized changes in yield components. They attributed heterotic higher rates of growth to more efficiently organized enzyme systems rather than to higher enzyme quantities or qualitatively different enzyme systems. Schrader (1985) built on this concept and assayed several key enzymes over the season finding substantial variability among some key inbreds but did not offer a firm selection plan to explain or improve heterosis at the enzyme level.

SUMMARY

Examples were given to illustrate how heterosis in grain fill length (developmental events) influences yield. How to proceed using metabolic events to enhance heterosis bears much thought because attempts to correlate yield with the reactions that drive it have been modest in number and even more modest successwise. That is not surprising in view of the complexity of the systems involved and the long time frame. Therefore, it seems that some sound developmental physiology should be pursued related to yield components in order to determine, over several years at a target location, when in time individual yield components detract from or contribute heavily to grain yield. Variation in seeds m^{-2} and g 100 seeds^{-1} and associated yield should suggest to a crop improvement scientist: (i) which yield components need to be bolstered on the average, (ii) when in time yield losses are suffered, and (iii) how to narrow the focus for metabolic research that is attempting to explain heterosis (such that levels of gain achieved by molecular geneticists might be similar to those being accomplished in pest and herbicide resistance). Better orientation of metabolic research seems essential if sound use is to arise out of newer biotechnology investments in crop improvement apart from simply inherited traits. Evans (1993) allowed that Engledow and Wadham (1923) would have considered it treasonable for someone to write a book on crop yield without a chapter on yield components while today's crop physiologists may consider it quite reasonable to ignore yield components (as do many if not most plant breeders). Engledow and Wadham (1923) make a compelling point. Simple, enlightening yield component analyses can offer closer, productive relationships between geneticists, whole plant physiologists and other metabolically oriented scientists which will, sooner or later, enhance productivity gains from heterosis.

ACKNOWLEDGMENTS

Published as Nebraska Agric Exp. Stn. Paper 12283 Journal Series. Research Supported by the Nebraska Grain Sorghum Development, Utilization and Marketing Board (NGSDUMB) and the International Sorghum/Millet Cooperative Research Support Program (INTSORMIL).

REFERENCES

Beevers, H. 1970. Respiration in plants and its regulation. p. 209–214. *In* Prediction and measurement of photosynthetic productivity. PUDOC, Wageningen.

Brewbaker, J.R. 1964. Agricultural genetics. Prentice Hall, Englewood Cliffs, NJ.

Cruz-García, F., V.A. González-Hernández, J. Molina-Moreno, and J.M. Vázquez-Ramos. 1995. Seed deterioration and respiration as related to DNA metabolism in germinating maize. Seed Sci. Technol. 23:477–486.

Dickinson, T.E. 1976. Caryopsis development and the effect of induced high temperatures in [*Sorghum bicolor* (L.) Moench]. M.S. thesis. Univ. of Nebraska, Lincoln.

Dhopte, A.M. 1984. Influence of night temperature on microsporogenesis and megasporogenesis in *Sorghum bicolor* (L.) Moench. Ph.D. diss. Univ. of Nebraska, Lincoln.

Eastin, J.D., J.H. Hultquist, and C.Y. Sullivan. 1963. Physiological maturity in grain sorghum. Crop Sci. 13:175–178.

Eastin, J.D., and C.Y. Sullivan. 1974. Yield considerations in selected cereals. p. 871–877. *In* R.L. Bielski et al. (ed.) Mechanisms of regulation of plant growth. Bull.12. Royal Soc. of New Zealand,Wellington, New Zealand.

Eastin, J.D., R.M. Castleberry, T.J. Gerik, J.H. Hultquist, V. Mahalakshmi, V.B. Ogunlela, and J.R. Rice. 1983. Physiological aspects of high temperature and water stress. p. 91–112. *In* C.D. Raper, Jr., and P.J. Kramer (ed.) Crop reactions to water and temperature stresses in humid, temperate climates. Westview Press, Boulder, CO.

Eastin, J.D., P.K. Verma, A. Dhopte, M.D. Witt, D.R. Krieg and J.L. Hatfield. 1988. Stress screening and sorghum improvement. p. 716–722. *In* B.W. Unger et al. (ed.) Challenges in dryland agriculture: A global perspective. Proc. Int. Conf. on Dryland Farming. USDA, Amarillo, TX.

Eastin, J.D., A. Dhopte, P.K. Verma, V. Gonzalez-Hernandez, M. Livera-Munoz, M.W. Witt, V.B. Ogunela, and L.Mendoza-Onofere. 1990. Sorghum development, stress mechanisms and crop improvement. p. 152–163. *In* S.K. Sinha et al. (ed.) Proc. Int. Congress of Plant Physiology. Soc. Plant Phys. and Biochem. Water Technology Centre and Indian Agricultural Res. Inst., New Delhi, India.

Engledow, F.L. and S.M. Wadham. 1923. Investigations on yield in the cereals. Part I. J. Agric. Sci. 13:390–439.

Evans, L.T. 1993. Crop evolution, adaptation and yield. Cambridge University Press, New York.

Farrar, J.F. 1985. The respiratory source of CO$_2$ Plant Cell Environ. 8:427–438.

Hageman, R.H., E.R. Leng, and J.W. Dudley. 1967. A biochemical approach to corn breeding. Adv. Agron. 19:45–86.

Lambers, H. 1985. Respiration in intact plants and tissues: Its regulation and dependence on environmental factors, metabolism and invaded organisms. p. 418–473. *In* R. Douce and D.A. Day (ed.) Encyclopedia of plant physiology: New series. Springer-Verlag, Berlin.

Lee, K.-W., R.C. Lommasson, and J.D. Eastin. 1974. Developmental studies on the panicle initiation in sorghum. Crop Sci. 14:80–84.

Lopez-Peralta, M., M.E. Garcia-Duran, and V. Gonzalez-Hernandez. 1991. Growth and development of young sorghum panicles cultured in vitro. p. 17. Proc. Sorghum Improvement Conf. of North America, Univ. of Georgia, Griffin, GA. Sorghum Newsletter Vol. 32.

Lund, A., P. Blum, D. Bhattranakki, and T.E. Elthon. 1998. Heat-stress response of maize mitochondria. Plant Phys. 116:1097–1110.

Manglesdorf, A.J. 1952. Gene interaction in heterosis. p. 320–329. *In* J.W. Gowen (ed.) Heterosis. Iowa State College Press, Ames.

Manjarrez, S.P., V.A. González H., L.E. Mendoza O., and E.M. Engleman. 1989. Drought stress effects on the grain yield and panicle development of sorghum. Can. J. Plant Sci. 69:631–641.

McWilliam, J.R. and B. Griffing. 1964. Temperature-dependent heterosis in maize. Aust. J. Biol. Sci. 18:569–583.

McWilliam, J.R., B.D.H. Latter, and M.J. Matheson. 1969. Enhanced heterosis and stability in the growth of an interspecific *Phalarus* hybrid at high temperature. Aust. J. Biol. Sci. 22:493–504.

Ogunlela, V.B. 1979. Physiological and agronomic responses of grain sorghum *Sorghum bicolor* (L.) Moench hybrid to elevated night temperatures. Ph.D. diss. Univ. of Nebraska, Lincoln (Diss. Abstr. 80:10871).

Ogunlela, V.B. and J.D. Eastin. 1984. Effect of elevated night temperature during panicle development on sorghum [*Sorghum bicolor* (L.) Moench] yield components. Cereal Crops Res. Commun. 12:245–251.

Ogunlela, V.B. and J.D. Eastin. 1985. After-effect of elevated night temperature and heat-preconditioning on net carbon dioxide exchange and grain development in *Sorghum bicolor* (L.) Moench. Z. Achker Pflanzenbau. 154:182–192.

Rice, J.R. and J.D. Eastin. 1986. Grain sorghum root responses to water and temperature during reproductive development. Crop Sci. 26:547–551.

Schrader, L.E. 1985. Selection for metabolic balance in maize. p. 79–89. *In* J.E. Harper et al. (ed.) Exploitation of physiological and genetic variability to enhance crop productivity. Waverly Press, Baltimore, MD.

Ziegler, H. 1990. Role of plant physiology in assessing productivity potential under stress environment. p. 10–17. *In* S.K. Sinha et al. (ed.) Proc. of the Int. Cong. of Plant Physiology. Soc. Plant Phys. and Biochem. Water Technology Centre and Indian Agricultural Res. Inst. New Delhi, India.

Chapter 20

Biochemistry, Molecular Biology, and Physiology of Heterosis

Discussion Session

QUESTIONS FOR C.W. STUBER

S. Smith, Pioneer Hi-Bred, International, USA: Why does the bottleneck hypothesis proposed by Mangelsdorf (1952) predict just one major bottleneck rather than many major ones plus many smaller ones?

Response: I was essentially repeating the Mangelsdorf hypothesis with the prediction by Rhodes et al. (1992) that for each inbred, there is one major bottleneck locus and potentially a series of minor bottlenecks that manifest themselves only when the major bottleneck is ameliorated. I tend to agree with Dr. Smith, and, when we begin to dissect QTLs, we may find a few major loci, some intermediate, and many minor loci.

J. Brewbaker, University of Hawaii, USA: Is complementation not the primary basis for the great heterotic groups in maize?

Response: That is probably true. We are probably just using a different term to describe what plant breeders have used to distinguish heterotic groups.

J. Neto, Universidad Federal Rio Grande do Sul, Brazil: Do you think that QTL cloning could lead to the establishment of the basis for heterosis?

Response: Cloning of QTLs that have been identified at this time will not be done any time in the near future, primarily because of their large size. Many involve 25 or more map units in length.

E. Arias, Pioneer Hi-Bred, International, Mexico: In reference to the bottleneck hypothesis, do you think if would be possible to find heterosis in F_1 hybrids among C-3 plants converted to C-4 plants?

Response: I would suspect that there are bottlenecks in both C-3 and C-4 plants, and I doubt that making such a conversion would have much effect on heterosis.

H. Geiger, University of Hohenheim, Germany: Many experiments have shown that heterosis is greater under stress conditions, whereas your QTL study did not show any effect of stress factors. What is your explanation for this discrepancy?

Response: In studies such as the one discussed by Dr. Reeves in his opening address, he indicated that selection under stress conditions had improved productivity.

However, the materials used were of tropical origin with a broad range of variability. In our study in which we evaluated QTLs under stress conditions, we used materials derived from the cross of inbred line B73 with Mo17 that have been highly selected for the temperate conditions in the U.S. Corn Belt. So, the kinds of plant materials used differ greatly.

D. Duvick, Iowa State University, USA: To reassemble the best epistatic combinations in one inbred line, would you prefer the use of markers or genetically defined physiological processes, or is it worthwhile to reassemble favorable epistatic combinations in one inbred?

Response: At this point in time, we can probably manipulate molecular markers easier than we can genes associated with physiological processes. In most cases, we have not identified or located those specific genes. It may be worthwhile to reassemble favorable epistatic complexes. In fact, if you consider the data presented by Dr. Duvick in this conference on the decrease in heterosis among commercial breeding lines and predict that inbreds may approach hybrids in performance in 20 or 30 years, then I believe that it would be necessary to reassemble favorable epistatic complexes.

S. Sanchez, Chapingo University, Mexico: Is the bottleneck model similar to the so-called enhancer-suppressor gene system?

Response: I believe the bottleneck model can encompass many different models and could possibly include that as well as several other models.

E. Redona, Philippine Rice Research Institute, Philippines: Given that QTL expression is influenced by both environment and genetic background, how can QTL information be best used in hybrid breeding?

Response: I did not have an opportunity to discuss a breeding scheme in my paper today that we have been promoting, and actually testing, that involves the development of a set of near-isogenic lines (NILs). Assume that you have an elite inbred line that you may want to improve for its heterotic response with some other line. From knowledge of other material or lines, you are able to identify a donor source of genetic factors. You then develop a set of NILs by crossing the donor to the recipient line, followed by two or three marker-facilitated backcrosses. The goal is to develop a series of NILs, each with a different chromosomal segment from the donor, with the total group of NILs encompassing the entire genome of the donor. This set of NILs would then be crossed to an appropriate tester(s) to determine which NILs have improved heterotic response. An important characteristic of this approach is that you do not need to map QTLs in advance. All that is required is a set of markers that is reasonably well distributed throughout the genome. QTL mapping is a fringe benefit that is accomplished by observing those NILs that perform well in the testcrosses. In addition, because each NIL has only a small segment introgressed from the donor source, the NIL has only a small change in its total genetic makeup when compared with the original elite line. Therefore, the field testing required for potential commercial use should be minimal.

QUESTIONS FOR M. LEE

C. Loeffler, Pioneer Hi-Bred, International, USA: What do you expect the first concrete, measurable contribution of genomic research to agriculture will be? When?

Response: It depends on your definition of genomics. In my opinion, DNA markers are part of genomics. For anyone who has been backcrossing with transgenes, it has already happened. I am also aware of some successful attempts to develop cultivars based on QTL mapping data. So, it has already started.

D. Duvick, Iowa State University, USA: Can you define genomics in one sentence? For how many years has the term been used?

Response: I would define genomics as an integrated assembly of approaches with the goal being a comprehensive assessment of the content of an expression of an entire genome, starting with DNA and the functional analysis of its products, and ending with the phenotype. I do not know how long the term has been used. I believe that the term *genomics* has been used widely since the 1980s in the wake of the Human Genome Project.

QUESTIONS FOR A.S. TSAFTARIS

A. Dogra, University of Missouri, USA: Are there differences in methylation between inbred lines? How comparable are the differences in methylation in inbreds and hybrids.

Response: Yes, there are differences in inbreds and in hybrids for total methylation as well as for methylation patterns.

G. Granados, CIMMYT, Mexico: Can methylation techniques differentiate between stress levels, and therefore, could the techniques be used for stress tolerance selection?

Response: We used only one kind of stress in our analysis. We have not attempted to differentiate among stresses associated with temperature, light, etc.; however, I can speculate that the CRED-RA technique that I described in my paper might be able to detect different stress responses. This assumes that you select the appropriate primer and that the stresses cause detectable responses.

A. Vanavichit, Kasetsart University, Thailand: Could DNA methylation in F_1 hybrids be a consequence, rather than a cause of heterosis?

Response: Yes, lower methylation in F_1 hybrids could be a consequence and not a cause of heterosis, but it is related to the phenomenon of heterosis. When studying the dynamics of the methylation pattern through cell division and DNA replication, cells can be found that are dividing faster than DNA is being duplicated and faster than the methyltransferases can perform their functions. I believe that most of the somoclonal variation observed in tissue culture results from fast cell divisions that do not allow adequate time for the methyltransferases to do their job.

J. Crow, University of Wisconsin, USA: Are there data showing a mutation rate increase at methylated sites?

Response: Yes. Methylation points are hot spots for mutations. When point mutations have been studied, the point frequently was methylated cytosine.

J. Ininda, Kenya Agricultural Research Institute, Kenya: If stress environments lead to increased methylation, will you comment on selection for higher yielding hybrids in stress condition?

Response: It is probably more difficult to evaluate heterosis in stress environments. There tends to be less genetic variation in more stressful environments, so if you want to uncover all of the variation, then spaced planting will provide a better condition for this purpose. As Dr. Duvick indicated to you in his paper, we should concentrate on stable yields as well as on high yields. I believe that today's popular hybrids that have high and stable yields are those that respond well under a number of different environments, including those with different kinds of stress.

B. Cukadar-Olmedo, CIMMYT, Mexico: You indicated that there is more methylation in inbred lines than in hybrids. How are the differences among inbred lines reflected in the hybrids? Do you get more methylation in hybrids which have parental inbreds with higher methylation?

Response: We have not measured this correlation in a significant number of pairs, so I cannot answer that question. However, we plan to get the answer.

QUESTIONS FOR J.D. EASTIN

C. Stuber, USDA-ARS, USA: Early in your presentation, you showed a graph with hybrids that were stress resistant but did not perform very well under optimum moisture conditions. Near the end of your presentation, you showed data from several hybrids that you had developed and that showed heterosis for both seed number and seed weight simultaneously. How did these later hybrids perform under stress conditions as well as under optimum moisture conditions?

Response: Most of these hybrids fall in the stress resistant-water responsive (SRWR) category, because they have done quite well in numerous stress tests, and they also did well in the high yield environment at Mead, NE. There are several commercial hybrids that are now is that category, also. Probably not a large number, but as least a few, so sorghum breeders are progressing.

R. Ortiz, INIFAP, Mexico: For an applied breeding program, what is a practical method for screening lines for rate of grain fill and length of grain fill?

Response: It is relatively easy to determine grain fill length in sorghum. When we conduct experiments for which we want very precise measurements, we usually use three replications with five or six plants per row or replication. An easy and very efficient method involves tagging the bloom date at the tip (upper one inch) of the head all in the same day in a row. That may not coincide with the average bloom date, but that does not make any difference. Then calculate a whole plot bloom date. When heads are tagged in a reasonably uniform line, the dark layers will become apparent within a day or two of each other, so that trait can be scored relatively quickly. If one is just comparing breeding lines, it should be adequate to tag five or six heads in a row. That will provide a very good comparative estimate of length of grain fill. Of course, it is then necessary to measure yield to calculate the grain fill rate.

When we measure field grain fill length, we take the tip grain from a panicle branch in the next to the last whorl from the bottom, so that a small amount of immature grain is left as a reference point. We generally score 15 to 10 plants whether in replicated tests or within breeding lines to determine dark layer. In maize, we check for dark layer about two inches from the base of the ear. Measurement of dark layer in maize is less accurate that in sorghum because approximate anthesis is harder to determine in maize.

Chapter 21

Selection Methodology and Heterosis

J. G. Coors

INTRODUCTION

The genetic mechanisms underlying heterosis are largely unknown. Competing genetic theories differ on the relative importance of overdominance, epistasis, and linkage and how they contribute to hybrid performance. In a practical sense, though, knowledge of the genetics of heterosis has not been essential for maize (*Zea mays* L.) improvement. In the early part of this century, the commercial potential of maize hybrids increased interest and funding for breeding far beyond the level of other crops, and this led to one of the most highly acclaimed genetic success stories of modern times. The ample financial and intellectual resources of the hybrid seed industry have been primarily dedicated to insuring high selection intensities, extensive wide-area testing, improvements in mechanization, efficient data collection, and thorough data analysis. There is nothing unique to the hybrid breeding system in this allocation of resources. High selection intensities and accurate, efficient testing strategies are required for breeders of all crop species.

The issue of overdominance has blurred our focus on breeding strategies, and the situation is only being slowly resolved. Initially, theoretical studies such as those of Crow (1948) strongly supported overdominance. Development of the North Carolina mating design III also led to pioneering studies in maize initially supporting genome-wide overdominance (Comstock & Robinson, 1948, 1952; Gardner et al., 1953; Robinson et al., 1949). However, if assumptions regarding genetic load, mutation, and extent of incomplete dominance are relaxed, theoretical support for overdominance is less convincing (Crow, 1993), and subsequent design III studies showed that linkage disequilibrium exaggerated the magnitude of overdominance in the original maize studies (Gardner & Lonnquist, 1959; Robinson et al., 1960). Genomic analysis using molecular genetic markers to identify quantitative trait loci (QTL) may provide new insight, but progress has been slow. Stuber et al. (1992) showed that many QTL contributed to grain yield in maize, and a large subset of these loci had either overdominant or pseudo-overdominant gene action. They could not distinguish the type of dominance gene action because their study was based on a mating scheme similar to the NC III design and involved nearly the same degree of linkage disequilibrium that misled the initial investigators in the 1940s and 1950s (Cockerham & Zeng, 1996). Continuing analysis, however, showed that a portion of the overdominance dissipated after additional generations of recombination (Graham et al., 1997).

Even if overdominance were not important, as long as dominance contributes significantly to total genetic variance, the hybrid breeding system may still be the most effective strategy for identifying productive genotypes and making them commercially available. If heterosis were due primarily to dominance gene action at linked, complementary loci, then prescreening of inbred progenies followed by evaluation of testcross hybrids may be a very efficient scheme to rapidly deliver improved germplasm to farmers. But other breeding systems can also rapidly improve germplasm. In fact, as many have pointed out, eventually the optimum genotype (if such a thing exists) under such a model may be an inbred, not a hybrid.

Recurrent selection has been underway in maize since humans first harvested seeds from desirable plants and used the remnant seeds to replant in successive seasons. It continues today and includes all types of selection where selected genotypes are interbred at some point to provide genetic recombination. Both classical recurrent selection and hybrid breeding follow the same general procedures. Both are cyclic processes of developing and evaluating families, selecting the better genotypes, and then recombining selected genotypes to form an improved population. The primary focus of classical recurrent selection is the continual development of improved populations. Hybrid breeders, on the other hand, face a more difficult task of first developing inbreds that can be used in commercial hybrids, and only then using the most elite inbreds to develop new breeding populations, commonly F_2s, backcrosses, or other relatively narrow-based populations, for the next round of inbred development. Hybrid breeders have not used classical recurrent selection to a great extent (Good, 1990), but the hybrid system is no different from any other breeding method in that long-term progress depends not only upon the correct identification of superior genotypes, but also upon the rapid cycling of germplasm.

Several investigators have evaluated yield gains in hybrid maize since the 1930s. Duvick (1992) presented results from his own research as well as results from 10 previous hybrid studies (Table 21–1). In Duvick's research, 41 hybrids introduced by Pioneer from 1934 through 1989 were evaluated at three planting densities, and genetic gain was calculated by using the yields from the highest yielding density for each era. Genetic gain was estimated to be 76 kg ha⁻¹ yr⁻¹ or 76% of the 100 kg ha⁻¹ yr⁻¹ total increase. The mean genetic gain of all eleven hybrid trials summarized by Duvick (1992) was 68 kg ha⁻¹ yr⁻¹, and this was 71% of the total gain of 96 kg ha⁻¹ yr⁻¹.

Lamkey and Smith (1987) estimated rates of gain, but used populations derived from intermating the hybrids within each era (S_0 populations) and inbred bulks from each S_0 (S_1 populations) (Table 21–1). Rates were positive and significant but were only 86% (S_0 populations) and 56% (S_1 populations) of the mean of the hybrid studies. Based on these data, not only have breeders significantly increased the frequency of favorable alleles over the last six decades, but also the hybrid system has effectively used heterotic interactions to increase response rate; however, there is no direct evidence that the amount heterosis has increased over eras of breeding. Even though the gain in yield of inbreds has been less than hybrids, single-cross yields expressed as a percentage of parental inbred yields have decreased slightly (Duvick, 1984).

Duvick's 1992 hybrid study also included two germplasm sets that were the products of classical recurrent selection: cycles C0 and C9 of BS10 × BS11 and cycles C0 and C5 of PIAA × PIABB. The former was the result of nine cycles of full-sib reciprocal recurrent selection (RRS) at Iowa State University (Eyherabide & Hallauer, 1991), and the latter was from a combined S_1 and half-sib selection program at Pioneer. The annual gains from the BS10 × BS11 and PIAA × PIBB programs were 77 and 59 kg ha⁻¹ yr⁻¹. These results indicated a high degree of efficiency for RRS because the effort spent on improvement of these populations, as reflected by selection intensity, was much less than that devoted to hybrid development. Duvick cautioned, however, that the standards for "all-round excellence" were higher for hybrids, thus mandating greater total effort for hybrid improvement."

Table 21–1. Estimates of the annual increase in grain yield of maize hybrids, populations, and inbreds in the U.S. Corn Belt (hybrid results adapted from Duvick, 1992).

Reference	Time span	Total gain	Genetic gain	% Genetic
		------ kg ha^{-1} yr^{-1} -----		%
Hallauer, 1973	1930–1970	99	33	33
Russell, 1974	1930–1970	78	63	79
Duvick, 1977	1935–1971	88	50	57
Duvick, 1977	1935–1972	88	53	60
Russell, 1983	1922–1980	91	71	79
Duvick, 1984	1930–1980	103	92	89
Duvick, 1984	1930–1980	103	73	71
Castleberry et al., 1984	1930–1980	110	82	75
Carlone and Russell, 1987	1920–1980	89	72	81
Duvick, 1990	1930–1986	103	82	80
Duvick, 1992	1930–1989	100	76	76
Hybrid means		96	68	71
Lamkey and Smith, 1987				
S_0 populations[†]	1930–1984	--	52§	--
S_1 populations[‡]	1930–1984	--	38§	--
Duvick, 1984 – inbreds	1930–1980	--	50§	--

[†] Populations derived from intermating hybrids within each era.
[‡] Populations derived from bulked inbreds from each S_0 population.
§ Rates are averages over plant densities, whereas genetic gains for hybrids were calculated on the basis of the yields at the optimum density for each era.

Unfortunately, germplasm from long-term recurrent selection studies has not been routinely included in other hybrid improvement studies, and it is not possible from published research to directly compare rates of gain via conventional recurrent selection with those achieved by the hybrid seed industry. The objective of this chaapter is to summarize the large number of recurrent selection studies that are now available in order to (i) examine whether empirical results from different methods of recurrent selection agree with theoretical expectations, (ii) compare breeding methods stressing different modes of gene action, and (iii) determine whether realized annual responses from classical recurrent selection are comparable to those achieved by the hybrid seed industry.

WHAT SELECTION STUDIES TO INCLUDE?

Selection studies in maize were reviewed, and 133 were chosen for this compilation (Table 21–2). These were chosen on the basis of several criteria. Recurrent selection studies were first categorized as using one of eight methods: mass selection (MASS), modified ear–to–row selection (MER), half-sib selection (HS), half-sib selection using an unrelated tester HS(SCA), full-sib selection (FS), reciprocal recurrent selection (RRS), S_1 selection (S1), or S_2 (S2) selection. Mass selection was defined as selection among individual plants within a population and involved methods using both uni-parental or bi-parental control. Method MER was

Table 21–2. References used for compiling recurrent selection results.

MASS selection
Compton et al., 1979
Darrah, 1986
Genter and Eberhart, 1974
Hallauer and Sears, 1969
Johnson and Geadelmann, 1989
Maita and Coors, 1996
Mareck and Gardner, 1979
Martínez-Barajas et al., 1992
Mulamba et al., 1983
Popi, 1997
Weyhrich et al., 1998a

MER selection
Compton and Bahadur, 1977
Darrah, 1986
de Léon and Pandey, 1989
Granados et al., 1993
Popi, 1997
Stojšin, 1994
Vasal et al., 1994

HS selection
Darrah, 1986
Haarmann et al., 1993
Horner et al., 1973
Popi, 1997
Stojšin, 1994
Weyhrich et al., 1998a

FS selection
Bänziger et al., 1997
Byrne et al., 1995
Darrah, 1986
Johnson and Geadelmann, 1989
Landi and Frascaroli, 1993
Moll et al., 1994
Moll, 1991
Pandey and Gardner, 1992
Pandey et al., 1986
Pandey et al., 1987
Singh et al., 1986
Stromberg and Compton, 1989
Vasal et al., 1994
Weyhrich et al., 1998a

HS(SCA)
Burgess and West, 1993
Darrah, 1986
Eberhart et al., 1973
Horner et al., 1973
Horner et al., 1976
Lonnquist, 1961
Russell et al., , 1973
Tanner and Smith, 1987
Walejko and Russell, 1977
Weyhrich et al., 1998a

RRS selection
Coors, 1998
Eyherabide and Hallauer, 1991
Galusha et al., 1997
Horner et al., 1989
Keeratinijakal and Lamkey, 1993
Lambert, 1992
Moll et al., 1994
Ochieng and Kamidi, 1992
Odhiambo and Compton, 1989
Popi, 1997
Rademacher et al., 1998
Stojšin, 1994

S1 selection
Burgess and West, 1993
Darrah, 1986
Edmeades et al., 1994
Galusha et al., 1997
Genter and Eberhart, 1974
Odhiambo and Compton, 1989
Stojšin, 1994
Sullivan and Kannenberg, 1987
Tanner and Smith, 1987
Tragesser et al., 1989
Weyhrich et al., 1998a
Weyhrich et al., 1998b

S2 selection
Helms et al., 1989
Holthaus and Lamkey, 1995
Horner et al., 1973
Iglesias and Hallauer, 1991
Lamkey, 1992
Weyhrich et al., 1998a

defined in the manner of Lonnquist (1964) where there is selection among half-sib families in replicated trials, and, in the same season, selection within half-sib families in a separate recombination block. All families are planted ear–to–row as females in the recombination block and pollinated by pollen from a bulk of all families. As with MER, HS uses the parental population as the tester and involves selection among HS families; however, there is no intentional within-family selection, and only selected HS families are used for recombination. The HS(SCA) method differs from HS selection because it uses unrelated germplasm as a tester. Methods FS, S1, and S2 involve selection among the designated family type and recombination of selected families (or in the case of S2 selection, recombination of related S_1 families). Reciprocal recurrent selection includes interpopulation improvement methods (both half and full-sib) designed to improve the cross between two specific populations based on procedures similar to those proposed by Comstock et al. (1949). For the purposes of this compilation, the selection methods can be classified on the intended direct response. The direct response to intrapopulation improvement methods MASS, MER, HS, and FS is population per se performance. Interpopulation improvement methods HS(SCA) and RRS are used to improve crosses between populations or crosses with testers. The direct response of inbred improvement methods S1 or S2 is the performance of inbreds derived from source populations.

Nearly all studies involved grain yield as a primary selection criterion. Many of studies also involved selection indices with various weights for traits relating to maturity, lodging, and pest resistance. Even those without formal selection indices often involved selection for traits in addition to grain yield. This is common for inbred and hybrid development as well, so no effort was made to chose among studies based on relative weights attached to grain yield per se. Several well-known recurrent selection programs were open-ended because new germplasm was added periodically to the germplasm pool being improved. For example, the HOPE program developed by Kannenberg (Popi, 1997) and many of the CIMMYT MER programs (de Léon & Pandey, 1989; Vasal et al., 1994) used this approach. These programs were included because hybrid breeding programs also can be considered open-ended.

There have been several very successful long-term studies that involved indirect selection for yield in unadapted germplasm by improving traits such as flowering date and plant or ear height (e.g., Johnson et al., 1986), but these were not included in the compilation because they are not directly comparable to selection studies in more adapted germplasm. The only studies involving per se selection for traits indirectly related to yield were those using prolificacy (Maita & Coors, 1996; Mareck & Gardner, 1979; Singh et al., 1986; Weyhrich et al., 1998a). Prolificacy is highly correlated with grain yield and was often mentioned as one of the desirable characteristics associated with yield in other studies, particularly those involving MASS or MER.

A large number of studies for each breeding method were needed so that mean values have relevance, but a selection study was included only if at least four cycles of selection were completed. Studies involving fewer than four cycles were more variable in response, many were several decades old, and the evaluations were often less rigorous than customary at the present time. In a few cases, even though four or more cycles were completed, the evaluation did not include germplasm representing the selection endpoints. Such studies were thought to be unreliable and were not used. Each study had to include either C0, or C1 and the most advanced cycle available at the time of the study.

Occasionally, more than one investigator evaluated the same selection program, or the same investigator evaluated the same selection program several times. In these cases, the most recent study involving the most advanced cycle was chosen for inclusion unless the study suffered from some deficiency. Several studies used several different selection protocols but started with the same initial germplasm. For example, Darrah (1986) initiated several mass selection programs in Kitale Composite A using different selection intensities and population densities. In these cases, where a variable influencing selection response was specifically designed to be part of the study, each separate selection scheme was considered as an independent program.

Data collected from each study included: (i) C0 and Cn grain yield (adjusted to 155 g kg⁻¹ moisture), and/or the linear regression coefficient corresponding to selection response, (ii) seasons required to complete each cycle, and (iii) selection intensity. Linear regression was used to calculate selection response, if appropriate data were available, using a model without a quadratic term. If the linear regression coefficient was not significant, the average response was calculated from the initial and final cycles. The number of seasons per cycle were those required to complete each phase of the selection protocol. Occasionally, protocols changed, resulting in a different number of seasons per cycle, and in such cases, gain per year calculations were based on the seasons required for the majority of the time the selection program was operating. It was assumed that yield trials could only be completed during one of two dissimilar seasons available each year and that seasons were used in the most efficient manner.

Selection intensities are a good reflection of the resources available to breeders because higher intensities are usually achieved by evaluating more materials. Selection intensities used for this compilation were either as reported by the investigator or calculated directly from the cumulative number of families evaluated and selected. No attempt was made to account for within-family selection intensity for MER or any other progeny-based method. Selection intensities in recurrent selection programs are probably quite a bit less than in most hybrid programs, but it is hard to precisely document what selection intensity is for the hybrid industry. One can estimate it from the number of early-generation inbred families that are first evaluated in testcrosses and the number of surviving inbreds that are then used to make the new populations (e.g., F_2s, or backcross populations) for the next round of inbreeding. During the initial years of hybrid breeding, selection intensity was probably on the order of 1%, i.e., one out of 100 inbreds evaluated in initial testcrosses would be used to create the next set of breeding populations. Now it may be approaching 0.01%. Theoretically, selection response, R, is a function of i, narrow-sense heritability, h^2, and the phenotypic standard deviation (Falconer & Mackay, 1996). Therefore, predicted gain from recurrent selection could be calculated at 1% and 0.01% selection intensities as long as the actual intensity used during selection was known.

Both direct and indirect responses were compiled for HS(SCA), S1, S2 and RRS. The direct response to HS(SCA) corresponds to the performance the cross between the population and the tester used during selection, while the direct response to RRS is measured by the performance of the cross between the two populations involved. The direct response of either S1 or S2 selection is reflected in the performance of the inbred populations, and all studies used bulked S1s for evaluation of direct response. For all four methods, the performance of the random-mated populations reveals the indirect response. While many studies included testcrosses to other unrelated germplasm not involved with selection, only the results reflecting direct and indirect responses as defined above were included in the compilation. This was difficult for several HS(SCA) studies where either the tester was no longer available or changed during selection. In such cases, the testcross results that most closely reflected the tester used during selection were used.

RESPONSE TO RECURRENT SELECTION

Mean responses for each recurrent selection method were calculated from at least 12 studies for all methods except S2 selection (Table 21–3). For S2 selection only six studies included direct response (i.e., performance of bulked

Table 21–3. Results from long-term recurrent selection for increased grain yield in maize.

Method	No. studies	No. cycles	Select. intensity	Response Cycle⁻¹	Cycle⁻¹	Year⁻¹	Predicted†
			-------- % -------	----------- kg ha⁻¹ -----------			
MASS‡	16	11.1	6.3	1.8	82	82	129–192
MER	25	11.8	22.4	2.1	83	83	157–234
HS	12	6.8	16.7	1.8	69	50	91–135
FS	27	6.9	27.9	3.1	154	87	178–265
HS(SCA)							
direct§	15	5.2	10.7	3.2	180	82	125–186
indirect	13	5.3	11.2	1.6	65	31	49–73
RRS							
direct§	14	6.7	14.6	4.6	270	116	197–292
indirect	28	6.7	14.7	1.3	56	21	37–56
S1							
direct§	13	5.6	16.0	7.4	182	93	154–229
indirect	18	5.3	14.3	1.9	86	43	67–100
S2							
direct§	6	4.5	17.0	4.6	132	43	78–116
indirect	6	4.8	17.0	2.0	78	26	49–72
Mean – Intra¶	80	9.2	20.2	2.3	105	79	149–221
Mean – Inter¶							
direct	29	5.9	12.6	3.9	223	98	160–237
indirect	41	6.3	13.6	1.4	59	24	41–61
Mean – Inbred¶							
direct	19	5.3	16.3	6.5	166	77	130–193
indirect	24	5.2	15.0	1.9	84	39	63–93

† Annual response adjusted for 1% (first number) and 0.01% (second number) selection intensity.

‡ Mean selection intensities and adjusted responses for MASS selection based on 13 studies because three of the 16 did not provide sufficient information.

§ Direct responses for HS(SCA) and RRS are those measured for testcrosses and population crosses, respectively. The direct response for S1 and S2 selection is that measured for bulked S_1 families. For all four methods, the indirect response is that measured for the random-mating populations.

¶ Means by either intrapopulation (MASS, MER, HS, and FS), interpopulation [HS(SCA) and RRS], and inbred (S1 and S2) recurrent selection methods.

inbreds), and eight studies included indirect response (i.e., performance of the random-mated population). The average number of cycles evaluated ranged from 4.5 to 11.8, and selection intensities ranged for 6.3% for MASS to 27.9% for FS. A large proportion of MER and FS programs were conducted by CIMMYT (12/25 and 16/27, respectively), and these programs used less intense selection than customary for most other studies. Percentage response per cycle ranged from 7.1% for the direct response from S2 selection to 1.4% for the indirect response from RRS selection.

Intrapopulation Improvement

Annual gains from intrapopulation recurrent selection using either MASS, MER, HS, or FS selection ranged from 50 kg ha^{-1} yr^{-1} for HS selection to 87 kg ha^{-1} yr^{-1} for FS selection. The mean annual response for the four intrapopulation programs was 79 kg ha^{-1} yr^{-1}. The relative ranking of selection methods in terms of percent response per cycle agreed with theoretical expectations, with one exception, HS selection. In the absence of within-family selection, HS should produce greater percent response and greater annual gains than MER because both can be completed·in one year, and HS allows greater parental control. In several HS studies, however, annual response was low because two or more seasons were needed for the particular protocol used by the breeder (Darrah et al., 1986; Haarman et al., 1993; Horner et al., 1973). It also was not possible to account for the intensity of within family selection used in the MER studies, and the results for both MER and MASS indicate that within-family or individual plant selection was quite effective.

Interpopulation Improvement

Gains from intrapopulation improvement have been substantial, but selection for specific combining ability has been especially effective. Direct responses for interpopulation improvement methods HS(SCA) and RRS (3.2 and 4.6%, respectively) were higher than MASS, MER, HS or FS methods on a cycle basis. On an annual basis, RRS was very effective and produced gains of 116 kg ha^{-1} yr^{-1}, which was the highest mean value recorded. Four RRS programs had responses in excess of 140 kg ha^{-1} yr^{-1} (Coors, 1998; Eyerabide & Hallauer, 1991; Horner et al., 1989; Keeratinijakal & Lamkey, 1993). Annual gains from HS(SCA) did not differ appreciably from the better intrapopulation improvement methods. However, the studies using HS(SCA) included four selection programs that required three years per cycle (Horner et al., 1973, 1976; Lonnquist, 1961; Weyhrich et al., 1998a). Furthermore, direct responses for HS(SCA) may be understated because in several instances the testers used during selection were not those used during evaluation.

The direct annual response for interpopulation improvement measured across all 29 studies was approximately 7 to 24% (adjusted and unadjusted for differences in selection intensity, respectively) higher than that achieved through intrapopulation improvement, even though a cycle of interpopulation improvement took an average of one additional year to complete. The direct annual response for the mean of both interpopulation improvement methods was 99 kg ha^{-1} yr^{-1}. Indirect responses of the random-mated populations for either HS(SCA) or RRS were as low or lower than other selection methods, and on average, the indirect response from interpopulation improvement was less than one third of the direct response.

As long as there is significant dominance variation, selection for both additive and dominance gene action in crosses should increase direct response over selection based on additivity alone. The degree of divergence between the two base populations determines, in part, the relative contributions of additivity and dominance to selection advance. If the base populations are unrelated, dominance may have a predominant role in selection, and consequently the indirect response may be appreciably less than the direct response. On the other hand, if the two base populations are closely related, indirect response may be closer to the direct response. Nearly all RRS programs started with two unrelated base populations in order to ensure that dominance contributed to selection response, and the relative magnitude of direct and indirect responses is as expected. However, not all of the RRS programs started with two distinct populations.

At the University of Wisconsin, RRS was initiated using two strains, GG(A) and GG(B), randomly chosen from the same initial seed lot of the Golden Glow maize population. The protocol has changed somewhat over cycles, but the current system involves developing S_1 families by selfing plants within each strain and then evaluating the interstrain $S_1 \times S_1$ crosses for grain yield. The GG(A) $\times$ GG(B) crosses were selected based on a performance index using grain yield and grain moisture. Selected S_1 families were then recombined to begin the next cycle. While the number of crosses evaluated in yield trials varied from cycle to cycle, the number of S_1 families recombined was always greater than 20 in order to maintain genetic variation and minimize genetic drift. The selection intensity has averaged 24% for the five cycles that have been completed.

Annual response for grain yield of the GG(A) $\times$ GG(B) population cross was 115 kg ha^{-1} yr^{-1}, which is nearly equal to the average direct response for all 14 RRS programs, 116 kg ha^{-1} yr^{-1} (Table 21–4). In contrast to other RRS programs, however, the indirect responses of strains GG(A) and GG(B) also increased appreciably, 70 and 90 kg ha^{-1} yr^{-1}, respectively. Even though the indirect responses were significantly less than the population cross response, they were considerably greater than the average indirect response for all RRS programs, 21 kg ha^{-1} yr^{-1}. The gene frequencies in the initial strains should have been equal, and, therefore, dominance interactions would not have contributed to the direct selection response. Initially, the program would have been equivalent to a FS intrapopulation improvement program. Once gene frequencies in the GG(A) and GG(B) strains began to diverge (either through drift or unintentional selection within the A and B strains), additional response was possible through utilization of dominance.

Only one other RRS program in maize is known to have started with two strains of a single base population. Population BS6 was developed at Iowa State University by intermating two populations, BSSS(R)C5 and BSCB1(R)C5, that had previously undergone five cycles of RRS (Eberhart et al., 1995, Smith, 1984). Two strains of BS6, BS6(RS) and BS6(RC), were then created by using BSSS(R)C5 and BSCB1(R)C5 as testers in a new RRS program. The rate of gain for the BS6(RS) $\times$ BS6(RC) population cross exceeded that of the original RRS program with BSSS(R) and BSCB1(R), and after three cycles, the grain yield of BS6(RS)C3 $\times$ BS6(RC)C3 population cross nearly equaled the yield of the BSSS(R)C8 $\times$ BSCB1(R)C8. However, in contrast to the per se yield increases observed for the GG(A) and GG(B) strains, the yield of the BS6(RS) and BS6(RC) populations did not increase. In fact, BS6(RC)C3 was significantly lower yielding than BS6C0. Unfortunately, the program was stopped after three cycles, and therefore, the results were not included in the compilation.

Inbred-Based Population Improvement

On a cycle basis, direct response to S1 and S2 selection was quite high, 7.4% and 4.6%, respectively. The annual response for S1 was also large, 93 kg ha^{-1} yr^{-1}; however, annual response for S2 selection was relatively poor, 43 kg ha^{-1} yr^{-1}, probably due to the excessive amount of time needed to complete a cycle. Although there are too few studies to be certain, it would appear that even the gains per cycle are disappointing for S2 selection since this method should be more effective than S1 on a cycle basis. As with RRS, indirect responses of the random-mating populations were relatively modest compared with intrapopulation methods. The indirect annual gain averaged for both inbred selection methods, 39 kg ha^{-1} yr^{-1}, was approximately one half of the direct gain of 77 kg ha^{-1} yr^{-1}.

Table 21–4. Grain yield, grain moisture, and ear number after five cycles of RRS for high grain yield and low grain moisture, and 21 cycles of biparental mass selection for prolificacy in the maize population Golden Glow. Grain yield and moisture expressed as means from three Wisconsin locations in 1995 and two Wisconsin locations in 1996 with three replications per location.

Entry	Grain yield	Grain moisture
	Mg ha^{-1}	g kg^{-1}
GG C0	5.00	277
GG(A) C1 × GG(B) C1	5.20	273
GG(A) C2 × GG(B) C2	5.87	264
GG(A) C3 × GG(B) C3	5.93	247
GG(A) C4 × GG(B) C4	6.47	253
GG(A) C5 × GG(B) C5	6.64	237
Response yr^{-1} [†]	0.11±0.01	-3±0.4
GG(A) C1	5.03	277
GG(A) C2	5.31	261
GG(A) C3	5.37	256
GG(A) C4	5.58	263
GG(A) C5	6.12	254
Response yr^{-1}	0.07±0.01	-2±0.5
GG(B) C1	5.32	271
GG(B) C2	5.40	257
GG(B) C3	5.86	252
GG(B) C4	6.05	241
GG(B) C5	6.36	236
Response yr^{-1}	0.09±0.01	-3±0.2
LSD(0.05)	0.57	12

[†] Based on linear regression coefficient and three years per cycle.

Duvick (1984) estimated that the annual increase in inbred yields from 1930 to 1980 was 50 kg ha^{-1} yr^{-1}, and Lamkey and Smith (1987) reported that S_1 populations spanning the same time period had an annual response of 38 kg ha^{-1} yr^{-1}. As would be expected, the direct response for the mean of S1 and S2 selection was greater than these estimates, but it was still less than anticipated. Inbred selection methods were originally designed to also increase population per se performance, but they have not been effective. Random genetic drift due to small populations size has been shown to be higher in S2 selection than other recurrent selection methods (Helms et al., 1989) but many of the other selection programs compiled in Table 21–3 used as small or smaller effective populations sizes than the inbred-based methods. Weyhrich et al. (1998b) used different effective population sizes (10, 20, and 30 selected families) across four cycles of S_1 selection in BS11. Only when the number of selects dropped to five was there a significant reduction in response rate. Many of the studies in Table 21–3 recombined 10 or more families each cycle, and most of those that did not were open-ended programs.

Hybrid Development vs. Classical Recurrent Selection

Response from recurrent selection was often greater than that reported for the hybrid seed industry, particularly when estimates were predicted at the 1% to 0.01% selection intensities (Table 21–3). If recurrent selection had been practiced using these intensities, apparently the direct annual responses would have been considerably higher than what has been achieved via hybrid breeding. One might even reach the unsettling conclusion that MASS selection has been more effective in increasing grain yield than the inbred-hybrid system, but this is an unreasonable conjecture because there are a number of important factors that contribute to the differences in response.

Publication of recurrent selection programs.   Unfortunately, it is not known how many unsuccessful selection programs have not been published or are otherwise unavailable for review. If results from the less successful programs have not been published, average responses would be less than reported in Table 21–3. However, there is little evidence that breeders have been hesitant in reporting small or negative gains. For example, of the 79 intrapopulation improvement programs, nearly 20% of the responses were 0.5% per cycle or less. Also, while it was anticipated that eliminating cycles with fewer than four cycles would bias the results because only the most successful programs would be continued past the third cycle, the opposite was observed. Initially, the compilation included all programs of three or more cycles, and the effect of eliminating studies with only three cycles was to decrease the compiled gains, quite substantially so for MASS and MER selection.

Genetic variation of source germplasm.   Using the initial and final estimates of selection intensity, 1% and 0.01%, used by the hybrid seed industry, one would predict that hybrid yield response should have increased by approximately 50% during the last six decades, but there is no evidence that this has happened. Some have proposed that an ever-narrowing genetic base is the limiting factor. Bauman (1981) estimated that over two-thirds of the germplasm sources used for inbred development are from populations where one inbred makes up at least 50% of the population. Most recurrent selection programs use relatively broad-based populations, and perhaps the greater predicted response from recurrent selection reflects the greater genetic variability in the source populations.

This is difficult to address from the compiled selection studies because very few involved populations were derived from single or three-way crosses among inbred lines that are more typical of the hybrid industry. Russell et al. (1973) used a population developed from Wf9 × B7 for five cycles of HS(SCA) using B14 as a tester with 13% selection intensity. The direct annual response was 66 kg ha^{-1} yr^{-1}. Moll (1991) conducted 15 cycles of FS selection to improve a population developed from the single cross NC7 × C121 using 10% selection intensity, and he achieved an annual responses of 130 kg ha^{-1} yr^{-1}. Landi and Fracaroli (1993) conducted four cycles of FS selection starting with the F$_2$ generation of the single cross A632 × Mu195 using 15% selection intensity, and the annual response was 211 kg ha^{-1} yr^{-1}. At the University of Wisconsin, we have completed five cycles of double-cross improvement using RRS with different two double-crosses, W577 and W03545. Each program used as parental populations, the F$_2$ generations of the two parental single crosses for each double cross hybrid (W577A = F$_2$ of A295 × W64A, W577B = F$_2$ of W374R × OH43; W03545A = F$_2$ of W64A × B46, and W03545B = F$_2$ of OH43 × A635). The selection protocol was similar to that described for the Golden Glow RRS program above. Selection intensities were 20% for W577 and 17% for W03545. After five cycles, the direct annual response from RRS (A × B crosses) using W577 and W03545 were 158 kg ha^{-1} yr^{-1} (6.2% per cycle) and 130 kg ha^{-1} yr^{-1} (4.5% per cycle), respectively (Fig. 21–1). By the fourth cycle, the population crosses significantly outyielded the original double-cross hybrids. By the fifth cycle, the A × B population crosses significantly outyielded the parental single crosses and the four possible nonparental single crosses.

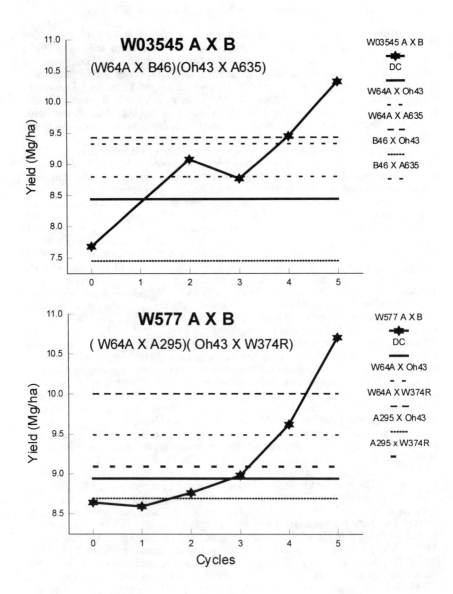

Fig. 21–1. Reciprocal recurrent selection using parental single crosses from double crosses as breeding populations. In the top graph, the two populations were W03545A and W03545B, developed from the F2 of W64A × B46 and OH43 × A635, respectively. In the bottom graph, the two populations were W577A and W577B developed from the F2 of A295 × W64A and W374R × OH43, respectively. Data are means from five Wisconsin locations collected in 1996 and 1997. LSD(0.05) = 0.63 and 0.61 Mg ha^{-1} for top and bottom figures, respectively.

Based on these studies, selection using narrow-based germplasm can still be quite effective, and there is no indication that selection response is any different in narrow than broad-based populations. While most populations used to develop hybrids are narrow-based, they are also open-ended in that new material is often added each cycle. To date, there is little objective evidence that the process of hybrid breeding has narrowed the genetic base to the extent that it is restricting gains.

Schedule slippage Annual responses in Table 21–3 rely on the most efficient use of seasons to most rapidly complete a cycle and, therefore, inflate responses above those actually achieved. Occasionally a season was lost due to some calamity such as drought, or, more likely, the project did not have the resources to efficiently use winter nurseries. Few in the public sector use winter nurseries for making large numbers of testcrosses, but rather limit activities to selfing or other tasks requiring fewer resources. Annual responses for hybrids from the studies of Duvick (1992) and others are realized responses that already account for accidental loss of generations or seasons. A more thorough compilation of recurrent selection would also estimate annual response from actual dates when selection programs are initiated and when the most recent cycles were completed. For example, the nine cycles of RRS selection in of BS10 × BS11 required 23 years, or approximately 2.6 years per cycle. The data in Table 21–3 assumed a two-year cycle, so the calculated gain of 155 kg ha^{-1} yr^{-1} should be reduced by about 23% to 119 kg ha^{-1} yr^{-1}. However, this might overcorrect the response rate since schedule slippage for the BS10 × BS11 might also have been due to lack of winter-nursery resources. The BS10 × BS11 example is probably typical of other RRS programs, and while the effect of schedule slippage may bias the results upwards, at least for the more complex selection systems, the magnitude of this bias is hard to determine without reviewing each step for each selection program.

Performance level of source germplasm. Yield gains are likely to be easier with relatively unimproved, low-performing germplasm. In unadapted germplasm, grain yield can be increased by altering traits such as daylength sensitivity and shelling percentage. As overall productivity increases, it becomes harder to maintain selection response. Populations used for recurrent selection can be relatively primitive, particularly so for MASS and MER selection. MASS and MER are often recommended to improve poorly-adapted germplasm for simply inherited traits that indirectly influence grain yield such as maturity, plant and ear height, and prolificacy, and selection response can be rapid. Nearly all of the MASS selection studies in Table 21–3 involved open-pollinated populations such as Iowa Ideal, Krug, Hays Golden, Golden Glow, and entry-level germplasm from open-ended breeding projects such as the Canadian HOPE program.

Germplasm in the commercial sector needs a full suite of performance attributes, not just high grain yield, to be successful. In particular, tolerance to relatively rare environmental stresses is an important criterion for selecting commercial hybrids, and maternal productivity and uniformity are essential for parental inbreds. Even though most recurrent selection programs involved some selection for standability, pest resistance, and maturity, performance requirements for hybrids and their inbred parents undoubtedly exceed what was demanded from most of the germplasm represented in Table 21–3.

Plant density, fertility, and moisture stress and estimate of response. Most of the gains reported in Table 21–3 are not true genetic gains, but are biased upwards because of the way the trials were conducted. Evaluating long-term selection results by conducting trials under management recommended at the time of the evaluation (particularly planting density) tends to inflate response. For example, Duvick's 1992 hybrid breeding study used three planting densities,   30 000, 47 000, and 64 000 plants ha^{-1}, spanning the range used since the introduction of hybrids, and the respective responses were 49, 77, and 91 kg ha^{-1} yr^{-1}. Because most recurrent selection programs have a shorter time span than the six decades of hybrid development, the bias would not seem to be great. However, for seven recurrent selection studies conducted at Iowa State University using at least three planting densities, Crosbie (1982) estimated that true genetic gains were approximately 25% lower than gains measured using recommended plant densities at the time of the evaluation. After 20 cycles of mass selection for prolificacy in Golden Glow, true genetic gains for grain yield, evaluated over four planting densities, were approximately 12% less than those measured at the highest (recommended) density (Maita & Coors, 1996). Duvick (1992) also used three plant densities of 30,000, 47,000, and 64,000 plants ha^{-1}, to measure response to nine cycles of RRS selection in BS10 × BS11, and reported respective gains of responses of 63, 89, and 79 kg ha^{-1} yr^{-1}. After five cycles, the responses for the Pioneer recurrent program involving PIAA × PIABB were 17, 38, and 123 kg ha^{-1} yr^{-1} at the same three planting densities.

Fertility and moisture stress also tend to reduce estimates of gain unless the program was specifically designed to improve yield under the particular stress. For example, Moll et al. (1994) reported a gain of 33 kg ha^{-1} yr^{-1} for the Jarvis × Indian Chief RRS program when the evaluation was conducted under low nitrogen applications, while the response was 74 kg ha^{-1} yr^{-1} under high nitrogen. Castleberry et al. (1984) showed that under drought stress, the rate of improvement for U. S. hybrids was 84 kg ha^{-1} yr^{-1}, while under irrigated conditions the rate was 123 kg ha^{-1} yr^{-1}. In the same study, rates were 51 versus 87 kg ha^{-1} yr^{-1} at low versus high fertility. Similar trends are seen in several other hybrid evaluation studies (Carlone & Russell, 1987; Russell, 1974, 1986; Duvick, 1977, 1984).

Selection in optimum versus poor environments.   Good performance in stressed environments is essential for the economic success of a hybrid. Therefore, most successful hybrid programs stress wide-area testing. Recurrent selection programs do not typically involve selection in both stressed and nonstressed environments, so the additional requirement of wide-area testing for hybrids might account for much of the difference between the responses from recurrent selection and the hybrid seed industry. It takes time to get to the stage where breeders have winnowed out the more unproductive inbred families and also have adequate seed supplies of testcrosses for multilocation yield trials. In Bauman's (1981) survey, most hybrid breeders did not initiate testcrossing until the S$_3$ generation. Extensive multilocation testing might not begin for several more inbreeding generations. Most hybrid breeders are reluctant to form new populations until wide-area trials are completed, and this reluctance slows the recycling of germplasm.

Selection in unproductive environments may also significantly reduce response rate. Selection can be less effective in poor environments than in good environments if the proportion of genetic relative to nongenetic variance decreases substantially (i.e., heritability decreases) in poor environments (Blum, 1988). Without additional resources, such as increased replication, response to selection in poor environments will be less than in optimal environments. For example, Johnson and Geadelmann (1989) completed five cycles of both MASS and FS selection in population AS-A under both irrigated and nonirrigated conditions. The response measured over all environments for the MASS program conducted under irrigated conditions was four times that of MASS under nonirrigated conditions. The response of FS under irrigation was nearly twice that of FS in moisture-stressed conditions.

Selection for performance across a diverse array of environments may require genetic compromises because superior genotypes in any given environment might not be the same as those with superior performance averaged across many different environments. This phenomenon is shown by results of Arboleda-Rivera and Compton (1974). They used two distinct environments in Columbia, wet and dry seasons, to select for prolificacy and grain yield in the variety Mezcla Varietales Amarillos. Response to mass selection in the wet season was greater than in the dry season, as expected, but selection in either the wet or dry seasons was nearly 50% higher than the response when selection occurred sequentially using both seasons.

If breeders can target specific stresses and achieve adequate experimental control in stressed environments, average selection response need not necessarily be less than under optimal conditions as shown by Byrne et al. (1995). Eight cycles of FS selection in Tuxpeño Sequía under drought-stress at Tlaltizapán, Mexico was more effective at increasing yields measured across 12 diverse environments in Africa, Asia, and Latin America than was multilocation selection by international cooperators in the same regions. However, breeders need good control of stress condition and also good understanding of the physiology of the particular stress to succeed with this strategy. Unfortunately, this is not often the case.

SUMMARY

Intrapopulation recurrent selection methods based solely on additive genetic variance have been successful in increasing the grain yield of maize, and comparisons among selection methods for rate of improvement agree with theoretical expectations in most instances. Interpopulation improvement has been especially productive. Breeders have been able to efficiently use dominance to increase annual response rates by approximately 24% above the mean of intrapopulation improvement methods. In fact, average annual gains from reciprocal recurrent selection substantially exceeded those achieved by breeders of commercial hybrids; however, the challenges faced by hybrid breeders are quite different, both in kind and extent, from those that confront breeders using recurrent selection. The most appropriate conclusion from the summaries provided in this paper is that selection for grain yield in maize has worked quite well, and selection for specific combining ability has been particularly effective.

In the absence of overdominance, one might argue that selection for combining ability in crosses is only a temporary phase in plant breeding. At some stage breeders should be able to fix the optimal allele at a large proportion of loci, and they would not have to rely on dominance to cover inferior alleles. It is apparent from the yield increase in maize inbreds that breeders of hybrids have succeeded in fixing some good alleles. The magnitude of heterosis, measured as the increase in hybrid performance above that of the mean of the inbred parents, has also decreased slightly over the decades (Duvick, 1984), so perhaps breeders are relying less on dominance interactions now than in the past (on a proportional basis at least). The notion of fixing the "optimal" allele, however, is probably an oversimplification, since breeders are always dealing with ever-changing conditions requiring new adaptations. Two of the most obvious examples in maize are the increase in planting densities and the change from hand to machine harvest. Both required new adaptations to increase competitive ability and ease of shelling. Without doubt, cultural procedures will continue to evolve, and environmental conditions will constantly and unpredictably change in the future.

The are two goals for breeders of hybrid crops; creating commercially-viable varieties, and increasing the genetic potential of germplasm so that future breeders will continue to make gains. While most individual breeders' careers stand or fall on the former, the sustained future success of the hybrid industry depends on the latter. In other words, the contrast between hybrid breeding and recurrent selection reflects, in part, the tradeoff the hybrid industry must make between short and long-term goals. The fundamental message from recurrent selection, which applies to nearly all breeding systems, is that rapid progression from evaluation to selection and finally recombination is essential for long-term germplasm improvement.

ACKNOWLEDGMENTS

The author is grateful to a number of colleagues including R. Bernardo, G. Drinic, D.N. Duvick, J.L. Geadelmann S.M. Kaeppler, L.W. Kannenberg, K.R. Lamkey, S. Pandey, O.S. Smith, M. Stojaković, W.F. Tracy, A.F. Troyer, and many others who shared their selection results and ideas.

REFERENCES

Arboleda-Rivera, F., and W.A. Compton. 1974. Differential response of maize to mass selection in diverse environments. Theor. Appl. Genet. 44:77–81.

Bänziger, M., H.R. Lafitte, G.O. Edmeades, J.F. Betrán, D.L. Beck, and A. Elings. 1997. Recent advances in breeding for tolerance to low nitrogen in tropical maize. April 21–25, 1997. West and Central African Regional Maize Workshop. Cotonou, Benin.

Bauman, L.F. 1981. Review of methods used by breeders to develop superior inbreds. p. 199–208. In H.D. Loden and D. Wilkinson (ed.) Proc. of the 36th Annual Corn and Sorghum Industry Research Conf. Chicago, IL, 9–11 Dec. 1981. Am. Seed Trade Assoc., Washington, DC.

Blum, A. 1988. Plant breeding for stress environments. CRC Press, Boca Raton, FL.

Burgess, J.C., and D.R. West. 1993. Selection for grain yield following selection for ear height in maize. Crop Sci. 33:679–682.

Byrne, P.F., J. Bolaños, G.O. Edmeades, and D.L. Eaton. 1995. Gains from selection under drought versus multilocation testing in related tropical maize populations. Crop Sci. 35:63–69.

Carlone, M.R., and W.A. Russell. 1987. Response to plant densities and nitrogen levels for four maize cultivars from different eras of breeding. Crop Sci. 27:465–470.

Castleberry, R.M., C.W. Crum, and C.F. Krull. 1984. Genetic yield improvement of U.S. maize cultivars under varying fertility and climatic environments. Crop Sci. 24:33–36.

Cockerham, C.C., and Z.B. Zeng. 1996. Design III with marker loci. Genetics 143:1437–1456.

Compton, W.A., and K. Bahadur. 1977. Ten cycles of progress from modified ear to row selection in corn (Zea mays L.) Crop Sci. 17:378–380.

Compton, W.A., R.F. Mumm, and B. Mathema. 1979. Progress from adaptive mass selection in incompletely adapted maize populations. Crop Sci. 19:531–533.

Comstock, R.E., and H.F. Robinson. 1948. The components of genetic variance in populations of biparental progenies and their use in estimating the average degree of dominance. Biometrics 4:254–266.

Comstock, R.E., H.F. Robinson, and P.H. Harvey. 1949. A breeding procedure designed to make maximum use of both general and specific combining ability. Agron. J. 41:360–367.

Comstock, R.E., and H.F. Robinson. 1952. Estimation of average dominance of genes. p. 494–516. *In* J.W. Gowen (ed.) Heterosis. Iowa State Univ. Press, Ames, IA.

Coors, J.G. 1998. Reciprocal recurrent selection using different types of base populations. p. 24 Report of the North Central Corn Breeding Research Committee (NCR–167). Ames.

Crosbie, T.M. 1982. Changes in physiological traits associated with long-term breeding efforts to grain yield of maize. p. 206–223. *In* H.D. Loden and D. Wilkinson (ed.) Proc. of the 37th Annual Corn and Sorghum Industry Research Conf.. Chicago, IL. 8–9 Dec. 1982. Am. Seed Trade Assoc. Washington, DC.

Crow, J.F. 1948. Alternative hypotheses of hybrid vigor. Genetics 33:477–487.

Crow, J.F. 1993. Mutation, mean fitness, and genetic load. Oxford Surv. Evol. Biol. 9:3–42.

Darrah, L.L. 1986. Evalaution of population improvement in the Kenya maize breeding methods study. p. 160–175. *In* B. Gelaw (ed.) To feed ourselves: A proceedings of the first eastern, central, and southern Africa regional maize workshop, Lusaka, Zambia. 10–17 Mar. 1985. CIMMYT, Mexico, D.F.

de León, C., and S. Pandey. 1989. Improvement of resistance to ear and stalk rots and agronomic traits in tropical maize gene pools. Crop Sci. 29:12–17.

Duvick, D.N. 1977. Genetic rates of gain in hybrid maize yields during the past 40 years. Maydica 22:187–196.

Duvick, D.N. 1984. Yield gains in U.S. hybrid maize. p. 15–47. *In* W.R Fehr. (ed.) Genetic contributions to yield gains in five major crop plants. CSSA Special Publ. 7. ASA, Madison, WI.

Duvick, D.N. 1990. Ideotype evolution of hybrid maize in the USA. 1930–1990. p. 557–570. *In* Proc. of the 2nd Nat. Maize Conf.: Research, Economy, Environment. Centro Regionale per la Sperimentazione Agrari per il Friuli-Venezia Giulia, Italy.

Duvick, D.N. 1992. Genetic contributions to advances in yield of U.S. maize. Maydica 37:69–79.

Eberhart, S.A., S. Debela, and A.R. Hallauer. 1973. Reciprocal recurrent selection in the BSSS and BSCB1 maize populations and half-sib selection in BSSS. Crop Sci. 13:451–456.

Eberhart, S.A., W. Salhuana, R. Sevilla, and S. Taba. 1995. Principles for tropical maize breeding. Maydica 40:339–355.

Edmeades, G.O., H.R. Lafitte, S.C. Chapman, and M. Bänziger. 1994. Improving the tolerance of lowland Tropical maize to drought or low nitrogen. p. 85–101. *In* S.K. Vasal, and S. Mclean (ed.). The lowland tropical maize subprogram. Maize Program Spec. Rep. CIMMYT, Mexico, D.F.

Eyherabide, G.H., and A.R. Hallauer. 1991. Reciprocal full-sib recurrent selection in maize: I. Direct and indirect responses. Crop Sci. 31:952–959.

Falconer, D.S., and T.F.C. Mackay. 1996. Introduction to Quantitative Genetics. Longman. Essex, England.

Galusha, D.D., S.M. Kaeppler, L. Nelson, and W.A. Compton. 1997. A valuation of eight cycles of replicated recurrent selection in maize. Ph.D. diss. Univ. of Nebraska, Lincoln.

Gardner, C.O., P.H. Harvey, R.E. Comstock, and H.F. Robinson. 1953. Dominance of genes controlling quantitative characters in maize. Agron. J. 45:186–191.

Gardner, C.O., and J.H. Lonnquist. 1959. Linkage and the degree of dominance of genes controlling quantitative characters in maize. Agron. J. 51:524–528.

Genter, C.F., and S.A. Eberhart. 1974. Performance of original and advanced maize populations and their diallel crosses. Crop Sci. 14: 881–885.

Good, R.L. 1990. Experiences with recurrent selection in a commercial seed company. p. 80–92. In D. Wilkinson (ed.) Proc. of the. 45th Annual Corn and Sorghum Industry Research Conf. Chicago, IL, 8–9 Dec. 1990. Am. Seed Trade Assoc. Washington, DC.

Graham, G.I., D.W. Wolff, and C.W. Stuber. 1997. Characterization of a yield quantitative trait locus on chromosome five of maize by fine mapping. Crop Sci. 37:1601–1610.

Granados, G., S. Pandey, and H. Ceballos. 1993. Response to selection for tolerance to acid soils in a tropical maize population. Crop Sci. 33:936–940.

Haarmann, R.J., D.G. White, and J.W. Dudley. 1993. Index vs. tandem selection for improvement of grain yield, leaf blight and stalk rot resistance in maize. Maydica 38:183–188.

Hallauer, A.R. 1973. Hybrid development and population improvement in maize using reciprocal full-sib selection. Egypt. J. Genet. Cytol. 2:84–101.

Hallauer, A.R., and J.H. Sears. 1969. Mass selection for yield in two varieties of corn. Crop Sci. 9:47–50.

Helms, T.C., A.R. Hallauer, and O.S. Smith. 1989. Genetic drift and selection evaluated from recurrent selection programs in maize. Crop Sci. 29:602–607.

Holthaus, J.F., and K.R. Lamkey. 1995. Response to selection and changes in genetic parameters for 13 plant and traits in two maize recurrent selection programs. Maydica 40:357–370.

Horner, E.S., H.W. Lundy, M.C. Lutrick, and W.H. Chapman. 1973. Comparison of three methods of recurrent selection in maize. Crop Sci. 13:485–489.

Horner, E.S., M.C. Lutrick, W.H. Chapman, and F.G. Martin. 1976. Effect of recurrent selection for combining ability with a single cross tester in maize. Crop Sci. 16:5–8.

Horner, E. S., E. Mogloire, and J.A. Morera. 1989. Comparison of selection for S_2 progeny versus testcross performance for population improvement in maize. Crop Sci. 29:868–874.

Iglesias, C.A., and A.R. Hallauer. 1991. Response to S_2 recurrent selection in exotic and semi-exotic populations of maize (*Zea mays* L.). J. Iowa Acad. Sci. 91:4–13.

Johnson, E.C., K.S. Fisher, G.O. Edmeades, and A.F.E. Palmer. 1986. Recurrent selection for reduced plant height in lowland tropical maize. Crop Sci. 26:253–260.

Johnson, S.S., and J.L. Geadelmann. 1989. Influence of water stress on grain yield response to recurrent selection in maize. Crop Sci. 29:558–564.

Keeratinijakal, V., and K.R. Lamkey. 1993. Responses to reciprocal recurrent selection in BSSS and BSCB1 maize populations. Crop Sci. 33:73–77.

Lambert, R.J. 1992. Evaluation of six cycles of maize reciprocal recurrent selection in a high yield environment. p. 101–111. *In* 28th Annual Illinois Corn Breeders School. Dep. of Agronomy, Univ. of Illinois, Champaign-Urbana.

Lamkey, K.R. 1992. Fifty years of recurrent selection in the Iowa Stiff Stalk Synthetic maize population. Maydica 37:19–28.

Lamkey, K.R., and O.S. Smith. 1987. Performance and inbreeding depression of populations representing seven eras of maize breeding. Crop Sci. 27:695–699.

Landi, P., and E. Frascaroli. 1993. Repsonses to four cycles of full-sib family recurrent selection in an F_2 maize population. Maydica 38:31–37.

Lonnquist, J.H. 1961. Progress from recurrent selection procedures for the improvement of corn populations. Nebraska Agric. Exp. Stn. Bull. 197. Lincoln.

Lonnquist, J.H. 1964. Modification of the ear–to–row procedure for the improvement of maize populations. Crop Sci. 4:227–228.

Maita, R., and J.G. Coors. 1996. Twenty cycles of biparental mass selection for prolificacy in the open-pollinated maize population Golden Glow. Crop Sci. 36:1527–1532.

Mareck, J.H., and C.O. Gardner. 1979. Response to mass selection in maize and stability of resulting populations. Crop Sci. 19:779–783.

Martínez-Barajas, E., C. Villanueva-Verduzco, J. Molina-Galán, H. Lopez-Travera, and E. Sánchez-de-Jiménez. 1992. Relation of rubisco to maize grain yield improvement: Effect of water restriction. Crop Sci. 32:718–722.

Moll, R.H. 1991. Sixteen cycles of recurrent full-sib family selection for grain weight in two populations of maize. Crop Sci. 31:959–964.

Moll, R.H., W.A. Jackson, and R.L. Mikkelsen. 1994. Recurrent selection for maize grain yield: Dry matter and nitrogen accumulation and partitioning changes. Crop Sci. 34:874–881.

Mulamba, N.N., A.R. Hallauer, and O.S. Smith. 1983. Recurrent selection for grain yield in a maize population. Crop Sci. 23:536–540.

Ochieng, J.A.W., and R. E. Kamidi. 1992. Response to eight cycles of reciprocal recurrent selection in Kitale Synthetic II, Ecuador 573 and their variety cross. J. Genet. Breed. 46:315–320.

Odhiambo, M.O. and W.A. Compton. 1989. Five cycles of replicated S1 vs. reciprocal full-sib index selection in maize. Crop Sci. 29:314–319.

Pandey, S., A.O. Diallo, T.M.T. Islam, and J. Deutsch. 1987. Response to full-sib selection in four medium maturity maize populations. Crop Sci. 27:617–622.

Pandey, S., and C.O. Gardner. 1992. Recurrent selection for population, variety, and hybrid improvement in tropical maize. Adv. Agron. 48:1–87.

Pandey, S., A.O. Diallo, T.M.T. Islam, and J. Deutsch. 1986. Progress from selection in eight Tropical maize populations using international testing. Crop Sci. 26:879–884.

Popi, J. 1997. A critical evaluation of the HOPE breeding system as a means for broadening the deployed germplasm base in maize. Ph.D. diss. Univ. of Guelph, Guelph, ON.

Rademacher, M.A.M, A.R. Hallauer, and W.A. Russell. 1998. Comparative response of two reciprocal recurrent selection methods in BS21 and BS22 maize populations. Crop Sci. (in press).

Robinson, H.F., R.E. Comstock, and P.H. Harvey. 1949. Estimates of heritablity and the degree of dominance in corn. Agron. J. 41:353–359.

Robinson, H.F., C.C. Cockerham, and R.H. Moll. 1960. Studies on estimation of dominance variance and effects of linkage bias. p. 171–177. In International Series of Monographs on Biometry. Pergamon Press.

Russell, W.A. 1974. Comparative performance for maize hybrids representing different eras of maize breeding. p. 81–101. In D. Wilkinson (ed.) Proc. of the 29th Annual Corn and Sorghum Industry Research Conf., Chicago, IL. 10–12 Dec. Am. Seed Trade Assoc., Washington, DC.

Russell, W.A. 1983. Agronomic performance of maize cultivars representing different eras of breeding. Maydica 29:375–390.

Russell, W.A. 1986. Contributions of breeding to maize improvement in the United States: 1920–1980s. Iowa State J. of Res. 6:5–34.

Russell, W.A., S.A. Eberhart, and U.A. Vega. 1973. Recurrent selection for specific combining ability for yield in two maize populations. Crop Sci. 13:257–261.

Singh, M., A.S. Khehra, and B.S. Dhillon. 1986. Direct and correlated response to recurrent full-sib selection for prolificacy in maize. Crop Sci. 26:275–278.

Smith, O.S., 1984. Comparison of effects of reciprocal recurrent selection in the BSSS(R), BSCB1(R), and BS6 populations. Maydica 29:1–8.

Stojšin, D. 1994. Genetic changes associated with different methods of recurrent selection in five maize populations. Ph.D. diss. Univ. of Guelph, Guelph, ON.

Stromberg, L.D., and W.A. Compton. 1989. Ten cycles of full-sib selection in maize. Crop Sci. 29:1170–1172.

Stuber, C.W., S.E. Lincoln, D.W. Wolff, T. Helentjaris, and E.S. Lander. 1992. Identification of genetic factors contributing to heterosis in a hybrid from two elite maize inbred lines using molecular markers. Genetics 132:823–839.

Sullivan, J.A., and L.W. Kannenberg. 1987. Comparison of S_1 and modified ear–to–row recurrent selection in four maize populations. Crop Sci. 27:1161–1166.

Tanner, O.S., and O.S. Smith. 1987. Comparison of half-sib and S_1 recurrent selection in Krug Yellow Dent maize populations. Crop Sci. 27:509–513.

Tragesser, S.L., W.C. Youngquist, O.S. Smith, and W.A. Compton. 1989. Drift vs. selection effects from five recurrent selection programs in maize. Maydica 34:23–32.

Vasal, S.K., S. McLean, and L. Narro. 1994. Lowland tropical germplasm development and population improvement at CIMMYT headquarters. pp. 1–13. *In* S.K Vasal, and S. McLean (ed.). The lowland tropical maize subprogram. Maize Program Spec. Rep. CIMMYT, Mexico, D.F.

Walejko, R.N., and W.A. Russell. 1977. Evaluation or recurrent selection for specific combining ability in two open-pollinated maize cultivars. Crop Sci. 17:647–651.

Weyhrich, R.A., K.R. Lamkey, and A.R. Hallauer. 1998a. Response to seven methods of recurrent selection in the BS11 maize population. Crop Sci. 38:308–321.

Weyhrich, R.A., K.R. Lamkey, and A.R. Hallauer. 1998b. Effective population size and response to S_1-progeny selection in the BS11 maize population. Crop Sci. 38:1149–1158.

Chapter 22

Recurrent Selection and Heterosis

C. L. Souza, Jr.

INTRODUCTION

Recurrent selection methods were designed to improve the agronomic value of populations by gradually increasing the frequency of favorable alleles, while maintaining genetic variability. To realize these objectives, progenies are developed, evaluated, selected, and recombined in a repetitive manner. Populations improved by recurrent selection are expected to be agronomically superior to unimproved versions and have enough genetic variability for the traits under selection to allow these populations be used for medium or long-term selection .

The performance of hybrids is associated with the level of heterosis, i.e., to the superiority of hybrids over their inbred parents. To exploit heterosis efficiently , populations are grouped into heterotic groups, where population crosses within and among groups produce low and high levels of heterosis, respectively. Hybrids are then produced by crossing inbred lines from different heterotic groups.

Recurrent selection has been effective in gradually improving population performance (Smith, 1983; Hallauer et al., 1988), as well as the performance of the hybrids developed from the succeeding cycles of selection in maize (*Zea mays* L.; Russell,1985; Betran & Hallauer, 1996). Therefore, recurrent selection programs should be integrated with hybrid breeding programs so that improved populations can be used as sources of inbred lines not related to that ones developed from recycled lines via pedigree breeding. The objectives of this paper are: (i) to present the effects of recurrent selection on hybrid breeding programs; (ii) to present the changes in heterosis following recurrent selection; and (iii) to compare intra- and interpopulation selection schemes .

RECURRENT SELECTION AND HYBRID IMPROVEMENT

Recurrent selection will be useful only if the hybrids developed from improved populations are superior to those developed from unimproved versions. To evaluate the effects of recurrent selection on hybrid improvement, consider two genetically divergent populations (P_1 and P_2) under recurrent selection, and assume that single-cross selection is carried out in the original (P_{1o} and P_{2o}) and in the populations improved by n cycles of recurrent selection (P_{1n} and P_{2n}). The expected means of the original populations and their crosses are P_{1o}, P_{2o} and $PC_o = [(P_{1o} + P_{2o})/2] + h_o$; and after n cycles of recurrent selection the corresponding means are P_{1n}, P_{2n}, and $PC_n = PC_o + n[(Rrs_1 + Rrs_2)/2] + n\Delta h$, where h_o and Δh are the initial heterosis and its change following selection, and Rrs_1 and Rrs_2 are the responses to recurrent selection on populations 1 and 2, respectively. The responses to single-cross selection in the original and in the improved populations are $RSC_o = i\sigma_{G_o}^2 / \sigma_{Ph_o}$ and $RSC_n = i\sigma_{G_n}^2 / \sigma_{Ph_n}$ where i, σ_G^2, and σ_{Ph}^2 are the standardized selection differential, the interpopulation genetic and phenotypic variances, respectively.

The expected means of the single crosses selected from original and improved populations are

$$SC_o = PC_o + RSC_o \text{ and } SC_n = PC_n + RSC_n$$

and the latter can be expressed as

$$SC_n = PC_o + n\left[\left(\frac{Rrs_1 + Rrs_2}{2}\right) + \Delta h\right] + RSC_n.$$

The contrast of the selected single-cross means from original and improved populations $(\Delta SC = SC_n - SC_o)$ is

$$\Delta SC = n\left[\left(\frac{Rrs_1 + Rrs_2}{2}\right) + \Delta h\right] + i\frac{\sigma_{G_n}^2}{\sigma_{Ph_n}} - i\frac{\sigma_{G_o}^2}{\sigma_{Ph_o}}$$

and by assuming $\sigma_{Ph_n}^2 \approx \sigma_{Ph_o}^2$, we have

$$\Delta SC = n\left[\left(\frac{Rrs_1 + Rrs_2}{2}\right) + \Delta h\right] + \left(\frac{i}{\sigma_{Ph_n}}\right)\left(\sigma_{G_n}^2 - \sigma_{G_o}^2\right).$$

Then, the improvement of single crosses through recurrent selection depends on the improvement of the performance of the populations per se, on the improvement of heterosis, and on the maintenance of the interpopulation genetic variance. For $\sigma_{G_n}^2 \approx \sigma_{G_o}^2$, we have $\Delta SC = n[(Rrs_1 + Rrs_2)/2] + n\Delta h$, and the improvement of the single crosses is directly proportional to the response from recurrent selection. Conversely, if the interpopulation genetic variance is reduced because of recurrent selection, i.e., if $\sigma_{G_n}^2 < \sigma_{G_o}^2$, we have $\Delta SC < n[(Rrs_1 + Rrs_2)/2] + n\Delta h$, and the total benefit of recurrent selection on hybrid breeding will not be achieved. Hence, the maintenance of the interpopulation genetic variance during selection is essential to have the improvement in hybrids be equal to that of the populations and their heterotic interactions. Breeders must recognize that a compromise between selection intensity and the effective size (Ne) of the improved populations is needed to avoid the reduction in the genetic variances due to selection and random genetic drift.

INTRA- AND INTERPOPULATION SELECTION

Recurrent selection methods are grouped into intra- or interpopulation schemes. The former emphasizes the improvement of populations per se, and the latter emphasizes population-cross improvement. Changes in the performance of population crosses and populations per se are considered as indirect responses to intra- and interpopulation recurrent selection, respectively.

It was shown above that improvement of single crosses is highly dependent on improvement of populations per se and heterosis, provided that genetic variances are maintained during selection. Thus, comparisons of intra- and interpopulation recurrent selection are necessary to evaluate their effectiveness in improving both heterosis and population per se performance.

Consider two genetically divergent populations (P_1 and P_2), both in Hardy-Weinberg and linkage equilibrium, under half-sib reciprocal recurrent selection (*RRS*; Comstock et al., 1949), and under simultaneous half-sib intrapopulation selection (*HSS*; Empig et al., 1972). Both selection methods use half-sib progenies (inter- and intra-) and S_1 progenies for evaluation and recombination, respectively. Assume also that intra- and interpopulation phenotypic variances of the progenies means are equal. The interpopulation genetic variances ($\sigma^2_{A_{12}}$ and $\sigma^2_{A_{21}}$) can be expressed as

$$\sigma^2_{A_{12}} = \sigma^2_{A_{11}} + \sigma^2_{\tau_{12}} + 4Cov_{(A_1\tau_{12})}$$

and

$$\sigma^2_{A_{21}} = \sigma^2_{A_{22}} + \sigma^2_{\tau_{21}} + 4Cov_{(A_2\tau_{21})}$$

and the additive genetic covariances of intra- and interpopulation genotypes ($\sigma_{A_1A_2}$ and $\sigma_{A_2A_1}$) are

$$\sigma_{A_1A_2} = \sigma^2_{A_{11}} + 2Cov_{(A_1\tau_{12})}$$

and

$$\sigma_{A_2A_1} = \sigma^2_{A_{22}} + 2Cov_{(A_2\tau_{21})}$$

where $\sigma^2_{A_{11}}$ and $\sigma^2_{A_{22}}$ are the intrapopulation additive variances, $\sigma^2_{\tau_{12}}$ and $\sigma^2_{\tau_{21}}$ are the variances of the deviations from intra- and interpopulation additive effects, and $Cov_{(A_1\tau_{12})}$ and $Cov_{(A_2\tau_{21})}$ are the covariances of intrapopulation additive effects with the deviations from intra- and interpopulation additive effects. For the one locus-two allele model where p and q, and r and s are the frequencies of the favorable and unfavorable alleles in populations 1 and 2, respectively; a and d are half of the difference of the genotypic values of the homozygotes and heterozygotes; and α_1 and α_2 are the additive effects of populations 1 and 2, respectively, the new variances and covariances can be expressed as

$$\sigma_{\tau_{12}}^2 = 8pq(p-r)^2 d^2$$

$$\sigma_{\tau_{21}}^2 = 8rs(p-r)^2 d^2$$

$$Cov_{(A_1\tau_{12})} = 2pq(p-r)\alpha_1 d$$

and

$$Cov_{(A_2\tau_{21})} = 2rs(r-p)\alpha_2 d .$$

Then, $Cov_{(A_1\tau_{12})} > 0$ and $Cov_{(A_2\tau_{21})} < 0$ for $p > r$, and vice-versa for $p < r$. Also, the σ_τ^2 parameters are positive and larger than the $|Cov_{(A\tau)}|$ parameters for traits whose levels of dominance are nearly 1.0, such as grain yield of maize, (Souza, 1993).

The direct and indirect expected responses to interpopulation (*RRS*) and intrapopulation (*HSS*) recurrent selection, as well as the expected change in heterosis, are formulated in terms of σ_A^2, σ_τ^2 and $Cov_{(A\tau)}$ in Tables 22–1 and 22–2. The contrasts between RRS and HSS responses for population cross, populations per se, and heterosis are presented below.

For the population cross:

$$Rs_{RRS} - Rs_{HSS} = i_1 \frac{\sigma_{\tau_{12}}^2 + 2Cov_{(A_1\tau_{12})}}{4\sigma_{Ph}} + i_2 \frac{\sigma_{\tau_{21}}^2 + 2Cov_{(A_2\tau_{21})}}{4\sigma_{Ph}}$$

For the populations per se:

$$Rs_{RRS} - Rs_{HSS} = i_1 \frac{Cov_{(A_1\tau_{12})}}{\sigma_{Ph}} \quad \text{for population } 1, \text{ and}$$

$$Rs_{RRS} - Rs_{HSS} = i_2 \frac{Cov_{(A_2\tau_{21})}}{\sigma_{Ph}} \quad \text{for population } 2$$

For heterosis:

$$Rs_{RRS} - Rs_{HSS} = i_1 \frac{\sigma_{\tau_{12}}^2}{4\sigma_{Ph}} + i_2 \frac{\sigma_{\tau_{21}}^2}{4\sigma_{Ph}}$$

Table 22–1. Direct and indirect responses to reciprocal recurrent selection.

Response for†	Rs
Population cross (PC)	$i_1 \dfrac{\sigma^2_{A_{11}} + \sigma^2_{\tau_{12}} + 4Cov_{(A_1\tau_{12})}}{4\sigma_{Ph_{12}}} + i_2 \dfrac{\sigma^2_{A_{22}} + \sigma^2_{\tau_{21}} + 4Cov_{(A_2\tau_{21})}}{4\sigma_{Ph_{21}}}$
Population 1 (P_1)	$i_1 \dfrac{\sigma^2_{A_{11}} + 2Cov_{(A_1\tau_{12})}}{2\sigma_{Ph_{12}}}$
Population 2 (P_2)	$i_2 \dfrac{\sigma^2_{A_{22}} + 2Cov_{(A_2\tau_{21})}}{2\sigma_{Ph_{21}}}$
Heterosis (h)	$i_1 \dfrac{\sigma^2_{\tau_{12}} + 2Cov_{(A_1\tau_{12})}}{4\sigma_{Ph_{12}}} + i_2 \dfrac{\sigma^2_{\tau_{21}} + 2Cov_{(A_2\tau_{21})}}{4\sigma_{Ph_{21}}}$

† $Rs_h = Rs[PC - (P_1 + P_2)/2]$.

Table 22–2. Direct and indirect responses to half-sib recurrent intrapopulation selection.

Response for†	Rs
Population cross (PC)	$i_1 \dfrac{\sigma^2_{A_{11}} + 2Cov_{(A_1\tau_{12})}}{4\sigma_{Ph_1}} + i_2 \dfrac{\sigma^2_{A_{22}} + 2Cov_{(A_2\tau_{21})}}{4\sigma_{Ph_2}}$
Population 1 (P_1)	$i_1 \dfrac{\sigma^2_{A_{11}}}{2\sigma_{Ph_1}}$
Population 2 (P_2)	$i_2 \dfrac{\sigma^2_{A_{22}}}{2\sigma_{Ph_2}}$
Heterosis (h)	$i_1 \dfrac{2Cov_{(A_1\tau_{12})}}{4\sigma_{Ph_1}} + i_2 \dfrac{2Cov_{(A_2\tau_{21})}}{4\sigma_{Ph_2}}$

† $Rs_h = Rs[PC - (P_1 + P_2)/2]$.

The expected responses to selection and the contrasts between inter- and intrapopulation selection show that *RRS* is expected to be more efficient than *HSS* for improving the population cross, one of the populations, and heterosis, but *RRS* is less efficient than *HSS* for improving the other population for traits whose levels of dominance are close to complete dominance $[(d/a) \approx 1.0]$ because the σ^2_τ parameters are positive and larger than the $|Cov_{(A\tau)}|$ parameters (Souza, 1993). The population with the highest frequency of favorable alleles will have

$Cov_{(A\tau)} > 0$ and will be more efficiently improved via *RRS*, but the other one will be more efficiently improved via *HSS* because $Cov_{(A\tau)} < 0$.

Notice that the expected response in heterosis via *RRS* depends on both σ_τ^2 and $Cov_{(A\tau)}$, and then heterosis will increase. However, with *HSS* the response for heterosis depends only on $Cov_{(A\tau)}$ and, therefore, could increase or decrease. If the populations are genetically divergent, e.g., $p > r$, we have $\left(\left| Cov_{(A_2\tau_{21})} \right| / Cov_{(A_1\tau_{12})} \right) > 0$, with the numerator being negative (Souza,1993), and the changes in heterosis via *HSS* will be zero or negative . Previous theoretical results have shown that *RRS* will increase heterosis, and intrapopulation selection will decrease heterosis for most situations (Moll et al., 1978; Souza & Miranda Fº, 1985; Jiang et al., 1990). The change in heterosis following recurrent selection depends on the differences between the rates of improvement of the population cross and of the populations per se. *RRS* will improve population crosses more efficiently than populations per se, and, therefore, heterosis increases. Conversely, *HSS* will improve populations per se more efficiently than population crosses, and heterosis decreases.

Most results from reciprocal recurrent selection for grain yield in maize have shown that the population cross and one of the source populations were improved at reasonable rates, but the other source population was not improved at acceptable rates, or the response was negative. Consequently, heterosis usually increased with RRS selection (Table 22–3). For example, Moll and Hanson (1984) reported that after 10 cycles of reciprocal recurrent selection, the population cross improved by 2.0% per cycle, the Jarvis population by 2.4%, and the Indian Chief population by -0.9%; Hallauer (1989) reported that full-sib reciprocal recurrent selection improved the population cross by 7.6% per cycle, the BS10 population by 3.0%, and the population BS11 by 1.6% after eight cycles of selection; and Hallauer et al. (1983) reported the population cross and the populations BSSS and BSCB1 were improved by 3.6%, 2.2% and 0.7% per cycle, respectively, after seven cycles of reciprocal recurrent selection. In these three programs, heterosis increased from 13.5 to 29.2%, from 2.5 to 21.1%, and from 29.3 to 46.1%, respectively. These experimental results are in agreement with the theoretical expectations of this paper.

For a hybrid breeding program, the improvement of both populations per se and heterosis should be as efficient as possible. As it was shown, neither intra- nor interpopulation recurrent selection can fulfill these requirements simultaneously. Therefore, a modification of these selection procedures should be utilized to overcome their drawbacks as follows: one of the two populations should be used as a tester for itself and for the other population. Then, interpopulation selection will be used for one of the base populations, and intrapopulation selection will be used for the other one. This modified selection procedure will be referred as Testcross Half-Sib Selection (*THS*) thereafter. The population to be used as a tester should be that one in which the $Cov_{(A\tau)}$ is negative, and it will probably be the lower yielding population because this covariance is a function of the differences between the frequencies of the favorable alleles in the two populations, i.e., $\sum_i (p_i - r_i)$. The expected responses for population crosses, populations per se, and heterosis for

Hallauer, A.R. 1989. Fifty years of recurrent selection in corn. Illinois Corn Breed. School. 25:39–63.

Hallauer, A.R., W.A. Russel, and O.S. Smith. 1983. Quantitative analysis of Iowa Stiff Stalk Synthetic. Stadler Genet. Symp. 15:83–104.

Hallauer, A.R., W.A. Russel, and K.R. Lamkey. 1988. Corn Breeding. p. 463–564. *In* G.F. Sprague and J.W. Dudley (ed.). Corn and corn improvement. Agron. Monogr. #18. ASA, CSSA, SSSA, Madison, WI.

Jiang, C., C.C. Cockerham, and R.H. Moll. 1990. Inter- and intracultivar effects of selection on heterosis. Crop Sci. 30:44–49.

Moll, R.H., C.C. Cockerham, C.W. Stuber, and W.P. Williams. 1978. Selection responses, genetic environmental interactions, and heterosis with recurrent selection for yield in maize. Crop Sci. 18: 599–603.

Moll, R.H., and W.D. Hanson. 1984. Comparisons of effects of intrapopulation vs. interpopulation selection in maize. Crop Sci. 24:1047–1052.

Russell, W.A. 1985. Comparison of the hybrid performance of maize lines developed from the original and improved cycles of BSSS. Maydica 30:407–419.

Smith, O.S. 1983. Evaluation of recurrent selection in BSSS, BSCB1, and BS13 maize populations. Crop Sci. 23: 35–40.

Souza, C.L. Jr., and J.B. Miranda F°. 1985. Changes in heterosis via intra- and interpopulation selection. Pesq. Agropec. Brasil. 20:1197–1201.

Souza, C.L. Jr. 1993. Comparisons of intra-, interpopulation, and modified recurrent selection methods. Rev. Brasil. Genet. 16:91–105.

West, D.P., W.A. Compton, and M.A. Thomas. 1980. Comparison of replicated S_1 per se vs. reciprocal full-sib index selection in corn. Crop Sci. 20:35–42.

Table 22–3. Responses for populations per se, population crosses, and heterosis following reciprocal recurrent selection for grain yield in maize.

Populations	Number of cycles	Response (% cycle)		Heterosis (%)	
		Population	Crosses	h_0	h_n
		------- % cycle^{-1} -------		--------- % ---------	
BS10†		3.0	7.6	2.5	21.1
BS11	8	1.6			
Jarvis‡		2.4	2.0	13.5	29.2
I. Chief	10	-0.9			
BSSS§		2.2	3.6	29.3	46.1
BSCB1	7	0.7			
NSS¶		4.1	6.0	13.8	27.4
NBS	2	-3.4			
NSS¶		4.1	2.6	11.2	26.4
NKS	2	-1.4			

†Hallauer (1989); ‡ Moll and Hanson (1984); §Hallauer et al. (1983); ¶West et al. (1980).

THS are the same as for interpopulation selection for one of the populations, e.g., population 1 (Table 22–1), and the same as for intrapopulation selection for the other population, e.g., population 2 (Table 22–2).

A numerical evaluation of the relative effectiveness $(RE\%)$ of *RRS* , *HSS* , and *THS* (interpopulation selection for population 1 and intrapopulation selection for population 2) was carried out as follows. First, the genetic variances and covariances related to the responses to selection were expressed as functions of a and d genotypic values assuming the population gene frequencies fit a Beta distribution .The phenotypic variances of the selection units were also assumed to be equal. Then, $RE\% = [(Rs_1 - Rs_2)/Rs_2]100$, where Rs_1 and Rs_2 are the expected selection responses for the two selection procedures being compared, was estimated for three levels of dominance and for two divergent populations whose average frequencies of the favorable alleles were $\bar{p} = 0.6$ and $\bar{r} = 0.4$ for populations *1* and *2*, respectively (Table 22–4).

The $RE\%$ evaluations showed that *RRS* was more efficient than *HSS* for improving the population cross (*PC*), population 1 (P_1), and heterosis (h), but it was nearly 30% less efficient than *HSS* for improving population 2 (P_2). *THS* was more efficient than *HSS* and as efficient as *RRS* for improving P_1, more efficient than *RRS* and as efficient as *HSS* for improving P_2, less efficient than *RRS* and

Table 22–4. Expected relative effectiveness $(RE\%)$ of three selection procedures for the population cross (PC), population 1 (P_1), population 2 (P_2), and heterosis (h) for three levels of dominance (d/a).

Comparisons	$d/a=$	$RE\% = \left[(Rs_1 - Rs_2)/Rs_2\right]100$		
		0.75	1.0	1.25
	PC	11.25	20.00	31.25
Rs_{RRS} vs.	P_1	18.56	20.00	18.56
Rs_{HSS}	P_2	-24.83	-32.00	-38.42
	h	240.00	240.00	240.00
	PC	7.22	7.14	6.12
Rs_{THS} vs.	P_1	0.00	0.00	0.00
Rs_{RRS}	P_2	33.03	47.06	62.39
	h	-71.43	-71.43	-71.43
	PC	19.29	28.57	39.29
Rs_{THS} vs.	P_1	18.56	20.00	18.56
Rs_{HSS}	P_2	0.00	0.00	0.00
	h	140.00	140.00	140.00

more efficient than *HSS* for improving heterosis, and more efficient than both methods for improving the population cross.

Hence, with *THS* heterosis will be improved at a slower rate than it would be with *RRS*, but the improvement of the population cross and of the populations per se will be greater than with *RRS* or *HSS* applied simultaneously in two divergent populations. For hybrid breeding programs, improvement of population crosses as well as the populations per se should be as efficient as possible, because their rates of improvement are approximately the same as for the derived hybrids and inbred lines per se, respectively. Therefore, the use of *THS* rather than *RRS* or *HSS* would be the suitable strategy to accomplish the requirements of hybrid breeding programs.

REFERENCES

Betran, F.J., and A.R. Hallauer. 1996. Hybrid improvement after reciprocal recurrent selection in BSSS and BSCB1 maize populations. Maydica 41:25–33.

Comstock, R.E., H.F. Robinson, and P.H. Harvey. 1949. A breeding procedure designed to make maximum use of both general and specific combining ability. Agron. J. 41:360-367.

Empig, L.T., C.O. Gardner, and W.A. Compton. 1972. Theoretical gains for different population improvement procedures. Nebraska Agric. Exp. Stn. Bull. MP26(revised). Lincoln.

Chapter 23

Molecular Markers and Heterosis

J. Moreno-Gonzalez

INTRODUCTION

Systematic exploitation of heterosis has played an important role in hybrid improvement. Classical methods for identifying heterosis include the diallel crossing system (Griffing, 1956; Gardner & Eberhart, 1966), the design II mating design (Comstock & Robinson, 1948) and testcrosses of the breeding material with testers of known heterotic performance (Hallauer & Miranda, 1981). Application of these methods to crops is extensively referenced.

The use of polymorphic marker genes to facilitate the process of selection was proposed early in this century. High quality molecular markers to be used as potential tools in plant breeding have been developed during the last three decades. The theoretical basis for identification of quantitative trait loci (QTL) associated with individual marker loci (McMillan & Robertson, 1974; Soller & Beckman, 1983) and flanking marker loci (Lander & Botstein, 1989; Knapp et al., 1990) has been developed. Some statistical techniques that have been proposed to estimate QTL effects associated with markers are: (i) comparison of marker class means (Soller & Beckman, 1983); (ii) maximum likelihood estimation of a putative QTL located between flanking markers, the so-called interval mapping method (Lander & Botstein, 1989; Luo & Kearsey, 1992); (iii) multiple linear regression (Knapp et al., 1990; Haley & Knott, 1992; Moreno-Gonzalez, 1992); (iv) composite interval mapping (CIM), which combines interval mapping and the regression approach (Jansen & Stam, 1994; Zeng, 1994).

Molecular markers have been used to investigate the presence of heterosis effects between inbred lines and to predict their hybrid performance (Lee et al. 1989; Dudley et al., 1991; Melchinger et al., 1992). The theoretical study of Charcosset and Essioux (1994) investigated the prediction of heterosis by molecular markers in different population structures. No conclusive method for hybrid prediction between inbreds of different heterotic groups seems to give convincing results.

Theory on molecular-assisted selection (MAS) was developed by Lande and Thompson (1990). Application of MAS to crop improvement in practical breeding programs is still very limited. Stromberg et al. (1994) reported similar results of testcross performance in MAS and conventional testcross selection of a maize population. Simulation studies on MAS have confirmed the theory predictions and have indicated in which situations MAS should be more beneficial (Edwards & Page, 1994; Gimelfarb & Lande, 1994).

The objectives of this study were to: (i) investigate the effect of some population parameters (i.e., progeny number, recombination frequency between markers, and heritability) on the estimation of marker-associated QTLs showing heterotic effects in F_2 and inbred testcrosses by stepwise regression; (ii) propose a strategy to create two heterotic F_2 populations to which marker-assisted reciprocal recurrent selection (MARRS) may be applied; and (iii) validate the theoretical findings with a simulation study on marker-assisted selection.

METHODS

Models

The regression approach was used for estimation of the marker-associated QTL effects because this technique (i) yields very similar results to the maximum likelihood method (Haley & Knot, 1992); (ii) allows for a simple analysis of the residual error structure; and (iii) permits use of the predicted regression phenotypic values based on molecular markers in the selection index of MAS.

Testcrosses of individual F_2 plants (F2T), recombinant inbreds (RIT) and doubled-haploid lines (DHT) with an inbred tester T can estimate the differential heterotic effect of marker-associated QTL's between the two parental lines from which the segregating populations are derived. A modification of the model proposed by Moreno-Gonzalez (1993) using flanking markers was introduced. The modification consisted in iterating the regression model for each significant QTL effect after the original model was first applied to the data. The iteration searched for the location of the QTL within the marker segment that yielded the least residual error. This approach is similar to the CIM method, but it uses regression instead of maximum likelihood techniques for locating the QTL. The equation of the proposed model resembles that of the CIM model reported by Lübberstedt et al. (1997) and will be called "iterative regression model" (IRM):

$$y_j = \mu_0 + h_k^* x_{kj}^* + \sum_i h_i x_{ij} + e_j \qquad for \ i \neq k \qquad [1]$$

where y_j is the phenotypic value of the jth testcross; μ_0 is the phenotypic mean; h_i is one-half the difference of the genotypic values of genotypes ${}^1Q_i{}^TQ_i$ and ${}^2Q_i{}^TQ_i$, where ${}^1Q_i$, ${}^2Q_i$, ${}^TQ_i$ are the alleles of the putative QTL i associated with the marker segment S_i (i.e., the chromosome interval between consecutive markers i and $i+1$) in the inbreds P_1, P_2 and T, respectively; x_{ij} are dummy variables associated with h_i for testcross j; e_j is the residual effect of testcross j; h_k^* and x_{kj}^* are similar to h_{ij} and x_{ij}, respectively, but they refer to QTL k which is being iterated for 20 different values of ρ_k (ρ_k = 0.025, 0.075.......0.975); ρ_k is the ratio of the recombination frequencies between QTL k and its left-hand side flanking marker to that between the two flanking markers M_k and M_{k+1} (r_k). Expected and initial assigned values of x_{ij} for the marker classes of generations F2T, RIT, and DHT are shown in Table 23-1.

Creation of Two Elite Heterotic F_2 Populations

Charcosset and Essioux (1994) found that prediction of hybrids between inbred lines from different groups on the basis of heterozygosity of marker loci was not effective; however, heterosis between lines of the same group can be predicted.

The following strategy is suggested for the creation of two heterotic F_2 populations if the objective were to improve the best selected hybrid between lines of two heterotic groups. MARRS can be later applied to the two F_2 populations:

Step 1. Assume that a number of N_A and N_B elite inbred lines are known to belong to heterotic groups A and B, respectively. A mating design II (Hallauer & Miranda, 1981) is carried out by crossing each line from group A with each line from

Table 23–1. Expected and initial assigned values of dummy variable x_{ij} for marker class of the F_2 (F_2T), doubled-haploid (DHT) and recombinant inbred (RIT) testcrosses.

Marker class	Expected values†			Initial assigned values		
	F_2T	DHT	RI	F_2T	DHT	RIT
$M_iM_iM_{i+1}M_{i+1}$	1	1	D‡	1	1	E§
$M_iM_iM_{i+1}m_{i+1}$	$1-\rho_i$			0.5		
$M_iM_im_{i+1}m_{i+1}$	$1-2\rho_i$	$1-2\rho_i$	$1-2\rho_i$	0	0	0
$M_im_iM_{i+1}M_{i+1}$	ρ_i			0.5		
$M_im_iM_{i+1}m_{i+1}$	0			0		
$M_im_im_{i+1}m_{i+1}$	$-\rho_i$			-0.5		
$m_im_iM_{i+1}M_{i+1}$	$2\rho_i-1$	$2\rho_i-1$	$2\rho_i-1$	0	0	0
$m_im_iM_{i+1}m_{i+1}$	ρ_i-1			-0.5		
$m_im_im_{i+1}m_{i+1}$	-1	-1	-D‡	-1	-1	-E§

† $\rho_i = r_{1i}/r_i$ where r_{1i}, and r_i are the recombination frequencies between M_i and Q_i, and M_i and M_{i+1}, respectively.
‡ $D = (1-4r_i^2\rho_i(1-\rho_i))/(1+4r_i^2\rho_i(1-\rho_i))$
§ $E = (1-r_i^2)/(1+r_i^2)$

group B. The numbers N_A and N_B are limited by the number of $N_A \times N_B$ crosses to be evaluated. Half-sib (HS) genetic variances of lines from group A when crossed with B ($\sigma_{(A)}^2$) and that of lines from group B when crossed with group A ($\sigma_{(B)}^2$) are estimated in the usual way for design II (Hallauer & Miranda, 1981).

Equations for $\sigma_{(A)}^2$ and $\sigma_{(B)}^2$) can be expressed as functions of some population genetic parameters in the two-alleles locus model, using an approach similar to that of Charcosset and Essioux (1994), as

$$\sigma_{(A)}^2 = \sum_{k=1}^{n_k} p_k(1-p_k)h_k^2 + 2\sum_{k=1}^{n_k}\sum_{f=1, f\neq k}^{n_f} D_{kf}h_kh_f \qquad [2]$$

where p_k is the frequency of allele A_k at locus k; h_k is the heterotic effect associated with locus k, measured as one-half the difference between the testcrosses of the favorable and unfavorable alleles with population B; n_k is the number of loci segregating for the quantitative trait; D_{kf} is the linkage disequilibrium between loci k and f defined as $qt - rs$; where q, r, s and t are the frequencies of gametes A_kA_f, A_ka_f, a_kA_f and a_ka_f, respectively.

Step 2. The criterion for choosing two lines i and j from group A and two lines u and v from group B to create the F_2(A) and F_2(B) populations is constructed in the following way:

(i) Assume that linkage disequilibria between QTL and marker loci exist within the A and B groups (Charcosset & Essioux 1994), and QTL effects are evenly distributed throughout the genome. The testcross genetic variance of the F_2(A) plants crossed with hybrid tester uv can be predicted as

$$\sigma_{F2(A)}^2 = \frac{n_{mij}\sigma_{(A)}^2}{n_{ms}} \qquad [3]$$
$$4\sum_{s=1} H_s$$

where $\sigma^2_{F2(A)}$ is the predicted testcross genetic variance of the F_2 population derived from the ij hybrid in group A; n_{mij} is the number of marker loci which have different alleles in both lines i and j; H_s is the heterozygosity of the group A of inbreds at marker locus s; n_{ms} is the number of polymorphic loci in group A. Only 1/2 of $\sigma_{(A)}^2$ contributes to prediction because the additive genetic variance of an F_2 population was predicted on the basis of the additive genetic variance of an inbred population.

(ii) The proposed criterion (C) is

$$C = \frac{1}{4} (Y_{iu} + Y_{iv} + Y_{ju} + Y_{jv}) + (\Delta G_A + \Delta G_B) \, t \qquad [4]$$

where Y_{iu}, Y_{iv}, Y_{ju} and Y_{jv} are the performances of hybrids iu, iv, ju and jv between the selected lines from groups A and B; ΔG_A and ΔG_B are the contributions from populations $F_2(A)$ and $F_2(B)$ to the genetic gain of the population cross when MARRS is carried out in both populations; and t is the number of cycles.

Expressions for ΔG_A and ΔG_B can be estimated according to MAS (Lande & Thompson, 1990) and selection index theory (Smith, 1936; Hazel, 1943) by combining the testcross performances and the marker-based predicted testcrosses into a selection index. The expression for ΔG_A is

$$\Delta G_A = i \mathbf{G} \mathbf{b} \, (\mathbf{b}' \mathbf{P} \mathbf{b})^{-\frac{1}{2}} \qquad [5]$$

where $\mathbf{G}$ and $\mathbf{P}$ are the genetic and phenotypic variance-covariance matrices, respectively; $\mathbf{b}' = [1, b_m]$ is the vector of relative weights; and i is the selection intensity. Furthermore (Lande & Thompson, 1990),

$$\mathbf{G} = \begin{vmatrix} \sigma^2_{F2(A)} & \sigma^2_m \\ \sigma^2_m & \sigma^2_m \end{vmatrix} \qquad [6]$$

$$\mathbf{P} = \begin{vmatrix} \dfrac{\sigma^2_{F2(A)}}{h^2} & \sigma^2_m \\ \sigma^2_m & \sigma^2_m \end{vmatrix} \qquad [7]$$

$$b_m = \frac{\sigma^2_{F2(A)} - h^2 \sigma^2_{F2(A)}}{h^2 (\sigma^2_{F2(A)} - \sigma^2_m)} = \frac{1 - h^2}{h^2 (1 - p)} \qquad [8]$$

where, $\sigma^2_{F2(A)}$ is the testcross genetic variance of the F_2 population derived from hybrid ij; σ^2_m is the genetic variance explained by the marker loci; h^2 is the heritability of the trait; and p is the portion of the genetic variance explained by the markers. The highest C value will select the inbreds to form the F_2 populations in groups A and B. If no estimate of σ^2_m is inicially available, then $\Delta G_A = i h \sigma_{F2(A)}$.

Step 3. MARRS is carried out in the two heterotic F_2 populations by crossing plants of population A with the hybrid tester uv from B, and reciprocally plants from B with the hybrid tester ij from A. A selection index is applied to populations A and B:

$$I_A = Y_A + b_{mA} \hat{Y}_{mA} \quad \text{and} \quad I_B = Y_B + b_{mB} \hat{Y}_{mB} \qquad [9]$$

where I_A, Y_A, $\hat{Y}_{mA}$ and b_{mA} are the selection index, the testcross performance, the predicted testcross based on molecular markers and the weight associated with $\hat{Y}_{mA}$ for population A, respectively; I_B, Y_B, $\hat{Y}_{mB}$ and b_{mB} are the same for population B.

Error Structure

Stepwise multiple linear regression analysis (Draper & Smith, 1981) was applied to the models. Significant QTL effects (h_k) can be detected at the α probability level with 50% chance (Moreno-Gonzalez, 1993), provided that

$$h_k > t_{f,\alpha}(S_e^2 c_{kk})^{\frac{1}{2}}$$ [10]

where $t_{f,\alpha}$ is the tabular t-value for the f degrees of freedom of the residual mean squares and the chosen α significance level; S_e^2 is the residual mean square of the regression; and c_{kk} is a diagonal term of the matrix $[\mathbf{X'X}]^{-1}$; where $\mathbf{X}$ is the design matrix of the multiple regression analysis (Draper & Smith, 1981). Expected values of c_{kk} are $2[N(1-r_i)]^{-1}$ and $[N(1-r_i)]^{-1}$ for the F_2 and doubled-haploid line testcrosses, respectively (Moreno-Gonzalez, 1993), where N is the progeny size.

The residual mean square S_e^2 of Eq. [10] has the components

$$S_e^2 = \sigma_e^2 + \sigma_{g'}^2 + \Phi^2$$ [11]

where σ_e^2 is the environmental error variance; $\sigma_{g'}^2$ is the component of the genetic variance (σ_g^2) accounted for by QTL not included in the model; Φ^2 is a component due to the deviations of the assigned genotypic values to marker classes in the model from their true genotypic values. The expected value of Φ^2 for several types of generations were derived by Moreno-Gonzalez (1992, 1993), when the assigned value for $\rho_k = 0.5$. Following the same approach (Moreno-Gonzalez, 1993) the expected value of Φ^2 for the F_2 and doubled-haploid line testcrosses in the iterative model will be:

$$\Phi^2 = 4r \sum_{k=1}^{n_k} (\rho_k(1-\rho_k) + (\rho_k'-\rho_k)^2) \sigma_{gk}^2 = rwp\sigma_g^2$$ [12]

where r is the recombination frequency between flanking markers; ρ_k and ρ_k' are the true and estimated values of the relative position of QTL k; w is the weighted average of $4\rho_k(1-\rho_k)+4(\rho_k-\rho_k')^2$ which is expected to be less than 1; σ_{gk}^2 is the contribution of QTL k to the total genetic variance (σ_g^2); n_k is the number of QTLs in the model; and p is the part of genetic variance explained by the model.

Taking into account Eq. [10], [11], and [12] and Moreno-Gonzalez (1993),

$$h_s > \sqrt{\frac{2F_{1,f,\alpha}[\frac{1}{h^2}-p(1-rw)]}{N(1-r_s)}}$$ [13]

where h_s is the smallest heterotic effect in genetic standard deviation units that can be detected in a group of estimated QTL's that account for a portion p of the total genetic variance σ_g^2; N is the number of evaluated testcrosses; $F_{1,f,\alpha}$ is the tabular F-value in the F distribution for 1 and f (residual) degrees of freedom at the α significant level; h^2 is the heritability of the trait.

Simulation Study

Sixty and 120 markers, each separated by $r = 0.2$ and $r = 0.1$ recombination frequencies, respectively, were simulated. Twelve chromosomes were involved in the simulations. Eleven QTLs (Q1 to Q11) at both independent and linked marker segments were arbitrarily located in the genome (Table 23–2). Random numbers from uniform distributions of the RANUNI SAS function (SAS Institute, 1985) were used to locate crossovers on chromosomes and produce gametes with random recombinant marker and QTL genotypes. Complete interference (Knapp et al., 1990) was assumed.

Iterative Regression Model

Monte Carlo data of doubled-haploid line testcrosses were generated to test whether the iterative regression model (Eq. 1) (i) located the QTL at a position close to the true location; and (ii) yielded less residual error and detected more QTL effects than the original model in which the assigned values to ρ_k were 0.5. Twenty sets of data were generated for testing the true location of Q4 and Q5 when the recombination frequencies between markers (r) were 0.2 and 0.1. One-hundred data sets were also generated for comparing the iterative model where the ρ_k was estimated for each significant QTL to the original model where the ρ_k was 0.5 for all QTLs. A random number from a normal distribution with mean zero and variance 2, $N(0,2)$, of the RANNOR SAS function (SAS, 1985) was added to the genotypic value of each generated testcross as an environmental error effect to simulate the phenotypic value.

Marker-Assisted Selection (MAS)

MAS was applied to simulation F_2 and doubled-haploid line testcrosses. Several situations were studied by combining three progeny sizes (125, 250 and 500 testcrosses), three selection intensities (selection of the top 5, 15 and 50 testcrosses) and three heritabilities (random environmental effects from normal distributions with zero mean and variances 2, 4, and 6 from the RANNOR SAS function were added to

Table 23–2. Location and effects of the simulated QTLs in the genome. Chromosome number and map distance from the chromosome end in cM for cases† 1 and 2†.

Location	Simulated QTLs										
	Q1	Q2	Q3	Q4	Q5	Q6	Q7	Q8	Q9	Q10	Q11
Chromosome number	1	1	1	2	3	4	5	8	9	11	12
Distance from the chromosome end						cM					
Case 1	2	36	77	57	35	17	35	59	94	53	72
Case 2	4	33	75	58	30	15	30	58	88	47	64
QTL effects‡	0.5	0.5	0.5	0.9	1.0	0.75	0.5	0.45	0.25	0.2	0.86

† Cases 1 and 2 refer to markers separated by $r = 0.1$ and 0.2, respectively.
‡ QTLs Q4, Q6, Q8, Q9, and Q11 had negative effects.

the genotypic values). One-hundred data sets were generated for each situation. The selection index (Lande & Thompson, 1990) was

$$I = Y + b_m \hat{Y}_m$$

where I, Y, $\hat{Y}_m$ and b_m are the selection index, the testcross performance, the marker-based predicted testcross, and the relative weight associated with $\hat{Y}_m$, respectively. $\hat{Y}_m$ was predicted from Eq. [1] by stepwise regression with probability level 0.001 for the effects entering the model. b_m was estimated from Eq. [8].

RESULTS AND DISCUSSION

Iterative Regression Model

The iterative regression model (Eq. [1]) was tested for 10 values of ρ_k ($\rho_k = 0.05, 0.1,......0.95$) at the marker segments of significant Q4 and Q5 effects. The least residual error of the model averaged over 20 simulation data sets occurred when the ρ_k value at the Q4 marker segment was 0.65 for both $r = 0.2$ and $r = 0.1$. The second minimum residual error resulted when the ρ_k value was 0.75. The iterative regression model located Q4 fairly well, because the ρ_k true value was 0.7. Likewise, the least residual error of the model appeared when the ρ_k values at the Q5 marker segment were 0.75 for $r = 0.2$ and 0.65 for $r = 0.1$. The iterative model gave an adequate location for Q5, because the true value of ρ_k was 0.75.

When the ρ_k values of all significant QTLs were estimated in the iterative regression model, the average residual errors were 2.8174 and 2.3913 for r = 0.2 and r = 0.1, respectively. In the original regression model, when the value assigned to ρ_k was 0.5 for all marker segments, the average residual errors were larger than in the iterative model, 2.9275 and 2.4342 for $r = 0.2$ and $r = 0.1$, respectively. The iterative model also detected the smallest QTLs Q9, Q10, and Q11 with a higher frequency and at a closer position to the true location than the original model. Thus the iterative regression model improved the original model.

The error component Φ^2 will be reduced in the iterative regression model, if the QTLs are close to the flanking markers (small ρ) and the estimates of ρ (ρ') are also close to the true values (Eq. [12]). If the location of the QTL is at the middle of the marker segment (ρ about 0.5) Φ^2 will not be reduced, and the iterative regression model and the original model will be similar (Eq. [12]).

Simulated Molecular-Assisted Selection

Marker-assisted selection (MAS) was superior to testcross conventional selection (CS) and molecular-based predicted regression selection (MPS) in all situations (Tables 23–3 and 23–4). MPS was superior to CS for high progeny size (250 and 500 testcrosses), but inferior for small progeny size (125 testcrosses). The disadvantage of MPS was more evident for low heritabilities, likely because the regression model was not able to detect genetic effects with small progeny size and low heritability (Eq. [12]).

The relative efficiency of MAS over CS increased as the progeny size increased and the heritability became smaller. There is not very much difference between MAS and MPS for the largest progeny size (500 testcrosses), probabily because a large portion (p) of the genetic variance was explained by the QTLs in the model. If p approximates 1, the contribution of the molecular-based predicted phenotypes (MPS) to MAS is very important (Eq. [8]). Thus the most significant requirement for MAS to be efficient is to detect QTLs accounting for a large portion of the genetic variance. This can be accomplished by increasing the progeny size and the heritability

Table 23–3. Relative genetic gain of phenotypic (PS), molecular-based predicted regression (MPS) and molecular-assisted (MAS) selections on F_2 testcrosses for several recombination frequencies between markers (r), selection intensities, evaluated testcross number and heritabilities (h^2). Average of 100 simulated sets.

r	Selected plants	125 Testcrosses			250 Testcrosses			500 Testcrosses		
		PS	MPS	MAS	PS	MPS	MAS	PS	MPS	MAS
	No.					$h^2 = 0.55$				
0.1	5	286†	286	312	316	380	390	354	448	449
0.1	15	233	229	248	266	322	324	308	392	394
0.1	50	136	136	146	194	232	234	244	307	309
						$h^2 = 0.39$				
0.1	5	239	201	262	262	347	358	307	433	438
0.1	15	193	167	209	223	287	295	261	372	375
0.1	50	113	105	127	163	205	212	205	294	296
						$h^2 = 0.29$				
0.1	5	220	161	229	238	282	307	265	398	406
0.1	15	171	134	185	204	245	265	228	351	354
0.1	50	100	82	108	146	177	193	181	277	279
						$h^2 = 0.55$				
0.2	5	289	272	307	323	379	386	355	432	444
0.2	15	233	212	248	272	314	325	310	379	385
0.2	50	134	125	143	195	222	230	244	297	304
						$h^2 = 0.39$				
0.2	5	243	170	259	283	328	350	298	413	420
0.2	15	190	142	203	232	281	294	263	360	365
0.2	50	113	85	120	164	200	209	203	282	288
						$h^2 = 0.29$				
0.2	5	211	122	224	240	278	312	261	392	402
0.2	15	168	111	176	202	247	263	232	346	352
0.2	50	100†	67	107	149	180	191	181	274	279

† Genetic gain 100 equals 0.833 simulation units and 0.528 genetic standard deviation units.

(Eq. [13]); however, a large heritability (h^2) has a negative effect on b_m (Eq. [8]). Thus, h^2 has opposite effects on QTL detection and b_m. A large h^2 allows detection of more and smaller QTLs, but also favors phenotypic selection.

The marker recombination frequencies ($r = 0.1$ vs. $r = 0.2$) did not have a large effect on MAS efficiency. MAS takes advantage of the effect of QTLs on the predicted value regardless of the place where they will be located, either at the true or neighbor marker segment. The so called *ghost effects* will contribute to MAS if they are a reflection of true QTLs. The theoretical advantage of $r = 0.1$ over I= 0.2 is small as indicated by Eq. [13].

Table 23–4. Relative genetic gain of phenotypic (PS), molecular-based predicted regression (MPS) and molecular-assisted (MAS) selections on doubled-haploid line testcrosses under several recombination frequencies (r), selection intensities, evaluated testcross number and heritabilities (h^2). Average of 100 simulated sets.

r	Selected plants	125 Testcrosses			250 Testcrosses			500 Testcrosses		
		PS	MPS	MAS	PS	MPS	MAS	PS	MPS	MAS
0.1	No.					$h^2 = 0.71$				
0.1	5	259†	264	275	288	320	325	302	352	354
0.1	15	207	216	222	246	273	277	273	311	313
0.1	50	125	128	133	179	197	200	220	249	251
						$h^2 = 0.55$				
0.1	5	223	225	253	253	299	305	279	349	351
0.1	15	182	185	201	216	255	260	244	306	309
0.1	50	109	110	119	159	187	189	193	244	245
						$h^2 = 0.45$				
0.1	5	208	184	228	231	286	300	248	338	342
0.1	15	166	158	188	200	244	252	220	299	301
0.1	50	100	99	109	143	178	181	174	239	241
						$h^2 = 0.71$				
0.2	5	258	242	272	288	303	313	309	369	345
0.2	15	208	199	219	243	257	267	273	299	305
0.2	50	126	120	131	178	187	193	220	239	244
						$h^2 = 0.55$				
0.2	5	229	186	243	252	286	298	274	330	336
0.2	15	184	163	195	215	245	254	243	292	298
0.2	50	110	104	118	156	178	184	195	234	238
						$h^2 = 0.45$				
0.2	5	209	149	220	233	268	280	250	323	327
0.2	15	164	135	179	197	228	238	222	288	291
0.2	50	100†	91	107	141	168	175	176	229	233

† Genetic gain 100 equals 1,463 simulation units and 0.656 genetic standard deviation units.

MAS relative genetic gain was higher in F_2T (Table 23–3) than in DHT (Table 23–4), but the absolute gain was larger in DHT than in F_2T, because the 100 relative figure was 1.463 simulation units for DHT and 0.833 for F_2T.

MARRS in Two Heterotic F_2 Populations

The criterion for the formation of two heterotic populations (Eq. [4]) includes the average of the four non-parental hybrids, the expected genetic gain of testcrosses in the two populations and the number of generations that are planned for MARRS. The tester of each population is single cross from the other population. This is a flexible criterion for the breeder who has to decide the number of generations of selections and the desired genetic gain over the best hybrid. The criterion score should be higher than the performance of the best hybrid plus the desired genetic gain, otherwise it would not be worthy to carry out selection. In some instances, it might be beneficial to make selection in one population using an inbred tester from the other heterotic group.

Response of simulated MAS to different selection intensities (Tables 23–3 and 23–4) agreed fairly well with expectations from selection theory. Since MARRS requires linkage disequilibrium in the F_2 populations, it is suggested to use a large progeny size, and a high selection intensity by recombining very few plants to attain maximum gain and keep the linkage disequilibrium of the populations for the first selection generations. Afterward, original RRS may be carried out.

CONCLUSIONS

Large progeny number was the most important factor for increasing the efficiency of molecular-assisted selection. For a given experiment size, it would be advisable to increase the number of entries and reduce the replication number. A large progeny number and a high selection intensity with MARRS is expected to increase the genetic gain and keep the linkage disequilibrium of the populations for a few generations. The effect of maker recombination frequency ($r = 0.1$ vs. $r = 0.2$) on selection was not very much important.

REFERENCES

Charcosset, A., and L. Essioux. 1994. The effect of population structure on the relationship between heterosis and heterozygosity at marker loci. Theor. Appl. Genet. 89:336–343.

Comstock, R.E., and H.F. Robinson. 1948. The components of genetic variance in populations of biparental progenies and their use in estimating the average degree of dominance. Biometrics 4:254–266.

Draper, N.R., and H. Smith. 1981. Applied regression analysis. 2nd edition. John Wiley & Sons, New York.

Dudley, J.W., M.A. Sagai Maroof, and G.K. Rufener. 1991. Molecular markers and grouping of parents in maize breeding programs. Crop Sci. 31:718–723.

Edwards, M.D., and N.J. Page. 1994. Evaluation of marker-assisted selection through computer simulation. Theor. Appl. Genet. 88:376–382.

Gardner, C.O., and S.A. Eberhart. 1966. Analysis and interpretation of the variety cross diallel and related populations. Biometrics 22:439–452.

Gimelfarb, A., and R. Lande.1994. Simulation of marker-assisted selection for non-additive traits. Genet. Res. 64:127–136.

Griffing, B. 1956. Concept of general and specific combining ability in relation to diallel crossing systems. Aust. J. Biol. Sci. 9:943–963.

Haley, C.S., and S.A. Knott. 1992. A simple regression method for mappingquantitative trait loci in line crosses using flanking markers. Heredity 69:315–324.

Hallauer, A.R., and J.B. Miranda. 1981. Quantitative genetics in maize breeding. Iowa State University Press. Ames.

Hazel, L.N. 1943. The genetic bases for constructing selection indices. Genetics 38:467–490.

Jansen, R.C., and P. Stam. High resolution of quantitative traits into multiple loci via interval mapping. Genetics 136:1447–1455.

Knapp, S.J., W.C. Bridges, Jr., and D. Birkes. 1990. Mapping quantitative trait loci using molecular marker linkage maps. Theor. Appl. Genet. 79:583–592.

Lande, R., and R. Thompson. 1990 Efficiency of marker-assisted selection in the improvement of quantitative traits. Genetics 124:743–756.

Lander, E.S., and D. Botstein. 1989. Mapping mendelian factors underlying quantitative traits using RFLP linkage maps. Genetics 121:185–199.

Lee M., E.B. Godshalk, K.R. Lamkey, and W.L. Woodman. 1989. Association of restriction length polymorphism among inbreds with agronomic performance of trait crosses. Crop Sci. 29:1067–1071.

Lübberstedt, T., A.E. Melchinger, C.C. Shön, H.F. Utz, and D. Klein. 1997. QTL mapping in testcrosses of European flint lines of maize: I. Comparison of different testers for forage yield trials. Crop Sci. 37:921–931.

Luo, Z.W., and M.J. Kearsey. 1992. Interval mapping of quantitative trait loci in an F_2 population. Heredity 69:236–242.

McMillan, I., and A. Robertson. 1974. The power of methods for detection of major genes affecting quantitative characters. Heredity 32:349–356.

Melchinger A.E., J. Boppenmaier, B.S. Dhillon, W.G. Pollmer, and R.G. Herrmann. 1992. Genetic diversity for RFLPs in European maize inbreds: II. Relation to performance of hybrids within vs. between heterotic groups for forage traits. Theor. Appl. Genet. 84:672–681

Moreno-Gonzalez, J. 1992. Genetic models to estimate additive and non-additive effects of marker-associated QTL using multiple regression techniques. Theor. Appl. Gen. 85:435–444

Moreno-Gonzalez, J. 1993. Efficiency of generations for estimating marker-associated QTL effects by multiple regression. Genetics 135:223–241.

SAS Institute. 1985. SAS user's guide: Statistics Basic. Version 5 ed. SAS Inst., Cary, NC.

Smith, H.F. 1936. A discriminant function for plant selection. Ann. Eugenics 7:24–250.

Soller, M., and J.S. Beckman. 1983. Genetic polymorphism in varietal identification and genetic improvement. Theor. Appl. Genet. 67:25–33.

Stromberg, L.D., J.W. Dudley, and G.K. Rufener. 1994. Comparing early generation selection with molecular marker assisted selection in maize. Crop Sci. 34:1221–1225.

Zeng, Z.-B. 1994. Precision mapping of quantitative trait loci. Genetics 136:1457.

Chapter 24

Best Linear Unbiased Predictor Analysis

R. Bernardo

INTRODUCTION

The choice of parental germplasm is crucial in a breeding program. Breeders continually have to decide which experimental single crosses to test, which advanced hybrids to recommend for further testing or commercialization, and which inbred parents to cross to form new base populations for inbred development.

Suppose a breeder has 100 inbreds from Heterotic Group 1 and 100 inbreds from Heterotic Group 1. There are 10 000 possible (Group 1 × Group 2) single crosses. For developing new inbreds, there are 495 000 possible (Group 1 F_2) × (Group 2 tester) combinations and 495 000 possible (Group 1 tester) × (Group 2 F_2) combinations. Due to limited resources, breeders are unable to test all of the above combinations but may test a limited subset of single crosses and F_2 × tester combinations. The crosses are evaluated in different sets of yield trials or in different environments, resulting in highly unbalanced data. Typically, <1% of the maize (*Zea mays* L.) single crosses tested by a breeder eventually become commercial hybrids (Hallauer, 1990). Methods for predicting the performance of single crosses would greatly enhance the efficiency of hybrid breeding programs.

Previous efforts to predict single-cross performance have been generally unsuccessful. Strong nonadditive genetic effects permit the exploitation of heterosis but prevent the prediction of single-cross yield from inbred *per se* performance (Gama & Hallauer, 1977). Correlations between single-cross performance and molecular marker diversity between unrelated parental inbreds have been too low to be of any predictive value (Godshalk et al., 1990; Melchinger et al., 1990; Dudley et al., 1991).

Best linear unbiased prediction, or BLUP for short, has been used for decades for evaluating the genetic merit of animals, especially dairy cattle (*Bos taurus*). Intrapopulation, additive genetic models have traditionally been used for BLUP in animal breeding (Henderson, 1975). During the last several years, I have attempted to use BLUP in maize breeding with interpopulation genetic models that involve both general and specific combining ability (Bernardo, 1994, 1995, 1996a, 1996b, 1996c). Results have indicated that BLUP is useful for routine prediction of single-cross performance. The predicted performance of single crosses may subsequently be used to predict the performance of F_2 × tester combinations, three-way crosses, or double crosses.

My objective is to describe the theory and application of BLUP in a hybrid breeding program.

BLUP ANALYSIS

The BLUP procedure requires two kinds of information that should be readily available to breeders: (i) performance data on single crosses that have been tested and (ii) genetic relationships among the parental inbreds. The principle underlying BLUP is the use of information from relatives. The classic example of using information from relatives is the prediction of milk yield performance of a dairy bull. If data are available on a bull's mother and paternal half-sister, the predicted milk yield of the bull is h^2 [0.50(milk yield of mother) + 0.25(milk yield of paternal half-sister)] (Falconer, 1981 p. 221). The bull is more closely related to its mother than to its half-sister and, consequently, more weight is placed on the mother's performance. Similarly, yield data may be available for hybrids X_1 and X_2 but not for hybrid Y. If X_1 and X_2 are genetically related to Y, the performance of Y can be predicted from the performance of X_1 and X_2. The weights placed on the performance of X_1 and X_2 will depend on the degree of relatedness of these hybrids to Y.

Not all single crosses are tested in the same trial at the same locations and in the same years. The unbalanced data are subject to different fixed, nuisance effects. For example, one set of hybrids may have been evaluated in high-yield environments whereas a second set of hybrids may have been evaluated in low-yield environments. Consequently, the observations need to be corrected for yield trial effects before they can be used in single-cross prediction. These fixed effects can be estimated by best linear unbiased estimation, which is used in conjunction with BLUP in genetic evaluation.

The theory and procedures for BLUP for single crosses have been described elsewhere (Bernardo, 1996a) and will not be discussed in detail in this paper. Briefly, the steps involved are:

1. Given a set of inbreds in Heterotic Group 1 and a second set of inbreds in Heterotic Group 1, retrieve all available data on all (Group 1 × Group 2) single crosses that have been evaluated in yield trials.
2. Calculate the coefficient of coancestry (Falconer, 1981 p. 80–82) among Group 1 and among Group 2 inbreds. The coefficient of coancestry quantifies the degree of relatedness among inbreds and can be determined from pedigree or molecular marker data (Bernardo et al., 1996).
3. With the so-called mixed-model equations, simultaneously (i) estimate residual and genetic variances, (ii) obtain best linear unbiased estimates of fixed yield trial effects, and (iii) obtain BLUP of general and specific combining ability.
4. Use the data on the tested single crosses, corrected for fixed yield trial effects, to obtain BLUP of the untested single crosses. The predictions are based on both general and specific combining ability.

Assume n single crosses are made between n_1 inbreds from Group 1 and n_2 inbreds from Group 2. The single crosses, along with n_C check hybrids, are evaluated in t different yield trials resulting in p total observations. The linear model in matrix notation for single-cross performance is

$$\mathbf{y} = \mathbf{X}\beta + \mathbf{Z}_0\mathbf{c} + \mathbf{Z}_1\mathbf{g}_1 + \mathbf{Z}_2\mathbf{g}_2 + \mathbf{Z}\mathbf{d} + \mathbf{e}$$

where: $\mathbf{y} = p \times 1$ vector of observed performance of single crosses (i.e., hybrid by multilocation trial means); $\beta = t \times 1$ vector of yield trial effects; $\mathbf{c} = n_C \times 1$ vector of check hybrid effects; $\mathbf{g}_1 = n_1 \times 1$ vector of general combining ability effects of Group 1 inbreds; $\mathbf{g}_2 = n_2 \times 1$ vector of general combining ability effects of Group 2 inbreds; $\mathbf{d} = n \times 1$ vector of specific combining ability effects; $\mathbf{e} = p \times 1$ vector of residual effects; and $\mathbf{X}$, $\mathbf{Z}_0$, $\mathbf{Z}_1$, $\mathbf{Z}_2$, and $\mathbf{Z}$ are incidence matrices of 1's and 0's relating $\mathbf{y}$ to β, $\mathbf{c}$, $\mathbf{g}_1$, $\mathbf{g}_2$, and $\mathbf{d}$, respectively.

Solutions to β, $\mathbf{c}$, $\mathbf{g}_1$, $\mathbf{g}_2$, and $\mathbf{d}$ as well as estimates of residual and genetic variances can be obtained iteratively by restricted maximum likelihood as described by Henderson (1985). After estimates of residual and genetic variances have been obtained, the performance of the $m = (n - n_1 n_2)$ untested single crosses can be predicted as

$$\mathbf{y}_M = \mathbf{C}_{MP}\, \mathbf{C}_{PP}^{-1}\, \mathbf{y}_P$$

where: $\mathbf{y}_M = m \times 1$ vector of predicted performance of untested single crosses; $\mathbf{C}_{MP} = m \times n$ matrix of genetic covariances between the untested single crosses and the tested single crosses; $\mathbf{C}_{PP} = n \times n$ phenotypic variance-covariance matrix among the tested single crosses; and $\mathbf{y}_P = n \times 1$ vector of average performance of tested single crosses, corrected for yield trials effects. Please refer to Bernardo (1996a) for details on the calculation of $\mathbf{C}_{MP}$, $\mathbf{C}_{PP}$, and $\mathbf{y}_P$.

APPLICATION

The BLUP procedure has been used for several years in the North American maize breeding program of Limagrain Genetics, a subsidiary of Groupe Limagrain (Chappes, France). A total of 404 private inbreds were assigned to nine heterotic groups, arbitrarily designated A, B, C, D, E, F, G, H, and I. Sixteen cross combinations (i.e., heterotic *patterns*) between the nine heterotic groups were considered. The hybrid performance data sets comprised the results from multilocation yield trials, conducted by Limagrain Genetics from 1990 to 1995, of 4775 single crosses. The data sets were highly unbalanced across multilocation trials but, disregarding occasional missing plots at individual locations, balanced within multilocation trials. The performance at each individual location was not considered in the data analysis. Instead, the elements of $\mathbf{y}$ were the average performance of a single cross or check hybrid at several locations in a multilocation trial. Data were available for grain yield (t ha^{-1} at 155 g H_2O kg^{-1}), moisture (g kg^{-1}), and the percentage of stalk and root lodging.

All the necessary computations were performed with *lgHYPER*, a proprietary [1996 Limagrain Genetics] FORTRAN PC program for predicting *hy*brid *per*formance. A breeder inputs lists of Group 1 and Group 2 inbreds. The *lgHYPER* program then (i) searches a pedigree file for progenitor inbreds and calculates the coefficient of coancestry among Group 1 and among Group 2 inbreds, (ii) searches a performance data file for the pertinent single cross and check hybrid data, (iii) performs best linear unbiased estimation of yield trial effects and BLUP, (iv) predicts the performance of untested single crosses, (v) finds the best F_2 populations for each tester based on predicted three-way cross performance (Gerloff and Smith, 1988), and (vi) outputs the general combining ability of each inbred for each trait.

The program ranks the check hybrids, tested single crosses, untested single crosses, and $F_2 \times$ tester combinations according to a multiple trait index specified by the breeder. In this study, the standardized (unitless) weights were 1.00 for yield, -0.25 for moisture, -0.30 for stalk lodging, and -0.20 for root lodging.

The effectiveness of BLUP was evaluated with a cross-validation procedure. Suppose 100 single crosses have data available. The *lgHYPER* program disregards the 1st single cross and predicts its performance from the remaining 99. The program then takes the 2nd, 3rd, 4th, . . . , 100th single cross and predicts its performance form the remaining 99. Lastly, the correlation between predicted and actual performance (sample size of 100) is calculated for each trait as an indication of the effectiveness of BLUP in predicting single-cross performance.

RESULTS AND DISCUSSION

Across the 16 heterotic patterns, the correlation between the predicted and observed performance ranged from 0.463 to 0.770 for yield, 0.868 to 0.936 for moisture, 0.466 to 0.685 for stalk lodging, and 0.164 to 0.518 for root lodging (Fig. 24–1). The cross-validation procedure is not perfect because predicted values are compared with observed values which are subject to error, i.e., h^2 <1. The square root of heritability is defined as the correlation between genetic and phenotypic value. Even if the correlation between the predicted and true genetic value is 1, the correlation between the predicted and observed value cannot be greater than the square root of h^2. Thus, the square root of h^2 (denoted by lines above the bars in Fig. 24–1) sets a theoretical upper bound on the correlation between predicted and observed values and should be the reference when interpreting the correlations obtained with the cross-validation procedure.

For yield, most of the correlations between predicted and observed performance were ≥0.60 and were 75 to 85% of the corresponding theoretical upper bounds. These correlations were not perfect but seemed sufficiently high (Johnson, 1989; Bernardo, 1992) for pick-the-winner schemes that characterize hybrid testing. For example, suppose a breeder desires to identify the best single cross (i.e., with the highest genetic value) out of 100. A correlation between predicted and true genetic value of 0.60 would allow a breeder to select the top 20 out of 100 single crosses while maintaining at least an 80% chance of retaining the best hybrid in the selected group (Bernardo, 1992).

Whereas the correlations between predicted and observed moisture were high (>0.85), such correlations were often low for root lodging. Maize breeders know that field trial results for root lodging are much less consistent than those for moisture. For root lodging, most of the correlations ranged from 0.30 to 0.50 and were 50 to 90% of their corresponding upper bounds.

For yield, the average ratio across heterotic patterns of specific combining ability variance (V_{SCA}) to total genetic variance (V_G) was 0.26 (results not shown). In contrast, average V_{SCA}/V_G was 0.11 for moisture. In agreement with published results (Hallauer & Miranda, 1981 p. 116), the importance of nonadditive effects relative to additive effects was less for moisture than for yield. Average V_{SCA}/V_G across heterotic patterns was also high for stalk lodging (0.24) and root lodging (0.39), indicating the importance of both additive and nonadditive effects for these traits.

When double-cross hybrids were grown in the USA in the 1930s to 1960s, Jenkins' (1934) Method B (i.e., average of four nonparental single crosses) was widely used for predicting double-cross performance. In a study conducted in a single year (and most probably at a single location near Ames, IA), the correlations between predicted and observed performance of 42 double crosses ranged from 0.42 to 0.78 for several traits (Jenkins, 1934). These correlations are comparable with those obtained in the present study. Hence, the BLUP procedure can be considered effective for predicting single-cross performance just as Jenkins' procedure was considered effective for predicting double-cross performance.

Several parameters such as the predicted three-way cross performance, μG (Dudley, 1987), an upper bound on μG (Gerloff & Smith, 1988), and the probability of net gain of favorable alleles (Metz, 1994) have been proposed for choosing parents to improve an elite single cross. Perhaps a main limitation to the routine use of these parameters has been the need for particular field experiments for their estimation. Much of the hybrid data required for estimating the above parameters can be obtained by BLUP with data already available from the breeder's hybrid testing program.

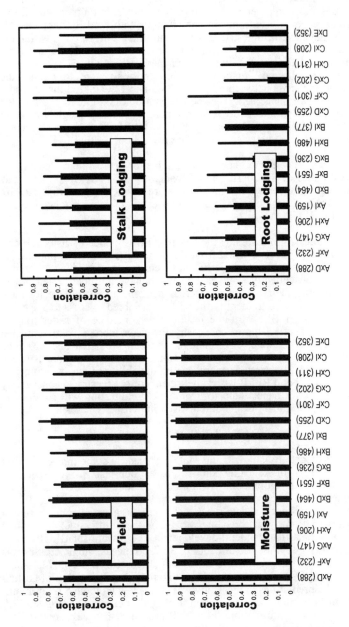

Fig. 24-1. Correlation between predicted and observed maize yield, moisture, stalk lodging, and root lodging for 16 heterotic patterns (AxD, AxF, AxG, ..., DxE). The number of tested single crosses in each heterotic pattern is in parentheses. The solid lines above the bars represent a theoretical upper bound on the correlations obtained with the cross-validation procedure.

SPECIAL SITUATIONS

1. Testcross data for an inbred are sometimes unavailable. For example, a maize breeder may be interested in crosses between early Iowa Stiff Stalk Synthetic (BSSS) inbreds and Lancaster Sure Crop (LSC) inbreds, but a particular BSSS inbred may have been tested only in combination with nonLSC inbreds. Or some inbreds may have been tested only in environments that are not within the breeder's target population of environments. The breeder may desire to predict the performance of the cross between an untested BSSS inbred and a tested LSC inbred. In extreme situations, the breeder may want to predict the performance of the cross between a BSSS inbred and LSC inbred that are both untested. The performance of a cross between an untested inbred and a tested inbred can be predicted effectively when the number of tested single crosses in the heterotic pattern is large (Bernardo, 1996c). But BLUP of the performance of the cross between two untested inbreds has given poor results.

2. Pedigree records may sometimes be unreliable or unavailable, leading to inaccurate estimates of the coefficient of coancestry. The BLUP procedure is robust with regards to erroneous measures of relatedness (Bernardo, 1996b). Resorting to the use of molecular markers to estimate genetic relatedness is unnecessary for BLUP.

3. Breeders are interested in assessing the testcross performance of sister inbreds, which are defined as inbreds selfed from the same F_2 or backcross population. Pedigree-based estimates of the coefficient of coancestry are identical among sister inbreds. Consequently, if the sister inbreds have not been tested in any hybrid combination, their predicted testcross performance would be identical and BLUP cannot be used to identify the best sister inbred to cross to a specific tester. In a study of BLUP of the testcross performance of sister inbreds (Bernardo, 1997), coefficients of coancestry for sister inbreds were estimated with random RFLP markers. Correlations between predicted and observed performance were erratic and mostly low. Correspondence was poor between ranks for predicted and observed general combining ability of the sister inbreds. The failure of BLUP and RFLP-based coefficients of coancestry to consistently predict testcross performance indicates that actual field testing will continue to be necessary for preliminary evaluation of sister inbreds.

CONCLUSIONS

The BLUP procedure is now being routinely used by some maize breeders in the USA, not only for selecting single crosses but also for choosing F_2 populations for inbred development. An attractive feature of BLUP is that no special experiments are needed for obtaining the predictions. Rather, BLUP exploits the massive amounts of data generated yearly in a commercial breeding program. Doing BLUP in hybrid breeding is extremely cost effective. The initial investments involve data management, development of software for BLUP, and a fast computer. In this study, the performance of 35 723 untested single crosses was predicted from the performance of 4775 tested single crosses. If all 35 723 single crosses were tested at five locations, each with one replicate and each plot costing $15, the value of the predictions in this study was equivalent to about $2.7 million.

The use of both trait and marker data in BLUP is currently being studied. But even without marker information, the BLUP procedure has vast potential for increasing the efficiency of present-day hybrid breeding programs.

REFERENCES

Bernardo, R. 1992. Retention of genetically superior lines during early-generation testcrossing of maize. Crop Sci. 32:933–927.

Bernardo, R. 1994. Prediction of maize single-cross performance using RFLPs and information from related hybrids. Crop Sci. 34:20–25.

Bernardo, R. 1995. Genetic models for predicting maize single-cross performance in unbalanced yield trial data. Crop Sci. 35:141–147.

Bernardo, R. 1996a. Best linear unbiased prediction of maize single-cross performance. Crop Sci. 36:50–56.

Bernardo, R. 1996b. Best linear unbiased prediction of maize single-cross performance given erroneous inbred relationships. Crop Sci. 36:862–866.

Bernardo, R. 1996c. Best linear unbiased prediction of the performance of crosses between untested maize inbreds. Crop Sci. 36:872–876.

Bernardo, R. 1997. RFLP markers and predicted testcross performance of maize sister inbreds. Theor. Appl. Genet. 95:655–659.

Bernardo, R., A. Murigneux, and Z. Karaman. 1996. Marker-based estimates of identity by descent and alikeness in state among maize inbreds. Theor. Appl. Genet. 93:262–267.

Dudley, J.W. 1987. Modification of methods for identifying inbred lines useful for improving parents of elite single crosses. Crop Sci. 27:944–947.

Dudley, J.W., M.A. Saghai Maroof, and G.K. Rufener. 1991. Molecular markers and grouping of parents in maize breeding programs. Crop Sci. 31:718–723.

Falconer, D.S. 1981. Introduction to quantitative genetics. 2nd ed. Longman, London.

Gama, E.E.G., and A.R. Hallauer. 1977. Relation between inbred and hybrid traits in maize. Crop Sci. 17:703–706.

Gerloff, J.E., and O.S. Smith. 1988. Choice of method for identifying germplasm with superior alleles. 1. Theoretical results. Theor. Appl. Genet. 76:209–216.

Godshalk, E.B., M. Lee, and K.R. Lamkey. 1990. Relationship of restriction fragment length polymorphisms to single-cross hybrid performance of maize. Theor. Appl. Genet. 80:273–280.

Hallauer, A.R. 1990. Methods used in developing maize inbreds. Maydica 35:1–16.

Hallauer, A.R., and J.B. Miranda, Fo. 1981. Quantitative genetics in maize breeding. Iowa State Univ. Press, Ames.

Henderson, C.R. 1975. Best linear unbiased estimation and prediction under a selection model. Biometrics 31:423–447.

Henderson, C.R. 1985. Best linear unbiased prediction of nonadditive genetic merits in noninbred populations. J. Anim. Sci. 60:111–117.

Jenkins, M.T. 1934. Methods of estimating the performance of double crosses in corn. J. Am. Soc. Agron. 26:199–204.

Johnson, B. 1989. The probability of selecting genetically superior S_2 lines from a maize population. Maydica 34:5–14.

Melchinger, A.E., M. Lee, K.R. Lamkey, and W.L. Woodman. 1990. Genetic diversity for restriction fragment length polymorphisms: Relation to estimated genetic effects in maize inbreds. Crop Sci. 30:1033–1040.

Metz, G. 1994. Probability of net gain of favorable alleles for improving an elite single cross. Crop Sci. 34:668–672.

Chapter 25

Prediction of Single-Cross Performance

O. S. Smith, K. Hoard, F. Shaw, and R. Shaw

INTRODUCTION

Maize *(Zea mays* L.) production in the U.S. has increased dramatically since the 1930s, as scientific knowledge in agronomy, genetics and plant breeding accumulated and this technology started to have a large effect. The change in production has been estimated by Duvick (1977), Russell (1974), and others to be roughly half due to increased inputs and agronomic practices and half due to genetic changes that allow modern hybrids to fully utilize increase inputs, i.e., chemical fertilizers, etc.

Maize breeding has also changed dramatically since the 1930s. During this period better maize hybrids have been developed, in terms of their grain production and agronomic performance. As superior hybrids have been developed, increased efforts have been needed to maintain the rate of genetic improvement. One of the results of this is that research funding associated with commercial breeding programs has increased almost logarithmically during this 60-year period. The increased research funding has supported larger and larger breeding programs that have been developed to identify new hybrids. Along with the development of larger breeding programs have come incentives for breeders to develop more efficient methods of identifying potential commercial hybrids.

The type of commercial hybrids grown has also changed during this 60-year period and has progressed from the use of open-pollinated varieties, or landraces, to double cross hybrids in the 1930s, and finally, single cross hybrids that have been used since the 1960s. The large amount of non-additive genetic variance associated with single cross hybrids and the amount of genotype by environmental interaction has made it difficult to identify high yielding hybrids without testing large numbers of potential hybrids over a relatively large sample of environments. Recently, Bernardo (1994,1995) has published on the adaptation of methods developed by animal breeders to predict single cross hybrids. Similar work has been done at Pioneer, and in this chapter I will present some of the objectives and details of this system.

EFFORT NEEDED TO DEVELOP A NEW HYBRID

Fig. 25–1 is a reminder of the gains in maize production in the U.S. since about the 1930s, when enough scientific knowledge had accumulated that technology could start to have a major effect on maize production. The following section lists some of these technologies that have allowed breeders to develop the highly productive hybrids farmers grow today.

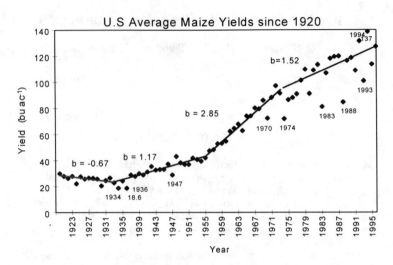

Fig. 25–1. U.S. average maize yields since 1920.

Maize breeding has changed dramatically since the 1930s. Much like the rest of society, the amount of effort needed to develop a commercial product has increased logarithmically since about 1970 when plant variety protection became available. This time period can be divided into six segments that correspond roughly to the following time periods.

Mid 1920s to about 1940

During this time, hybrid maize was rapidly being developed and adopted in the U.S. Corn Belt. Breeders were developing inbred lines, and with a limited amount of testing, found combinations of these that would produce superior products.

1940s to 1950s

More effort was invested in testing new hybrid combinations. A few good inbred lines had been identified by both public and private breeders, and limited numbers of these combinations were being tested in replicated field trials. The double cross prediction methods developed by Jenkins (1934) were being used. The types of gene action (additive vs. nonadditive) were being debated as well as how best to use nonadditive genetic variance.

1950s to 1960s

Industry starts to use multi-station testing (wide area testing) to identify superior hybrids. The concepts of general and specific combining ability (GCA and SCA) were being used in the development of superior inbred lines. In about the mid 1960s, the industry switches from double cross to three-way crosses and finally to single cross hybrids. Based on theoretical studies (Cockerham, 1961), the breeders observed that the highest yield hybrids are single crosses and the increase in inbred line performance made production of single cross hybrids commercially feasible.

1970s to 1980s

The multi-stage hybrid development programs currently in use are developed. Wide area research testing and large amounts of on farm strip testing are used in hybrid development and commercialization. Adjusted GCAs and the ability to use data from across research experiments and strip trials are developed. Winter nursery decisions become data driven.

1990s

Rapid cycling of inbreds and wide spread use of data driven winter nurseries allow reduction in time to introduce new genetics into hybrids. The number of hybrids that are in wide area testing explodes. Hybrid prediction system is developed to refine techniques to use data across the breeding departments. Performance predictability and product life cycle management becomes key concepts in product development.

The previous introduction shows how the amount of effort needed to develop a commercial hybrid has increased over time. A large part of this effort is involved in performance testing. So the question becomes, "Why do we need so much performance testing?"

There are Several Reasons

We are trying to measure a 5 to10% performance advantage with traits that have a significant measurement error associated with them. The standard deviation on a location basis is equal to one to two times the difference we are trying to measure.

Genotypes interact with environments, therefore, which genotype is best, is environment dependent, and no one genotype is best in the full range of environments.

Environments sampled in a given year are not a random sample of all possible environments. The sample of environments realized in any given year is usually a biased sample of the environments of interest, therefore, getting a representative sample of the environments of interest is very difficult.

This point is very important and needs some further explanation. Genes that affect quantitative traits, the traits that are the most valuable to our customers, interact with environments. Currently, we cannot characterize these interactions to any great extent. Therefore, we cannot predict performance in unsampled environments, or in a sample of environments for which the distribution of the environments is different, from the distribution in the sampled environments. The strategy is to test in such a large sample of environments that we find those hybrids that are likely to perform well NO MATTER WHAT THE FUTURE DISTRIBUTION OF ENVIRONMENTS. Is this difficult and inefficient? Yes, it is comparable to inserting a gene at random in a genome and predicting the effect of this random insertion on the rest of the genome!

Table 25–1 presents one of the advancement schemes that have been used to advance hybrids to commercial status. The R1 stage consists of early generation topcrosses, and most of the testing is done at the breeding station level. The R2 stage is where hybrids are made among lines that have survived the initial testing and have been inbred to near homozygosity. At this stage the hybrids are grown in wide area tests, i.e., they are grown in multiple locations across the

Table 25–1. Typical yield testing effort for #2 yellow hybrids.

Stages	Research Total locations	On farm Locations
R1	5	–
R2 wide area	20	–
R3 Y number	50	–
R4 X number	75	200–500
R5 commercial number	75	300–1500

geographic area where the hybrid is thought to be adapted. In the R1 to R3 stages the topcrosses and hybrids are grown in two row plots about six meters in length. Those hybrids that advance to the R3 stage become precommercial hybrids, and enough seed is made during the R3 stage so that those hybrids that advance to the R4 stage can be grown in strip trial or on farm trials. The on farm trials consist of 6 to12 row strips, generally, several hundred meters long. They are harvested using commercial equipment and grain yields are determined using weigh wagons.

Table 25–2 lists the numbers of new wide area hybrids tested in North America over the years 1990 to 1994. The numbers have grown dramatically during this time as has the number of inbred parents coded by the breeders each year. The number of inbreds generate a huge number of potential single cross hybrids. The number being on the order of 25 000 for about 700 new inbreds with about ten potential established inbred parents and about 25 new inbreds as potential parents. This represents a huge amount of potential yield testing. In the past breeders have made potential commercial hybrid combinations based on their knowledge of the germplasm, knowledge that had been acquired through years of experience, and in working with the parents and grandparents of the material being evaluated; however, as the private seed business consolidates and becomes global, and as the number of new inbred parents and potential new hybrids increase, new technology is needed to help breeders efficiently use the large amount of germplasm developed.

Now, let's look at a hybrid development process with the objective of looking at how we might improve the efficiency of this process. A simplified overview of a pedigree inbred line and hybrid development process are presented in Fig. 25–2.

Table 25–2. North America #2 yellow hybrids by year †

Year	R2	R3	R4	R5
1990	800	65	20	15
1991	1,100	50	25	15
1992	1,500	55	25	15
1993	2,400	70	20	20
1994	3,800	80	25	10

† Source: annual hybrid release books.

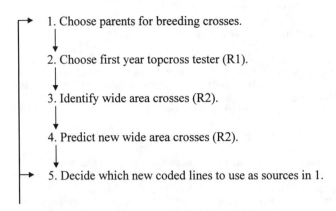

1. Choose parents for breeding crosses.

2. Choose first year topcross tester (R1).

3. Identify wide area crosses (R2).

4. Predict new wide area crosses (R2).

5. Decide which new coded lines to use as sources in 1.

Fig. 25–2. Overview of steps involved in a pedigree inbred line/hybrid development process.

This process consists of the following steps:

1. Choose parents for breeding crosses

This step is still part of the art of plant breeding. The object in this step is to find combinations of parents that complement each others' yield, or agronomic shortcomings, and result in progeny that are high yielding in hybrid combination. Experience has shown that parents that do well in hybrid combination, make the best parents for breeding crosses. Sprague and Tatum (1941) developed methods for estimating general combining ability using balanced designs and these methods have been extended over time to include unbalanced datasets. Selection of parents based on general combining ability estimates is widely used for parental selection.

2. Tester choice for the initial topcross evaluations

The choice of the initial tester is based on experience, with most commercial hybrid development programs using inbred parents with proven hybrid performance. Breeders use information on the pedigree of the genotypes being tested, along with the knowledge of the performance of the tester with the parents of these genotypes in making this choice. In practice, the newest elite's parents are used as testers. There is a desire to use the newest lines available as testers; as the information generated using these lines will speed the process of hybrid development; however, based on limited amounts of data, the performance of the tester with the new lines being tested, or the parents of the new lines being tested, is unknown. This lack of information on the new lines being tested, coupled with a lack of information on how well the testers combine with the new lines, leads to inefficiencies in the initial topcross test.

3. Identify potential commercial hybrids based on topcross performance with other previously identified elite parents

This is similar to choosing the first year topcross tester, with the exception that there are now many potential parents for each new line. Since the number of potential crosses is very large, the breeders can only grow a subset of the potential single crosses involving newly developed lines. This step could be greatly facilitated if the breeder could predict hybrid performance.

4. Predict new potential hybrid combinations based on results from step 3

Again, the number of potential hybrid combinations that the breeders are interested in is very large. At this point the breeder has limited information from steps 2 and 3 on potential new parents, and wants to test only the best crosses among these parents as well as these new parents and other previously identified elite's parents. This step has been addressed by Bernardo (1994, 1995). Efficient use of these new parents would be facilitated if an estimate of the performance of single crosses, with lines being developed, could be predicted.

5. Predict future performance based on limited testing

After a limited amount of wide area testing is done on experimental hybrids, they are advanced to precommercial status. The advancement decisions are primarily based on the performance as an experimental hybrid. Generally, due to the costs involved, a limited amount of testing is done for a very large number of experimental hybrids. Breeders base the advancement decision on this information. Because of the number of experimental hybrids involved, all experimental hybrids generally cannot be directly compared. Since the data are generated from a limited number of environments, the superiority of those hybrids advanced is to some extent environmentally dependent. Those hybrids that advance are those that performed well in the sample of environments they were tested in, and as breeders well know, this does not guarantee superior performance in future environments.

6. Decide which new parents developed in steps 2 to 4 to use in step 1. The quicker the genetic merit of inbreds can be determined the faster they can be used as parents in the development process

In the past, several years were needed to accurately determine which parents had a high enough genetic merit to warrant recycling in the breeding programs; however, in practice, new lines would be used in breeding crosses before complete information was available on their genetic merit. Since the genetic merit of the new parents was not well estimated, breeders could not focus their resources on using the truly superior parents in breeding crosses. If they chose to use only the elite lines in their breeding programs, then they had to wait until several years of data had been collected on new lines to determine which of these new lines were truly elite.

The point of this overview of an inbred line/hybrid development program, is that, there are several steps in the process where information is needed on the genetic merit of the lines (steps 2 and 5), or where the process could be made more efficient if hybrid performance could be predicted (steps 3, 4, and 5).

As indicated above, Sprague and Tatum had developed the concept and some methodology for estimating the general combining ability of a line (GCA), and Jenkins (1934) had developed methods of double-cross prediction. Although both of these concepts and methods were used once the industry moved into the single-cross era, there were no generally available methods to use direct prediction of single cross performance. The number of lines developed for each cycle became too large for the methods developed by Sprague and Tatum to be economically feasible for directly predicting the genetic merit of these new lines.

Animal breeders starting with concepts developed in the 1950s by C.R. Henderson (1975) developed methods that, in theory, could be used to predict individual genotypic performance. These methods were very computationally intensive and were adopted and refined over a period of about 30 years. The general use of these methods became possible as both computing hardware and efficient algorithms were developed to the point where the capability to do the computations required by the methodology were realized. These methods were first used by breeders of large animals where performance information was very expensive and time consuming to generate. These methods, which have become known as Best Linear Unbiased Estimation (BLUE) and Best Linear Unbiased Prediction (BLUP), are based on the linear mixed model. This model was developed to handle two aspects of performance data that confronted animal breeders. First, that the data are very unbalanced, i.e., the information of interest was not generated experiments where all genotypes of interest could be directly compared; and second, because the trait of interest was milk production, information from relatives was the only way the genetic merit for the males (bulls) could be estimated.

By the late 1980s and early 1990s, computing technology had advanced to the point where the new RISC (Reduced Instruction Set CPU) machines were as fast as the supercomputers of a decade ago. At about this same time, commercial plant breeding programs started to seriously look at the tools the animal breeders had developed for animal breeding programs. The key reason for this was the recognition that commercial plant breeding programs were essentially pedigree breeding programs, and that the information from relatives would be useful in determining genetic merit. The potential advantages of the BLUE and BLUP methodologies were recognized; and furthermore, it was recognized that the computing tools were available to extend the animal models to include non-additive variances. Commercial companies also started to develop the necessary databases needed to bring the pedigree data and phenotypic data together, both of which are used in this methodology. Given that the computing tools are now available and that the breeders recognize the need for rapid, accurate estimates of genetic merit—what are the features of a hybrid prediction system and how does it impact breeding programs?

As stated previously, many commercial breeding programs are so large that not all hybrid combinations can be tested in the same environments in a manner that will generate simple, 'clean' estimates of genetic merit for the genotypes of interest. This is, in fact, by design. Breeders generate hybrids that are of different maturities or adapted to different environments. In addition, the number of potential crosses is too large to test all possible combinations. What the BLUE/BLUP methodologies do is allow combining data over many experiments, involving many different hybrids. The uses of these techniques allow better separation of the major environmental effects, from the genotypic effects, for genotypes grown in different samples of environments.

Genotypes of interest can only be economically tested in a limited number of environments, and not all hybrids of interest can be tested. However, by combining data from relatives through the use of genetic relationships, much more information can be used to predict genetic merit. The current limitation of the systems developed is that genotype by environmental interactions is not accounted for. Since only a sub-sample of the environments of interest is sampled (which is a practical limitation associated with realized weather sequences for which the breeder has no control), the genetic effects cannot be totally separated from environmental effects.

If we refer to Fig. 25–2, again, we can see how information from a hybrid prediction system would be used in breeding programs.

1. Choice of parents to use for breeding crosses

The objective is to identify the genetic merit of new lines rapidly and then use the superior ones in breeding crosses. The breeding value estimates combine the performance of the line of interest in hybrid combination, with the performance of close relatives in hybrid combination. This should allow breeders to identify lines with superior genetic merit faster than they have in the past.

2. Choice of tester(s) for initial progeny test

In this case, the hybrid prediction system allows breeders to use the information from the parents of lines being developed to predict the performance of untested lines, as well as providing better estimates of the genetic merit of the new elite lines being used as testers.

3. Which combinations to make for advanced testing

Again, based on the performance of the parents of the lines that survive initial testing, plus the results of initial testing, breeders can predict which hybrid combinations involving new lines to make for advanced testing. Once the advanced testing results are available, breeders can predict hybrid performance based on the observed performance of the hybrid of interest, plus data from close relatives. This allows use of information from close relatives tested in environments, besides those actually sampled for the hybrid of interest. The implication here is that the predicted performance of a hybrid based on BLUP methods should be, on the average, a better predictor of future performance than the observed mean performance. This is, in part, due to combining larger sample of environments in estimating the hybrid performance.

4. Prediction of new combinations among new and/or established elite lines

Prediction of hybrid performance of untested combinations, involving newly developed lines, will allow the identification of potential new hybrids faster and involve fewer testing resources. The performance of a large number of potential new hybrids can be predicted and only the best of these would then be tested.

Use of a single cross hybrid prediction system is not going to revolutionize plant breeding; but it will provide another tool that breeders can use to develop superior products. Another advantage of this approach is that it provides plant breeders with a quantitative model to use in single cross prediction. This is a genetic-mathematical model that they can continue to build upon. This technology helps breeders make maximum use of the performance data being collected. The methodology is flexible; in that, information provided by newer technologies can be added to improve the predictions. An obvious use of newer technology in this model is the use of molecular marker information in the prediction process. There are at least two ways that molecular marker information can be used in the prediction process. The first is to calculate *specific genetic relationships* between the genotypes being tested vs. the *average genetic relationships* calculated from pedigree information alone. Calculation of genetic relationships based on pedigree information does not account for the effects of sampling (i.e., the random fixation of alleles during the inbreeding process) nor of selection during line development. Considering pedigree, all individuals that have the same pedigree have the same genetic relationships; however, we know that although this is the expectation on the genome basis, given no selection during the inbreeding process, it is not always true for the selected individuals. Therefore, we can modify the genetic relationships we calculate based on pedigree with those based on molecular marker information. Second, if marker-trait associations are known (i.e., which markers are linked to loci that affect traits of interest), then we have a much more powerful tool to use for hybrid prediction. Here the genetic relationship information based on markers linked to alleles affecting the traits of interest can be used to predict performance. Until we identify and understand the actual genes of interest, this can be a very useful tool to predict hybrid performance. Using this information we can predict performance conditional on alleles that affect the trait of interest; much the way we do for very simply inherited traits like disease resistance.

These new tools coupled with the tools being developed in other areas of biology are truly exciting, and they are being developed at a very opportune time. The cost of additional genetic progress has become very expensive over the last decade. **However, in light of this, I would like to leave you with a quotation that I find to be just as true today as it was 46 years ago, when H.A. Wallace sent it to Raymond Baker.** *"As you know I have always been more interested in superior genetic material than in methods as such. THERE IS NO SUBSTITUTE FOR THE MAN WHO CAN OBSERVE AND WHO LIVES SO CLOSELY WITH HIS MATERIAL THAT HE CAN RECOGNIZE A LUCKY BREAK WHEN HE SEES IT."*

REFERENCES

Bernardo, Rex.   1995. Genetic models for predicting maize single-cross   performance in unbalanced yield trial data. Crop Sci. 35:141–147.

Bernardo, Rex. 1994. Prediction of maize single-cross performance using mixed model analysis. Proc Annu Corn Sorghum Res. Conf. 49:104–116.

Cockerham, C.C. 1961. Implications of genetic variances in a hybrid breeding program. Crop Sci. 1:47–52.

Duvick, D.N. 1977. Genetic rates of gain in hybrid maize yields during the past 40 years. Maydica 22:187–196.

Jenkins, M. T. 1934. Methods of estimating the performance of double crosses in corn. J. Am. Soc. Agron. 26:199–204.

Henderson, C.R. 1975. Best linear unbiased estimation and prediction under a selection model. Biom. 31:423–447.

Russell, W. A. 1974. Comparative performance for maize hybrids representing different eras of maize breeding.   Proc Annu. Corn Sorghum Res. Conf. 29:81–101.

Sprague, G.F., and L.A. Tatum. 1941. General vs. specific combining ability in single crosses of corn. J. Am. Soc. Agron. 34:923–932.

Chapter 26

Selection Methodology and Heterosis

Discussion Session

QUESTIONS FOR J.G. COORS

M. Cooper, University of Queensland, Australia: To what extent can the results of experimental studies be biased upwards by the compounding of G × E interactions relating to gains realized by industry given the small samples of environments used in experimental studies?

Response: That is a good question, and I would think that gains from experimental evaluations of recurrent selection are biased upwards because of limited numbers of environments used for evaluation.

H. Geiger, University of Hohenheim, Germany: Was the gain from inbred line selection determined for inbred performance per se or for GCA?

Response: I summarized direct response from inbred response per se, and indirect response from the performance of the populations. I did not attempt to measure GCA response.

C. Stuber, USDA-ARS, USA: Most public programs are focused primarily on grain yield. The hybrid industry has to take into account many other traits such as lodging. Could this not also account for the lower gains for the hybrid seed industry?

Response: Most definitely. I think that this is almost entirely the reason for the differences in rates of gain. The hybrid industry sector has to deal with factors such as maternal productivity and uniformity, and the agronomic standards are much more stringent.

T. Wehner, North Carolina State University, USA: A comment. Multi-environment testing provides needed information on G × E, and that should increase the estimate of narrow-sense heritability. The issue is whether large G × E reduces the possibility of finding one elite genotype suited to all environments.

D. Duvick, Iowa State University, USA: Was mass selection before 1930 less precise than in current times?

Response: Both Irwin Goldman and Don Duvick pointed out at the beginning of this symposium that many of the biometrical concepts that we now use in selection and experimental design were developed by R.A. Fisher and others and were not really used until after 1930. Perhaps this is largely why mass and other forms of selection were relatively ineffective initially.

M. Fakorede, Obafemi Awolowo University, Nigeria: One reason for the difference in gains from recurrent selection in hybrid programs, which you did not comment on, is the time lag between development of the last population and utilization of inbreds from that population. Obviously, lines from cycle n populations are not used in hybrids until much later, perhaps several years.

Response: I agree that there is a time lag. I make the assumption, though, that the response using inbreds derived from populations roughly parallels the response depicted by the populations from which they were derived. Perhaps this is a mistaken assumption, but one we often make.

R. Sevilla, Universidad Nacional Agraria La Molina, Peru: Since recombination is required for future gain, what is the best recurrent selection method for both near-term and future gains?

Response: Based purely on response rates, intrapopulation improvement via full-sib recurrent selection works well, and for interpopulation improvement, reciprocal recurrent selection also works quite well; however, I do not know how to answer the question relative to the amount of recombination and ultimate gain.

F. Troyer, USA: Can you calculate gain per dollar, using data for hybrids from commercial companies?

Response: Good question, and I think it is possible if someone wants to look into this. Gains from the public sector have probably been cheaper than in the private sector because they have been easier to attain. As mentioned earlier, the requirements for commercial hybrids are more stringent.

B.S. Dhillon, Punjab Agricultural University, India: Did you compare gains obtained from half-sib reciprocal recurrent selection and full-sib reciprocal recurrent selection? Which approach would be more suitable for exploitation of heterosis?

Response: I did not separate the two types of reciprocal recurrent selection. There are relatively few studies in each category, and I am uncomfortable with splitting up the studies any more than I already have. So, I have no real answer to that question.

H. Geiger, University of Hohenheim, Germany: Did you find any indication on how selection changed the proportionate contribution of heterosis to cross performance?

Response: This is a very interesting issue, and I think it relates to the fact that indirect response, i.e., that of the populations per se, to reciprocal recurrent selection has been quite disappointing. This may be due to genetic drift, but also due to the fact that we often stack the deck in favor of dominance interactions by choosing parental populations that are quite divergent.

A. Melchinger, University of Hohenheim, Germany: Based on the outcome of your review, what would you recommend for the maize program at CIMMYT? Less emphasis on hybrids, and if so under what conditions?

Response: There are several other advantages to the hybrid breeding system other than the biological issues that I have addressed. It is very important to foster an active seed industry, and I think that the hybrid system does that very well. I whole-heartedly endorse CIMMYT's hybrid approach, and I think it is biologically sound as well.

D. Duvick, Iowa State University, USA: Have you any suggestions for how to increase genetic diversity in the inbred-hybrid method without lowering performance level?

Response: I honestly don't know. This is the issue isn't it for breeders? And I would welcome anybody to come up with suggestions because it's something we're all worried about.

M. Olsen, University of Minnesota, USA: Some studies show that there is little evidence for genetic plateaus in long term selection studies. If you convert genetic gain to a cost per unit gain, do you think we may be approaching yield plateaus? Please comment on the sort of costs of making gains 60 years ago at the selection intensities used then compared to current intensities.

Response: Selection intensities for the hybrid system initially were probably on the order of 1%, and a good guess now is that they are approaching 0.01%. So the fact that the rate of gain has not changed appreciably, even though the amount of effort has increased substantially, indicates that some sort of plateau may be approaching.

Lin Xiaochuan, Chinese National Rice Research Institute, China: Why does selection response increase with higher plant density?

Response: The question is really why do we see greater selection responses when hybrids are evaluated at higher plant densities. It is the result of evaluating hybrids that have been developed over the last sixty years. The most recent hybrids have been bred under very high planting densities, and the older hybrids weren't exposed to such stresses. So, evaluating older and newer hybrids together at high plant densities will favor the most recent hybrids, i.e., bias the response upwards.

QUESTIONS FOR C.L. SOUZA

S. Pandey CIMMYT, Mexico: If P2 (population 2) is lower yielding , we may have trouble deriving good lines from it. For an applied breeding program, how lower yielding should P2 be in relation to P1 (population 1)?

Response: It is not possible to answer your question precisely. The theoretical and experimental results showed that with reciprocal recurrent selection one of the populations (the higher yielding, say P1) will be improved, whereas the other one (the lower yielding , say P2) will not be improved at acceptable rates or will not be improved at all. Then, after a few cycles of RRS the difference in performance between the two populations will increase. Consequently, the trouble in deriving outstanding inbred lines from P2 will also increase. The Test-cross Half sib Selection (THS) was proposed to overcome these difficulties and to integrate recurrent selection effectively with applied hybrid breeding programs.

A. Cardinal, Iowa State University, USA: You started your derivations assuming that both populations have the same alleles in both populations but different gene frequencies. Will your conclusions hold if the two populations have different alleles? What about some of the covariances?

Response: In the theoretical derivations I assumed a simple model, but I think it worked very well because the theoretical expectations agreed very well with the experimental results. That is a good question, and maybe it is necessary to investigate what happens if we consider different alleles at the same locus in the two populations. I am not sure, but if a simple model worked well one, can expect that a more sophisticated model may not provide additional information.

A. Gallais, INRA, France: To combine inter- and intrapopulation improvement do you think it would be possible to use an index of intra- and interpopulation performances?

Response: I think we will have two problems if we use your suggestion. First, as you know, intra- and interpopulation selection make use of different gene effects; also, the genetic correlation between intra- and interpopulation progenies stemming from the same plant is usually high for one of the populations and low for the other one. Therefore, it will be very difficult to establish an index that could account for these differences. Second, we will have to evaluate intra- and interpopulation progenies simultaneously from the same populations; thus, the amount of resources needed for this type of procedure could limit its use for applied breeding programs.

H. Geiger, University of Hohenheim, Germany: Wouldn't it be more effective to use line selection rather than THS selection for improvement of the seed parent pool?

Response: For hybrid breeding programs the improvement of the population cross as well as the populations per se performance should be made as efficiently as possible. This is because the development of new outstanding hybrids from inbred lines depends on the improvement of the population cross; and the development of new elite inbred lines depends on the improvement of the population per se performance. The results presented showed that THS should be a suitable strategy to fulfill these requirements.

F. Sanvicente, CIMMYT, Mexico: How does your proposed THS method differ from the classical half-sib selection for SCA?

Response: Half-sib selection for SCA uses elite inbred lines from different heterotic groups as testers, in which population and tester line are from different heterotic groups. THS uses one of the populations as a tester for itself and for the other population. Inbred lines are not used in this procedure.

F. Sanvicente, CIMMYT, Mexico: Has your method been tested and how does it compare to Russell's results?

Response: THS has not been tested yet, whereas theoretical expectations agreed very well with experimental results reported by several researchers. THS was not compared to half-sib selection for specific combining ability.

QUESTIONS FOR J. MORENO- GONZÁLEZ

A. Melchinger, University of Hohenheim, Germany: When you calculated the selection gain for marker assisted selection (MAS), what were your assumptions concerning the proportion of the genetic variance explained? I am asking this question because based on the theory of Lande and Thompson, p, the proportion of the genetic variance explained by markers, is a critical parameter in the genetic gain expected from MAS.

Response: The portion of genetic variance explained by the markers, p, is certainly a crucial parameter for determining the expected genetic gain from MAS. When constructing the selection index for MAS, the estimate of weight b_m associated with the molecular-based predicted testcross from regression, $\hat{Y}_m$, includes $1-p$ in the denominator (Eq. [8] in the paper). As p approaches 1, the relative weight of $\hat{Y}_m$ rapidly increases in the selection index (Eq. [9]). Thus, it will be important to control the false QTLs entering in the stepwise multiple regression model. If false QTLs are retained in the model, p will be inflated, the weight of the regression predicted testcrosses (b_m) will be overestimated, and the realised genetic gain will be reduced. Therefore, it would be advisable to choose a small probability level for controlling the type I errors when performing the F-test in the stepwise regression. A probability level of 0.001 was used in our simulation study, which included 60 and 120 markers (dummy variables) for the multiple regression.

C. Jiang, CIMMYT, Mexico: Can you comment on the possible effect of epistasis on the relative efficiency of MAS and conventional selection?

Response: There are not very good models to detect QTLs with epistatic effects. One of the problems of including digenic epistatic effects is that models become very complicated because they add nonlinear terms. One possible strategy to deal with this difficulty is to approximate the nonlinear models to linear models (Moreno-Gonzalez 1992, TAG 85:435-444). After the monogenic QTLs with significant heterotic effects have been identified and retained in the model, digenic interactions between QTLs could be sequentially added to the stepwise regression model to test whether they account for significant additional epistatic variance. Later on, the methodology described in this paper can be also applied to models including epistatic effects. If significant epistatic variance is retained by the models, the predicted regression testcrosses, $\hat{Y}_m$, will be more accurate, and the expected genetic gain from MAS will increase. Therefore, the additional efficiency of MAS relative to conventional selection will depend on the ability of the models for detecting additional epistatic variance, if it really exists. Epistatic variance has been found in many experimental situations, however it has been very small compared to the additive and dominance variances.

Liu Xiaochuan, China National Rice Research Institute, China: Have you tested the efficiency in predicting heterosis with molecular marker-assisted selection (MAS) using your methods?

Response: No, I haven't. Unfortunately I do not have molecular facilities in my maize breeding program at home. The efficiency of MAS has been tested in several (not many) breeding programs for traits controlled by additive and dominance effects. The models developed in this paper for heterotic effects are similar to the current models for additive effects. Thus, I do not expect more difficulties for analysing heterotic effects in F_2 testcrosses than for additive effects in F_2-F_3 families. The obvious comment is that heterotic effects are less common than additive effects. Therefore, heterotic effects should be tested for those traits where they are presumably present.

QUESTIONS FOR R. BERNARDO

D. Johnson, Dekalb Genetics, USA and Alain Charcosset, INRA, France: How much predictive accuracy is lost if specific combining ability is ignored? Or, in other words, what would happen if you simply include the general combining ability effects?

Response: My experience has been generally that the correlations between predicted and observed performance would decrease by about 0.03 to 0.05 for grain yield. You have to decide whether that is a significant decrease or not. In my opinion, the data are there. Let's use all the available data we have.

N. Rajanaidu, Palm Oil Research Institute, Malaysia: Can the model be expanded to include epistasis?

Response: Yes, but there is a problem. It makes your model bigger. Computing time is longer. Another problem is that the coefficients for the additive by additive epistatic effects, for example, are squares of the coefficients for additive effects. That introduces a lot of dependency or multicolinearity in the model, and in the end I don't think you gain that much by doing so. If we formulate epistasis in some other way, perhaps there could be an advantage.

A. Melchinger, University of Hohenheim, Germany: Was there coancestry between different heterotic groups?

Response: Yes, the different heterotic groups in many instances were related with each other, and often I had to ask a breeder, for example in corn, is this more of a Mo17 or Oh43 inbred? But the two groups in a heterotic pattern were not related.

A. Melchinger, University of Hohenheim, Germany: What is the explanation for the high specific combining ability variance in some of the heterotic patterns?

Response: I really don't know. I know if it's not very good data or failure of the model, but I can't explain it, and I've quit trying to explain it.

J. LeDeaux, North Carolina State University–USDA, USA: The first question is on the recommended minimum sample size to predict hybrid performance. And also, how do the correlations between observed and predicted values change as that inbred line is tested against more and more lines?

Response: I think the key here is to simply use as much data as you have available. Sometimes you might not have enough data to obtain good predictions, but as your breeding program expands, you should eventually have enough data. If I were to put a figure on number of hybrids, I'd say 200 or 250 would be a good number to start with.

M. Lee, Iowa State University, USA: You presented predicted versus observed correlations for each trait separately. What would happen if you consider the traits simultaneously?

Response: Yes, the predictions were made for yield, then moisture, then for stalk lodging, then for root lodging. It is certainly possible to make the predictions simultaneously for all four traits. What that would do is two things; you would have a much larger matrix to worry about, but also you might get better predictions if you have genetic correlations between traits. If yield is genetically correlated to moisture, then the moisture information could improve yield predictions. I don't have any specific answers, but it would be a function of how genetically correlated the traits are.

QUESTIONS FOR O.S. SMITH

B. Hunter, Novartis, Canada: Do you see a role for GIS in the selection of great hybrids?

Response: I think so. Right now we are certainly using GIS to try to characterize hybrid performance and try to get some handle on stability of hybrids, or at least identify those areas where the hybrids have stable performance.

D. Beck, CIMMYT, Mexico: Do you see value in using managed environments to better understand stress tolerance and related stability of maize hybrids? If so, do you think this is adequately exploited in the seed industry?

Response: I do see the advantage of managed environments. I know of cases in the western cornbelt where there is dryland and limited irrigation testing, as well as full irrigation testing to try to get a feel for how hybrids perform under drought stress. We certainly also use disease nurseries and the like. Yes, I think these sorts of tools are being used in industry to advantage.

R. Pratt, Ohio State University, USA: The R-1 phase of testing used few no-till or reduced till environments ten years ago. Has this changed?

Response: The question relates to how well do our test environments match our target environments. I think the breeders review that on a constant basis, and certainly they are always looking for ways to identify test environments that adequately measure performance in target environments. So, the use of no-till environments has increased in our testing procedures.

B. Dhillon, Punjab Agricultural University, India: What is the number of replications and environments in R-2 and precommercial tests on research stations?

Response: We're doing practically all one rep testing at each location. We're trying to maximize the number of different environments that we test this material in. The number of environments per station has gone down over time. What we're trying to do is spread those environments out over a larger area. I would say that we're trying to sample the target environment as extensively as we can, as early in the testing program as we can.

D. Hess, CIMMYT, Mexico: How can the BLUP techniques be used to identify those genotypes to be used in new populations?

Response: I think that is a particularly good question. what you're trying to do with BLUP is figure out which are the best genotypes. And what you can do with those genotypes is start using the information not only on that genotype itself but on the relatives of that genotype. So what you're trying to do is look at the genes that are in that individual across a population and evaluate that. The breeding values that you can then estimate will give you a better estimate, the best estimate that you can get on that line's performance. What you put into your populations are the lines that have the best performance that you can identify. I'd like to build on this a little bit and say that you can use these techniques in recurrent selection programs. Right now al you have for estimates of genetic relation is whatever family structure you have in developing those progenies. But once we start getting marker information, you can see that those progenies, for example half sibs, or full sibs, don't really have the same relationship.

And they don't have the same genetic relationship for genes affecting the traits of interest. I think eventually that these techniques can be incorporated into population improvement schemes, and they will provide more information than we now use.

P. Sun, Dairyland Research, International, USA: What is the minimum number of locations and years needed to assess yield advantage and stability?

Response: Why don't we call it a stable yield advantage. You can see from the testing program that we feel that three years in the minimum number that we need. We have found that genotype by year interactions is something that we really have to deal with. Locations don't completely substitute for years. The number of locations is a real hard question to answer. What you want to have is a large enough sample to get a distribution of tested locations that in some way matches your distribution of target locations. And so each year you end up choosing, particularly as you get into the later stages of testing, a much larger sample of locations than you actually need. But you're forced to do that such that you hope to get a distribution that approximates the target distribution.

Chapter 27

Commercial Strategies for Exploitation of Heterosis

D. N. Duvick

INTRODUCTION

Private firms are attracted to the hybrid seed business because of the built-in plant variety protection of hybrids. Customers need to buy new seed for every planting season. But the breeding, production, and sale of hybrid seed—the commercialization of heterosis—can be successful only if it meets the following criteria: (i) The hybrids must satisfy the needs of the customer for all important traits. Simply to be hybrid, or simply to exhibit heterosis, is not enough. (ii) The price of hybrid seed must be low enough to enable the customer to make substantial profits from annually recurring investments in expensive hybrid seed. A rule of thumb is that a first time use of hybrid seed should enable the farmer to earn an extra profit equal to at least three times the added cost of the seed. (iii) The price of hybrid seed must be high enough to enable the seed company to make substantial profits from its investments in research, production, and sales. A successful seed company needs to realize a 10 to 15% return on equity. Its investments in research—one of the essential business expenditures for a research-based seed company—should be equivalent to 5 to 10% of sales income.

Two other criteria underpin all other requirements for success in the hybrid seed business: (i) Farmers will risk investment in improved seed only when they have some assurance of a fair price — a dependable market — for their crop. (ii) Government regulations, formal and informal, must give minimal hindrance to honest and prudent business operations. These two requirements apply to all seed firms, not just hybrid seed companies. They have particular significance in many developing countries.

DISCUSSION

To satisfy the three primary criteria for success in the hybrid seed business, companies must integrate a host of variables such as: (i) the pollinating system of the crop, (ii) options for manipulation of the pollinating system, (iii) supply and cost of labor for emasculation or other requirements for hybridization, (iv) the yield of the crop in the farmer's field, (v) the commercial value of the crop per unit of land area, (vi) the seeding rate of the crop, (vii) the seed yield in the seed production field, (viii) the extra yield to be expected from heterosis, (vix) the implications of hybrid uniformity, (x) the most important traits to improve in the crop, and their genetics, (xi) the ease of demonstrating improvements in new hybrids, (xii) availability of inbred parents and other breeding materials in either public or private institutions. The following examples illustrate, for three different crop species, some of the many ways in which these 12 variables can be integrated.

Hybrid Maize

Hybrid maize (*Zea mays* L.) was introduced in the USA in the late 1920s and early 1930s. Hybrids were well received by the farmers and they rapidly replaced open pollinated maize varieties in the major maize growing areas of the country. The first maize hybrids yielded only about 15% more than the better open pollinated varieties (OPVs), but they had much better resistance to root and stalk lodging. U.S. farmers were beginning to use mechanical maize pickers in the 1930s. The mechanical pickers were inefficient at gathering lodged maize, and so farmers often chose to plant hybrids because the hybrids lodged less, and, therefore, were better adapted to machine harvest. Some of the pioneering maize breeders have said that the very first hybrids might not have been accepted so readily if their higher yield had not also been accompanied by superior resistance to lodging. Superior drought tolerance of hybrids compared with OPVs also helped sell the next generation of hybrids; they were introduced just at the time of two exceptionally severe drought seasons (1934 and 1936) in the U.S. Corn Belt.

Maize is a naturally outcrossing species. The complete separation of male and female flowers ensures ease of emasculation (called detasseling), and the plant sheds copious amounts of pollen for hybridization, as a byproduct of its natural adaptation for outcrossing. Although hand detasseling is relatively easy and gives precise results, cytoplasmic male sterility eventually was developed (starting in the 1950s) as an option for hand detasseling. It reduced dependence on expensive hand labor and lessened the problem of interruptions of detasseling by rainy weather. Despite failure of one cytoplasmic system (Texas cytoplasm) due to disease susceptibility, other systems are available and in use. Machinery also has been developed to mechanically remove tassels. Thus, three options are available for emasculating the maize plants in the crossing fields. Each has its limitations but seed producers can choose an optimal mix of the three. (Several variations of a fourth option, involving genetically engineered male sterility, also are in process of development.)

When hybrid maize was introduced in the USA, the commercial value of the maize crop per unit of land was not particularly high, compared with high value crops such as tomato (*Lycopersicon esculentum* Mill.) but seeding rates were low for a field crop (one seed gave about 300 seeds in return), and seed yields in the hybridization field (to make double cross hybrids) were relatively high. Therefore, seed companies could set prices at a level low enough to be attractive to farmers but high enough to allow comfortable profit margins for the seed companies.

Yield and standability were the prime traits in need of improvement in maize. Both traits were susceptible to improvement by use of the inbred–hybrid method, and improvements in both traits (especially standability) could be demonstrated to the farmer with relative ease. Both traits were inherited in quantitative fashion, and were governed by many genes, but genetic variability for both traits was high, and replicated yield trials at a few locations easily differentiated the poorest from the best hybrids. To identify the most superior hybrids for yield and standability required several years of performance trials at multiple locations, but this requirement was not onerous, once a well-organized maize breeding program was in operation. Minimal application of scientific method, and rudimentary statistical design and analysis, ensured reliable decisions about hybrid performance.

Breeders had some difficulty in making further improvements in hybrid yield and standability, once the first cycle of selfing in OPVs was completed. Inbreds derived from a second round of selfing OPVs did not give improved hybrids. But breeders soon found that progress could be made by developing new inbreds from crosses of the best first cycle inbreds (Hallauer & Miranda, 1988). Although exact inheritance of yield and standability was not known, and still is not known, breeders were able to establish, through trial and error, the procedures and population sizes that were needed to ensure satisfactory breeding progress.

The uniformity of hybrids, as compared with heterogeneous OPVs, allowed farmers to distinguish hybrids from OPVs, and made it easier for them to make critical comparisons between the two classes. Uniformity of hybrids helped the maize breeders also, in their efforts to develop cultivars with specific product qualities, or to fit unique ecological niches. Differences among uniform hybrids were more clear-cut than among heterogeneous OPVs.

But hybrid uniformity increased the dangers of susceptibility to unforeseen disease or insect problems. Widespread use of a few hybrids, or of hybrids based on a small number of inbred lines, gave opportunity for explosive multiplication of specifically adapted disease or insect pests. In the early days, only a few hybrids were available, and farmers tended to concentrate on an even smaller number, those they judged to be the best of the lot. In the 1940s, a severe outbreak of Northern Corn Leaf Blight (*Exserohilum turcicum*) occurred in the eastern Corn Belt of the USA, largely because of over-dependence on a few inbreds that had been developed in the western Corn Belt, where climatic conditions are less conducive to development of the disease. Breeders responded rapidly. They replaced susceptible hybrids with tolerant ones, and initiated new breeding programs to develop blight resistant inbreds for use in future hybrids.

This cycle of: (i) concentration on a few hybrids, (ii) pest epidemic, and (iii) introduction of new hybrids, was only the first of many. These cycles continue to occur in various regions of the country, and with various pest organisms. None have been catastrophic, except for the anomalous epidemic involving T cytoplasm in 1970. Hybrid choices within and among the various seed companies are numerous enough that farmers can switch to more resistant hybrids from one season to the next, if necessary, and breeders have sufficient strength in their breeding pools to bring out new resistant hybrids in a relatively short time. Collectively, the hybrid maize seed industry provides the farmers with genetic diversity in time (Duvick, 1984).

In the critical early years of development of the hybrid maize seed industry in the USA, inbreds were bred and supplied by public breeding institutions at the universities and in the USDA (Duvick, 1997). Commercial seed companies combined the inbreds into hybrids that they produced and sold. Sometimes public institutions produced single cross seed for use as parents of double cross hybrids. The commercial seed companies needed to make only the final double crosses, in hybrid combinations recommended by the public institution. Some of the public institutions multiplied and sold the parent seed that they had developed. This practice continued for a decade or two but gradually was dropped by all public institutions.

Private firms produced their own inbred lines also, but in the early years they could not produce enough to fill their needs. For many years they depended on public breeders for most or all of their inbred lines. The large seed companies became relatively self-sufficient in inbred lines by about the 1950s, although they continued to use public inbreds whenever they gave superior hybrids. The smaller seed companies depended on public inbred lines until about the 1970s and 1980s, when private foundation seed companies began to lease out their own privately developed inbreds, on large scale. Their proprietary inbreds were available to all seed companies but the primary targets were the small companies that had few or no proprietary lines of their own.

Most of the public institutions reduced or completely stopped efforts to develop commercially useful inbred lines in the 1980s, and devoted their maize research programs to investigation of maize genetics and breeding techniques. A small number of public programs still develop and release new inbred lines for use in commercial hybrids, but most public germplasm releases now are of basic breeding materials and genetic stocks, rather than finished inbred lines.

The record shows, therefore, that hybrid maize in the USA was commercialized successfully as a joint public–private enterprise, and that the roles of public and private entities have undergone continuous evolution over the years. The contribution of public breeders was absolutely essential in the start-up years, but as the seed industry matured, the seed companies gradually assumed responsibility for all functions of research and development, except for long-range fundamental studies in genetics and breeding.

Changes continue. Use of intellectual property rights for plants has stimulated public entities such as the USDA and land grant universities to encourage their researchers to patent or obtain plant variety protection for their inventive products or processes. In the universities, royalties from the protected materials and processes typically are divided between the researcher and the parent institution. Thus public maize research in the USA has gone from partly commercial, to noncommercial, and then back to partly commercial practices, in the course of the past 70 years.

Hybrid maize was introduced in Canada shortly after its introduction in the USA, and was adopted at about the same rate and intensity as in the USA.

Hybrid maize was successfully introduced in Europe in the 1950s, following World War II. Use of hybrids started in the southern countries, those best able to use germplasm from the U.S. Corn Belt. Hybrids gradually moved north as new inbreds and hybrids were developed with adaptation to needs and growing practices of the northern parts of Europe. Subsidized prices for maize grain stimulated farmers to make annual investments in high yielding hybrid maize seed, once adapted hybrids were available. Public breeding played an important role in establishing maize hybrids in Europe, but, as in the USA, private seed companies gradually assumed dominant roles in breeding as well as in seed production and sales.

Hybrid maize was introduced on a limited scale in the tropics and subtropical regions (mostly developing countries) in about the 1960s. Progress in using hybrids was slow at first in most of the developing countries, with a few exceptions. During the past decade, however, interest and planting of hybrid maize has increased, perhaps due to an increased market demand for feed grain to produce meat and eggs demanded by the rapidly urbanizing populations of many developing countries. A second change in many developing countries is encouragement of the development of a private seed industry, in contrast to earlier emphasis on development of public seed enterprises, sometimes known as *parastatals*.

The country of Zimbabwe was an outstanding early exception to the rule of slow progress in adoption of hybrid maize in developing countries. Hybrid maize was introduced in the late 1940s and its area expanded rapidly, reaching nearly 100% concentration in about 25 years' time. Single cross hybrids were successful, and semi-subsistent smallholders as well as large scale commercial farmers adopted hybrid maize (CIMMYT, 1994). Hybrid success was due in part to the fact that the hybrids were purposely bred to fit a new niche in dryland farming; they were early flowering and drought tolerant, traits not found in the existing OPVs.

In each of these examples of establishment and growth of a hybrid maize seed industry, success has depended on strong farmer demand for hybrid maize. Farmer demand in turn was predicated on a strong and reasonably dependable commercial market for maize grain, and on financial and technical capability of the farmers themselves to supply extra inputs to allow hybrids to reach their yield potential. Farmers bought hybrid maize seed from the private seed companies because of proven ability of the companies to deliver quality products on time, in needed amounts, and at affordable prices. Also, in these examples, the public sector has led the way in research and development, but in time the private sector has taken over most of the applied aspects of research and development, in addition to performing its original function of production, sales, and delivery of hybrid seed.

Apomictic Maize Hybrids

A notably different approach to use of heterosis and hybrids has been advocated by researchers who propose use of apomixis to generate self-reproducing maize hybrids. Farmers who cannot afford to buy hybrid maize seed could plant apomictic hybrids and save part of their grain production as replant seed (Anonymous, 1996; Jefferson, 1994). Such farmers typically might be poor semisubsistent smallholders in developing countries. Two parallel systems have been proposed for making apomictic plant hybrids; each has potential pluses and minuses; neither of them is ready to use.

Simply to develop genetic systems for production of apomictic hybrids will not be the end of the task, however. As smallholders replace their heterogeneous OPVs with homogeneous apomictic hybrids, they will be confronted with the potential dangers of genetic uniformity. The familiar cycle of narrow genetic base, pest epidemics, and hybrid replacement could easily be instituted for the poor smallholders, just as has happened with farmers in other parts of the world, when they adopted conventional maize hybrids. Maize breeders will need to ensure that genetic diversity in time and genetic diversity in place are available for poor smallholders using apomictic maize hybrids, in the same manner as they have done, successfully, for commercial farmers who adopted conventional maize hybrids in temperate regions of the world.

Some important differences should be noted, however. Disease and insect pressure is greater in the tropics than in the temperate zones, and hybrid lifetimes may be shorter; replacements may be needed more frequently. (This will be a problem for conventional as well as for apomictic hybrids.) The poorest smallholders are heavily dependent on genetic diversity within and among their crop varieties, for security against problems with disease and insects, unfavorable weather, or variability in soil type. Unlike larger scale commercial maize farmers, they cannot purchase chemical or mechanical aids to control insects and diseases, or to correct nutrient imbalance.

Maize breeders, therefore, must be prepared to provide the needed kinds of genetic diversity to smallholders who plant apomictic hybrids. They can do so by bringing out replacement apomictic hybrids at frequent intervals (diversity in time), and they can release large numbers of genetically dissimilar apomictic hybrids for each adaptation zone, and encourage farmers to plant all of them (diversity in place), rather than to concentrate on planting one or two favorites.

Alternatively, the farmers themselves, rather than professional maize breeders, might take responsibility for providing necessary amounts and kinds of genetic diversity in place and in time, much as they have done through the millennia with their own farmer varieties. Genetically heterogeneous populations of apomictic hybrids, or heterogeneous facultative apomictic populations, could be furnished to the smallholders; they could select desired apomictic hybrids and grow them in mixtures that seemed best to them. But professional breeders still would have the ultimate responsibility of furnishing base populations to smallholders in needed amounts, at suitable intervals, and with appropriate kinds of pest resistance and environmental adaptation. Explanation and instructions for use probably should accompany the releases, and a reliable system of delivering them would be needed. Thus, even though they selected and saved seed of their own hybrids, smallholders would not be self-sufficient; they would depend on the professionals.

One need not assume that only poor smallholders would appreciate the potential savings from saving seed of apomictic maize hybrids. Commercial maize farmers may wish to save money, also, by growing publicly available apomictic hybrids and replanting their own seed. Whether or not they do so will depend primarily on whether the apomictic hybrids are competitive with standard hybrids. For example, in the U.S. Corn Belt, a yield reduction of about 5% would cancel out any savings from not buying hybrid seed.

A second financial consideration might influence the more specialized commercial farmers, as they consider whether or not to plant apomictic maize hybrids. Maize seed harvest and conditioning—harvesting with special equipment, drying, shelling, sizing, treating with insecticide and fungicide, packaging, and labeling—is a highly technical operation. If done improperly, it can severely damage the yielding ability of hybrids with good genetic potential. Some commercial producers might prefer to leave this specialized and important operation to the seed companies. But less specialized growers might decide that it was worth their time to save and prepare seed for planting.

Seed companies some day may produce and sell apomictic maize hybrids. An easily manipulated apomictic system might improve opportunities to develop new hybrids. The method might make it possible to bypass multi-generation selfing for inbred development, and it could eliminate the need for large-scale and expensive cross-pollination blocks to produce hybrid seed. Intellectual property rights would be used to ensure that the companies, as well as the farmers, could profit from the results of the companies' self-financed research and development.

Such an outcome, of course, would not provide products for farmers who are too poor to buy seed; it primarily would serve commercial maize farmers. Public breeding programs would have the major and continuing responsibility to provide apomictic hybrids and breeding materials to the poor smallholders. One can expect that farmers would abandon their own OPVs once they converted to use of apomictic hybrids, just as has happened wherever farmers have switched to new, professionally bred varieties such as the Green Revolution wheat (*Triticum aestivum* L.) and rice (*Oryza sativa* L.) varieties. The trade-off for improved incomes has been dependence on professional plant breeders.

Time will tell just what niches can be filled by apomictic maize hybrids. One can be certain that, as skill develops in manipulating apomixis in maize, new uses for apomictic hybrids will be devised. As with all plant breeding, the breeding of apomictic maize hybrids will be an evolutionary art and science.

Hybrid Wheat

Wheat hybrids can yield up to 30% more than their parents, but hybrids with heterosis at these levels usually are the product of crosses between different classes of wheat, such as a cross of hard red winter wheat by soft red winter wheat. Commercially useful wheat hybrids must be made within a class, to maintain milling and baking quality. Crosses within a quality class typically have less heterosis, only about 5 to 15% more than their parents. The lower amount of heterosis may be because of relationship among members of a relatively closed gene pool.

Wheat is a self-pollinated crop. It has perfect florets, limited supplies of pollen, and a relatively brief period of stigma receptivity. Hand emasculation is impractical for commercial production of hybrid seed, but cytoplasmic male sterility allows production of hybrid wheat seed on a field scale. Limited pollen production by male lines (in comparison to maize, for example) means that the ratio of male rows to female rows must be relatively high, and seed yield per hectare is reduced correspondingly.

Value of the wheat crop per unit of land is similar to that for maize. (Both crops are commodities; wheat yields less than maize but it commands a higher price.) Seed yield in the crossing field is low and seeding rates for commercial grain production are high for wheat, relative to maize. One kg of wheat seed will produce 30 to 50 kg of grain, compared to the maize ratio of 1 to 300 or more. Therefore, if a seed company prices its hybrid wheat seed safely above cost of production, the seed cost from the farmer's point of view could be very high in relation to the expected extra income from the hybrid. If a hybrid has only a small yield advantage over the best pure line cultivars, the expected gain in the farmer's income from increased yield of the hybrid could be less than the cost of the hybrid seed (assuming the company priced the seed to cover cost of research, production, and sales plus profit). Such a hybrid would be unacceptable, despite its yield advantage.

Yield, standability, and pest resistance are important traits for wheat varieties, just as with maize, but acceptable wheat varieties also must meet rigorous milling and baking standards. A cross with high heterosis for yield may be unusable if it lacks needed levels of milling and baking quality, or is out of bounds for protein percentage.

Wheat hybrids are not more uniform than standard inbred cultivars, thus they do not introduce new dangers due to genetic uniformity, nor do they introduce new opportunities based on an increase in uniformity. But wheat hybrids, although uniform from plant to plant, are heterozygous at many loci, in contrast to homozygous inbred cultivars. Therefore, wheat hybrids can carry useful combinations of dominant disease or insect resistance genes in heterozygous form. Two inbred parents, neither of which has all needed genes for resistance, can be crossed to make a hybrid with acceptable resistance. Further, by crossing new combinations of inbred parents, one can quickly produce new hybrids with needed new forms of resistance. This process can be much faster than the usual backcrossing or selfing process for placing pest resistance genes into a new inbred variety.

In the early years of hybrid wheat breeding, widely used elite wheat cultivars could be used as female parents with no change in their nuclear genotypes, since they nearly always lacked fertility restoration genes. This fact allowed rapid access to the high general combining ability of these varieties. Through simple backcrossing, their nuclear genomes were placed in sterility-inducing cytoplasm, and they then could be used as the female parent of a hybrid.

But because of this same circumstance, breeders had to develop entirely new male lines, by inserting nuclear genes for fertility restoration into nonrestorer genotypes. Restorer lines usually were made by introgressing dominant fertility restorer genes from widely divergent germplasm into elite wheat lines. Typically, the strongest and most useful restorer genes came from different species, sometimes weedy or wild species. This introduced problems of linkage to undesirable traits from the alien species. Breeders devoted years of time and energy to backcrossing with selection for strong fertility restoration, and, as a consequence, they spent less time on breeding for increased yield and general performance. Also, in contrast to the early years of hybrid maize development, publicly employed wheat breeders gave very little input to hybrid wheat breeding. This led to under-investment in development of germplasm and breeding methods (particularly for restorer males) in the important start-up period. The private sector had to carry the load.

Breeding work with hybrid wheat in especially the 1970s coincided with a period of rapid improvement in yielding ability of standard inbred wheat cultivars, as high yielding semi-dwarf germplasm came to dominate the U.S. wheat germplasm pool. The rapid increase in yield and performance of standard inbred cultivars meant that hybrids, in spite of a yield advantage from heterosis, could not compete with standard cultivars. The hybrid parents lagged behind, in improvement of non-heterotic traits for yield and general performance.

Several research-based seed companies in the USA invested heavily in research to develop hybrid wheat, starting in the 1960s (Knudson & Ruttan, 1988). A few wheat hybrids were developed and released, but most of them did not succeed in the marketplace, primarily because farmers decided that the hybrids' performance did not justify the increased cost of the hybrid seed. Seed companies gradually dropped their hybrid wheat programs, and by the end of the 1980s only a few programs were still in operation.

Interest in hybrid wheat is still present, however, particularly in regions where wheat yields and commercial value of the crop are relatively high (Edwards, 1997, personal communication). Two companies are marketing wheat hybrids in France. Both companies use chemical hybridizing agents (CHAs) to produce the hybrids. (CHAs, applied at appropriate stages of development, prevent pollen development.) Several wheat hybrids, both hard red winter and soft red winter, are bred and sold at the present time by a seed company in the USA. Four wheat hybrids are marketed in Australia by a private company. The University of Sydney is a shareholder in the company. A cooperative in South Africa is selling wheat hybrids. The American, Australian, and South African hybrids are made with the cytoplasmic male sterile method.

The French and also the Australian hybrids are targeted for high yield production areas, where the farmers use high levels of management. Such producers, it is expected, can make best use of added investment in hybrid seed. In sharp contrast, the South African hybrids are sold in dryland production areas with low yield expectation; however, the hybrids are planted at very low seeding rates, thus keeping seed cost in line with expected return.

Research is in progress in several seed companies, on new ways to produce hybrid wheat seed, using new sterility systems, some of which are introduced into wheat via genetic transformation. The goal is to build systems that are reliable, easy to manipulate, and that interfere as little as possible with routine wheat breeding programs aimed at making improvements in yield and general performance.

These examples show that seed companies and breeders still believe that hybrid wheat can succeed on large scale, and perhaps more importantly they show that there are numerous ways to produce the hybrids, and then to manage them for profit in the farmers' fields. The examples also demonstrate that small companies as well as large can participate in the hybrid wheat seed business.

Hybrid Tomato

Tomatoes are a self-pollinating inbred crop with perfect flowers. Although genetic sterility is available in tomato, hand emasculation and hand pollination are preferred for making hybrids. Crossing is performed in countries where labor costs are low. Seed number per pollination is high. Tomatoes are a high value crop, grown for fresh market or for processing, and seeding rates are very low compared with the value of the commercial crop. In the USA, 100% of fresh market and 80% of processing tomatoes are F_1 hybrids.

Although tomato hybrids can exhibit heterosis for yield, the amount of yield increase in absence of stress is small or even nonexistent. The unique utility and attraction of hybrid tomatoes is that they allow breeders to assemble, in one cultivar, complementary genes for disease resistance as well as for traits affecting product quality such as shelf life. Breeders of hybrid tomatoes do not need to place all desirable resistance genes in one inbred cultivar, which accentuates problems with linkage drag; they instead can hybridize two complementary inbred lines to produce a hybrid with the desired full set of resistance genes. Hybrids are essential for expression of the slow ripening trait governed by the gene *nor*. The homozygous wild type, +/+, ripens too fast; homozygous *nor/nor* does not ripen at all, but the heterozygote *nor*/+ ripens slowly, as desired by the market. Tomato hybrids also exhibit increased yield stability, perhaps because they have a better balance of genes for disease resistance (Janick, 1996).

The success of hybrid tomatoes shows that hybrids can be commercially successful in an inbred crop. Expensive means of seed production, such as hand pollination, are feasible with tomatoes because of the high value of the commercial crop, the relatively low seed requirement, and the large number of seeds produced per pollination. This example also points up the fact that heterosis for yield need not be the major factor in determining whether hybrids will be successful. Hybrids can provide many advantages over non-hybrid cultivars, in addition to heterosis for yield.

Large vs. Small Seed Companies

In recent years, the size of many seed companies has increased, and total numbers have been reduced, due to consolidations and buyouts. This phenomenon has been true for seed companies of all kinds, not just those specializing in hybrids. In part this may be because of the growing need to incorporate expensive biotechnology research into the seed breeding process, which means that only large companies can support biotechnology research on a meaningful scale. The consolidations and buyouts also may be just one more part of the current global trend of corporate enlargement.

For whatever reason, to an outsider it may seem that opportunities no longer exist, for small hybrid seed companies. In actuality, small seed companies are still numerous in all parts of the world, and they account for a large amount of the seed business, including the hybrid seed business. In the USA, for example, small firms account for about 25 to 30% of hybrid maize seed sales (Duvick, 1997). Small hybrid seed firms tend to depend heavily on parental lines developed by public institutions. They often sell seed in niches not conveniently reached by large firms, or perform specialized contract services. They fill a vital role in the hybrid seed economy, and will continue to do so, especially in countries where the hybrid seed industry is in early stages of development. As time goes by, some small firms become large firms, and in their place new small companies arise. This cyclic phenomenon has been documented for hybrid maize in the USA (Norskog, 1995), and general observation indicates that it occurs in other crops in other countries, as well.

The three criteria for success in the hybrid seed business: selling good hybrids, providing profit for farmers, and providing profit for seed companies, can be met by small companies as well as by large ones.

COMMENTARY AND CONCLUSIONS

Demonstration of heterosis for yield and other traits in many crops has prompted efforts to commercialize the breeding, production, and sale of hybrids of cross-pollinated field crops, self-pollinated field crops, and numerous vegetable and bedding floral crops. As a general rule, hybridization has been commercially successful with cross-pollinated field crops, relatively unsuccessful with self-pollinated field crops, with the exception of sorghum and rice (Doggett, 1988, Virmani, 1994), and successful with many kinds of high value vegetable and bedding crops (Janick, 1996). Heterosis is only one of several determinants to the success of hybridization.

Cytoplasmic male sterility has been the method of choice for hybridizing field crops. (Maize also can be detasseled.) Vegetable and ornamental crops are hybridized in a variety of ways, including cytoplasmic male sterility, hand emasculation, genetic male sterility, self incompatibility, and production of gynoecious or highly pistillate monoecious plants. For all crops, research is in progress on use of chemical male sterilants, or new ways of manipulating pollen sterility with genetic engineering, to produce new systems for hybridization. And finally, new knowledge about the genetics and manipulation of apomixis someday may open up entirely new ways for commercial exploitation of heterosis and hybrids, in many crops.

Commercial development of the hybrid seed business has been most successful when public breeders led the way, providing not only breeding technology and genetic knowledge, but also the breeding materials needed to make hybrids. Private firms, in the early years, primarily produce and deliver hybrid seed of materials developed by the public sector. They then begin to develop their own proprietary germplasm and proprietary hybrids, and gradually take over much or even all of the public sector's responsibility for applied research and development. The rate of change and the amount of change varies with the crop species, as well as with the economy and organization of agriculture in a particular country.

In recent years, reduction in public funding for plant breeding research has impelled public researchers to seek funds from private industry, and to produce products to be marketed to private industry. Industry-oriented research naturally tends to be pointed towards short-term goals that can help industry fulfill its function of producing improved seeds. Over-concentration on such goals by the public sector may lead to neglect of pioneering research for the long term, and neglect of research for the public good, i.e., research on needed food production practices that cannot be commercialized. Underfunding of such long-range and public good research eventually will hamper success of commercial seed companies as well as limit progress in improving vital noncommercial aspects of sustainable food production.

When farmers and seed companies simultaneously can profit from production and use of hybrid varieties, a hybrid seed industry can flourish. But the industry as a whole is based on a complex interweaving of public research, private research, small local seed companies, and large national or international companies. Each crop species calls for a slightly different mix of ingredients. And farmers—those who grow the crop—are at the base of it all.

REFERENCES

Anonymous. 1996. Apomixis: Ensuring equity in the use of hybrids: A pre-proposal for a CGIAR systemwide initiative. CIMMYT, México, DF México.

CIMMYT. 1994. 1993/4 World maize facts and trends. CIMMYT, México, DF, México.

Doggett, H. 1988. Sorghum. Longman Scientific & Technical, Harlow, Essex, England.

Duvick, D.N. 1984. Genetic diversity in major farm crops on the farm and in reserve. Econ. Bot. 38:161–178.

Duvick, D.N. 1997. The American maize seed industry: An historical overview. In M. Morris (ed.) Maize seed industries in developing countries. México, DF, México.

Hallauer, A.R., and J.B. Miranda, Fo. 1988. Quantitative genetics in maize breeding. Iowa State Univ. Press, Ames.

Janick, J. 1996. Hybrids in horticultural crops. p. 45–56. In K.R. Lamkey and J.E. Staub (ed.) Concepts and breeding of heterosis in crop plants. CSSA Spec. Publ. 25. CSSA, Madison, WI.

Jefferson, R.A. 1994. Apomixis: A social revolution for agriculture? Biotech. Dev. Monit. 19:1994.

Knudson, M.K., and V.W. Ruttan. 1988. The R&D of a biological innovation: The case of hybrid wheat. Food Res. Inst. Stud. XXI:45–67.

Norskog, C. 1995. Hybrid seed corn enterprises: A brief history. Curtis Norskog, Willmar, MN.

Virmani, S.S. 1994. Hybrid rice technology: New developments and future prospects. In S.S. Virmani (ed.) Selected papers from the International Rice Research Conf. Int. Rice Res. Inst. Manila, Philippines.

Chapter 28

Heterosis for the Development and Promotion of the Seed Industry

R. B. Hunter

INTRODUCTION

The commercial exploitation of heterosis has been one of the driving forces behind the rapid and extensive development of privately funded crop breeding efforts around the world; however, the decision of private industry to invest in plant breeding for a given species has not been based solely on the presence and degree of heterosis. The major factors influencing privately funded breeding investment will be discussed followed by examples of development in the USA and in Thailand.

FACTORS AFFECTING PRIVATE INVESTMENT IN THE SEED BUSINESS

There are many factors involved in the decision by private individuals or companies to invest in crop improvement research for a particular segment of the seed industry. The factors include:
1. Market value
2. Competitive structure
3. Intellectual property protection
4. Productive operating environment
5. Ability to differentiate products
6. Opportunities for synergy

The right mix and balance of the above factors provide the rational for private investment in crop improvement research.

Market Value

Market value is a key factor affecting the willingness of individuals or corporations to invest in crop improvement research. For example in the USA, the area devoted to maize (*Zea mays* L.) production is about 30 million ha, seed market value of the U.S. seed industry is more than two billion dollars. Soybeans (*Glycine max* Merr.), another major U.S. crop, are grown on approximately 26 million ha. However, the seed value is considerably less than maize since seed costs per ha are much lower and only a portion of the area planted to soybeans is seeded with seed purchased directly from seed companies. The value of the soybean seed market in the USA is estimated at about 800 million U.S. dollars. In contrast, the area planted to sorghum [*Sorghum bicolor* (L.) Moench] in the USA is about five million ha, and the market value of the seed is about 180 million U.S. dollars. The privately funded breeding effort is impacted by these values. The science person years (SY) of private breeding effort devoted to maize, soybeans, and sorghum are 510, 101, and 41 respectively (Frey, 1996). In contrast to maize, soybeans and sorghum, crops such as onions (*Allium cepa* L.) and muskmelon (*Cucumis melo* L.) have much smaller seed market value because of the lower area devoted to these crops. The result is less privately funded breeding effort with these crops.

Competitive Structure

For a given species, many factors affect the competitive structure of the seed industry. A major factor is the competitive structure within the private sector component of the seed industry. In mature seed markets, such as maize in the USA, it is very difficult for a new company to become a significant player without purchasing an existing company. In recent years there has been a lot of activity in the purchase of existing seed businesses. Monsanto, ELM/ Seminis, and DuPont have all recently made significant purchases in existing seed companies.

Other aspects of the competitive structure include the presence of a significant publicly funded research effort that is in direct competition with the privately funded crop improvement effort. In North America, wheat (*Triticum aestivum* L.) is a major crop species with considerable publicly funded breeding research. Depending on the specific location and type of wheat, privately funded wheat breeding may compete directly with the public sector breeding effort. It can be much more difficult to justify private investment in breeding when it must compete against the public purse. For wheat breeding in the USA, there are 76 publicly funded SYs compared with 54 privately funded SYs (Frey, 1996).

The transition from publicly funded to privately funded research can be difficult and result in considerable tension between the two groups. When private investors initially invest in breeding research for a crop that has traditionally benefited from abundant public support, the private breeders are competing with strong public programs. In some cases the publicly funded breeders are reluctant to give up their dominant position in crop improvement. On the other hand the newly established private breeding effort will take time to become productive. This is particularly difficult if the public sector does not provide support to these new programs. If it is decided by the country or region to try to encourage the development of privately funded breeding for a given species, then an accommodating approach is necessary, especially during the early establishment years.

Examples of a very accommodating and cooperative approach to the establishment of private maize breeding are the USA in the 1920s and 1930s and more recently Thailand. More will be presented on these two examples in a later section of this chapter.

An example of a more difficult start up of the private seed industry was with canola (*Brassica napus* L.) in Western Canada. The private sector established breeding programs in competition with a well established and very successful publicly funded breeding effort. Today there is a strong private sector canola breeding effort that compliments the public efforts; however, there were some strenuous years during the transition. In Canada there is a government registration system for many crop species. All new canola cultivars must be registered in Canada prior to the sale of any seed. The tension between private and public breeders has often been manifested at committee meetings where public breeders play a major role in deciding what varieties will be eligible for registration. In countries that do not have registration systems, such as the USA, this may facilitate entry of the private sector into the market and certainly provides the private sector with more freedom to operate.

In order to commence breeding activity for a given crop in a given region, the private sector is often prepared to preinvest, but there are limits to the time and amount of preinvestment that are compatible with good business practices and are in the best interest of the owners.

Intellectual Property Protection

Historically there were little if any specific systems established to provide protection to the inventor of new cultivars resulting from an investment in crop improvement research. This reflected the fact that almost all plant breeding was initially carried out by publicly funded programs with free distribution of the cultivars resulting from this research effort.

Partially as a result of pressure from the seed industry and partially as a result of government desire to stimulate more private plant improvement research, there has been an effort made to establish vehicles for the protection of intellectual property resulting from plant breeding. During the past 60 years, systems of Intellectual Property Protection (IPP) specific to variety protection have evolved. The International Union for the Protection of New Varieties of Plants (UPOV) is an organization that serves as the international forum for countries interested in plant variety protection. UPOV's primary objective is to provide a framework for the protection of the intellectual property of plant breeders. There are currently 32 member countries. In addition 14 countries have their regulations in compliance but have not deposited their instruments with UPOV. The current convention was established in 1978 but a revision was made in 1991. The latest convention has not yet come into effect because sufficient countries have not yet ratified it. In order for a country to become a signatory to the UPOV convention, the Plant Variety Protection Act of that country must be in compliance with either the 1978 or 1991 convention.

Prior to the development of Plant Variety Protection (PVP) legislation, the sale of hybrids formed a means of Intellectual Property Protection (IPP). For example, the hybrid maize seed industry of North America was well developed prior to the development of any IPP systems for the protection of cultivars and products of plant breeding. Prior to 1963 the hybrid maize seed industry was selling three-way and double-cross hybrids. The fact that the purchase of seed by the grower did not result in the grower gaining access to the inbred lines that could be used to reconstitute the hybrid, provided a biological form of intellectual property protection. In addition, farmers could not, from an economic stand point, save seed of the hybrids they planted since the loss of heterosis and the increase in variation (segregation) in the F_2 generation resulted in considerable reduction in performance and an increase in plant to plant variability. Experiments in the 1930s (Richey et al., 1934) indicated that about 15% reduction in yield can be expected from planting second generation double–cross hybrid seed of maize.

In the past few years there have been some new developments in the seed industry that have changed the IPP situation. The introduction of products resulting from the efforts of biotechnology has resulted in patents becoming a major factor in IPP. In addition several companies are introducing a new concept in IPP. For example, Monsanto has introduced a Technology Use Agreement and is licensing seed directly with the grower. The license signed by the grower at the time of seed purchase, stipulates conditions of sale including aspects of IPP. These new approaches to protection may stimulate increased investment by the private sector in self pollinated crops such as wheat, barley (*Hordeum vulgare* L.), and soybeans.

Productive Operating Environment

Successful plant breeding is a long term and costly endeavor. Prior to the decision of privately held companies to invest in plant breeding in any given region of the world, the expectation must be that the investment will not be negated by government policies, country or regional internal structural constraints, or political unrest. Private plant breeding has been limited in several regions of the world because of investor concerns about operating stability. In parts of Africa, concerns about stability have limited and in some cases reduced or eliminated investment by the private sector in plant breeding. Today, investment in some of the newly created counties resulting from the demise of the USSR also may carry risks due to structural disruptions that, for now, will limit investment by private industry.

In contrast, investment in plant breeding has been strong in many countries that it is believed will provide a stable environment conducive to realizing a reasonable return on the investment. Many countries in Latin America and Southeast Asia provide examples. In a subsequent section of this chapter, Thailand is cited as an example with a rapidly developing private sector breeding efforts coexisting in harmony with a strong publicly funded effort.

Ability to Differentiate Products

Successful marketing and selling of cultivars by private seed companies requires that the company be able to differentiate its products from other cultivars available on the market. In order for a cultivar to become truly successful it is necessary to be able to demonstrate that the product and/or the service associated with the product add value for the user.

For several species the higher yields and plant to plant uniformity associated with hybrids clearly differentiates them from the traditional cultivars. For example, in Thailand, where both maize hybrids and maize OPVs coexist in the market, the top five hybrids out-yielded the top OPV by 37% (Table 28–1). In addition, hybrids, and particularly single crosses, because of their uniformity, can be defined and positioned more precisely. Every plant in a single cross hybrid will carry the same insect resistance, disease resistance, plant height, ear height, maturity, etc.
Hybrid crop breeding allows the breeder to bring in components from multiple sources (inbred lines). The breeder can rely on a particular high performing line (high general combining ability and good agronomics) to produce a range of hybrids over time. In addition, with rDNA based traits, a dominant gene is needed in only one parent of a single cross hybrid. This allows for rapid adoption across a range of maturities and hybrids when a trait such as corn borer resistance (Bt) is incorporated into a widely used inbred line. With self pollinated crops such as soybeans, a trait such as Roundup resistance needs to be backcrossed into every variety. This leads to a slower introduction of varieties across a range of maturities and markets and greatly increases the effort needed to accomplish the backcrossing.

Table 28–1. Estimates of yield differences (%) between the top five hybrids and a top OPV in Thailand in 1977, 1987, and 1997.

Variety	1977	1987	1997
Top five hybrids	Nil	126†	137‡
OP Variety	100	100	100
Difference (%)		26	37

† Mostly three-way crosses   ‡ Mostly single crosses

Opportunities for Synergy

Today, more than in the past, privately funded crop improvement activities are part of a broader investment strategy. Recently a number of specialty chemical companies have entered the seed business because they believe seed will become the dominant delivery system for crop improvement inputs or specialty chemical outputs. The input side includes insect resistance, disease resistance and herbicide tolerance. The outputs include specific nutritional quality traits such as oil content and quality, carbohydrate content and quality, protein content and quality, and reduced phytic acid. In addition, new and very different compounds will be plant based synthesized. These will include pharmaceuticals, neutraceuticals, specialty organic compounds, and raw materials for processing into down stream compounds. Many private corporations are now associating themselves with the seed industry either through agreements or through investments. In the past few years, nearly seven billion U.S. dollars have been invested by chemical industries to purchase or align with aspects of the seed industry. In recent years, Monsanto has been a major purchaser including more than one billion U.S. dollars for Holden's, a successful foundation seed company with a strong maize breeding effort. DuPont has just made a major investment in Pioneer Hi-Bred (1.7 billion U.S. dollars). Companies such as AgrEvo, Dow, and Novartis have all made investments in seed businesses because of the interactions with other aspects of their businesses.

ROLE OF HETEROSIS IN THE SEED INDUSTRY, SOME EXAMPLES

Maize in the USA

As pointed out in the previous section, the commercial exploitation of heterosis has been one of the key factors in the rapid and extensive development of privately funded plant breeding efforts around the world. No where is this more apparent than in the development of the privately funded maize breeding efforts in the USA. Hybrid maize was the first major food and food crop to move largely from publicly funded to privately funded breeding. Private industry entered into maize breeding very early in the history of the development of hybrid maize.

In 1923, Funk Brothers Seed Company in Bloomington, IL, was the first company to market hybrid maize seed (Crabb, 1993). In the 1920s and 1930s the hybrid seed industry was in its infancy. By the early 1940s the use of hybrids in the USA had become well established. It was during this period of rapid increase in the use of hybrid seed that the USA sustained the nations first major increase in maize production. The increase had begun in the 1930s following a period of 60 years (1870 to 1930) during which there had been no real annual gain in average yield in the United States. During these 60 years, the average yields varied between 1.67 to 2.00 t ha^{-1} (Crabb, 1993). From the time of their wide spread use, hybrids have been the major factor contributing to maize yield improvements in the USA. It is estimated that fully two thirds of the increase in maize yields in the USA from the 1940s to the present are the result of improvements in hybrid performance resulting from a massive plant breeding effort (Duvick, 1984; Russell, 1986).

During the early years of the establishment of the hybrid maize seed industry in the USA there were very few privately funded maize breeding programs. Those that existed were clearly tied to publicly funded programs, and they relied very heavily on materials and expertise from these programs. During this period, several commercial programs were strongly linked with government-funded programs. An example was the establishment of the Funks Farm Federal Field Station in conjunction with the Funk Seed Company. A similar relationship was established between what is now Iowa State University and the forerunner of Pioneer Hi-Bred.

Today in the USA almost all of the breeding effort aimed at the development of maize hybrids has shifted from publicly to privately funded research. In a recent survey by Frey (1996) reported that there were 510 private sector SYs devoted to dent maize breeding research and product development in the USA in 1994. This represents 94% of the total breeding effort in dent maize. Of the 27 state budgeted SYs working on field maize, only 3 SYs were devoted to cultivar development. There was no Federally funded maize research devoted to cultivar development.

In addition to private sector dent maize breeding in the USA, private industry investment in sweet corn and popcorn hybrid breeding accounts for an additional 46 SYs. The total publicly funded effort was reported to be 7 SYs.

A key conclusion of the study by Frey (1996) was that privately funded research in the USA has concentrated on breeding crops that use hybrid cultivars in agricultural production. The private industry has invested heavily in species with entirely hybrid cultivars or a mixture of hybrids and pure line cultivars. Major crops with entirely hybrid cultivars are field maize, sorghum, sunflowers (*Helianthus annuus* L.), sweet corn, sugar beets (*Beta vulgaris* L.) and muskmelons. Three crops, tomatoes (*Lycopersicon esculentum* Mill.), peppers (*Capsicum annuum* L.), and onions, with both hybrids and pure line cultivars accounted for an additional 119 SY's of private sector investments.

In contrast to crops with hybrid cultivars, there is still considerable publicly funded breeding effort devoted to pure line cultivar development in such crops as soybeans, wheat, rice (*Oryza sativa* L.), and cotton (*Gossypium hirsutum* L.). For these four crops the ratio of public to private plant breeding research effort is 2:3.

In the USA, the decision of private industry to invest in crop improvement for a given species has not been based solely on the presence of and the degree of heterosis; however, for species like maize and sorghum, the production of hybrid cultivars has provided an approach to IPP as well as an ability to clearly differentiate products from competitors. In the case of maize, the value of the seed market has been another major factor. In addition, the USA supplied a favorable operating environment for private investment. In the past five years the opportunities for synergy between seed and other aspects of the agricultural input and output industries have become a strong force in stimulating private investment in crop breeding and its extension, biotechnology. A disadvantage to investing in maize hybrid improvement in the USA is the very strong competitive nature of the business. The 510 SY's devoted to commercial breeding give some indication of the degree of competition. Overall, the investment in private crop improvement research in the USA, although not totally driven by heterosis, has been strongly influenced by this phenomenon.

Thailand Maize Research Development

Thailand offers a good example of a country that has provided an environment for the rapid development of a privately funded plant breeding industry. The private sector has been actively involved in plant breeding in Thailand since the late 1970s. Their efforts have focused primarily on hybrid maize and vegetables. The private seed sector started as a result of the Thai government's policy to encourage the private sector to participate in crop improvement and in the production of good quality seed for the farmers. Such participation was considered as an efficient way to further develop Thai agriculture while at the same time reducing the Governments need to invest in crop improvement research. The Board of Investment Committee offered interested investors from the private sector seed industry promotional incentives. Between 1978 and 1981, Thailand was able to attract five major seed companies to establish maize breeding programs in Thailand. The investment was attractive because Thailand offered a stable, progressive operating environment with a strong agricultural base and a farm population that had demonstrated it would rapidly adopt new technology. The newly established seed companies also were able to hire competent and experienced scientists.

In Thailand there has been and still is considerable cooperation between the public and private seed sector. The overall objective has been to better serve the growers with improved cultivars and high quality seed. The seed companies working with maize are actively involved in screening elite germplasm provided by the national Thai programs such as Kasetsart University and with international organizations such as the Asian Regional Program of CIMMYT. In addition to providing elite material for the private sector breeding activities, Kasetsart University has provided trained staff, technical services (for example, DNA fingerprinting) and training for technical staff already employed by the industry. CIMMYT also has been a major collaborator of the seed industry.

Using the maize seed industry as an example, let's look at what has happened to maize breeder numbers in just 20 years. In 1977 there were nine public maize breeders in Thailand, four with the Department of Agriculture and five at Kasetsart University. At that time there were no privately funded maize breeders. By 1987 there were still nine publicly funded maize breeders but in addition there were 12 maize breeders in the private sector. Today there are seven public and 14 private sector breeders serving the needs of the Thai farmers (statistics for Thailand used throughout this chapter have been graciously supplied by Kriangsak Suwantaradon and Tawatchai Prasastsrisupab of Novartis (Thailand) Ltd., Field Crop Seeds).

What has been the result of the Thai Government policy to provide an operating environment conducive to stimulating private investment in crop improvement? Once again using maize as an example, the number of maize breeders has more than doubled. In 1977 the effort devoted to improvement of OPVs was 100% of the total maize breeding effort. In 1997, 90% of the maize improvement effort (100% in the private sector and 70% in the public sector) is devoted to hybrid production. In 1977, 100% of the maize seed planted was OP or retained seed. In 1997, hybrids account for 75% of the total area planted to maize. Table 28–1 illustrates the yield advantage associated with the development of hybrids for the Thai market. The gains have been impressive and have not required large expenditures from the public. The investment by the private industry would not have been possible without the strategy put in place by the Thai government; however, the private investment in maize breeding would not have taken place if the strategy had been to develop OPVs rather than hybrids.

REFERENCES

Crabb, R. 1993. The hybrid corn makers. 2nd ed. West Chicag Publ. Company, Chicago, IL.

Duvick, D.N. 1984. Genetic contributions to yield gains of U.S. hybrid maize 1930 to 1980. p. 15–47. *In* W.R. Fehr (ed.) Genetic contributions to yield gains of five major crop plants. CSSA Spec. Publ. 7. CSSA, Madison, WI.

Frey, K.J. 1996. National plant breeding study-1 human and financial resources devoted to plant breeding research and development in the United States in 1994. Spec. Rep. 98. Iowa State Univ., Ames.

Rickey, F.D., G.H. Stringfield, and G.F. Sprague. 1934. The loss in yield that may be expected from planting second generation double cross corn seed. J. Am. Soc. Agron. 26:196–199.

Russell W.A. 1986. Contribution of breeding to improvement in the United States, 1920s–1980s. Iowa State J. Res. 61:5–34.

Chapter 29

Logistics of Seed Production and Commercialization

A. B. Maunder

INTRODUCTION

An obvious need of developing country agriculture is technology transfer, such as with improved planting seed, from national or international research improvement programs to the producer, both subsistence as well as the large operator. Unfortunately, major improvements at the experiment station level often only show up in annual reports and producer yields remain low. Lack of the proper infrastructure for effective multiplication and distribution of improved cultivars, such as from hybridization, which requires input from the private sector, most likely explains this constraint. For any crop the existence of dependable markets, relatively large areas of cultivation, and a desire on the farmers' part to increase yields through hybridization indicate that farmers might benefit from the presence of commercial seed firms. Advantages and likely reasons for success would be: (i) dependable supply, (ii) acceptable quality/purity, and (iii) improved level of performance (Maunder et al., 1994).

To develop a seed industry means providing a reasonable return on investment without undue government restrictions since capital risk will be required. Private but indigenous seed enterprises are a logical first step but their inability to cope with monetary fluctuations and lack of sufficient funds for research investment often has put them at a serious disadvantage. Historically, public institutions have not been effective or efficient as providers of hybrid seed nor have they been structured to do production and marketing. To the contrary, on-going technical support and encouragement by the public breeder until the seed producer becomes adequately experienced is highly desirable. Those new to the hybrid business must anticipate a more complex procedure with outside training early on but every country possesses the type of individuals capable of such activity. A supply of trained agriculturalists will be indispensable in operating the seed firms.

Over the past seventy years, since the hybridization of maize (*Zea mays* L.), numerous field and vegetables crops have benefited from the phenomenon coined by Shull in 1917 as heterosis (Crabb, 1947). Examples include plants, characterized by both perfect and imperfect flowers, with genetic and cytoplasmic-genetic-sterility, protogyny, incompatibility, and genetic markers allowing for wind, insect, of hand pollination (Table 29–1). Even the oil palm (*Elaeis guineensis*), which accounts for the major cooking oil in much of the world, is in fact now derived from a wide F_1 hand-cross of *dura* x *pisifera*, which brings not only higher yield of oil but better seed characteristics into the end product.

Table 29–1. Specific field and vegetable crops currently hybridized and the method
of pollination.

Field crops				Vegetables			
Maize	w†	Cotton	i h	Onions	i	Squash	i
Sorghum	w	Alfalfa	i	Broccoli	i	Beets	w
Wheat	w	Oil palm	h	Cauliflower	i	Cantaloupe	h
Canola	w i	Millet	w	Cabbage	i	Cucumber	i h
Sunflower	i	Castor	w	Tomatoes	h	Eggplant	h
Rice	w h	Sugarbeet	w	Watermelon	h	Pepper	h
				Carrots	i	Spinach	w

† w, by wind; i, insect transmission; h, hand pollination.

LOCATION OF PRODUCTION CRITICAL

Success in seed production often relates to the specific eco-geographic area chosen. Location should be determined by region of crop adaptation that will determine parental line performance in relation to frost-free days, temperature means and extremes, relative humidity, adequate and timely moisture, often by irrigation, well drained soils, and minimal abiotic and biotic stresses. Certainly freedom from weedy relatives of the specific crop will be essential. Heat units to enable normal growth and pollen viability are just as critical as a relatively dry harvest period to provide adequate purity as well as quality of the seed crop. Areas of heavy disease pressures on the seed crop will not only reduce yield but may very well prevent movement of seed into commercial channels. Weedy relatives not only cause high costs of roguing but may cause whole lots or segments of the production to be eliminated from the market.

Also, the availability of temporary labor for field activity such as roguing, detasseling, and harvesting can affect cost of the finished product as well as quality. This unskilled labor still requires a reasonable amount of training. Proximity of the seed production and conditioning to its intended market further relates to unit cost. Most likely because the parental material as well as the end product generally adapt to the same area, production will take place reasonably near the greatest demand and will become a component of more intense growing areas of a specific crop to reduce freight costs.

EQUIPMENT REQUIRED

Field equipment required for seed production will be dependent on that available by the contract growers with some more expensive or complex items at times furnished by the seedsman, examples being a pesticide applicator and a stationary, or preferably mobile, harvester/combine. Depending on the level of inputs, even some conventional farming equipment may of necessity have to be provided. Seed conditioning equipment would be the principal investment along with a reasonably useable building for this equipment as well as seed storage. Cost of such equipment based on U.S. prices for such crops as maize, sorghum [*Sorghum bicolor* (L.) Moench], and sunflower (*Helianthus annuus* L.) may approach $75,000 U.S. (Table 29–2) where season long conditioning would readily handle 1000 tons. Harvest bags, holding bins, scales, and finished product sacks of paper or cloth as well as chemical treatments would be additional as would be the need for a dryer. Some level of useable equipment could very well already be in place as with Niger's effort to produce NAD-1, a first sorghum hybrid for the country (L.R. House, 1997, personal communication). Also developed seed operations constantly upgrade, putting much useful conditioning equipment on the market at more

Table 29–2. Equipment to condition 1000 tons of maize, sorghum, or sunflower seed based on U.S. cost estimates.

Component for conditioning	Estimated cost U.S. export†
Seed cleaners and sizer	$32,000
Ideal length grades	4,500
Pre-mix tank	2,700
Seed treater	3,300
Elevator	8,000
Conveyors (2)	8,000
Total	$58,500

† FOB prices plus packing

favorable prices. Finally, consultation with experienced seedsmen will prove invaluable at the time of putting together an efficient and cost effective seed conditioning plant.

GERMPLASM, FOUNDATION, AND HYBRID PRODUCTION

Without preliminary germplasm research to develop and determine best parents for hybrid production, parental lines may be obtained from IARCs, NGOs, NARS, or US–AID CRSPs (Collaborative Research Support Programs). As reiterated often by L.R. House from experiences in India, Sudan, and Southern Africa, the importance of a significant yield increase for the first hybrid or varietal release cannot be over emphasized. In fact, N. Borlaug believes that under some cultures yield increases of as much as 50% may be required to achieve movement away from traditional landrace cultivars (Maunder, 1997). Yield gains of 20 to 100% for hybrids over varieties are common in crops like sorghum, along with enhancement of seedling vigor, 5 to10 days quicker flowering, and improved test weight, not to mention the greater ease of handling abiotic and biotic stress, so often associated with this crop.

Foundation

Ready availability of the parental seed stocks will determine the need for foundation maintenance. Should foundation seed be a component of the seed industry smaller blocks for inbred increase with even greater isolation to conform to certification–purity restrictions must be established. These may require increase of all inbreds used in large scale production as well as single cross production where hybrids are three-way or double cross types. Male sterility will require A-line or female production by the single cross of A × B where the B-line is identical genetically but in normal cytoplasm allowing for pollen production. This requirement to maintain seed stocks in sufficient quality and quantity when no other source is available becomes a significant component of a successful seed venture. Where the developing or research entity has personnel and equipment to handle foundation, new seed ventures will have a higher probability for success.

Hybrid Seed Production

The process of production requires many variables, some controllable, others not, making it impossible to be crop specific within the framework of this chapter. Reference to Table 29–1 made earlier is critical in that the transmission of pollen during hybridization significantly effects many of these variables.

First, growers, often on a contract arrangement, must be located for the actual crop production. These individuals should be the more progressive farmers in their community. The more interested and prepared they are to cooperate, the less the risk for a successful crop. If at all possible they should have access to irrigation and be able to provide reasonably uniform soils without toxicity or water logging problems. Some agreement within the contract may be necessary to insure sufficient chemical inputs for a successful seed crop. These seed growers may be the supplier of labor necessary to handle such field activities as detasseling or rouging. Payment can vary but is often based on market price plus premium per unit weight of harvested seed in an amount sufficient to attract the best possible growers.

Successful hybrid production depends on high purity and quality parental (foundation) seed planted in the correct sequence to assure pollination nicking and in the correct ratio female to male for an adequate volume of pollen. Examples of ratio by crop would be sorghum 3:1, maize 2:1 to 4:1, sunflower from 2:1 to 7:1, and wheat (*Triticum aestivum* L.) from 1:1 to 3:1 (Wright, 1980). Vegetable crop ratios also range from 1:1 to 4:1 (D. Homes, 1997, personal communication). Row width, ease and volume of pollen movement are other important factors. Where nicking is environmentally affected, even from year to year, two or more planting dates of male are practiced to spread the pollen over more days to assure a good seed set. Good nicking best reduces off-types in the final product and allows sterile plants to escape such problems as sorghum ergot. Planting date studies to predict most likely splits are encouraged for all parental or most likely parental lines. Roguing before, during, and following pollination is essential to maximize purity and to meet inspection requirements. Another complexity of dealing with hybridization relates to adequate isolation from nonmale pollen such as commercial cultivars and or weedy species some of which may result in off-types of an extremely objectionable nature such as rhizomes or seed shattering with sorghum.

Harvest must stress seed purity as well as quality. Generally, the male parent should be removed first to avoid contamination. Often a preharvest roguing will take place depending on season-long challenges with the specific field. Where and when possible harvest will take place which avoids use of a dryer or need for a desiccant. High moisture at harvest with too much heat from artificial drying can induce dormancy, reduce germination, or lead to rapid aging, a condition often associated with light frost damage. Every effort at this stage should be made to carefully label and store separately by seed lots, often quantities which represent a specific area of the field. These lots then allow for purity checks affecting only a part of a field and can avoid excessive discard should a problem with grow-outs occur.

With the overview of production requirements the complexity of hybrid production suggests outside technical input at the beginning of a hybrid seed industry. Many experienced with the crop could serve as short term consultants. Where government or international center research has developed the specific hybrid cultivar, the breeder or related agronomist would greatly enhance success by being available for guidance and season-long problem solving. Additionally, the risks of low seed parent yields, inclement weather, and failure to meet purity standards all favor input from the private sector where investors are willing to take on risk for a potential profit, an undertaking not common to the public sector.

MARKETING, DISTRIBUTION, AND SERVICE

Once the seed is produced, the cultivar's attributes must be communicated to the commercial producer and the seed be made readily available for purchase. This may take place (i) directly from the seed operation itself, (ii) through a store or other retail type operation, or (iii) through farmer dealers, the approach most commonly used in the early history of the U.S. hybrid industry. Larger distributors, capable of providing more services, may become dominant in some markets. This trend has developed to better provide financing and service after the sale. The larger seed companies, however, also are expanding their efforts to provide agronomic support to customers. These services attempt to provide product information as well as information on likely crop production problems and their control such as from abiotic and biotic stress. At the time of purchase the grower should additionally receive information on germination, purity, seed size, and even specific lot numbers.

Product advertising is especially essential in a competitive environment. Perhaps one of the most effective and inexpensive approaches is through demonstration or show plots either done independently or in conjunction with government extension efforts. Producer meetings are often held at these plots to allow for discussion of the specific hybrid attributes or limitations with harvest data helpful in convincing growers to make the transition to high technology seeds from their common varieties. These meetings also may take place between cropping seasons where an update on agronomic practices, insect, disease, and weed problems and their control as well as nutrient requirements can be discussed.

Pricing of the seed also must be given high priority along with collection of sales. No specific figures can be given here but cost of seed from the contract grower, sales and advertising expense, and where applicable research cost are critical to determine price that still must take into account a reasonable return on the investment by the seed industry and a competitive range to other companies offering the same or similar performing products. Remember the grower will do his own calculation and a critical piece of information will be the opportunity for return on the potentially extra cost of his seed. In most countries utilizing hybrids the seed investment is relatively small in relation to the increased potential for both higher yield and extra earnings.

CONCLUSIONS

Success as a seedsman, once an infrastructure is in place, will relate to many variables among which will be the volume to produce to adequately meet demand. This may require a 25 to 30% additional production to account for distribution logistics, change in cropping systems, possible quality discards, and hopefully competitive gains. As stated earlier, likely reasons for success not only include a dependable supply but also credibility and conformance to purity/quality standards. In developing countries this assurance of a trueness-to-type or expectation can be an adequate advantage to develop a successful business. The third and most significant reason for a commercial seed venture would be to produce and distribute to an adequate customer base a product of sufficient yield, quality or other agronomic advantage to provide the consumer or producer a more profitable business whether it be selling the seed or growing a superior crop. The profit incentive promotes improved quality and performance, a characteristic not natural to the public sector.

To start up a successful operation requires the ability to secure adequate financing by an individual or organization, be willing to study and learn the business and then become a primary component of the agribusiness community. At times the best of planning will still result in too little or a carryover of supply which will require storage facilities to maintain an adequate quality to the seed. There will always be challenges with collection on sales but this is a business and these challenges are part of the risk. Depending on the quantity and quality of competition, the product performance must develop a customer base adequate in size to support the available seed industry. Success at this entry level can then lead to a form of in-house research activity to provide an improved and proprietary product. India has no doubt been the recent best example where in sorghum and millet the private sector, value-wise, has captured 60% of the formal seed market (Maunder, 1997). Now, according to a survey by Pray et al., (1991) some 17 firms had Research and Development programs on these crops, spent an average 4% of seed sales on research and employed 31 Ph.D graduates and 45 with MS degrees.

To realize benefit from the billions of dollars of agricultural investment for improvement of developing country agriculture will require a strong private sector seed industry. Only with this transition joined by those providing other critical inputs such as chemicals and machinery can the basic principles of agronomic research, already developed, return the necessary impact to mankind.

REFERENCES

Crabb, R.A. 1947. The hybrid-corn makers, prophets of plenty. Rutgers Univ. Press, New Brunswick, NJ.

Maunder, A.B., F. Bidinger, L. Busch, J. Hulse, and M. Shepard. 1994. INTSORMIL five year EEP review. US–AID, Washington, DC.

Maunder, A.B. 1997. Role of private sector. *In* Genetic improvement of sorghum and pearl millet. INTSORMIL., Lincoln, NE.

Pray, C.E., S. Ribeiro, R.A.E. Moeller, and P.P. Rao. 1991. Private research and public benefit: the private seed industry for sorghum and millet in India. p. 315–324. Res. Policy 20, Elsener Science Publishers B.V.

Wright, H. 1980. Commercial hybrid seed production. p. 161–176. *In* W.R. Fehr and H. Hadley (ed.) Hybridization of crop plants. ASA. Madison, WI.

Chapter 30

Exploitation of Heterosis: Uniformity and Stability

J. Janick

INTRODUCTION

Crop uniformity is considered a desirable character in modern agriculture because product uniformity is essential in marketing; uniformity in maturity permits crop scheduling; and uniformity in plant structure and maturation permits efficient mechanical harvest. Furthermore, crop uniformity is essential for maximizing yield, a little understood feature of uniformity that will be expanded in this chapter. With the increasing importance of world urban markets, product uniformity becomes an essential feature of crop quality especially for horticultural commodities.

Crop uniformity is associated with a lack of diversity but diversity can be increased by planting a number of different homogeneous cultivars. Crop diversity has been considered desirable in some environments and situations because nonuniform populations are assumed to produce population buffering under stress and diversity spreads risk. Consider the planting of sweet corn (*Zea mays* L.) in a garden situation. The planting of a uniform cultivar with asynchronous silk emergence and pollen shed could result in pollination problems while uniformity in maturity causing the entire crop to ripen at once would produce a feast or famine situation. Diversity could be increased by the use of open pollinated seed, by planting F_1 hybrids with different maturity dates, or by staggered plantings. Variability in development and ripening has been considered a virtue under primitive or subsistence (low input) agriculture in order to hedge risks but carries the penalty of decreased yield and quality under optimum conditions.

Production of F_1 hybrids of seed-propagated crops is a successful breeding technique because it exploits heterosis, promotes homogeneity in allogamous species, and is a way for commercial breeders to control their product. The uniformity of hybrids has been considered one of their special benefits. There are two dimensions to the uniformity of hybrids: (i) genetic homogeneity and (ii) genetic stability. Genetic homogeneity refers to the presence of identical genotypes while genetic stability refers to phenotypic uniformity (homeostasis) in different environments.

GENETIC HOMOGENEITY

Phenotypic homogeneity can be achieved by various methods. These include vegetative propagation or special breeding techniques.

320 JANICK

Vegetative or Clonal Propagation

Vegetative propagation, the increase of plants via mitotic divisions and differentiation, is the basic method of propagation for perennials, such as most fruit crops, sugar cane (*Saccharum officinarum* L.), and many ornamentals. These crops are typically allogamous and thus highly heterozygous and show genetic segregation when planted from seed. Heterozygous genotypes are cloned by various techniques including cuttage, graftage (includes budding), special vegetative structures (such as runners, bulbs, and corms), and various tissue culture techniques (involving shoot proliferation, organogenesis, or somatic embryogenesis). Genetic homogeneity in heterozygous clones can also be achieved by seed propagation through apomixis, the development of seed without the sexual process, and thus a form of vegetative propagation. A common type of apomixis in nature is the occurrence of somatic embryogenesis from nucellar tissue which is common in crops such as citrus (*Citrus* sp.) and mango (*Mangifera indica* L.). In citrus culture, apomixis is exploited to propagate commercial rootstocks such as Rough lemon and Trifoliate orange which are highly polyembryonic but cannot be used for commercial propagation of scion cultivars because of the problems of juvenility. (Nucellar seed has been used to free plants from virus because virus is normally not transmitted through either nucellar seed or zygotic seed.) Apomixis is under genetic control and has been suggested as a means to clonally propagate heterozygous genotypes in cereals.

Selection

In seed propagated crops it is possible with selection to achieve homozygosity of important morphological traits and yet maintain enough residual heterozygosity to avoid inbreeding decline in open-pollinated populations. Thus, selected open-pollinated populations or land races may appear relatively uniform for a number of obvious morphological traits after many generations of mass selection. Uniformity for various traits in open-pollinated crops can be enhanced by a combination of mass and pedigree selection (selfing and intercrossing of many selected genotypes). This is the breeding system that is carried out in Texas for improvement of open-pollinated onions (*Allium cepa* L.).

Inbreeding

Selfing of heterozygous lines will increase homozygosity and this is the standard breeding system for autogamous crops; however, while genetic homogeneity may be achieved by selfing allogamous crops, high inbreeding depression usually makes such inbred lines weak and low yielding, so that in many cases they may be difficult to maintain, much less produce an adequate seed crop. Furthermore, although these weak inbreds may be genetically homogeneous, they often show high variability for many traits despite the fact that they appear morphologically uniform. A number of studies have shown that in maize, variability of inbreds for some traits is greater than that found in F_1 hybrids and even greater (!) than in open-pollinated populations (Table 30–1). This extra variability of inbreds when it exceeds the variability of homogeneous hybrids has been difficult to explain. It is not genetic variability (since the plants in lines are highly homozygous after five or six generations of selfing), but must represent *general environmental* (or *microenvironmental*) variability (V_{Eg}) the environmental variance contributing to the between-individual component arising from permanent or non-localized events, and is similar to developmental or *special environmental variability* (V_{Es}), the variability that occurs between organs of a single plant arising from temporary or localized circumstances operating during development (Falconer, 1989). Developmental variability would be greater for quantitative

Table 30–1. Yield performance and uniformity of various types of hybrids involving inbred lines HY, L317, WF9, and 38–11. Adapted from Jugenheimer (1976).

Hybrid type	CV ($F_1 = 100$)						
	Yield (t ha^{-1})	Ear weight	Ear length	Ear circumference	No. kernel rows	Ear height	Mean
4 inbred lines	2.57	150	156	122	102	160	138.0
6 single-crosses	6.57	100	100	100	100	100	100.0
12 three-way-crosses	6.26	106	109	115	108	126	112.8
12 single-backcrosses	6.40	103	109	116	122	133	116.6
12 backcrosses	6.57	101	114	115	129	136	119.0
3 double-crosses	6.16	120	120	131	138	140	127.8
6 top-crosses	6.08	110	121	129	132	186	135.6
1 open-pollinated cv.	5.73	117	130	127	149	178	140.2

characters that are influenced throughout development. Thus, of five maize characters (ear height, kernel rows, ear weight, length, and circumference) investigated by Jugenheimer (1976) the character with the least variability in inbreds was row number, a qualitative character probably determined very early in development. Ear height, which is related to the node of first flower and, therefore, fixed early in development also shows great variability in inbreds but this might be due to its co-dependence on internode length which must be affected by microenvironment. However, uniformity of maize inbreds has been greatly increased as inbred yields have increased with cycles of selections (D. Duvick, 1997, personal communication). In conclusion, genetic homogeneity does not insure uniformity in non-adapted material where instability is high unless coupled with incorporation of genes contributing to buffering.

Production of F_1 Hybrids

F_1 hybrids are created by inbreeding to create homozygous lines, followed by intercrossing divergent inbred lines to create heterozygous but homogeneous hybrids. This breeding system insures uniformity in seed propagated, allogamous species, where open-pollinated populations consist of a mixture of genotypes. Genetic homogeneity combined with high vigor can be achieved by selecting within and among inbred lines.

The genetic variability of the F_1 hybrid is a function of the homozygosity of the parents. This can be demonstrated with a single gene. With a gene frequency of 0.5, 1/4th of random crosses in a random mating population (S_0 generation = 1st segregating generation consisting of 1AA:2Aa:1aa) can be expected to be nonsegregating (AA × AA, aa × aa, AA × aa, aa × AA). After one generation of selfing (S_1), the expected population would consist of 3AA:2Aa:3aa and 9/16 of random crosses would be nonsegregating. With n segregating generations, $[(2^n-1)/2^n]$ random crosses would be nonsegregating. With m number of segregating genes the number of completely nonsegregating crosses would be $[(2^n-1)/2^n]^{m+1}$ (Table 30–2).

As genetic homogeneity in single crosses is a function of the degree of homozygosity of the parents, F_1 homogeneity can be increased by increasing the homozygosity of the inbred parents through inbreeding. However, inbreeding for

Table 30–2. Percentage of random crosses that are completely nonsegregating following self fertilization for n segregating generations with m pairs of factors assuming an original random mating population with gene frequency of 0.5.

Self generation	Segregating generation	No. factor pairs				
		1	2	5	10	m
S_0	1	25.00	6.25	0.10	0.00	$(1/4)^m$
S_1	2	56.25	31.64	5.63	0.32	$(9/16)^m$
S_2	3	76.56	58.62	20.14	6.92	$(49/64)^m$
S_3	4	87.89	77.25	52.44	27.51	$(225/256)^m$
S_4	5	93.85	88.07	72.80	52.99	$(961/1024)^m$
S_9	10	99.80	99.60	99.00	98.06	$(1046524/1048576)^m$
S_{n-1}	n	$\left[\dfrac{2^n-1}{2^n}\right]^2$	$\left[\dfrac{2^n-1}{2^n}\right]^3$	$\left[\dfrac{2^n-1}{2^n}\right]^6$	$\left[\dfrac{2^n-1}{2^n}\right]^{11}$	$\left[\dfrac{2^n-1}{2^n}\right]^{m+1}$

five generations still maintains 3.125% heterozygosity. Although homozygosity of inbred lines is expected to increase 50% on average with each generation of selfing, actual heterozygosity may be greater than expected for the following reasons: (i) natural selection for heterozygosity, (ii) mixtures and open pollinations, and (iii) mutations. In most allogamous species, it is often difficult or impractical to produce homozygous inbreds because the decline in vigor is so severe that seed yield of inbreds is uneconomically low and in some cases the inbreds are too weak to survive. In addition, when cytoplasmic male sterility is required to produce all female lines, the male-sterile line is maintained by crossing two lines, S cytoplasm *ms ms* (male sterile) × N cytoplasm *ms ms* (male fertile). Consequently, different strategies are used to produce hybrids when completely homozygous lines are unattainable or unavailable. One is to produce hybrids using inbreds that may be only selfed one or two generations. Selection can increase phenotypic uniformity of the inbreds; in onions these parents are called *heterozygous uniforms*. The other strategy is to use an F_1 hybrid as one parent and either inbred, or cultivar, or another F_1 hybrid for the other parent (known as three-way crosses, top-crosses, or double-crosses). There are other combinations (Table 30–3). All of these systems increase the genetic variability of the hybrid as compared to single-cross hybrids; however, single crosses display maximum heterosis and thus have become the hybrid of choice in maize where homozygous inbreds can be maintained. Single-crosses were always used in sweet corn because the window of quality is very short and uniformity is critical, especially in processing. In autogamous (self-pollinating) species, genetic homogeneity of parental populations and hybrids are equivalent because each population consists of identical genotypes, one homozygous and the other heterozygous.

GENETIC STABILITY

Genetic stability (homeostasis) refers to the smaller genotype by environmental (G×E) interaction in response to environmental variation, particularly to stress. When a series of inbreds, clones, or hybrids are evaluated across a wide range of locations or years, the rankings for any character may differ.

Table 30–3. Kinds of hybrids based upon the number and arrangement of the parental inbred lines (Jugenheimer, 1976).

Type of hybrids	Pedigree
Top-crosses	(A) × op cultivar, or (A × B) × op cultivar
Single-crosses	(A × B)
Modified single-crosses	(A × A') × B)
Sister line crosses	(A × A') × (B × B')
Three-way crosses	(A × B) × (C)
Modified three-way crosses	(A × B) × (C × C')
Double-crosses	(A × B) × (C × D)
Double backcrosses	[(A × B) × A] × [C × D) × C]
Single backcrosses	[A × B] × [(C × D) × C]
Multiple crosses	[(A × B) × (C × D)] [(E × F) × (G × H)]
Synthetics and composites	Many lines

The change in rank and the relative differences over a range of locations is defined statistically as G×E interaction. The environmental effects may be due to differential response to stress including abiotic factors (e.g., temperature gradients, photoperiods, insulation, soil moisture, length of growing season, soil characteristics, and farming practices) and biotic stress (e.g., plant density, response to virus, bacteria, nematodes, fungi, insects, birds, and weeds). In the U.S. Corn Belt, the most critical factor is weather, especially moisture stress at pollination (A.F. Troyer, 1997, personal communication). The optimum hybrid would be one that consistently gives above average yields, especially in situations of stress because, if widespread, prices would be expected to be highest.

Open-pollinated populations of allogamous crops have been considered to have genetic stability by the virtue of genetic heterogeneity, which creates *population buffering*. Population buffering must be absent in homogeneous single-crosses that consist of a single genotype. Thus, there would appear to be a dilemma in the breeding strategy involving hybrids. The highest yields (maximum heterosis) would be expected to be found with single-cross F_1 hybrids, but the uniformity of F1 hybrids precludes population buffering; however, this dilemma is more apparent than real because it can be demonstrated that genetic stability is an inherited trait and can be incorporated into heterotic combinations (Russell & Eberhart, 1968; Eberhart, 1969; Eberhart & Russell, 1969). Furthermore, population buffering for yield may be an artifact except in the most extreme environments (see below). The selection of single crosses with high yield and high stability in high density plantings, is responsible for the consistent increase in maize yields in the USA since their introduction in the early 1960s (Fig. 30–1). Duvick (1984, 1997) has demonstrated that the best modern single cross hybrids are on average not more productive than the best hybrids of previous eras in widely spaced plantings. Yield increases have come about principally because of increased stress resistance, particularly the ability to produce under the increased stress of high plant populations (Fig. 30–2). A small part of the yield ability has come about from morphological changes (small tassels, upright leaves) and reduced grain protein, but further increases in yield based on these characters is unlikely.

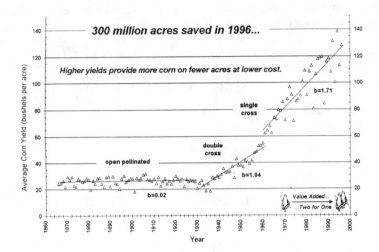

Fig. 30–1. Average U.S. maize yields from 1866 to 1996. The regressions (b) indicate gain in bushels per acre per year (1 bushel = 25.4 kg). Note that 1996 production with prehybrid era yields would require 151 M ha. Source: A.F. Troyer (1997, personal communication).

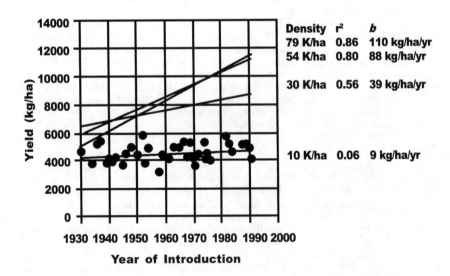

Fig. 30–2. Grain yield per hybrid regressed on year of hybrid introduction at four densities (10 000, 20 000, 54 000, and 79 000 plants ha^{-1}) at three locations in 1994. Source: Duvick (1997).

Characterizing Stability

Eberhart (1969) has summarized ways to characterize the yield stability of hybrids. This is achieved by comparing trials of the same entries conducted over many locations or years. The basic technique is to regress yields of each entry on an environment index, based on the deviation of each environment from the grand mean over all environments (Fig. 30–3). The deviation from regression (deviation mean square) provides additional information concerning yield stability.

The regression of the mean yield of all hybrids on the environmental index must necessarily have a slope of 1.0. An entry with a greater mean and a slope close to 1 will have higher average yielding ability over all environments, while an

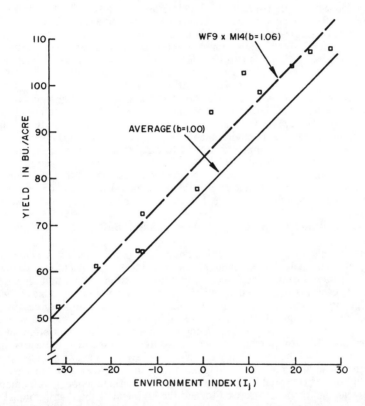

Fig. 30–3. Examples of stability in maize trials. Response of WF9 × M14 to varying environments in south-central Iowa from 1948 to 1951. The hybrid would be considered high yielding and stable because the slope (b_1) was near 1.00, and the mean performance was greater than the average of all entries under each environment, while the deviations from regression were not significant. Examples of instability are shown in Fig. 30–4. When the regression coefficient is >1, the hybrid yield increases under non-stressful environments; when the regression coefficient is <1, the relative increase over the mean is greater in stressed environments. Note that if the range in environments is very great, high-performing nonstable cultivars in nonstressed environments can be expected to perform worse than the average in extreme environment. Source: Eberhart (1969).

entry with a lesser mean and a slope of 1.0 will yield less over all environments. Entries with slope >1 perform relatively better as the stress levels decline, while entries with a slope <1 perform relatively better as stress levels increase. Large deviations from regression in particular environments suggest the involvement of some other factor at specific locations. Let us assume that one hybrid is higher yielding at all environments but that it contains susceptibility to a particular disease. In an environment where the disease was prevalent and where most of the other entries were tolerant, the entry would perform very poorly. This would affect the deviation from regression but not necessarily the slope of the regression.

Genetics of Stability

Evaluations of single-cross and double-cross hybrids by Eberhart and Russell (1966, 1969) indicate genetic differences among single-crosses for stability. Diallel studies of single-crosses from 10 inbreds indicate additive gene action for stability as measured by the regression coefficient but less so for mean square deviations. Thus, stable hybrids can be selected from stable inbreds. Fortunately, there appears to be a positive correlation between stability and yield.

An example of a character that may affect stability is the propensity for one- vs. two-eared hybrids (Russell & Eberhart, 1968). In an extreme example, two hybrids were compared in different environments (Hy × C103) × B37 (all one-eared inbreds), and (B60 × R71) × B59 (all two-eared inbreds). While both hybrids had similar average yields over all environments, the one-eared type was superior in the high yield (low stress) environment and the two-eared type was superior in the low yield (high stress) environment (Fig. 30–4). Furthermore, the single-eared hybrid has a greater deviation mean square. The effect appears to be associated with the increase in barrenness of single-eared genotypes exposed to stress, i.e., as stress increases not only does the second ear fail to develop but growth of the top ear is inhibited.

Competition and Stability: Does Population Buffering Exist?

In a recent review of competitive ability and plant breeding, Fasoula and Fasoula (1997a) indicate that population buffering, at least in respect to maximizing yield, is in fact an illusion. They consider that maximum yields per unit area, only occur where resources are shared equally by plants, a situation only achieved with identical genotypes and uniform environments. In unequal sharing of resources, due to environmental, genetic variability, or both, the yield gain from the winners does not compensate for the yield reduction of the losers. The direct effect of interplant competition for resources can be measured by plant-to-plant variability (measured by CV). When variability is high (CV>33%), defined as *negative competition*, the mean and mode of the population are skewed toward the low end of the distribution; when variability is low (CV<33%), defined as positive competition, the mean and mode are skewed toward the high end of the distribution (Fig. 30–5). Genotypes that display *positive competition* are associated with genetic stability (well buffered plants); they resist barrenness and respond to increased plant density with higher yields.

The concept of plant-to-plant competition may be easier to understand using the analogy of aggressive and nonaggressive chickens in a confined space. The presence of aggressive chickens interferes with performance so that productivity per unit area decreases. Thus, the best strategy to increase yield of chickens per unit area would be to select for nonaggressiveness or to use group selection (Muir, 1996).

Factors that enhance the scarcity of resources magnify existing plant-to-plant differences and increase competition. Thus, maximum yield is obtained by genetic and cultural practices that reduce plant-to-plant variability (e.g., through genetic homogeneity and cultural practices that increase uniform germination and

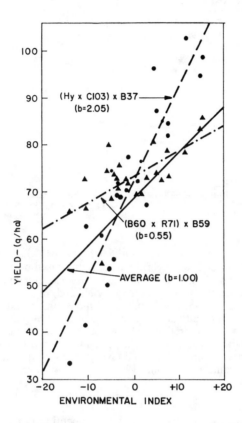

Fig. 30–4. Average response of two three-way crosses in maize to varying environments at two locations. (Hy × C103) × B37 is derived from one-eared inbreds whereas (B60 × R71) × B59 is derived from two-eared inbreds. Note that the two-eared hybrid yielded better under stress conditions. Source: Russell and Eberhart (1968).

uniform growth). Extreme cases of stress (as in high density plantings) result in self thinning and barrenness, when interplant competition is high.

There are no convincing experimental examples that unequal sharing of resources increases yield per unit area. Evidence for the superiority of identical genotypes derives from results of mixed vs. unmixed plantings. When high yielding, high stability genotypes are involved, mixed plantings do not exceed yields of individual genotypes planted alone and are usually less. The increased yielding ability and thus adoption of single-crosses over double-crosses in maize confirm the desirability of uniform genotypes as does the increased performance of F_1 over F_2 populations.

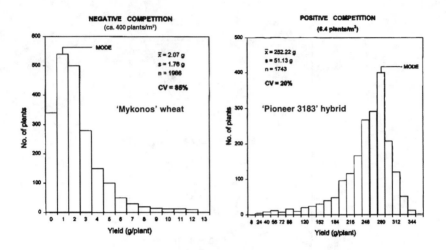

Fig. 30–5. Examples of negative competition ('Mykonos' wheat, left) and positive competition ('Pioneer 3183' hybrid maize, right) based on yield per plant frequency distribution. Negative competition is associated with a CV > 33% with the mean and mode transposed to the low yield end; positive competition is associated with a CV < 33% with mean and mode transposed to the high yield end. Since these two cultivars are homogeneous due to genetic uniformity, the variation is due to acquired differences among plants during development. Positive competition maximizes yield per unit area and is associated with increased resistance to barrenness and a favorable response to high plant population. Source: Fasoula and Fasoula (1997a).

The consequences of this response to competition in plants leads to plant breeding strategies that not only are not self evident but are counter intuitive. For example, it is now well established that the ability of genotypes to compete in a stress environment does not predict their yield performance in uniform stands. It is a truism that genotypes that are high competitors will produce poorly in monoculture; poor yielders in mixed stands out perform the high yielders when planted alone. (High competitors prevent equal sharing of resources and increase interplant variability even when genetically homogeneous because they exploit environmental variation.) This means that maximum yields cannot be achieved with heterogeneous genotypes. It also has serious implications for selection. Selection in heterogeneous genotypes for yield under interplant competition is counterproductive because the high competitors are selected which perform poorly in monoculture (In our chicken example, selection of high performing chickens in a stressed competitive situation would select primarily for aggressiveness.) This may explain why yield has such a low heritability.

MAXIMIZING GENETIC PROGRESS

Techniques to maximize genetic progress for crop yield must include selection for both yield performance and stability from the earliest segregating generations. Fasoula and Fasoula (1997a,b) present five points that need to be considered:

1. Select on a single-plant basis under the target environments with spacing that precludes plant interference. This condition mimics the ideal environment encouraging equal sharing of resources. Under this situation the phenotype comes to maximum expresion, resulting in increased heritability of the trait under selection.

2. Enhance gene fixation either by selfing or by intercrossing selected plants and applying high selection pressure.

3. Practice multi-environment screening, including the target environments, with a high number of replications in order to expose entries to diversity of environmental and cultural practices.

4. Use suitable experimental designs that sample effectively for the diversity encountered across the target environments, allow for effective selection for stability among entries from the earliest segregating generations, and cope with the confounding effects of spatial heterogeneity on single plant yields to allow effective selection within entries. Honeycomb selection designs have been proposed for this purpose (Fasoulas & Fasoula, 1995).

5. The criteria for selection must combine yield potential (mean) and yield stability (phenotypic standard deviation). Fasoula and Fasoula (1997b) provide a prediction criterion (PC) that selects for all categories of cultivars:

$$PC = \bar{x}(\bar{x}_S - \bar{x})s_P^2$$

namely, the product of the standardized entry mean ($\bar{x}/s_P$) as an estimate of yield stability (tolerance to stresses), and the standardized selection differential,

$$(\bar{x}_S - \bar{x})s_P^2$$

as an estimate of yield potential (responsiveness to inputs), where $\bar{x}$ is the entry mean; $\bar{x}_s$ is the mean of the selected genotypes; and s_P^2 is the phenotypic variance of the selected entry. Note that the PC is very similar to Falconer's general equation for the response to selection (R):

$$R = s_g^2(\bar{x}_S - \bar{x})s_P^2$$

except that the mean, $\bar{x}$, is substituted for the unavailable genetic variance, s_g^2. This is made possible by the fact that conditions that maximize the mean, primarily the absence of interplant competition, are those that allow maximum expression of the genetic variance and maximize its contribution to the phenotypic variance s_P^2. In non-mathematical terms, the most effective selection occurs when high performance and low variability are considered at the same time, in the complete absence of interplant competition, from the earliest segregating generations.

STRATEGIES FOR ACHIEVING UNIFORMITY AND STABILITY

Maize

At present, the maize industry in the USA has shifted to single-cross hybrids to maximize yields and increase uniformity. Historical U.S. maize yields from 1865 to the present (Fig. 30–1) indicate that the introduction of the double-cross hybrids and the use of single-crosses, in concert with advances in cultural practices, have each had a major effect on increasing average yields. At the present time, the introduction of stable, single-crosses that produce well under high plant populations rather than increased performance of inbreds is responsible for this increase. This has been achieved by extensive multi-environment testing of single crosses with attention to incremental genetic improvement of hybrids and combining ability.

Although stability has been shown to be an inherited character, stability has not been selected for directly but rather as a side effect of single-cross evaluation. Stability in inbreds can be inferred either from an analysis of their hybrid performance in diallel crosses or directly. Because the number of inbreds is always much less than the potential single-crosses, it would seem that direct selection for inbred stability might be a more efficient procedure. At the present time the yield performance of inbreds has not proceeded as rapidly as the performance of hybrids (Meghji et al., 1984), especially under conditions of stress (D. Duvick, 1997, unpublished data). A more appropriate future strategy may be to concentrate efforts on inbred productivity combined with stability.

Tomato

In the past decade a remarkable shift has occurred in the processing tomato industry where a shift to hybrids has occurred despite their relatively high cost. A kilogram of hybrid tomato seed (352 000 seeds) has a retail value of $704 in California as compared with $88 for open-pollinated seed. The seed costs per hectare using hybrid seed (direct seeded) is about $286 (2.8 ha kg^{-1}) as compared to $88 for open pollinated seed; however the use of transplants significantly brings down the seed costs per hectare because growers can plant about 24 ha kg^{-1} of seed, but this does not change establishment costs. (Note that the cost of hybrid seed for fresh market tomatoes may be $11,000 kg^{-1} while the cost for greenhouse tomato, a much smaller market, may be $22,000 or more. Thus, hybrid seed may be more valuable than gold!)

The value of the hybrids is not explained by yield differences for, in well conducted studies, standard cultivars and hybrids show similar yields (M.A. Stevens, 1997, personal communication); however, the hybrids are more stable than standard cultivars under stress. As a result the yields per unit area in California are now very consistent. Some of this advantage may be due to disease resistance which is easier to combine in hybrids than by pyramiding genes using conventional breeding strategies. As the disease resistance is found in unadapted material, combining multiple resistance results in genetic drag that reduces yield. Future breeding strategies in tomato breeding should include selection for tolerance to stress at the inbred line level.

Onion

Onion is an allogamous species that shows extensive vigor decline when selfed (Pike, 1986). Hybrids are produced using cytoplasmic male sterility, a trait first discovered and analyzed in onion by H.A. Jones and A.E. Clarke (1943). Hybrids have taken over long day onions but open-pollinated cultivars are still prominent in short day onions such as Bermuda, and the Texas Grano types. The open-pollinated cultivars in short day onions are fairly uniform (known as heterozygous uniforms) for traits such as disease resistance, bulb shape, color, and size. Breeders of open pollinated onions, such as Leonard Pike, feel that it is possible to recombine sib lines so as to retain sufficient heterozygosity to maintain vigor while creating homozygosity, and thus uniformity, for selected morphological traits. Texas Grano 10-15, an important short day onion, and the only onion marketed by cultivar name, derives from a single bulb of Texas Grano 951. However, hybrids are increasing in short day onions because yield is higher (M.A. Stevens, 1997, personal communication).

The F$_1$ hybrids on the market are not genetically homogeneous because most inbred lines are not homozygous. The high inbreeding depression in onion greatly reduces seed yield so most hybrid onions are not single-crosses but three-way-crosses in which each parent is only partially inbred. Hybrid onions have taken over the hybrid seed business, in part, because they provide protection to seed producers, but also because yields have been high.

There are some new techniques that promise to increase uniformity. One is rapid cloning of the male sterile parent by tissue culture, which could increase uniformity of the hybrid provided the clone is inbred (L. Pike, 1997, personal communication). Another is the use of doubled haploids from ovule culture to produce inbreds. These doubled haploids are more vigorous than the highly inbred types, probably because the haploid process eliminates deleterious genotypes (M.A. Stevens, 1997, personal communication).

UNIFORMITY VS. STABILITY IN SUBSISTENCE AGRICULTURE

Hybrids have been considered an unwise choice for subsistence farmers for two reasons. The first reason is the requirement for cash outlay for seed. In modern agriculture this argument is specious because the extra cost of hybrid seed provide a yield return that makes the extra cost a good investment. However, subsistence farmers have no resources to invest. The second reason is the supposed fear of higher risk that uniformity imposes in response to severe stress, a risk that would be catastrophic for those existing on the margins. This risk is real if the available hybrids are unadapted but the risk may be negligible with adapted hybrids. Furthermore, risk could be alleviated by planting a range of hybrids to avoid the yield penalty of open-pollinated populations but here the first reason (investment cost) comes into play.

Are modern cultivars inferior in stability to long-established land races? Ceccarelli and Grando (1996) report that this is indeed the case for genotypes in very highly stressed environments. They conclude that new entries selected under well-managed conditions have been superior to local cultivars only under conditions of improved management, but not under extreme low input conditions as in the droughty areas of Africa and the Mideast. This result is that despite the introduction of new cultivars, few are actually grown in these areas. Poor farmers in extreme environments maintain genetic diversity in the form of different crops and heterogeneous cultivars to maximize adaptation over time. Survival in bad years, not yield in good years, is key. The failure of new cultivars under the extreme stressed environments is explained on the basis of "cross over" (nonstable) cultivars, as shown in Fig. 30–6. It assumes that genotypes that perform well under high inputs perform low under low inputs and vice versa. It further assumes that cultivars cannot be transferred across the "box." However, the authors show evidence that yield of selections from local land races [barley, (*Hordeum vulgare* L.)] when selected vigorously under high stress can show genetic improvement providing careful testing is carried out with designs adapted to stress conditions (border effects, large plots, check entries). This is strong evidence for the possibility of incorporating individual buffering into inbred line cultivars.

The issue of where selection should be carried out is contentious. The issue is between those who favor "selection in favorable environments where genetic differences are maximum and environmental noise minimum and those who believe that selection has to be done in the target environment or close to it." The theory for the latter, paraphrased from Falconer (1952), is "to improve performance in environment A, select in environment A." These opposing views on selection may not be contradictory because all would agree that testing must include the target environment

Perhaps the take home lesson is that strategies to increase yields of subsistence farmers in extreme environments cannot be restricted to providing better performing genotypes, especially those adapted only to improved environments, but must be to improve cultural practices at the same time. The unavailability or inability to use superior germplasm by subsistence farmers must be considered another factor that traps subsistence farmers in a downward cycle of poverty relegating

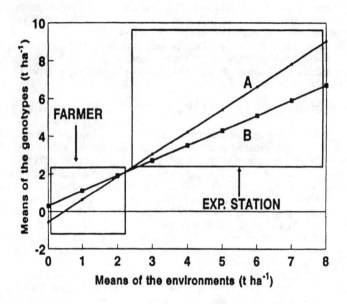

Fig. 30–6. Hypothetical G×E interaction of cross-over genotypes derived from experiment stations and farmers fields under highly stressed, drought environments. A is a typical genotype selected in high yielding environments and B is a typical genotype selected in low yielding environment. Source: Ceccarelli and Grando (1996).

them to the worst of all possible worlds, a poverty from which there is no escape. Potential solutions to this problem must be economic (the better availability of credit) and social (education) as well as technical (improved genotypes and cultural practices).

ACKNOWLEDGMENTS

I thank Donald Duvick, Apostolos C. Fasoulas, Dionysia A. Fasoula, Leonard Pike, and M. Allen Stevens for stimulating discussions.

REFERENCES

Ceccarelli, S., and S. Grando. 1996. Drought as a challenge for the plant breeder. Plant Growth Regul. 20:149–155.

Duvick, D., 1984. Genetic contributions to yield gains of U.S. hybrid maize, 1930 to 1980. p. 15–47. *In* Genetic contributions to yield gains of five major crop plants. CSSA Spec. Pub. 5. CSSA, Madison, WI.

Duvick, D. 1997. What is yield? *In* G.O. Edmeades et al., (ed.) Developing drought and low N-tolerant maize. Proc. Symp., March 25–29. 1996. CIMMYT, El Batan, Mexico.

Eberhart, S.A. 1969. Yield stability of single-cross genotypes. p. 22–35. *In* J.I Sutherland and R.J. Falasca (ed.) Proc. of the 24th Annual Corn and Sorghum Industry Research Conf., Chicago, IL. 9–11 Dec. Am. Seed Trade Assoc., Washington, DC.

Eberhart, S.A., and W.A. Russell. 1966. Stability parameters for comparing varieties. Crop Sci. 6:36–40.

Eberhart, S.A., and W.A. Russell. 1969. Yield and stability for a 10 line diallel of single cross and double cross maize hybrids. Crop Sci. 9:357–361.

Falconer, D.S. 1952. The problem of environment and selection. Amer. Nat. 86:293–298.

Falconer, D.S. 1989. Introduction to quantitative genetics. 3rd ed. John Wiley & Sons, New York.

Fasoula, D.A., and V.A. Fasoula. 1997a. Competitive ability and plant breeding. Plant Breed. Rev. 14: 89–138.

Fasoula, D.A., and V.A. Fasoula. 1997b. Gene action and plant breeding. Plant Breed. Rev. 15: 315–374.

Fasoulas, A.C., and V.A. Fasoula, 1995. Honeycomb selection designs. Plant Breed. Rev. 13:87–177.

Jugenheimer, R.W. 1976. Utilization of inbred lines. p. 375–385. *In* Corn improvement, seed production, and uses. John Wiley & Sons, New York.

Jones, H.A., and A.E. Clarke. 1943. Inheritance of male sterility in in the onion and the production of hybrid seed. J. Am. Soc. Hort. Soc. 43:189–194.

Meghji, M.R., J.W. Dudley, R.J. Lambert, and G.F. Sprague. 1984. Inbreeding depression, inbred and hybrid grain yields, and other traits of maize representing three eras. Crop Sci. 24:545–549.

Muir, W.M. 1996. Group selection for adaptation to multiple-hen cages: Selection program and direct responses. Poultry Sci. 75:447–458.

Pike, L.M. 1986. Onion breeding. p. 357–394. *In* M. Bassett (ed.) Breeding vegetable crops. AVI, Westport, CT.

Russell, W.A., and S.A. Eberhart. 1968. Test crosses of one-and two-ear types of corn belt maize inbreds. II. Stability of performance in different environments. Crop Sci. 8:248–251.

Chapter 31

Apomixis and Heterosis

W. Hanna, P. Ozias-Akins, and D. Roche

INTRODUCTION

Apomixis is a genetically controlled reproductive mechanism that can be used to vegetatively propagate a plant genotype through seed. This reproductive mechanism introduced into a hybrid could make possible a true-breeding hybrid. Through various gametophytic apomictic mechanisms (Asker & Jerling, 1992), an egg cell with an unreduced chromosome number develops into an embryo without fertilization by a sperm cell; however, fertilization of one or more nuclei of the central cell by a sperm cell is usually necessary for endosperm development (a mechanism referred to as pseudogamy). Endosperm development is especially important in grain crops valued for their endosperm that includes some of the major world food crops. It is important to note that although megasporogenesis does not occur (or is not completed) in apomicts, microsporogenesis does occur in apomicts and produces chromosomally reduced sperm cells in the pollen that makes possible the transfer of this mechanism, especially if apomixis is controlled by a dominant gene(s).

Researchers have been aware of apomixis since the mid-1800s (Asker & Jerling, 1992). However, it has been in the past 35 years, as more information has been generated on the genetics and manipulation of this asexual reproductive mechanism, that interest has increased in using it to produce true-breeding hybrids propagated by seeds. The discovery of sexual plants in apomictic species to produce apomictic hybrids (Bashaw & Hussey, 1992), new information on the genetics of apomixis (Nogler, 1984; Asker & Jerling, 1992), progress in transferring apomixis from wild to cultivated species (Asker & Jerling, 1992), progress made in molecular mapping of the gene(s) controlling apomixis (Hanna et al., 1996; Koltunow et al., 1995; Ozias-Akins et al., 1993), and the rapid advances being made in molecular techniques have all contributed to renewed interests and efforts to use apomixis to produce superior hybrids. There are numerous advantages to using apomixis in breeding programs to produce superior cultivars and to produce commercial seed. Many of these have been previously discussed (Hanna, 1995; Hanna & Bashaw, 1987).

FACTORS FAVORING USE IN PLANT BREEDING

Apomictic mechanisms are relatively wide-spread in plants, having been reported in >300 species in at least 35 different plant families (Hanna & Bashaw, 1987). At least three major apomictic mechanisms have been documented: apospory, diplospory and adventitious embryony (Bashaw, 1980).

Apospory is the easiest to identify by the absence of antipodal cells and by embryo sac orientation , shape, and number, especially at anthesis. Megaspores or the products of meiosis degenerate and one or more nucellar cells enlarge to produce embryo sacs with one or more chromosomally unreduced nuclei.

Diplospory is more difficult to identify because mature embryo sacs in ovaries of both sexual and diplosporous plants look identical, with an egg, one binucleate central cell, two synergids, and three antipodal cells (or a proliferation of antipodals). Nuclei in embryo sacs of sexual plants are chromosomally reduced whereas nuclei in embryo sacs of diplosporous apomicts are chromosomally unreduced. Diplospory can be identified by the absence of both meiosis and a linear tetrad of megaspores.

In adventitious embryony embryos develop through mitotic division of a cell in the ovule, integuments, or nucellus and begin as bud-like structures. Endosperm is derived from fertilization of the central cell in sexual embryo sacs of the same ovule.

Apomixis is found in the tertiary gene pools of a number of the major grain crops such as maize (*Zea mays* L.; Dewald et al., 1992); wheat (*Triticum aestivum* L.; Carman & Wang, 1992); and pearl millet (*Pennisetum glaucum* (L.) R. Br.; Hanna et al., 1992). In addition, it is also found in forage and turf genera such as *Paspalum* (Burton, 1992), *Panicum* (Savidan, 1981), *Poa* (Pepin & Funk, 1971) and *Brachiaria* and in *Citrus* and other fruits and nuts (Asker & Jerling, 1992). Facultative apomixis has been reported in a number of genotypes of sorghum [*Sorghum bicolor* (L.) Moench; Schertz, 1992].

It is not necessary to develop inbred lines with the gene(s) for apomixis (by backcrossing) before hybrids can be developed. In fact, the more heterozygous the genotypes that are crossed, the greater the opportunity to develop many new hybrid genotypes from a particular cross. In traditional inbred development for hybrid production, one is testing combining ability between two inbreds(each of which produces gametes with a single genotype). In a heterozygous sexual × heterozygous apomict cross, each parent is producing thousands of genetically diverse gametes. Therefore, combining ability in such a cross is tested between thousands of diverse gametes from each cross. It would be necessary to incorporate the gene(s) controlling apomixis (whether by traditional crossing or by molecular methods) into genotypes with desirable agronomic characteristics. Any single superior plant from a sexual × apomictic cross that reproduces by apomixis has the potential to become a true-breeding cultivar. Any superior apomictic genotype can be crossed as male parent with a superior [or one with desired gene(s)] sexual genotype (preferably from another cross combination or pedigree) to develop new apomictic hybrids and/or to pyramid genes or blocks of genes in a new cultivar (Hanna, 1995).

CHALLENGES

Apomixis provides unlimited opportunities for maximizing crop production through the development of unique true-breeding gene combinations; however, transfer of a gene(s) controlling apomixis into commercial cultivars of our important food, feed, fiber, and oil crops is not without its problems and challenges.

In traditional backcrossing programs, problems in both male fertility [to allow transfer through the pollen of gene(s) controlling apomixis] and female fertility (to allow economic levels of commercial seed production) need to be overcome. Part of the male and female sterility is due to the alien chromosomes (from a tertiary gene pool) introduced into crosses with the cultivated species. The negative effect of this alien germplasm can be reduced if (i) partial homology to allow recombination is present for the chromosomes and/or (ii) translocations can be induced to transfer small segments of the chromosomes between chromosomes from the cultivated and wild species. Inbreeding from backcrossing to a single recurrent genotype from the cultivated species appears to cause problems in maintaining seed set in *Pennisetum* (Dujardin & Hanna, 1989; Hanna et al., 1993). This inbreeding effect can possibly be overcome by using diverse genotypes from the cultivated species in the backcrossing process.

Although information on the genetics of apomixis has been reported in the past 35 years from sexual × apomictic crosses, the lack of F_2 data and the inability to use apomicts as a female parent in crosses,especially when apomixis is dominant, limits to some degree the genetic and linkage data that can be obtained. This information is important in understanding the genetic control of apomixis so that it can be successfully manipulated.

Apomixis is mainly found in polyploids. Many of the agronomic crops are diploid. Will apomixis be expressed at the diploid level? Future experimentation will provide the answer. However, Dujardin & Hanna (1986) and Leblanc et al. (1996) showed that obligate apomixis was expressed in polyhaploids with the gene for apomixis in the simplex condition. Apomixis may be easily used in polyploid crops such as wheat, peanut, soybean and cotton. For diploid crops, it may be necessary to explore the possibility of using the gene(s) for apomixis in induced tetraploids if polyploidy will be necessary for the expression of apomixis.

Molecular markers linked to apomixis are being identified in *Pennisetum* (Ozias-Akins et al., 1993), *Tripsacum* (Leblanc et al., 1995), and *Brachiaria* (Pessino et al., 1997; Miles et al., 1994). In *Pennisetum*, 12 PCR-amplifiable markers strictly cosegregate with the trait and define an apospory-specific region. The lack of recombination between the trait and these markers in a large population could indicate that the gene(s) controlling apomixis are part of a complex locus. Physical mapping methods are being used in *Pennisetum* for characterizing the locus so that it can be cloned and inserted into other species.

USING APOMIXIS TO IMPROVE PRODUCTION

Apomixis is in the early stages of making a contribution to crop improvement. Bashaw and Hussey (1992) reported on the use of apomixis to produce apomictic forage hybrids in *Cenchrus* by traditional crossing methods. Hanna et al. (1993, 1996) reported on the progress being made in transferring apomixis from *Pennisetum squamulatum* to cultivated pearl millet. Major efforts are also underway to transfer gene(s) controlling apomixis from a wild to cultivated species in maize (Kindiger et al., 1996; Leblanc et al., 1996; Savidan et al., 1993) and wheat (Carman & Wang, 1992). The citrus industry uses apomixis to produce virus-free rootstock.

Apomixis has the most immediate potential to improve production in crops where efficient systems for producing hybrids are not available. In crops (such as maize, sorghum, and pearl millet) with efficient cytoplasmic-nuclear male sterility systems, apomixis would simplify hybrid seed production and allow breeders to produce and use unique gene combinations to maximize yield (Hanna, 1995). It would even make a major contribution in (i) vegetatively propagated crops (such as potato) where the seed generation could help eliminate viruses and simplify transportation of seed-stocks and (ii) tree species where unique genotypes could be fixed and easily propagated. In developing countries apomixis would make a major contribution in most crops.

High-yields and uniformity are two characteristics that come to mind when the term hybrid is used. Although high-yield is a characteristic that may be universally needed or desired, uniformity may not be necessary or even desirable for most crops grown around the world. It is possible to use apomixis to maximize yield and produce any level of uniformity that may be needed or desired.

Apomixis allows one to maximize yields because any superior genotype from a sexual × apomictic cross that reproduces by apomixis (we will refer to obligate apomixis unless otherwise indicated) can be maintained and is ready for performance testing. Maintaining the superior genotype is not dependent on the homozygosity or heterozygosity of the genotype. The main way to release the genetic variation in an apomictic genotype is by crossing the apomict as pollinator with a sexual female (with outstanding performance, when possible) to produce new superior gene combinations or by using it to pollinate a facultative (reproduces by both apomixis and sexually) apomict.

Levels of Apomixis

Obligate apomixis is a term used to describe plants that reproduce only by apomixis. Technically, there are few obligate apomictic plants because off-type plants, due to some sexuality and/or fertilization of an unreduced egg (referred to as B_{III} hybrids), can be found if large enough populations are observed in most apomictic species (Asker & Jerling, 1992). A few off-type progeny may prevent a plant from being classified as obligate, but practically, such a plant can function as an obligate apomict with little or no detrimental effect on the cultivar. Obligate apomixis is generally considered more desirable if uniformity is important. Additionally, it preserves more of the vigor of the hybrid.

Facultative apomicts reproduce by both apomixis and sexuality. The frequency of maternal progeny produced by facultative apomicts can range from low to high. Facultative apomixis can be effectively used to fix various levels of heterosis (to be discussed later) when complete genetic homogeneity and morphological uniformity are not necessary. Pepin and Funk (1971) were able to increase the levels of apomixis in Kentucky bluegrass by crossing selected facultative apomicts.

Breeding Approaches

Obligate apomixis allows a breeder to rapidly produce numerous stable genotypes, regardless of heterozygosity, that are ready for performance evaluation. A sexual × apomictic cross will segregate for method of reproduction when apomixis is controlled by a dominant gene because the apomict will always be heterozygous in most natural crossing situations. If apomixis is controlled by a recessive gene(s), a sexual × apomictic cross will segregate for method of reproduction if the sexual plant is heterozygous for method of reproduction, but all progeny from such a cross will be sexual if the sexual plant is homozygous for sexuality.

A number of breeding procedures have been described for utilizing apomixis to produce superior cultivars (Bashaw & Funk, 1987; Burson et al., 1984; Burson & Hussey, 1996; Hanna, 1995; Nakajima, 1990; Savidan, 1981; Taliaferro & Bashaw, 1966). Hanna (1995) and Hanna and Bashaw (1987) have previously discussed the benefits of apomixis in commercial seed production. In this chapter, we want to elaborate on the use of apomixis to increase crop production in developing countries.

It is well documented that use of hybrid vigor is an effective way to maximize production; however, one of the arguments for not using hybrids in developing countries is that they reduce the genetic diversity in farmers' fields, which increases the vulnerability to changes in environmental conditions. Another argument is that farmers cannot afford to purchase seeds of hybrids each year. In our opinion, apomixis provides a viable solution to both of these situations.

Genetic Diversity-Obligate Apomixis

Maintaining genetic diversity, which may help to overcome environmental stress (such as drought) and various pests, is an important consideration in cultivar development, especially in developing countries. For example, variation for maturity may help escape the effects of drought at a critical time in the development of the plant.

Apomixis allows one to develop a true-breeding hybrid population consisting of one to an unlimited number of genotypes. An apomictic population could be considered a variety with unique hybrid genotypes. This can simply be accomplished by mixing equal quantities of seed of the apomictic hybrids that are to make up the population before the population is commercially increased. Adjustments for reproductive potential (e.g., size and number of inflorescences produced, and seed size) would need to be made if one wanted each genotype equally represented in the hybrid population. One also could develop populations with higher frequencies of certain genotypes represented, by blending more seeds of certain genotype(s) before commercial increase. Alternately, one also could commercially increase the individual apomictic hybrids and blend the seeds of the hybrids before the hybrid population is sold to the farmer.

Where do all of these apomictic genotypes come from? Can they be produced fast enough? They should be easily produced from sexual × apomictic crosses. The number of superior apomictic genotypes produced will mainly be limited by the frequency of desirable genes in the parents and the size of the populations that are evaluated from each cross. We can foresee gene banks (private and public) with thousands of apomictic hybrids for a crop where each hybrid has been evaluated for agronomic characteristics such as maturity, height, seed traits, pest resistance, response to environment, etc, and can be readily identified from data bases. Diverse hybrid populations can be developed that have any combinations of hybrids that are diverse for certain characteristics and uniform for others.

These apomictic hybrid populations would maintain genetic diversity and at the same time allow farmers to reap the benefits of hybrid vigor. This approach would work for both self- and open-pollinated crops. The frequency of the hybrids in the population in the farmers' field, for a locality or for an environment may shift over time. This would be advantageous to the farmer if the genotypes with a performance advantage for a particular situation increase in frequency. If the population shift of the hybrids is undesirable for some reason, the farmer could purchase the original population again.

If genetic diversity is important in farmers' fields, another approach to using apomictic hybrids, especially in open-pollinated species (but also in self-pollinated species with some cross-pollination or with an introduced genetic male sterile to enhance cross-pollination), would be to introduce the gene(s) controlling apomixis into a diverse population or landrace that reproduces sexually. Sexual × apomictic crosses would produce sexual and apomictic hybrids that probably would have a selective advantage in the population or landrace. The best sexual and apomictic hybrids adapted to the particular environment would increase in frequency and interpollinate to produce a superior population or landrace. Higher seeding rates could be used in any cycle in the population development to eliminate weaker and/or non-hybrid plants. This approach would maintain the genetic diversity of the population or landrace and increase its performance through continued recombination and fixation of superior genotypes. A dominant gene(s) would fix genotypes at a more rapid rate than a recessive gene (Hanna, 1995).

Genetic Diversity-Facultative Apomixis

In general, a dominant gene controlling obligate apomixis would be the most efficient to use in maximizing the opportunity to develop superior cultivars, especially if sexual counterparts are available for crossing with the obligate apomict. In reality, various levels of sexuality can be found in most apomicts. Although the benefits of apomixis decrease as the level of facultative apomixis increases, any level of apomixis that increases production is beneficial if morphological and maturity variation can be tolerated.

Less than 95% apomixis in commercial production of cultivars may be undesirable because it could cause production and cultivar identification problems; however, a gene(s) controlling facultative apomixis introduced into a farmer's field in a developing country may be welcome because some hybrid vigor is better than none.

SUMMARY

Of the major world agronomic food crops, possibilities for transferring apomixis from wild to cultivated species exists for maize, wheat and pearl millet.   Research programs on a number of crops are attempting to map and clone genes controlling apomixis.   Success in cloning genes controlling apomixis will open the opportunities for widely using this reproductive mechanism in many crops.

There is probably no single characteristic that could have a greater impact on increasing food, feed, oil and fiber production around the world than apomixis.   The opportunities and possibilities that apomixis provides for developing superior cultivars are almost beyond imagination and its advantage in effectively utilizing heterosis is quite apparent.

ACKNOWLEDGMENTS

The authors gratefully acknowledge partial support by U.S. Department of Energy Contract DE-FG05-93ER20099, Rockefeller Foundation grant GA-AS-9016, and USDA-NRICGP grant 93-37304-9363.

REFERENCES

Asker, S.E., and J. Jerling. 1992. Apomixis in plants. CRC Press, Boca Raton, FL.

Bashaw, E.C. 1980. Apomixis and its application in crop improvement. p. 45–68. *In* (ed.) W.R. Fehr, and H.H. Hadley Hybridization of crop plants, ASA, Madison, WI.

Bashaw, E.C., and C.R. Funk. 1987. Breeding apomictic grasses. p. 42–82. *In* W.R. Fehr (ed.) Principles of cultivar development: crop species. MacMillan, New York.

Bashaw, E.C., and M.A. Hussey. 1992. Apomixis in *Cenchrus*. p. 1–4. Proc. Apomixis Workshop, Atlanta, GA. 11–12 Feb. 1992. Nat. Technical Inform. Serv. Springfield, VA.

Burson, B.L., P.W. Voigt., and E.C. Bashaw. 1984. Approaches to breeding apomictic grasses. p. 14–17. *In* Proc. 40th Southern Pasture and Forage Crop Improvement Conf., Baton Rouge, LA.

Burson, B.L., and M.A. Hussey. 1996. Breeding apomictic forage grasses. p. 226–230. *In* Proc. 1996. American Forage and Grassl. Council, Vancouver, British Columbia.

Burton, G.W. 1992. Manipulating apomixis in *Paspalum*. p. 16–19. *In* Proc. Apomixis Workshop, Atlanta, GA. 11–12 Feb. 1992. Nat. Technical Inform. Serv. Springfield, VA.

Carman, J.G., and R.R.C. Wang. 1992. Apomixis in the Triticeae. p. 26–29. *In* Proc. Apomixis Workshop, Atlanta, GA. 11–12 Feb. 1992. Nat. Technical Inform. Serv. Springfield, VA.

Dewald, C.L., P.W. Voigt, and B.L. Burson. 1992. Apomixis in *Tripsacum*. p. 43–48. *In* Proc. Apomixis Workshop, Atlanta, GA. 11–12 Feb. 1992. Nat. Technical Inform. Serv. Springfield, VA.

Dujardin, M., and W.W. Hanna. 1986. An apomictic polyhaploid obtained from a pearl millet × *Pennisetum squamulatum* apomictic interspecific cross. Theor. Appl. Genet. 72:33–36.

Dujardin, M., and W.W. Hanna. 1989. Developing apomictic pearl millet-characterization of a BC$_3$ plant. J. Genet. Breed. 43:145–151.

Hanna, W.W. 1995. Use of apomixis in cultivar development. Advances in Agron. 54:333–350.

Hanna, W.W., and E.C. Bashaw. 1987. Apomixis: its use and identification in plant breeding. Crop Sci. 27:1136–1139

Hanna, W.W., M. Dujardin, P. Ozias-Akins, and L. Arthur. 1992. Transfer of apomixis in *Pennisetum*. p. 30–33. *In* Proc. Apomixis Workshop, Atlanta, GA. 11–12 Feb. 1992. Nat. Technical Inform. Serv. Springfield, VA.

Hanna, W.W., M. Dujardin, P. Ozias-Akins, and L. Arthur. 1993. Transfer of apomixis in *Pennisetum*. p. 25–27. *In* K.J. Wilson (ed.) Proc. International Workshop on Apomixis in Rice, Changsha, PRC, 13–15 Jan. 1992, CAMBIA, Canberra, Australia.

Hanna, W., M. Dujardin, P. Ozias-Akins, Ed. Lubbers, and L. Arthur. 1993. Reproduction, cytology and fertility of pearl millet × *Pennisetum squamulatum* BC$_4$ plants. J. Hered. 84:213–216.

Hanna, W., D. Roche, and P. Ozias-Akins. 1996. Use of apomixis in crop improvement- traditional and molecular approaches. *In* Proc. Third International Symp. on Hybrid Rice, Hyderabad, India, 14–16, Nov, 1996.

Kindiger, B., D. Bai, and V. Sokolov. 1996. Assignment of a gene(s) conferring apomixis in *Tripsacum* to a chromosome arm: Cytological and molecular evidence. Genome 39:1133–1141.

Koltunow, A.M., R.A. Bicknell, and A.M. Chaudhury. 1995. Apomixis: molecular strategies for the generation of identical seeds without fertilization. Plant Physiol. 108:1345–1352.

Leblanc, O., D. Grimanelli, D. Gonzalez-de-Leon, and Y. Savidan. 1995. Detection of the apomictic mode of reproduction in maize-Tripsacum hybrids using maize RFLP markers. Theor. Appl. Genet. 90:1198–1203.

Leblanc, O., D. Grimanelli, N. Islam-Faridi, J. Berthaud, and Y. Savidan. 1996. Reproductive behavior in maize-*Tripsacum* polyhaploid plants: implications for the transfer of apomixis in maize. J. Hered. 87:108–111.

Miles, J.W., F. Pedraza, N. Palacios, and J. Tohme. 1994. Molecular marker for the apomixis gene in Brachiaria. P. 51 *In* Abstract in Proc. Plant Genome II. San Diego, CA. 24–27 Jan. 1994.

Nakajima, K. 1990. Apomixis and its application to plant breeding. p. 71–92, *In* Proc. Gamma Field Symp. No. 29. Ohmija-machi, Ibaraki-ken, Japan.

Nogler, G.A. 1984. Gametophytic apomixis. p. 475–518. *In* B.M. Johri (ed.) Embryology of Angiosperms. Springer-Verlag, New York.

Ozias-Akins, P., E.L. Lubbers, W.W. Hanna, and J.W. McKay. 1993. Transmission of the apomictic mode of reproduction in *Pennisetum*: Co-inheritance of the trait and molecular markers. Theor. Appl. Genet. 85:632–638.

Pepin, G.W., and C.R. Funk. 1971. Intraspecific hybridization as a method of breeding, Kentucky Bluegrass (*Poa pratensis* L.) for turf. Crop Sci. 11:445–448.

Pessino, S.C., J.P.A. Ortiz, O. Leblanc, C.B. do Valle, C. Evans, and M.D. Hayward. 1997. Identification of maize linkage group related to apomixis in *Brachiaria*. Theor. Appl. Genet. 94:439–444.

Savidan, Y. 1981. Genetics and utilization of apomixis for the improvement of guineagrass (*Panicum maximum* Jacq.). p. 182–184, *In* Proc. XIV Int. Grassl. Congr., Westview Press, Boulder, CO.

Savidan, Y., O. Leblanc, and J. Berthaud. 1993. Progress in the transfer of apomixis in maize. p. 101. *In* Agronomy Abstracts, ASA, Madison, WI.

Schertz, K.F. 1992. Apomixis in sorghum. p. 40–41, *In* Proc. Apomixis Workshop, Atlanta, GA. 11–12 Feb. 1992. Nat. Technical Inform. Serv. Springfield, VA.

Taliaferro, C.M., and E.C. Bashaw. 1966. Inheritance and control of obligate apomixis in breeding buffelgrass, *Pennisetum ciliare*. Crop Sci. 6:473–476.

Chapter 32

Commercial Strategies for Exploitation of Heterosis

Discussion Session

QUESTIONS FOR D.N. DUVICK

T. Wehner, North Carolina State University, USA: Some groups are complaining about hybrid maize as an example of corporations forcing their systems on captive growers. How do you explain their negative view of private maize breeders?

Response: Well I could say, facetiously, maybe they are jealous of the success of the seed companies. Seriously, I believe that this point of view is an example of how some people believe the agricultural economy (or perhaps any economy?) operates. They believe that when a for-profit organization succeeds in making a profit, the profit must have been extracted from its customers, who now have less of their own because the company made a profit. I have tried to point out in my talk that unless the customer also makes a profit, that is, unless the customer makes extra income because of hybrid seed purchases, there will be nothing in it—no more sales—for the commercial seed companies, assuming a competitive market for seeds. But this fact of life seems to be ignored by those who argue against the commercialization of the seed business. I suppose there may be other reasons for their concern, also, but this is the one that comes to mind.

A. Dogra, University of Missouri, USA: Can you explain why the apomictic hybrids are expected to have a shorter life span?

Response: I am rather glad that question was asked because perhaps I did not make myself clear. The shorter life span is not because the hybrids are apomictic, but rather because if current plans are followed, they are most likely to be grown in the tropical and subtropical regions; these regions have many more problems with disease and insects than do the temperate regions, where hybrid maize (with conventional hybrids) has had its greatest success to date. I pointed out that even in the temperate regions there are cycles of reduction in number of hybrids, increase in pests that attack those hybrids, and then need for replacement of the hybrids, and I would expect that this problem would happen more in the tropical regions. And I would say that this would be a problem for not only apomictic hybrids; it also will be a problem—and I think it already is a problem—for the conventional maize hybrids in the tropics.

Yitbarek, Institute of Tropical Crop Science, Ethiopia: What do you think are the best commercial strategies for exploitation of heterosis in the sub-Saharan African countries? The majority of areas of crop production are less than one hectare, and these areas are subject to unfavorable environmental conditions, especially rainfall.

Response: I would guess that there will be many places for which hybrid seeds will not be developed for a long time, if ever, particularly in special environmental niches with a small demand for seed. It simply will not be within the capabilities of a relatively small number of professional breeders, whether public or private, to develop hybrids that fit each of those many environmental niches. I think that if there really is a need for genetic adaptation to these niches (and some niches may be more imagined than real) local varieties are going to be needed for those areas for a long time to come. I am speaking, now, of small environmental niches, not of small farms. How about small farms? I have seen areas in other parts of the world where the family operated a very small area of land, one hectare or less, and yet they successfully bought seed and grew hybrid maize. Here, I think, the biggest problems will be on-time delivery of needed amounts of the right kind of seed, and then, of course, finding a market for the crop when it is harvested, whether for home consumption or sale. I think that in regions like sub-Saharan Africa, seed supply and market stability will be more important as limiting factors, than farm size or yield potential.

M. Goodman, North Carolina State University, USA: Should public US institutions get out of inbred line release work?

Response: Personally, I do not think so - but I will qualify that statement. I think that a certain amount of work to carry the breeding process to its end—developing and releasing inbreds that work well in hybrids—is necessary to keep breeders aware of exactly what is needed in the final product as well as in the basic research that underlies product development. This need exists even for those whose mission is primarily to develop breeding technology, or to conduct genetics research related to hybrid work. I will qualify my statement even further, and say that, in my opinion, not every publicly employed maize breeder and geneticist needs to develop and release inbreds. But I am pleased to see at least some of the breeders and geneticists in the public institutions developing and releasing inbred lines. You know, they might even teach commercial people a thing or two, once in a while!

V. Ahuja, Indian Agricultural Research Institute, India: Is there any male sterility being used in the USA for maize seed production? If yes, what kind and what type?

Response: Many of you may know, I have been retired for seven years now, and I am increasingly getting behind on what is going on in the real world. But as far as I know, yes there is cytoplasmic male sterility being used, and I presume it is of the two kinds that were not susceptible to the race T epidemic: the S cytoplasm and the C cytoplasm and variations of them. I have never heard of any entirely new type of cytoplasm being used. I see one of the commercial people nodding his head, so I guess I have not been too far off, in my reply.

S. Pandey, CIMMYT, Mexico: What are the main reasons for the large difference in the rate of adoption of hybrids during the first thirty years in developing and developed countries, 1960 to 1990 in the LDCs, 1930 to 1960 in the USA?

Response: This became a little more clear to me recently as I thought about it when writing a review paper on growth of the hybrid seed industry in the USA. First, why did hybrids come in when they did in the USA? I will give you a personal example. I can remember when my father first switched to hybrid seed in the early 1930s. This was the height of the Great Depression, as we called it, and like most farm families we were very poor, maybe not compared to developing country poor, but we were poor. And yet my father decided to spend money on hybrid seed. I wondered why. Well, for one thing, the hybrid yielded grain in a severe drought the first year that he grew it when his open pollinated variety did not yield any grain at all. That was a fairly convincing yield trial. The hybrid showed much more than a 15% gain. But more importantly, at just about that time, a subsidy of the maize price was introduced by law in the USA. And I have an idea that he and a lot of farmers decided it was worth it to invest in hybrid seed to grow more corn because they were now guaranteed some sort of a price for their product. And I have an idea that the inverse of this may be the reason in many developing countries why farmers decided it was not economically worthwhile to invest in hybrid seed. They could not depend on a decent price for their product, or for any surplus over their immediate needs for food. That is one thought. In some of the developing countries, of course, the tremendous instability of the country due to war, revolution, so forth, certainly destabilized the market, so the farmers were hesitant to invest in anything, including hybrid seed. Perhaps the recent rise in popularity of hybrid maize in many developing countries is a barometer of better financial stability for the farm sector in those countries, as well as of greater demand for maize.

B. Bowen, Pioneer Hi-Bred International, USA: Do you think heterosis could be improved in the future through genetic manipulations in genes responsible for its manifestation?

Response: Well, I will put it this way, I hope so. First, of course, one has to find some genes that act on heterosis as such. And I have indicated, a couple of days ago, that there is confusion in the minds of most of us as to when we are talking about heterosis itself, and when we are talking about something that increases yield but is not what we define as heterosis. But I really do think that as we learn more about some of the major genes for yield and yield stability (and I think yield stability is in some ways more important than just yield per se) —as we learn more about those genes, we will learn how to manipulate them, and some of them may cause this magical thing, heterosis. We must continue to search, and continue to experiment.

A. Ortega, CIMMYT, Mexico: Could you please comment on what may be appropriate differential prices in double, three-way, and single-cross maize hybrids?

Response: Well here I am afraid I simply cannot. I am too many years away from the business to remember those numbers. Perhaps some of the other people in our panel might tackle that question later on. I cannot even remember what the differences were, except that I remember when single crosses came in, there was a very large price increase that was instituted. The higher price was established partly because it cost more to produce the seed of inbreds. But also, it was more risky to produce seed on inbreds, so an extra amount was added to the total as sort of an insurance policy for the seed companies. The seed companies thought the farmers surely would not pay the high price—and then we could go back to the more comfortable practice of producing double cross or three-way cross seed—but the farmers did pay, and the seed companies, in the end, were happy that it happened. I suppose the farmers gained more than the extra yield of single crosses than they paid out in higher seed prices. But other than that recollection, I am sorry to say that I cannot answer this question.

A. Ismail, Maize Program, ARC, Egypt: Small farm size should not limit the use of hybrid maize. In Egypt using single and three way-crosses, maize productivity increased to 7.1 tons per hectare. The only limiting factor would be the market price. Would you agree?

Response: I would agree with that statement. I think it is interesting to speculate on why some places where the farms are very small, hybrids, or improved seeds for that matter, have not been successfully adopted, and in other places, as in Egypt, they have. I have seen the same sort of things in very small farms in India for example, and there is the example that Dr. Geiger mentioned, about Zimbabwe. He pointed out there that the very small farmers, under low yield conditions, moved to hybrids successfully and quickly.

G. Granados, CIMMYT, Mexico: You mentioned in your closing remarks that private seed companies will eventually take over some of the responsibilities of public breeding programs. In the case of maize, could you elaborate on this statement?

Response: Well, I think Dr. Hunter and others could talk about this as well, or better than I, but I will say that I did point out the probability of considerable shift from public to private research in developing countries, as the seed maize industry matures. The development of a commercial hybrid seed industry will depend to a large extent on the economics, but also on many other factors, in any particular country. In general, it seems that as the use of hybrids increases, and if the volume of hybrid use is large enough, private companies that previously had been exclusively producing and selling seed of hybrids produced by the public sector, will begin to feel it worthwhile to invest in their own research and development. This effort can give them a competitive advantage and they also can more precisely tailor hybrids to their own needs. This seems to be a gradual process, and I am quite sure that it will vary from country to country in how fast it goes and, for that matter, how far it goes.

L. Machida, Crop Breeding Institute, Zimbabwe: You expect hybrids from apomixis to have a shorter life span in the tropics due to biotic pressures. Are you aware that the hybrids SR-52 and R-201 each had a lifespan of at least 20 years under wide cultivation? Do not you expect this to occur with apomictic hybrids? If it occurs would seed production of apomictic hybrids be sustainable?

Response: Well, it is very interesting to know that in some conditions there have been very long lasting hybrids and varieties in tropical countries. I have heard that too. I would suppose that this depends on the particular type of environment where they are grown. I do not pretend to be enough of an expert to go into that. But there is no reason to think that this cannot happen sometimes. On the other hand, I know of other examples where a disease pressure changes so rapidly that it is very difficult for breeders to keep up. So it will doubtless be a mixture of both kinds of longevity, depending on the region. Regarding the last part of the question, "would production of apomictic hybrids be sustainable?", I am not sure I understand the meaning of that, but I think perhaps the idea is there must be some places where the breeders will not have to move fast to stay ahead of diseases and insects with regard to populations of apomictic hybrids that they have released. Apomictic hybrids (and therefore, also conventional hybrids) would have longer lifetimes before specially-adapted diseases and insects caught up to them. Of course, we must remember that if new, more productive hybrids of either type are developed and placed on the market, they will replace the older hybrids even in the absence of disease and insect pressure. The chief threat to old hybrids is new ones, in many parts of the world.

B. Rana, AICSIP, National Research Centre for Sorghum, India: Fixation of heterosis through apomixis is amazing. How much private sector investment is expected to be protected if they endeavor to develop apomictic hybrids?

Response: I am not sure I understand this question. On the one hand the question may be will the private sector use intellectual property rights to protect their investment in breeding apomictic hybrids? And I did say that I would expect that they would; they would have to. One might ask, however, "Is the private sector threatened by the introduction of apomictic maize hybrids?" And I think the answer to that would be "No they are not", because in the areas where commercial sales of apomictic hybrids might succeed, intellectual property rights would be used by industry, as they now do for their soybean and wheat varieties in certain farming economies. On the other hand, there is a vast area where the private sector has no place in the foreseeable future. These are the regions with very poor farmers that cannot afford to buy seed of any sort. And in this area, if apomictic hybrids work, they will not interfere with the business of the hybrid sector, since the private sector does not - and cannot - do business in those places. In fact, successful apomictic hybrids in such places probably will help private industry in the long run, because they will raise the general level of agricultural productivity and well-being, particularly among poor farmers. As these farmers become more prosperous, they will be natural customers for the commercially available hybrids, either apomictic or conventional, if the commercial hybrids provide needed superior traits.

A. Hernandez, Instituto de Investigation y Capacitation Agropecuaria, Mexico: To make attractive the wheat hybrid seed production, how much should this seed cost compared with the seed of pure varieties?

Response: I would suppose that hybrid wheat should be priced at a level where its extra expense (compared to the cost of pure-line variety seed) would equal not more than one-third of the extra income that the farmer could expect to get from planting the hybrid seed. In other words, on a per hectare basis, the extra yield that could be expected when the hybrid is harvested should be worth about three times the extra cost of the hybrid seed.

R. Paliwal, CIMMYT, USA/India: I assume from your talk that apomictic hybrids are likely to become a possibility sooner than generally believed or expected. Is this correct? Are any private companies involved in research on apomictic hybrids other than with forage and lawn grasses?

Response: I do not think apomictic hybrids will be available sooner than expected. In fact, recent research discoveries indicate more, rather than less, time will be required. But if funding and enthusiasm by researchers continue, the genetic problems look like they can be solved. I do not know of any seed company in-house research programs on apomixis, but I have been told that some of them are interested in its possibilities, long-range.

QUESTIONS FOR R.B. HUNTER

M. Lee, Iowa State University, USA: Do the commercial strategies for exploiting heterosis include considerations for preparing/developing the human resources in-house, for example, in-house graduate assistantship, internships, etc. Is there an industry wide strategy or plan? Are there any concerns about the sources of new persons?

Response: It is a concern I think. I have never heard anybody in the industry yet say that they do not strongly support the need for training. And the industry can do some training, but we certainly prefer the good graduate students that earn their Ph.D. and M.S. degrees from universities. And I think the universities can certainly count on industry support to try and maintain that, because I do not think we are in a position, nor do we want to have to train our own breeders. We will obviously do some training, continual training as people join any organization.

R. Paliwal, CIMMYT, USA/India: In several countries the private sector involvement in crop research is encouraged to reduce the public sector investment in crop research. Do you consider a reduction in public sector research investment a justifiable cause for promotion of private research?

Response: Well I think it is a difficult question. I think there are lots of opportunities for long periods of time for both to coexist and perform a function. I think if the public sector continues to try and service the seed part, you know the cultivar part of the market, for a long time with a major effort, it will discourage private sector involvement, but there is lots of room for special trait developments and breeding technology approaches. In Thailand, for example, the public sector helped supply some of the technology that can be used by the private sector to enhance their programs, genotyping, for example. So I think there are lots of opportunities for cooperative research, and I do not think it is necessary for the public sector to disappear for the private sector to be successful. I think they just need to work together. How they are going to work together is going to depend on the phase of the time frame of the development of the private sector. When it is starting off, it is going to need a lot of help from the public, and later on, it is going to need a different kind of help.

N. Singh, IARI, India: Most of the developing countries are located in tropical parts of the world. About 60 to 80% of the areas of these countries are prone to abiotic and biotic stresses, coupled with the early maturity resulting in very low yields. The only sector left to address these areas is the public sector. What kind of linkages between public and private sector do you envision in the coming years?

Response: First of all, I do not think all segments of crop improvement can be handled satisfactorily by the private sector. There are some segments where the private sector just cannot do a good job and survive. And it is better that they stay out of those areas. That may be small area crops, or those with special requirement, whatever it may be. I do think though that we have a major challenge ahead in terms of increased productivity, and I think there could be all kinds of linkages between the public and private sectors. It is a terrible waste of resources to not complement each other. I think there is lots of room for complementarity, and I think it is a shame not to pursue those opportunities, because resources will be wasted if we do not. We just cannot afford that in the future.

N. Singh, IARI, India: Do you think PVP is feasible in countries where farmers have very small holdings, and it will be impossible to stop them from replanting seed or supplying them to other farmers, no matter how much governments try to stop them? Farmers' rights, in this regard, should not be ignored. Governments must frame suitable laws to protect the interests of both farmers and the seed sector.

Response: In essence, the concept of plant variety protection is a balance between farmers' rights and the inventors' rights. And that is why, instead of just using the normal patent process, which was not developed specifically for the protection of intellectual property of plant breeders, we have developed plant protection laws. For example, the Plant Variety Protection Act in the USA allows a farmer to plant back seed of the crop he grows. What he cannot do is sell it across seven states on a half a million hectares. So he cannot get into the seed business with someone else's variety. The protection systems have evolved from limited protection to more protection because of the much larger investment in private sector research today. Twenty years ago a grower could buy seed of a soybean variety, and he could sell seed from the crop he produced to his neighbors. And that was not really outside the law. The point is that we must take into account plant breeders' rights and the farmers' rights.

A. Charcosset, INRA, France: Public research has provided private companies with elite germplasm and still does so in a number of cases. Is there a general reflection about the possible feedback to allow public research to develop methodologies using germplasm of actual agronomic value?

Response: I think the idea here is that the private sector supply germplasm for the public sector to work on in terms of developing methodologies. And I think that is something that several groups are struggling with. As I understand it, that is one of the aspects of the GEM project in the USA, that is taking some open-pollinated material, but also getting material from private industry and using both as part of the base of new material. I think that the private industry has been quite receptive to it. It has been receptive, obviously, with the proviso that a procedure is established so their lines do not get into the hands of the competition. The private germplasm, in certain forms, can be passed back to the public sector so they can do research with it. I think that most companies would be willing to support that concept.

Response (A.B. Maunder): If I could add to that, it also could affect genetic resources. There are many cases in which the private sector for one reason or another may have access to some especially good early generation genetic resource material, and I think that should very definitely be shared, not only with the public sector, but also deposited in national seed storage facilities.

Response (D.N. Duvick): I think the GEM project indeed is a good example of germplasm sharing among private and public institutions. I think GEM stands for "Genetic Enhancement for Maize." It is centered at Iowa State University, but I believe the USDA has initiated it. Essentially the private industry crosses some of its better inbred lines to good exotic open-pollinated varieties, then makes resulting breeding material available to all participants in the program, private and public.

QUESTIONS FOR A. B. MAUNDER

R. Velasquez, Guatemala: What priorities would you give to these three factors for the development of the seed industry: strategies and politics of the official or government sector, capital needed for inventory, and genetic or technical information?

Response: I would say the capital to get started, and then the problems that might be internal within the government situation, and finally the genetic or technical information.

P. Wilson, Hybrid Wheat Australia, Australia: It is nearly 40 years since extensive breeding commenced on hybrid wheat, yet the area grown is insignificant. This contrasts with other self-pollinating crops such as sorghum or rice. Failure to produce seed at a competitive price has been frequently suggested as the main reason for the failure to develop successful hybrid wheat. Can you suggest methods to improve seed production in wheat—a self pollinating crop with a high seeding rate and a low "multiplier" ratio?

Response: Good question, Peter. I think that I would like to look at it from the other way around. I think the better chance would be to get a better level of heterosis somehow in the cross, or by adding value-added traits that are not easy to obtain through conventional breeding. Maybe you need to go from a C-3 to a C-4 plant. The work at CIMMYT on spring by winter shows some real exciting things and may be one way to eventually get the female yield up. I do not think we have to worry about male parent yield as much as we did in the early days, because we are really looking more for hybrid vigor than we are for that pollinator to be high yielding. Maybe another direction we can go is to look more for the combining ability as opposed to the yield on the male side. But I am afraid that just to answer your question *per se*, I really do not have a real good suggestion right now.

P. Peterson, Iowa State University, USA: What percentage of sales of hybrid seeds end up in litigation for collection?

Response: I think that depends on how good a person you have chasing down accounts. In the USA I would say it is a very low percentage. Anybody else have a thought of percent of sales ending up in litigation, but not being paid?

Response (R.B. Hunter): You mean in terms of collecting? Pretty small I think....very, very small.

QUESTIONS FOR J. JANICK

W. Zhang, CIMMYT, Mexico: It is an interesting statement that we should not select for aggressive chickens which may reduce the final production because the aggressive ones will take advantage of the non-aggressive ones and finally give low yield. But if we select aggressive ones and put them together to form a uniform community, would yield be higher?

Response: A cage full of all aggressive chickens will still lead to reduced yield because a pecking order is established. Similarly, two equally aggressive plants may not be desirable because environmental variation may give the advantage to one. In any case, the more aggressive plant in a heterogeneous population may not be the best yielder when planted alone.

K. Coot, Iowa State University, USA: Would you recommend selecting at higher plant densities than are used in current cultural practices?

Response: Yes, I predict that the next increase in yields in temperate climates will come about by still higher densities.

A. Oyervides, Universidad Agraria Antonio Narro, Mexico: I understand the high CV in two-way and three-way crosses, but how do you explain the high CV in finished inbred lines?

Response: The high CV is a function of non-adaptability. In the example of Jugenheimer, all the characters were ear related. A high frequency of barrenness would cause a high CV for ear characters.

T. Wehner, North Carolina State University, USA: Professor Brim shows an increase in yield of soybeans using blends of two or three inbreds. Does this support buffering?

Response: Yes, I would concede that if blends increase yields this would support buffering. I would like to know the extent of yield increase in soybean blends. However, mixtures cannot be expected to increase yields over the separate components grown alone when genotypes with high stability are involved.

I. Alvarez, Asgrow Mexicana, Mexico: Would you like to comment more about homeostasis, its origin and how it interacts with the genotype, advantages and disadvantages?

Response: At the present time, homeostasis is as mysterious as heterosis. However, the indication that two-eared maize is homeostatic while one-eared maize is not suggests the kind of characters that may be involved. Professor Athanasios Tsaftaris, University of Thessaloniki, suggests that DNA methylation may be involved.

L. Machida, Crop Breeding Institute, Zimbabwe: Maize yield improvement in the USA is largely due to improving tolerance to high plant densities. Do you think the same approach would lead to yield improvements in the tropical and subtropical environments?

Response: I would concur that density increases would increase yields, but because of the large plant size in tropical maize, I would expect optimum populations to be lower in the tropics than in temperate climates. I suggest a good strategy would be to reduce plant size and increase density.

QUESTIONS FOR W. HANNA

M. Lee, Iowa State University, USA: Given the lack or absence of recombination in the region of apomictic gene(s), how do you know there are so many linked genes (presented in linear order on your slide)?

Response: Actually, those were not linked genes, but were molecular markers that were linked to apomixis. It is just a group of 12 molecular markers that were linked to apomixis. There is no representation of genes at all.

R. Velasquez, Guatemala: How will apomixis contribute to the maintenance of genetic purity of cultivars?

Response: Apomictic hybrids planted side by side will not intercross because the pollen will not fertilize the embryo. So it just clones itself through the seed. The pollen does not enter into the reproduction cycle.

R. Phillips, University of Minnesota, USA: How qualitative is the genetics of apomixis? Is it a major gene with modifiers, or is it a cluster of genes?

Response: In a traditional backcrossing it acts like a major gene with modifiers. But the molecular work would indicate that it is a complex gene, or possibly a cluster of genes.

B. Rana, AICSIP, National Research Centre for Sorghum, India: As apomixis in cultivated sorghums is not presently at a useable level, how can the frequency of it be enhanced to bring it to a useable level?

Response: Levels of facultative apomixis were increased in *Poa* by crossing facultative apomicts (Pepin & Funk, 1971). Perhaps this would work with sorghum.

H. Cordova, CIMMYT, Mexico: How many years will it take until apomictic hybrids become a reality, and how will this event affect the conventional seed industry?

Response: Well, it depends on which crop you are talking about. I do not think you have to be changing your seed production techniques in the near future, but I would say that by the use of traditional backcrossing, pearl millet will probably be the first apomictic crop to be commercially cultivated. An apomictic maize looks like it is next. It is a little bit further down the road, because you have to get rid of a lot of wild things yet and get some fertility. Wheat would be third. And then the molecular aspects of transferring a gene to species or genera that are not apomictic, that is a long way down the road because we do not even have a gene map. I am evading specifics, because I do not know specifics.

R. Bernardo, Purdue University, USA: Do you think it might be possible to have chemically induced apomixis? This would be advantageous to seed producers, not necessarily to farmers.

Response: Yes, I believe most things are possible, and I believe this is one that is possible. But, we are going to have to understand a lot more about the apomictic process, the genes, and their products. We have discussed this for 25 years and need a lot more people working on it to work out the processes.

R. Kumar, Indian Agricultural Institute, India: Once an "obligate apomictic variety" of wheat becomes susceptible to some race of a rust pathogen, how do you suggest its improvement?

Response: Cross the apomict as a male parent to sexual germplasm with rust resistance.

S. Sanchez, UAA, Antonio Narro, Mexico: Do you know if Dr. Ozias-Akins is working on apomictic peanuts? Could you tell us something about this scheme?

Response: Peggy is not working on apomixis in peanuts. However, she has been successful in the transformation of peanut. When we clone the gene(s) for apomixis, peanut would be a good candidate because of its polyploidy.

R. Paliwal, CIMMYT, USA/India: Could you please comment on the source of apomixis that could be used in potato for apomictic seed production?

Response: I am not familiar with all of the potato relatives. If potato does not have apomixis in wild relatives, it will need to come from cloned gene(s) from another genus.

M. Ahmed, Hybrid Rice Laboratory, India: It is reported that the Chinese have found/incorporated apomixis in rice. What are your views/comments?

Response: I have not seen any report or documented evidence of apomixis in rice.

Chapter 33

Temperate Maize and Heterosis

A. R. Hallauer

INTRODUCTION

Interest in the potential of hybrid vigor from crosses of maize (*Zea mays* L.) cultivars was studied after the report by Beal (1880) that hybrid vigor was expressed in the cross of two open-pollinated cultivars. Maize was an important feed grain in the temperate areas of the USA, but similar yield levels (1.88 t ha^{-1} or less) were experienced from 1865 to 1935 (Fig. 33–1). Because of the interest and the desire to increase maize yields in the USA, hybrid vigor in cultivar crosses was studied, but the use of cultivar crosses by the producers was not extensive. Richey (1922) summarized the reported data for 244 cultivar crosses, and he stated: "In such more or less haphazard crossing, therefore, the chances seem about equal of obtaining a cross that is or is not better than the better parent." Hybrid vigor was expressed in some crosses, but it seems the choice of cultivars used to produce the variety crosses was not always conducive to producing a cross that was significantly better than the better parent. Crossing itself did not always result in hybrid vigor. The genetic composition of the open-pollinated cultivars and the genetic basis of hybrid vigor were not understood to consistently produce greater yielding cultivar crosses. With the rediscovery of Mendel's laws of inheritance in 1900 and the research that was summarized by Shull (1952), an alternative method was suggested to exploit hybrid vigor in maize crosses. The inbred–hybrid concept provided the impetus to develop methods for increasing maize yields in temperate areas.

INBRED–HYBRID CONCEPT

The studies reported by Shull (1908, 1909, 1910) and the interpretations Shull provided from his inbreeding and crossing experiments provided insights for the genetic composition of a maize cultivar, for the genetic basis of inbreeding effects, and for an interpretation of hybrid vigor in crosses among pure lines. Inbreeding within a maize cultivar produced pure lines that were homozygous and homogeneous. Hybrid vigor occurred in crosses of pure line because heterozygosis (i.e., heterosis) would occur in crosses of pure lines homozygous for different alleles. Although Shull's (1914) nonMendelian interpretation that hybrid vigor was because of physiological stimulation, Shull (1909, 1910) correctly suggested the future course of maize breeding. Initially, enthusiasm for Shull's suggested methods was limited. Because of the poor vigor and productivity of the inbred lines then available, the method did not seem cost-effective. Jones (1918) suggestion for

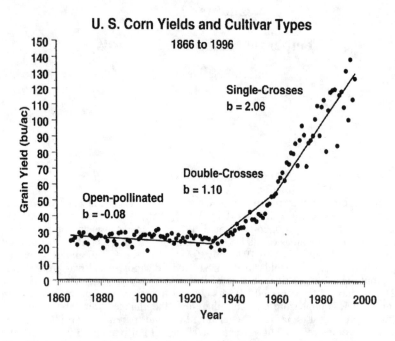

Fig. 33–1. Average U.S. maize yields for three cultivar types from 1865 to 1996 inclusive.

use of single crosses as parents to produce double crosses removed the restriction for use of inbred lines to produce hybrid seed for the growers.

Because maize yields in the USA remained stagnant (Fig. 33–1), concerted plans were made to test the suggestions of Shull (1910) and Jones (1918). Maize breeding programs were initiated in the 1920s by state agriculture experiment stations and the U.S. Department of Agriculture to develop inbred lines and evaluate the inbred lines in crosses. Because of the number of combinations of possible double crosses that could be produced and tested for n inbred lines, it became evident that methods were needed to determine the relative breeding values of inbred lines in crosses. It was obvious that the ultimate value of an inbred line in hybrids had to be determined in comparative yield trials. Hayes and Olson (1919) indicated that the use of first-generation crosses between inbred lines was a means of increasing yield of maize although such crosses are not equally productive, some being of no value. Research was conducted to determine relations between traits of inbred lines and their hybrids, test crossing to determine relative combining ability of lines, prediction of double crosses based on single-cross and testcross data, generation of inbreeding to test for combining ability, relative importance of general and specific combining ability, and extent of testing required to determine relative differences among hybrids. An experimental basis of maize breeding had evolved by the 1940s (Sprague, 1946). Except for the change from use of double crosses to single crosses and the types of germplasm emphasized in breeding programs, the breeding methods presently used are similar to those suggested by Shull (1910), which were enhanced by the research conducted before 1950.

Development of inbred lines by self-pollination was initiated in the more productive and locally grown open-pollinated cultivars. Reid Yellow Dent and strains developed from Reid Yellow Dent (e.g., Osterland Dent, Walden Dent, McCulloch Dent, Krug, Black's Yellow Dent, and Iodent), Lancaster Sure Crop, Leaming, Illinois High Yield, Illinois Two-ear, and others were sampled in the development of inbred lines. Inbred lines rapidly became available to test in hybrids. Comparisons of the newly produced hybrids were made with the open-pollinated cultivars to determine if the hybrids had performance levels that exceeded the currently grown open-pollinated cultivars.

The Iowa Corn Yield Test was initiated in 1920. The goal of the Iowa Corn Yield Test was to provide objective data on the relative performance of the cultivars available to the producers. Until 1924, all of the entries were open-pollinated cultivars. In 1924, 121 entries were entered, two of which were classified as hybrids. The two hybrids were designated as F_1 Hybrids (mixture of Holberts A and B lines with Iodent lines) entered by the U.S. Department of Agriculture and as Copper Cross (C.I.226 × Conn.1-6) entered by Kurtzwell of Des Moines. For the one district that included F_1 Hybrids (3.21 t ha^{-1}) and Copper Cross (2.82 t ha^{-1}), their yields ranked first and second in the district for the top six cultivars, including Ioleaming (2.58 q ha^{-1}), Leaming × White Pearl (2.52 q ha^{-1}), Bloody Butcher × White Pearl (2.51 q ha^{-1}), and Black's Yellow Dent (2.44 q ha^{-1}). Relatively to the average yield (2.51 q ha^{-1}) of the two open-pollinated cultivars and the two cultivar crosses, F_1 Hybrids had 0.70 t ha^{-1} (or 27.9%) and Copper Cross had 0.31 t ha^{-1} (or 12.4%) greater yield. The number of hybrids entered in the Iowa Corn Yield Test increased rapidly after 1924: 15 (3.1% of total entries) in 1925, to 230 (44.6% of total entries) in 1929, to 391 (82.0% of total entries) in 1934, to 1,240 (93.2% of total entries) in 1937. For the seven years of 1926 to 1932 inclusive, the hybrids averaged 8.8% more yield than the average of the open-pollinated cultivars, which was similar to the advantage of cultivar crosses relative to the cultivars themselves (Richey, 1922). But improved hybrids were being identified. For the five years of 1933 to 1937 inclusive, the hybrids averaged 17.6% more yield than the average of the open-pollinated cultivars. There was one year (1936), however, that convinced many growers that hybrid maize was the cultivar of the future. The year of 1936 experienced extremely hot, dry weather, and the average yield of the 769 hybrid entries was 30.8% more than the average yield of 110 open-pollinated entries.

The growers were convinced that the double-cross hybrids were superior to the traditionally grown open-pollinated cultivars and rapidly changed to growing hybrids when seed became available. But growers were cautioned in the 1937 Iowa Corn Yield Test "… that the mere fact that a strain of maize is a hybrid does not in itself give it any superiority." More double-cross hybrids continued to be tested, and adequate quantities of good quality seed were being made available by commercial seed producers. Jenkins (1942) stated that U.S. maize growers had harvested a "bumper crop": 3.132 billion bushels were produced, which exceeded the previous record crop (3.071 billion bushels) produced in 1920 by 61 million bushels. Average yield in 1942 was 2.19 t ha^{-1} (35 bu acre^{-1}), which was 0.21 t ha^{-1} (3.3 bu acre^{-1}) greater than previous U.S. high average yield of 1.98 t ha^{-1} (31.7 bu acre^{-1}) in 1906. Jenkins also reported that 16.6 million hectares (45%) were planted with hybrid maize. In the U.S. Corn Belt, 15.4 million hectares (72%) were planted with F_1 hybrid seed. Jenkins concluded 300 million more bushels of maize were produced than if open-pollinated maize cultivar seed had been used. The acceptance of the new double-cross hybrids continued until nearly 100% of the Illinois and Iowa maize area was planted to hybrid seed by 1945 and in the USA by 1950. Average U.S. maize yields during the era of double-crosses increased about 1.10 bu acre^{-1} yr^{-1} (0.07 t ha^{-1} yr^{-1}; Fig. 33–1). Although maize yields seemed to plateau in the 1950s, improved maize production practices and management skills, recycling of elite inbred lines, and replacement of double-cross hybrids with single-cross hybrids have contributed to greater rates of gain in the past 30 years (Fig. 33–1). Average rates of increase of U.S. maize yields has been about 2.06 bu acre^{-1} yr^{-1}

(0.13 t ha^{-1} yr^{-1}) since 1960. The impact of maize hybrids on average U.S. maize yields was obvious for two distinct 30-year eras: average yields increased two-fold after double crosses replaced the open-pollinated cultivars (1935–1965) and average yields have increased two-fold after single crosses replaced the double crosses (1966–1996; Fig. 33–1).

HETEROTIC GROUPS

The experimental methods that evolved with the development of improved inbred lines and hybrids also included germplasm sources and their management. Germplasm sources for extraction of lines have changed from the use of open-pollinated cultivars to pedigree selection within F_2 populations of elite line crosses (Jenkins, 1978). Whereas, the more productive open-pollinated cultivars were the original germplasm source, open-pollinated cultivars presently have a very minor role as breeding germplasm. The recycling of elite lines has provided consistent, incremental improvement of the lines and their hybrids. The changes in germplasm sources, their management, and the performance of the lines in crosses also suggested that some germplasm sources and combinations were better than others for how inbred lines performed in crosses.

Heterotic groups, and their patterns, have been empirically developed by maize breeders within most of the world's major maize production areas. Heterotic groups initially were identified by how lines performed in crosses; i.e., A × B crosses were superior to either A × A or B × B crosses where A and B represent different germplasm sources. Heterotic groups generally represent broad, but distinct, sources of germplasm; e.g., Reid Yellow Dent-Lancaster Sure Crop and U.S. Dents–European Flints.

Heterotic groups were not identified until extensive yield test data of different combinations of inbred lines in double crosses became available. The inbred lines used to form the first double crosses included lines with good general combining, with limited consideration given to their origin. Pedigrees of some earlier hybrids illustrate the trends for the arrangement of lines in double crosses. Iowa 931 [(L289 × C/447)(Os420 × Os4260] included L289 developed from Lancaster Sure Crop and three lines from two different strains of Reid Yellow Dent; Iowa 931 was tested in 1930–1935. Iowa 939 [(L289 × I205) (Os420 × Os426)] was a widely grown hybrid released in 1934, and Iowa 939 included L289 and three lines from two strains of Reid Yellow Dent. Promising new hybrids in 1938 included Iowa 3110 [(I159 × I224A2)(L289 × L317 B2)], Iowa 3480 [(I159 × I224)(KB397 × WD456)] and Illinois hybrid [(WF9 × 38-11)(I159 × I224)]. By this time, the single-cross parents tended to originate from the same open-pollinated cultivar source: Iowa 3110 includes two lines from Iodent, a strain of Reid Yellow Dent in one parent, and two lines from Lancaster Sure Crop in the other parent; Iowa 3480 includes two lines from Iodent, one from Krizer Brothers Dent, and one from Walden Dent, all three sources being strains of Reid Yellow Dent; and Illinois hybrid also includes four lines that include Reid Yellow Dent germplasm.

Eckhart and Bryan (1940) reported results that indicated inbred lines that are more similar genetically should be included in the parental single crosses when making double crosses and that the inbred lines that are more diverse genetically should enter in the final double cross from the opposite single-cross parent. Hayes (1939) reported that good double crosses where all four inbred lines came from the same cultivar had not been obtained in Minnesota. Evidence suggested that, on the average, greater productivity occurred when divergent lines were included in single-cross parents as (A1 × A2) and (B1 × B2) and not as (A × B). Inbred lines that included Reid Yellow Dent germplasm were prominant in all hybrids and the other source of lines in the earlier hybrids was Lancaster Sure Crop. Hence, the origin of the Reid Yellow Dent–Lancaster Sure Crop heterotic pattern.

Because of the seemingly important need to maintain diversity between the single crosses included in the double crosses, a committee, Grouping of Inbred Lines for Breeding Purposes, was appointed at the 9th Corn Improvement Conference of the North Central Region to consider the development of groups of inbred lines. G.H. Stringfield (1947) urged the advisability of grouping inbred lines and that crosses for the improvement of lines (i.e., recycling of elite lines) should be made only among lines of the same group. The goal for grouping of inbred lines would be to maintain genetic diversity and to avoid relations among inbred lines that are to be used in the production of double crosses. Inbred lines within the North Central Region were to be assigned to two groups, which were to be kept distinct for breeding purposes; i.e., crosses for breeding purposes are made only between lines included in the same group. The reasons and justifications for identifying and maintaining two distinct breeding groups seemed valid, but a very arbitrary method of assigning inbred lines to one of the two groups was used. The first assignment of lines to the two breeding groups was included in a report presented by Stringfield (1950), and the method for assignment of lines was not based on origin of lines: lines having odd entry numbers in the 1948 uniform inbred tests were assigned tentatively to Group A and lines having even entry numbers were assigned tentatively to Group B. Because of previous discussions and data from double-cross trials for combinations of lines used in double crosses, it is surprising that origins of the inbred lines were not considered in assigning them to the two groups. Os420 and Os426, for example, were used as one single-cross parent of a widely grown double-cross hybrid, Iowa 939, but Os420 was assigned to Group A and Os426 was assigned to Group B. Some revisions of the group assignments were made in 1953 (e.g., Os426 moved to Group A with Os420), but the main addition was the assignment of Iowa Stiff Stalk Synthetic lines to Group B. The policy of grouping inbred lines in the North Central Region has continued with inbred lines considered to have primarily Reid Yellow Dent germplasm assigned to Group B and those inbred lines considered to have non-Reid Yellow Dent germplasm assigned to Group A, but it is the breeders' responsibility to assign newly developed lines to a heterotic group. Heterotic groups were desired because of the need to include related lines in the same single-cross parents of the double crosses, but considerations of heterotic groups are equally useful in pedigree breeding programs for developing inbred lines as parents of single-cross hybrids.

Heterotic groups, or patterns, are frequently used in maize breeding discussions, but it is obvious that some overlapping of germplasm has occurred in assignment of lines to the original heterotic groups. The original assignment of inbred lines for the North Central Region of the USA was certainly without consideration in the origin of lines. Heterotic groups in recent years have become more distinct and refined because of emphasis of selection within elite line crosses for developing parents of single crosses and with the assistance of molecular markers to resolve questionable assignments. Each organization or group has their own heterotic pattern(s) which include families of lines developed by recycling important elite lines. Heterotic groups are used to classify groups of germplasm that will have maximum expression of heterosis in crosses of lines from the distinct groups. Heterotic groups have evolved within breeding programs. Newer heterotic groups could be developed, but because of extensive recycling of elite lines and extensive testing, it will become difficult to replace presently used heterotic groups. Present heterotic groups have evolved by intensive breeding efforts, and similar efforts would be required to develop other heterotic groups competitive with those currently used.

SELECTION FOR CROSS PERFORMANCE

Interest in improving germplasm resources that could be used in applied breeding programs and the types of genetic effects important in the expression of heterosis led to the development of different recurrent selection methods. At the time when the ordering of lines in double-cross hybrids and grouping of inbreds for breeding purposes were being considered, Comstock et al. (1949) proposed the use of reciprocal recurrent selection for two genetically broad-based maize populations, P1 and P2, in which P1 served as tester for P2 and P2 served as tester for P1. The method was expected to emphasize selection for complementary alleles in the two populations. Comstock et al. (1949) demonstrated that reciprocal recurrent selection would be as effective as methods that emphasized selection either for specific combining ability (Hull, 1945) or for general combining ability (Jenkins, 1940), and would be more effective than either method if both specific and general combining effects were important.

Direct response to reciprocal recurrent selection is measured in the population crosses of the two populations undergoing selection. Falconer (1981) indicated midparent heterosis could be expressed as $\Sigma y^2 d$, where y is the difference in allele frequency of the two parents of the cross, d is level of dominance, and summing over loci affecting the trait. Partial to complete dominance has been shown to exist for most maize traits (Hallauer & Miranda, 1988). Evidence from reciprocal recurrent selection studies conducted in temperate areas suggests the method is effective for improvement of grain yield of the population crosses (Table 33–1). Average direct response to selection in the population crosses was 5.2% cycle^{-1} of selection. Average midparent heterosis increased from 8.9% for the original parental crosses (C0 × C0) to 42.5% for the last cycle of selection (Cn × Cn). Because level of dominance probably did not change during the limited number of selection cycles, the selection method must have increased the divergence of allele frequencies between the two populations. The effects of indirect response among the populations themselves were erratic and only averaged 1.2% cycle^{-1}.

Indirect evidence from the reciprocal recurrent selection programs suggests maize breeders can, and have, improved the cross performance of their important heterotic groups. The midparent heterosis for the genetically broad-based population crosses would not include the unique combinations of alleles that surely exist within and between families of highly selected, elite inbred lines. The heterosis expressed in the population crosses includes only the modal or average of possible interactions (i.e., epistasis) that exist in highly selected inbred line families, which would be specific for the pairs of inbred families emphasized in modern breeding programs. It seems that selection for heterosis can continue to be enhanced with the assistance of molecular markers to identify and to maintain unique allele combinations during selection for potential expression in hybrids.

SUMMARY

The development and the application of the concept of hybrid maize have impacted temperate maize production areas in many ways. Hybrid maize's immediate impact was increased yields area^{-1}. In 1932 when open-pollinated cultivars were being used primarily, it required 110.6 million acres to produce 2 billion bushels of maize grain; average yield was 26.5 bu acre^{-1}. By 1955, most of the maize area was planted with double-cross hybrid seed, 2.9 billion bushels of maize grain was produced on 79.5 million acres with an average yield of 40.6 bu acre^{-1}; 45% more grain was produced on 28% less area. Total U.S. maize grain production in 1975 was 5.8 billion bushels which was produced on 66.9 million acres for an average yield of 86.2 bu acre^{-1}. New, improved hybrids continued to be developed and made available to the growers. In 1994, about 11 billion bushels of grain

Table 33–1. Responses to reciprocal recurrent selection conducted in temperate area maize populations with emphasis on selection for greater grain yields.

Populations	Reference	Types of families	Cycles of selection	Response cycle^{-1}		Midparent heterosis	
			no.	Direct	Indirect	C0 × C0	Cn × Cn
				—%—	—%—	—%—	—%—
BSSS and BSCB1	Keeratinijakal and Lamkey, 1993	Half-sib	11	7.0	0.0	25.4	76.0
Jarvis and Indian Chief	Moll and Hanson, 1984	Half-sib	10	2.7	3.1 / -0.7	6.6	28.9
BS10 and BS11	Eyherabide and Hallauer, 1991	Full-sib	8	6.5	3.0 / 1.6	2.5	39.6
BS21 and BS22	Menz, 1997	Half-sib	6	4.5	-1.0 / 1.6	1.0	25.4
Average			8.8	5.2	1.2	8.9	42.5

was produced on 79 million acres for an average yield of 138.6 bu acre^{-1}. Total U.S. maize acreage for 1955 and 1994 was similar, but average yield per acre increased about 3.5 times. Hybrid seed maize also has led to the development of new industries and companies to develop, produce, and market a reliable supply of high quality seed for each growing season, development of alternative uses because of a reliable supply of grain, and for exporting maize grain and products derived directly or indirectly from maize. Today, hybrid maize is an important component of the economies of countries located in temperate areas.

REFERENCES

Beal, W.J. 1880. Indian corn. Rep. Michican Bd. Agric. 19:279–289.

Comstock, R.E., H.F. Robinson, and P.H. Harvey. 1949. A breeding procedure designed to make maximum use of both general and specific combining ability. Agron. J. 41:360–367.

Eckhart, R.C., and A.A. Bryan. 1940. The effect of the method of combining the four inbreds of a double cross upon the yield and variability of the resulting hybrid. J. Am. Soc. Agron. 32:347–353.

Eyherabide, G.H., and A.R. Hallauer. 1991. Reciprocal full-sib selection in maize: I. Direct and indirect responses. Crop Sci. 31:952–959.

Falconer, D.F. 1981. Introduction to quantitative genetics. 2nd ed. Longman, New York.

Hallauer, A.R., and J.B. Miranda, Fo. 1988. Quantitative genetics in maize breeding. 2nd ed. Iowa State Univ. Press. Ames.

Hayes, H.K. 1939. Sources of inbreds. p. 6. *In* Rep. 3rd Corn Impr. Conf. North Central Region. Columbia, MO. 27 Nov. 1939.

Hayes, H.K., and P.J. Olson. 1919. First generation crosses between standard Minnesota corn varieties. Minnesota Agri. Exp. Stn. Tech. Bull. 183:5–22.

Hull, F.H. 1945. Recurrent selection and specific combining ability in corn. J. Am. Soc. Agron. 37:134–145.

Iowa Corn Yield Tests. 1920–1937. Iowa Corn and Small Grains Assoc. cooperating with Iowa Agric. Exp. Stn. and Bureau of Plant Industry. USDA. Ames.

Jenkins, M.T. 1940. The segregation of genes affecting yield of grain in maize. J. Am. Soc. Agron. 32:55–63.

Jenkins, M.T. 1942. The corn breeding program in relation to the war effort. p. 16–17. *In* Rep. 6th Corn Impr. Conf. North Central Region. St. Louis, MO. 12 Nov. 1942.

Jenkins, M.T. 1978. Maize breeding during the development and early years of hybrid maize. p. 13–28. *In* D.B. Walden (ed.) Maize breeding and genetics. John Wiley & Sons, New York.

Jones, D.F. 1918. The effects of inbreeding and crossbreeding upon development. Connecticut Agric. Exp. Stn. Bull. No. 207:5–100.

Keeratinijakal, V., and K.R. Lamkey. 1993. Responses to reciprocal recurrent selection in BSSS and BSCB1 maize populations. Crop Sci. 33:73–77.

Menz, M.A. 1997. Comparative response to selection of two reciprocal recurrent selection procedures in BS21 and BS22 maize populations. Ph.D. diss. Iowa State Univ., Ames.

Moll, R.H., and W.D. Hanson. 1984. Comparisons of effects of intrapopulation vs. interpopulation selection in maize. Crop Sci. 24:1047–1052.

Richey, F.D. 1922. The experimental basis for the present status of corn breeding. J. Am. Soc. Agron. 14:1–17.

Shull, G.H. 1908. The composition of a field of maize. Am. Breeders' Assoc. Rep. 4:296–301.

Shull, G.H. 1909. A pure line method of corn breeding. Am. Breeders' Assoc. Rep. 5:51–59.

Shull, G.H. 1910. Hybridization methods in corn breeding. Am. Breeders' Mag. 1:98–107.

Shull, G.H. 1914. Duplicate genes for capsule form in *Bursa bursa pastoris*. Zerts. Ind. Abst. Ver. 12:97–149.

Shull, G.H. 1952. Beginning of the heterosis concept. p. 14–48. *In* J.W. Gowen (ed.) Heterosis. Iowa State College Press. Ames.

Sprague, G.F. 1946. The experimental basis for hybrid maize. Biol. Rev. 21:101–120.

Stringfield, G.H. 1947. Grouping of inbred lines for breeding purposes. p. 25. *In* Rep. 9th Corn Impr. Conf. North Central Region. Cincinnati, OH. 16 Nov. 1947.

Stringfield, G.H. 1950. Grouping of inbred lines for breeding purposes. p. 8–9. *In* Rep. 11th Corn Impr. Conf. North Central Region. Chicago, IL. 3–4 Feb. 1950.

Chapter 34

Tropical Maize and Heterosis

S. K. Vasal, H. Cordova, S. Pandey, and G. Srinivasan

INTRODUCTION

Tropical environments are diverse and harsh, where maize (*Zea mays* L.) is grown on approximately 58 million hectares and of which 36 million ha are in the tropical lowland, 16 million ha in the subtropics and the mid-altitude ecologies, and about 6 million ha in the highlands. Except in China, Argentina, and Brazil, area planted to hybrids does not exceed 19% of the total maize area (CIMMYT, 1994). Also, about 85% of the total seed is produced and sold by multinational and national private companies. The 1970s witnessed a sharp decline in the number of hybrids released in the tropics. This could partly be attributed to a shift toward population and composite breeding and to the phenomenal success of improved OPVs during that decade. The hybrid release during the 1980s and 1990s shows an upward trend. The activity by the private sector has increased resulting in a larger share of hybrids in the total maize releases and more area under hybrids as percentage of the total maize area.

HISTORICAL TRENDS IN HYBRID MAIZE DEVELOPMENT

Hybrid development in tropical maize dates back to the 1940s. In the initial stages, the Rockefeller Foundation played a vital role by stimulating hybrid maize research in Mexico, Colombia, India, and several Central American countries. Several other countries initiated hybrid research on their own. These efforts were relatively more successful in Zimbabwe, Kenya, Malawi, Mexico, Egypt, Turkey, India, Pakistan, Venezuela, El Salvador, Colombia, Brazil, and China. But the sustainability of these efforts varied even in these countries, due to shifting balance in population improvement vs. hybrid research, lack of seed production infrastructure, difficulties in producing parental and hybrid seed, insufficient hybrid superiority, and reduced public funding for research (Pandey & Gardner, 1992). Countries that can be cited as success stories for hybrid use today include Zimbabwe, Kenya, Thailand, El Salvador, Guatemala, Egypt, and Vietnam.

Kenya released a number of hybrids beginning 1964. The first three hybrids, H611, H621, and H 631, were a variety cross (VC), a double cross (DC), and a three-way hybrid (TWC), respectively. Subsequently, the Kenyan program released better hybrids, such as H622 (DC), H632 (TWC), H612 (TC), H511 (VC), H512 (VC), H611 (VC), H613 (TC), H614 (TC), and H625 (DC). Maize hybrids have gained popularity in Kenya and most farmers in the favored environments grow only hybrids. In 1992, 74% of the total maize area was under hybrids. In Zimbabwe, almost 100% of the maize area is planted to hybrids. SR52, released in

1960, was the world's first single-cross (SC) hybrid for commercial planting. It dominated the commercial scene for more than two decades in Zimbabwe and some neighboring countries. Two other early hybrids, R200 and R201, occupy 15% of the total area. R200 (TWC) was released in 1970, followed by R201 in 1975 and R215 in 1976. The R200 series hybrids have played a significant role throughout Eastern and Southern Africa and enabled Zimbabwe to become the leading exporter of hybrid maize seed in Africa. Zambia has used SR52 and it also released MH752, MM501, MM504, MM602, MM603, MM604, and MM606 in 1984. In Western and Central Africa, countries growing small areas (<5% of total maize area) of hybrid maize are Nigeria, Cameroon, and Ivory Coast. Commercial IITA hybrids, 'Oba-Super-1' and 'Oba-Super-2' are adapted to savanna and forest zones.

Egypt has released new and superior hybrids on a continuous basis. At present, approximately 40% of the total maize area is planted to 16 hybrids from public programs, and Pioneer, Dekalb, Agro Seed Co., and Agri. Seed (Syndicate) are marketing their own products.

In Asia, China, Thailand, and Vietnam have made significant advances in hybrid maize technology. In Thailand, 80% of the area is mostly planted to SCs. In Vietnam, from no hybrid maize in 1990, 40% of its maize area now is planted to hybrids. Hybrid use in India, the Philippines, and Indonesia is expected to grow during the next few years.

In Latin America, the countries with significant maize area planted to hybrids are Brazil, Mexico, Venezuela, El Salvador, Guatemala, Panama, and Cuba. Other countries such as Costa Rica, Nicaragua, Honduras, Bolivia, Colombia, Ecuador, and Peru also plant about 10% of the total maize area to hybrids.

Source Germplasm for Hybrid Development

A wide array of maize germplasm is available in tropical maize for use in hybrids. It has evolved in environments with specific agro-ecological conditions and stresses. Several useful racial complexes have been identified for hybrid maize research (Wellhausen, 1978; Brown & Goodman, 1977; Goodman, 1985). These are Tuxpeno, Cuban flints, Coastal tropical flints (Caribbean flint), ETO, Tuson, Chandelle, Haitian yellow, and Perla. Tuxpeno, Cuban flints, Coastal tropical flints, Tuson, and ETO have been extensively used worldwide. Any germplasm can be used for hybrid development but its effectiveness will depend on its ability to provide inbreds with acceptable frequency and with superior agronomic performance.

Germplasm Development: The Changing Scenario

Breeding efforts in the early 1960s and 1970s emphasized formation of populations disregarding heterotic patterns and combining ability. Their mixed genetic constitution makes the task of characterizing and placing them in specific heterotic groups difficult. Occurrence of stresses such as downy mildew in Asia, stunt complex in Central America, Maize Streak Virus (MSV) in Africa, and soil-acidity tolerance in South America has changed focus of germplasm development and improvement strategies. More resources are devoted to stress-related traits. CIMMYT, Kasetsart University, and Department of Agriculture in Thailand have developed several downy mildew resistant populations such as Suwan-1, Suwan-2, Suwan-3, Population 101, Population 145, Population 300, Population 345, and NS-1. Several populations now combine resistance to both downy mildew and maize streak virus or downy mildew and acidity tolerances a result of shuttle breeding among CIMMYT programs in Thailand, Colombia, and Ivory Coast. In case of MSV, a large volume of resistant germplasm has become available through the collaborative efforts of IITA and CIMMYT outreach locations in Harare, Zimbabwe and Ivory Coast in West and Central Africa. Several normal and QPM populations now have resistance to MSV. CIMMYT outreach program in Cali, Colombia, has developed populations tolerant to soil acidity and aluminum toxicity. Initially six population, SA-3, SA-4, SA-5, SA-6, SA-7, and SA-8, were developed but at a later stage SA-3 and SA-8 were merged with others to reduce the germplasm volume to two yellow populations, SA-3, SA4, and two white populations, SA-6 and SA-7. Efforts in forming two early acidity tolerant populations are nearing completion. Excellent research done by EMBRAPA in Brazil has been equally successful in providing sources of resistance for millions of hectares needing this type of germplasm.

Drought and low N are encountered practically in all production environments where tropical maize is grown. This has led to extensive efforts by CIMMYT, IITA, and several national programs to develop germplasm suitable for the two stresses. Recurrent selection has been used to improve drought tolerance of several lowland tropical maize populations at CIMMYT. Such populations include TS6, La Posta Sequia, Pool 16 Sequia, Pool 26 Sequia, DTP-1, and DTP-2 for drought and AC 8328BN for improved N-efficiency. The insect resistant germplasm is equally important in several production environments. Because of the complexity of insect resistance work, progress has been slow in developing tolerant germplasm. Nevertheless populations, Multiple Borer Resistant (MBR) and Maize Insect Resistant Tropical (MIRT), have been developed at CIMMYT. Inbred lines such as CML 67 and 139 also have good level of resistance but unfortunately lack agronomic performance.

The work on controlling parasitic weed *Striga hermonthica* and *S. asiatica* through genetic resistance is in progress at IITA and CIMMYT and several sources of resistance have been either identified and/or developed (Kim & Adetimirin, 1994). Work on water logging tolerance is a more recent initiative but heterotic pairs of populations have already been formed at Pantnagar, India, in the early maturity group, and in the full season white and yellow groups at Crop Research Institute, Tak Fa, Thailand.

Heterotic Patterns and Recurrent Selection in Tropical Maize

Characterization of maize germplasm for combining ability and heterotic patterns is important in hybrid breeding. Several useful heterotic patterns have been described by Wellhausen (1978) and Goodman (1985). Tuxpeno combines well with Cuban flint, Coastal Tropical flints (Caribbean flint), Tuson, and ETO. Cuban flints combine well with Tuxpeno, Tuson, Coastal Tropical Flints, and Perla. Coastal Tropical Flints combine with Tuxpeno, Cuban flints, and Chandelle. Heterosis among CIMMYT germplasm is not high considering their broad genetic base. However, several parents with high CGA have been identified (Table 34–1; Beck et al., 1990; Crossa et al., 1990; Vasal et al., 1992a, b, c).

Through earlier efforts of the Rockefeller Foundation in Mexico, followed by inception of CIMMYT in Mexico in 1966, a large amount of maize germplasm has been developed for most agro-ecologies of tropical lowlands, subtropical and mid-altitude, transition highland, and highlands (Vasal et al., 1982). Many of these populations were improved using intrapopulation selection initially. Interpopulation recurrent selection is now receiving greater emphasis. The lowland tropical program is emphasizing five pairs of heterotic populations: Tuxpeno (P21) × ETO (P32), Mezcla Tropical Blanca (P22) × La Posta (P43), Amarillo Dentado (P28) × Cogollero (P36), Amarillo Cristalino-1 (P27) × Ant. Ver.181 (P24), and Blanco Cristalino-1 (P23) × Blanco Dentado-2 (P49). The subtropical program is using interpopulation selection to improve Amarillo subtropical (P33) × Amarillo Bajio (P45), A.E. Dent-Tuxpeno (P44) × ETO Illinois (P42), and SIW-HG88A (P501) × SIW-HT88B (P502). And the highland program has only one pair of populations, P902 × P903, for interpopulation improvement.

Table 34–1. Heterotic partner(s) for some CIMMYT maize populations.

Population	Possible heterotic partner(s)
21	POP25, POP32, POOL 23
22	POP25, POP32
23	POP49, POOL 20
24	POP27, POP36, Suwan-1, POOL25
25	POP21, POP22, POP43, POOL24
26	POOL21
27	POP24, POP36, POOL26
29	POP25, POP32
31	POP26, POOL22
32	POP21, POP22, POP29, POP44
33	POP34, POP45
34	POP44, POP47, POP501, POOL32 '
36	POP27, POOL25, Suwan-1
42	POP43, POP44, POP501, POP502, POOL32
43	POP42, POP44, POOL32
49	POP33, POOL19

The populations undergoing improvement at CIMMYT outreach sites are SA-4 × SA-5 and SA-6 × SA-6 for soil acidity tolerance in Colombia, P351 × P352 for downy mildew and other traits in Thailand, and INT-A × INT-B, LAT-A × LAT-B and DR-A and DR-B in Zimbabwe.

Several national programs in recent years have also adopted interpopulation selection. The DOA program in Thailand practices reciprocal recurrent selection with two pairs of populations for excess water tolerance work and one pair for low N tolerance. Interpopulation improvement is also used at Pantnagar, India, for improvement of two populations for tolerance to waterlogging.

Shift to interpopulation improvement efforts fulfills the objectives of deriving superior OPVs and hybrids. Many comprehensive population improvement schemes are available (Hallauer & Eberhart, 1970; Eberhart et al., 1967, 1995). Number of populations to be handled in a program, particularly if the resources are limited, could be a concern and perhaps in that scenario one can handle two superior heterotic populations (Eberhart et al. 1995); however, when the germplasm product needs are diverse in terms of grain color, maturity, and adaptation, one must improve more than two populations.

Developing Hybrid-Oriented Source Germplasm

Germplasm should not show large inbreeding depression if inbreds are to be developed for hybrid formation. Since most tropical maize germplasm has been improved using recurrent selection based on noninbred progenies, selection has not helped much in improving inbreeding depression in such populations. Since the inception of hybrid program at CIMMYT, tolerance to inbreeding has been emphasized for all types of germplasm. Studies conducted in both tropical and subtropical maize germplasm using S_3 recurrent selection have shown that it is possible to improve this trait (Vasal et al., 1995).

Hybrid-oriented source germplasm must possess attributes needed in the inbreds. It should have low inbreeding depression, high GCA, and high SCA with one or more other populations. Since CIMMYT, IITA, and several other institutes are now emphasizing inbreeding and testcross progeny evaluation, more populations suited for hybrid work should become available in the near future.

Based on previously known patterns or, alternatively, using two or more testers, separation of S_1 to S_3 lines has been made, followed by recombination of the two opposing fractions to form new populations (Table 34–2). Use of such heterotic populations would greatly facilitate hybrid development work, as lines developed from one population would cross well with lines from the opposite heterotic population. CIMMYT has formed several such populations, e.g., IBP-1 (Late White Dent), IBP-2 (Late White Flint), IPB-3 (Late Yellow Dent), and IPB-4 (Late Yellow Flint). Similarly, 92 tropical maize inbreds were crossed with four testers. Also, the two testers that provided better separation of more advanced lines were used to select two fractions opposite in heterotic response. The fractions were recombined separately to provide two populations, THG-A and THG-B. Several synthetic populations with lodging resistance, stay green character, long ears, and rust resistance have also been developed.

Types of Hybrids Grown

At present, many types of conventional and nonconventional maize hybrids are grown in different countries. Brazil (BR 300, BR 301), Colombia (ICA-H353, ICA-H556), Mexico (HV-313, HV-373), Guatemala (ICTA T101, HA-44), Egypt (VC80, VC69), Kenya (H611, H511, H611C, H512), Madagascar (Plata 383, Plata 375), Reunion (IRAT-143, IRAT 279), Senegal (BOS-101, HVG-1), Swaziland (PNR95), Tanzania (H6303, H614), India (Sangam), and the Philippines (AMC-101, SMC-102) are still growing varietal hybrids. A few topcross hybrids also are being marketed in Brazil (BR-302), Kenya (H612, H613, H614), China (Kueiding), and Thailand (CP-1). And some double topcross hybrids are under cultivation in Malawi (MH-17, MH18), China (Kueidin, Kueidin No.4), and India (Ganga 4, Ganga 11, Ganga 5, Hi Starch, Ganga Safed 2). Nonconventional hybrids are grown in those areas where productive inbreds and infrastructure for seed

Table 34–2. Formation of heterotic populations based on testcross data.

Country	Institution	Populations formed	
		Group-A	Group-B
Colombia	CIMMYT	SA-4 (Y)	SA-5 (Y)
	CIMMYT	SA-6 (W)	SA-7 (W)
Thailand	CIMMYT-ARMP	P. 351 (Y)	P. 352 (Y)
India	Pantnagar	Yellow Early Pool-1	Early Yellow Pool-2
	P.A.U.	LL Pool (C_4)	LSS (C_4)
	Peninsular	X_1 (Y)	X_2 (Y)
Thailand	KU	Suwan-1	KS6
	DOA	WLTFY-1	WLTFY-2
		WLTEY-3	WLTEY-4
		KK (DR) C_5	NS (DR) C_5
Mexico	CIMMYT	IPB-3	IBP-4
		THG-A	THG-B
		P902	P903
		P501	P502
		P401	P402
		P601	P602

production are lacking. The current trend, however, is to grow three-way and sin-gle-cross hybrids. Some countries like Thailand are shifting rapidly to single-cross hybrids.

Intrapopulation vs. Interpopulation Hybrids

Intrapopulation maize hybrids have been developed and commercialized in tropical maize. Some of the earlier hybrids were intrapopulation hybrids (Aekata-sanawan et al., 1997). The first hybrid (Suwan 2301) developed by Kasetsart University in Thailand was an intrapopulation interline hybrid from Suwan-1. Data suggest that interpopulation interline hybrids are generally superior (Han et al., 1991). Even when the populations are not heterotic, the interpopulation interline hybrids give superior performance. However, populations like AED × Tuxpeno (P44), Tuxpeno (P21) and La Posta (P43) have produced outstanding intrapopula-tion interline hybrids. It is possible to produce good hybrids from the same popu-lation provided it has high per se performance and high GCA.

Testers in Tropical Maize

Testers are essential to the success of a hybrid breeding program. At pres-ent, all types of materials including population(s), synthetics, hybrids, and inbreds are used as testers. Little effort has been devoted in developing countries to de-velop good inbred testers. In recent years, use of inbred testers has increased sig-nificantly. CIMMYT has an ongoing research to find new and better testers. The data generated so far indicate that it is desirable to use tester from the opposite population, that inbred testers are as efficient as other types, and generally one tester is enough to identify superior lines.

In the lowland tropical maize program at CIMMYT, the inbred testers used are CML 247 and CML 254 for full season white populations, CML 287 and CL -00331 for full season yellow populations, CL-G15 and CL-G16 for early white populations, and CL-03101 against CL-G1805 and CL-G1704 for yellow early populations. The subtropical program is using CML 321 and CML 320 (CML 78, CML 313) for white materials and CML 323 and CML 327 for yellow materials. The highland program uses CLM 242 and CML 246 as testers. The KU program in Thailand uses Ki 44 and Ki 21 (Ki 45) and INIFAP in Mexico uses T39 and T 40 (T43) as testers. The Philippines uses Pi-17 and Pi 23, Vietnam DF-2 and DF-8, India CM202 and CM111, and in many Eastern and South African countries N3 and SC or their modified version are used as testers.

Breeding for Stress Tolerant Hybrids

Breeders now pay more attention to stress tolerance today than in the past. Important stresses include drought, acid-soil tolerance, low N, tolerance to parasitic weed *Striga hermonthica*, and a whole array of diseases and insects. Vasal et al., (1997) outlined strategies, which could be used to develop maize germplasm and hybrids tolerant to stresses. Several lines and hybrids have been identified which are tolerant to drought and low N (Table 34–3). CIMMYT has a strong stress breeding program to develop and refine methodologies and produce sources of germplasm tolerant/ resistant to stresses (Edmeades et al., 1994).

Drought and low N are also receiving emphasis in Latin America, Asia, Af-rica, and at IITA. Extensive work is in progress developing hybrids tolerant to downy mildew, stunt virus, MSV, and soil-acidity tolerance. Breeding tolerance/ resistance to *Striga* sp. has received considerable attention at IITA and at CIMMYT's regional program in Western and Central Africa. Several good lines, hybrids, and other sources of resistance to Striga are now available (Kim & Adet-imirin, 1994; Diallo, 1997).

Table 34–3. Inbred lines announced by CIMMYT (1991-1996).

Year	Adaptation	CML no.	No. of lines
1991	Tropical	1-74	74
	Subtropical	75-116	65
1992	QPM Tropical	140-172	33
	QPM Subtropical	173-194	22
	Mid-Altitude (Streak)	195-216	22
	Tropical (Streak)	217-238	22
1993	Highland	239-246	8
1994	Tropical	247-281	37
	Tropical	282-308	27
	Tropical	309-310	2
1995	Subtropical	311-329	19
1997	Subtropical	330-338	9
	Tropical	339-348	10
	Highland	349-356	8
	Tropical (Soil acidity)	357-366	10

Hybrid Performance

Considerable data on hybrid performance are now available from various international, regional, and national maize trials (Table 34–4). At present CIMMYT is distributing several international hybrid trials with tropical, subtropical and highland germplasm (Table 34–5). Similarly, regional trials are also conducted in Central America (PCCMCA trial) and in Asia (TAMNET trial). IITA conducts regional trial in Central and West Africa. In addition, some countries have their own trials where hybrids from the public and private sector are tested.

Development and Evaluation of Tropical Maize Inbreds

In tropical maize, frequency of good lines is usually low because of high inbreeding depression. The lines at high level of inbreeding do not have high seed yield and tolerance to stress. Generally, ear-to-row or pedigree method is used for developing lines. More attention should be paid to whether a line will be used as a seed or pollen parent. High population densities should be used during inbred development. Lines can be modified through backcrossing and recycling. Line performance should receive greater attention during the development process. Development of pedigree populations and extracting new lines from them needs to receive more emphasis. Several programs have now initiated recycling among superior or elite lines to develop better lines. Breeders also self in the commercially available hybrids to develop lines. In developing lines, CIMMYT scientists generally plant higher densities and depending on the row length, leave half of the row for observation purposes including visual evaluation for yield at harvest.

Breeders evaluate lines for combining ability at different stages of inbreeding. Support for early evaluation of lines has accumulated over the years. The stage at which combining ability should be evaluated has to be determined by inbreeding minimum score, inbred survival rate, and inbreeding level necessary to realize hybrid potential. It may be more appropriate to choose an intermediate inbreeding stage for combining ability evaluation. If the survival rate is extremely low it is better to test for combining ability after S_3 or S_4 stage of inbreeding. In

Table 34–4. Five top performing hybrids in different regional and national trials.
Source: K.U. (Chock); CIMMYT (Jorge Bolaños).

Trial name	Locs.	Hybrid	Institution/ company	Grain yield
	no.			t ha⁻¹
Cooperative hybrid corn yield trial (Thailand)	9	KS × 3851	K.U.	9.5 (122) †
		5124002R	CARGILL	9.3 (119)
		3012	PIONEER	9.2 (118)
		KS × 3852	K.U.	9.1 (116)
		PAC417	PACIFIC	8.9 (114)
PCCMCA white hybrid trial-96	30	CML9 × CML47	CIMMYT	7.1
		A-7573	ASGROW	7.0
		CML247 × CML254	CIMMYT	6.9
		HE-59	CENTA - El Salvador	6.8
		A-7530	ASGROW	6.5
PCCMCA Yellow hybrid trial-96	14	CML287 × CL00331	CIMMYT	6.2
		CB-X HS-8GH3	CRISTIANI BURK	5.9
		CB-X HS-8GH2	CRISTIANI BURK	5.8
		DK888A	DEKALB	5.8
		HS-6	CRISTIANI BURK	5.6

† Numbers in parentheses are yields relative to check hybrids in %.

Table 34–5. Performance of CIMMYT tropical and subtropical hybrids in international trials and under mild stress.

Trial	Pedigree	Int. testing		Drought stress		
		Yield	Rank	Yield	Rank	ASI
		t ha⁻¹		t ha⁻¹		
CHTT-W	CML258 × CML 264	6.7	2	5.5	4	-0.1
	CML258 × 270	6.5	3	5.8	2	0.0
	CML 251 × 267	6.4	4	5.7	3	0.0
CHTS -W	CML 78 × CML321	8.0	2	5.9	1	1.0
CHTT-Y	CML287 × CL00331	5.8	1	4.7	6	0.8
	CML287 × 298	5.5	2	5.1	2	0.1
	CL 299 × CL00331	5.4	3	4.8	5	0.6
	CML297 × 295	5.4	3	4.8	5	1.4

institutions with strong population improvement programs, lines will be tested for combining ability at S_1 or S_2 as part of the on-going population improvement program.

For single-cross hybrids, it is imperative to have lines that are productive and have good seed yield. Lines identified promising after evaluations for combining ability and per se performance, should be tested in line evaluation nurseries for agronomic performance, yield, and diseases and other stresses. Characterization of lines for many attributes helps build data base which can be subsequently used in crossing program and in recycling of lines.

CIMMYT has conducted line evaluation trials for the past few years. Some of lines have yielded as much as 3 to 4 tons/ha (Table 34–6). Lines have also been identified with tolerance–resistance to stress traits such as drought, low N, acid soil, maize streak virus, Striga, and several foliar, ear rot and stalk rot diseases.

Inbred Line Availability

During the past decade, availability of tropical maize inbreds with different adaptation, grain color and texture has greatly increased. A few of lines are of Quality Protein Maize (QPM) while others are tolerant to different biotic and abiotic stresses. The tropical inbreds are available from CIMMYT, IITA, University of Hawaii, Kasetsaart University, Institute of Plant Breeding Philippines, Maize Directorate, India, and from USA (Florida & Texas A&M). CIMMYT has announced/ released 366 lines to date (Table 34–3). New and diverse lines will continue to accelerate hybrid maize production in the developing world.

Use of Exotic Germplasm

Exotic germplasm, particularly of temperate adaptation, has been used quite extensively by CIMMYT, IITA and many national programs. Several pools and populations of subtropical adaptation have varying proportions of temperate germplasm. Several released lines from IITA involve temperate lines such as N28, Oh43, B73, H15, F44, and others. Lines Ki 21 and Ki 44 also contain temperate germplasm. India, Pakistan, Vietnam, and China are using temperate germplasm in their breeding programs. MIR lines from the University of Hawaii contain temperate germplasm. More systematic efforts are, however, needed to make the best use of exotic germplasm.

Table 34–6. Performance of maize inbreds from different CIMMYT populations.

Population	Lines	Grain yield	Female flowering	Plant ht
	no.	t ha^{-1}	d	cm
24	12	3.30	60	157
28	8	3.05	63	158
36	5	4.05	59	175
21	24	2.54	63	161
43	20	2.68	62	169
21(drought)	3	3.11	62	165
Sint. An. TSR	30	3.37	61	162

OUTLOOK AND PERSPECTIVE

Hybrid area is growing in several countries and a shift from multiparent to two-parent hybrids is rapidly taking place. It is, therefore, timely that strategies emphasizing two-parent hybrids be considered and various issues and concerns relevant to this technology receive emphasis in breeding programs. Seed yield of the female parental line is extremely important in single-cross production. Issues relating to economics of seed production will require allocating more resources to inbred line development. Biotic and abiotic stresses are important and some efforts in their direction have already been made. Private sector involvement and investment has increased and will probably grow further. Private-public partnerships are becoming strong every day, promising a bright future for hybrid maize technology for tropical maize.

REFERENCES

Aekatasanawan C., S. Jampatong, N. DChuichoho, C. Balla, and C. Chutkaew. 1997. Exploitation of heterosis in maize at Kasetsart University, Thailand. CIMMYT 1997. p. 252–253. *In* Book of abstracts. The Genetics and Exploitation of Heterosis in Crops: An International Symposium, México, DF. México.

Beck, D.L., S.K. Vasal, and J. Crossa. 1990. Heterosis and combining of CIMMYT's tropical early and intermediate maturity maize (*Zea mays* L.) germplasm. Maydica 35: 279–285.

Brown, W.L., and M.M. Goodman. 1977. Races of corn. p. 49–88. *In* G.F. Sprague (ed.) Corn and corn improvement. 2nd ed. Agron. Monogr. 18. ASA, CSSA, and SSSA, Madison, WI.

CIMMYT. 1994. CIMMYT 1993/94 World Maize Facts and Trends. Maize seed industries, revesited: emerging roles of the public and private sectors. CIMMYT, México, DF. México.

Crossa, J., S.K. Vasal, and D.L. Beck. 1990. Combining ability estimates of CIMMYT tropical late yellow maize germplasm. Maydica 35:273–278.

Diallo, A.O. 1997. p. 9–11. Annual research report 1997. West Central Africa, Lowland Tropical Sub-Program. CIMMYT, México, DF. México.

Eberhart, S.A., M.N. Harrison, and F. Ogada. 1967. A comprehensive breeding system. Der Zuchter 37:169–174.

Eberhart, S.A., W. Salhuana, R. Sevilla, and S. Taba. 1995. Principles of tropical maize breeding. Maydica 40:339–355.

Edmeades, G.O., S.C. Chapman, J. Bolanos, M. Banziger, and H.R. Lafitte. 1994. p. 94–100. *In* Recent evaluations of progress in selection for drought tolerance in tropical maize. IV Eastern and Southern Africa Regional Maize Conf..

Goodman, M.M. 1985. Exotic maize germplasm: Status, prospects, and remedies. Iowa State J. Res. 59:497–527.

Hallauer, A.R., and S.A. Eberhart. 1970. Reciprocal full-sib selection. Crop Sci. 10:315–16.

Han, G.C., S.K. Vasal, D.L. Beck, and E. Elias. 1991. Combining ability analysis of inbred lines derived from CIMMYT maize (*Zea mays* L.) germplasm Maydica 36:57–64.

Kim, S.K., and V.O. Adetimirin. 1994. p. 255–262. *In* Overview of tolerance and resistance of maize to *Striga hermonthica* and *S. Asiatica*. Fourth Eastern and Southern Africa Regional Maize Conference..

Pandey, S., and C.O. Gardner. 1992. Recurrent selection for population, variety and hybrid development in tropical maize. Adv. Agron. 48:1–87.

Vasal, S.K., A. Ortega, and S. Pandey. 1982. CIMMYT's maize germplasm management, improvement and utilization program, CIMMYT, El Batan, México, DF. México.

Vasal, S.K., G. Srinivasan, D.L. Beck, J. Crossa, S. Pandey, and C. De Leon. 1992a. Heterosis and combining ability of CIMMYT's tropical late white maize germplasm. Maydica 37:217–223.

Vasal, S.K., G. Srinivasan, J. Crossa, and D.L. Beck. 1992b. Heterosis and combining ability of CIMMYT's subtropical and temperate early maturity maize germplasm. Crop Sci. 32 884–890.

Vasal, S.K., G. Srinivasan, S. Pandey, H.S. Cordova, G.C. Han, and F. Gonzalez C. 1992c.Heterotic patterns of ninety two white tropical CIMMYT maize lines. Maydica 37:259–270.

Vasal, S.K., G. Srinivasan, and N. Vergara. 1995. Registration of 12 hybrid-oriented maize germplasm tolerant to inbreeding depression. Crop Sci. 35:1233–1234.

Vasal, S.K., H.S. Cordova, D.L. Beck, and G.O. Edmeades. 1997. Choices among breeding procedures and strategies for developing stress tolerant maize germplasm. *In* G.O. Edmeades et al. (ed.) Proc. Developing drought- and low N-tolerant maize. 25–29 Mar. 1996. CIMMYT, El Batan, México.

Wellhausen, E.J. 1978. Recent developments in maize breeding in the tropics. p. 59–89. *In* D.B. Walden, (ed.) Maize breeding and genetics. John Wiley & Sons, New York.

Chapter 35

Heterosis in Sorghum and Pearl Millet

J. Axtell, I. Kapran, Y. Ibrahim, G. Ejeta, and D. J. Andrews

INTRODUCTION

Of the 700 million ha planted to cereals in the world, 45 million are planted to sorghum [*Sorghum bicolor* (L.) *Moench*] and about 80% of this is grown in developing countries (Dendy, 1995). Approximately 70 million tons of sorghum grain is produced annually as a dietary staple for some 500 million people in 30 countries. It has developed into an important feed grain in the USA, South and Central America, Mexico (Table 35–1), and Asia where yield increases (mainly through hybrid use) have been impressive (B. Maunder, personal communication). Sorghum also is used as a forage crop in the USA and several other countries. Pearl millet [*Pennisetum glaucum* (L.) R. Br.] is planted on about 28 million ha, mainly in Africa and India, to produce 10 million tons of grain per year for food for 70 million people. It also is an important summer annual forage crop in the USA, Australia, and South America (Hanna, 1996).

Sorghum and the millets (including finger millet and foxtail millet) are essential to diets of poor people in the semi-arid tropics where droughts cause frequent failures of other crops. They are most important in West Africa (70% of total cereal production) and North Africa where they provide from 8 to 91% of the total staples for >216 million people. In East Africa (30% of total cereal production), central and southern Africa they provide from 9 to 85% of the total staples for >126 million people. They are important crops for >2.1 billion people in Asia (mainly China and India).

Sorghum is second in importance to maize in Africa, south of the Sahara. However, it is probably more important than maize in Africa as a subsistence crop rather than as a commercial crop (Dendy, 1995; Maunder, 1990). The yields may be low under the area's environmental conditions but they are relatively the most dependable. Four countries in the Sahel of Africa with a population of 38 million depend on the millets (mainly pearl millet) for >1000 calories per person per day (Dendy, 1995).

Sorghum and pearl millet are, therefore, the two most important cereal crops grown to feed people living in the semi-arid low input dry-land agriculture regions of Africa and southeast Asia (Andrews and Bramel-Cox, 1993). In the last 35 years the area harvested to sorghum in Africa has nearly doubled, but yields averaging 800 kg ha^{-1} have not increased. A similar trend exists for all millets in Africa where the area planted has increased by 50% but yields, averaging 620 kg ha^{-1} and showing more fluctuation than sorghum, have changed little. By contrast in India during the same period sorghum area has declined 37%, but yields

Table 35–1. Area (thousand ha) planted to sorghum in USA, Argentina, and Mexico.

Period	USA	Argentina	Mexico
1991 – 1992	3994	823	1115
1992 – 1993	4876	810	540
1993 – 1994	3608	670	612
1994 – 1995	3609	622	868
1995 – 1996	3350	671	1412
1996 – 1997	4816	802	1353

have increased 80% (now about 880 kg ha^{-1}). Likewise area planted to millets has decreased 28%, but yields have increased 75% (now ±720 kg ha^{-1}; USDA, 1997). Hybrid use, and a consequent increase in genetic research has been a major factor in the yield increases recorded in India.

HYBRID DEVELOPMENT

Sorghum

Hybrid sorghum was first introduced commercially in the USA in 1956, although researchers at the Texas Experiment Station were interested in hybrid vigor in sorghums as early as 1920. R.E. Karper and A. B. Conner, in 1921, reported observed steriles (= male steriles) in the segregating generation of milo × kafir crosses; however, J.C. Stephens (Stephens & Quinby, 1952) must be credited for isolating a sterile-restorer system that would allow for the production of a hybrid in a plant with a perfect flower. Steriles found in 1929 and 1935 were discarded in favor of the Day Milo male sterile discovered by Glenn Kuykendall. N.W. Kramer in 1951 used sterile plants from a milo × kafir F$_2$ for backcrossing by Combine Kafir-60. These sterile Combine Kafir-60 plants were then crossed in 1953 by numerous males for testing in 1954. Previously, Stephens had tested hybrids made with the genetic male sterile and reported that crosses by common varieties yielded 40% more than parentals. It should be mentioned that Stephens and R.F. Holland in 1952 determined from milo × kafir crosses that cytoplasmic male-sterility (CMS) existed in sorghum. This ability to have 100% male steriles resulting from a cross quickly interested other agencies and the private industry to begin sorghum hybrid breeding programs.

J. Roy Quinby reported that after two years of testing, seven hybrids were worthy of production. These were given Texas Station or Regional Sorghum (RS) designations. To determine the economics of hybrid seed production, one kg of male sterile seed was put into the hands of about 200 seed growers along with suitable males in cooperation with the Extension Service. These small blocks suggested that pollination by windblown pollen was effective and that hybrid seed could be produced at a reasonable cost. By 1956, enough foundation seed was available from state foundation seed programs to plant > 4000 ha of seed fields. Early comparisons using the first hybrids suggested yield advantages of 15 to 25%.

Acceptance of sorghum hybrids occurred rapidly and approximated an S-curve with >1% of the acreage in 1956, 10 to 15% in 1957, 75% in hybrids in the 1959 crop, and >90% in hybrids in the early 1960s, with a present figure of above 95% (Maunder, 1969). National yield levels in excess of 50 bushels per acre are significantly higher than yields of <20 bushels just prior to hybrids.

Pearl Millet

Pearl (cattail) millet is the best annual summer grazing crop in the southern USA (Burton, 1995). A series of steadily improved hybrids replaced open pollinated varieties or synthetics since 1965 have raised pearl millet forage yields 50% in 20 years

In 1956 Burton (1958) discovered cytoplasmic male sterility (CMS) in crosses made in 1954 between a line (556) derived from Russian introductions, with inbreds (#12, 18, 23, and 26) used to make complex forage hybrid Gahi 1. In 1955 male sterile plants were noticed only in the F_2 of 556 × 23. These steriles, when pollinated with line 23, set seed and gave an all sterile backcross population, proving the 556 had carried the cytoplasm (designated A_1) that, interacting with the maintainer genes in line 23, supplied the CMS system that has been used to make hybrid pearl millet worldwide.

Gahi 3 (Tift $23A_1$ × Tift 186), the first forage hybrid was released in 1972. It yielded 10 to 19% more dry matter and gave 50% better animal gains than Gahi 1 (a complex of hybrids produced from first generation intercrosses in a bulk of four inbreds).

Hybrid Gahi 1 (which replaced the synthetic 'Starr') is 2 to 3 m tall when mature. A high percentage of the tall forage is unpalatable, low-quality stem. From a study of the inheritance of 5 dwarfs (Burton & Fortson, 1966) came the d_2 gene that shortened the stem 50% but did not dwarf other plant parts. Backcrossing the d_2 gene into Tift 23B and Tift 186, produced Tift $23DB_1$ and Tift 383. Tift $23A_1$ backcrossed to Tift 23DB made Tift $23DA_1$ with Tift 23DB its maintainer. Tift $23DA_1$ × Tift 383 produced a dwarf forage hybrid, Tifleaf 1, that possessed 60% more leaves and gave 15% better daily gains when grazed.

Hanna et al. (1985) discovered the dominant rust resistant gene, Rr_1, that was introduced into Tift $23DA_1$ and Tift 23DB to develop Tift $85D_1A_1$ and Tift $85D_1B_1$ that crossed with Tift 383 made Tifleaf 2 that yielded 22% better quality forage than Tifleaf 1 when exposed to rust.

Hanna (1993) released inbreds Tift $90D_2E_1$ A_1/B_1 (the female parent) and inbred Tift 8677 (the male fertility restorer) in 1991 and produced HGM100™, the first commercial pearl millet grain hybrid in the USA. HGM100™ (and similar pearl millet grain hybrids) have potential for producing high quality grain for poultry, livestock, and wildlife.

HB1 (Tift $23A_1$ × Bil 3B), the first CMS grain hybrid bred by D.S. Athwal in India, was released in 1965 because it had yielded 88% more grain than the best varieties in the 12 million ha pearl millet belt. With the help of HB1 and the next hybrid HB3 (Tift $23A_1$ × J104), India was able to increase pearl millet grain production from 3.5 million metric tons in 1965 to 8 million tons in 1970. Downy mildew soon became a major problem on the initial hybrids, and because of the adherence to the single cross dictum, progress in breeding resistant hybrids in India was slower than it might have been (Andrews & Bramel-Cox, 1993). Durable resistance to downy mildew existing in pearl millet—a naturally cross-pollinating species where traditional cultivars are heterozygous populations—proved very difficult to incorporate into inbreds. It was finally done, after 20 years, but the reliable supply of hybrids to Indian farmers still partly depends on the diversity that has now been created between parental lines, and hybrids, developed by the public and private sector. As in sorghum, other CMS systems have been discovered in pearl millet with various advantages and disadvantages. The A_4 system discovered by Hanna (1989), has shown numerous advantages (Andrews & Rajewski, 1994) over previous systems, facilitating more rapid breeding of parental lines, and better stability of male fertility restoration in hybrids.

BREEDING

Andrews et al (1998) reviewed current breeding procedures for sorghum and pearl millet hybrids. In both grain sorghum and pearl millet, single-cross hybrids are made with CMS seed parents and pollen parents that restore male fertility. In pearl millet, hybrids also can be made with an inbred (or an F_1) seed parent pollinated by a variety that probably are the best types for African conditions for reasons of disease resistance, adaptation and ease of production. Such topcross hybrids can be made either with or without CMS. For forage hybrids, a male sterile F_1 hybrid is used as a seed parent as this greatly increases seed yields, with little reduction in hybrid forage yields.

Improving levels of combining ability for yield together with desired grain qualities, phenotype, and wide adaptation are the keys to breeding hybrid parents. Early generation testing to select for combining ability is an accepted need, but a certain level of per se eliteness together with a good B or R reaction in the CMS system must be obtained before testing can begin. The concept of heterotic groups is useful in breeding hybrid parents, even if the classification is mostly by sterile (indicating B or maintainer) or fertile (R or restorer) reactions on a CMS sterile system. According to its reaction, new germplasm is added to either the B or R pool. Germplasm giving an incomplete reaction is normally avoided. However, new germplasm chosen for its per se values does not necessarily add to the heterotic value of the B or R pool.

Balanced breeding programs have both short term and long term aims. Better parents are most frequently obtained from elite × elite crosses, however, for sustained improvement new variability must be found and moved into elite backgrounds. Conventionally such introgression programs have backcrossed to elite lines. This may be satisfactory in pearl millet (because unadapted germplasm is usually a population), and for partly adapted sorghums but research on sorghum has shown distinct advantages of using adapted random mating populations (with genetic male sterility) over lines as recurrent parents to facilitate the discovery and the introgression of useful traits from very unadapted cultivated or even wild species (Menkir et al., 1994).

Hybrid testing programs are based on the principles of first testing in ideal conditions and against the known major specific constraints before extensive multilocation testing across 20 to 100 locations that will adequately expose new hybrids to the range of environmental variations expected in the target domain. In the USA in sorghum these are known as strip tests and are almost all placed in farmers fields and managed by farmers. This permits a good estimation of the G × E interaction of the new hybrids. Performance data (including grain qualities), extensive visual evaluations, farmers opinions and seed production aspects are all considered in the decision to release.

Existing breeding methodology should continue to raise yields at around 1% per year. Sorghum grain yields in the USA, since hybrids were introduced, have increased, from all causes, at an average of 1.25% per year during the last 36 years (USDA, 1997). There is very adequate untapped genetic variability in cultivated germplasm of both crops (for example in sorghum <10% of this has contributed to commercial U.S. hybrids). Breeding for quality and defensive traits, except perhaps for drought tolerance, may be accelerated by better methods of gene identification and transfer, which will almost certainly involve biotechnology. But the ability of biotechnology to help improve very complex traits such as yield potential still is uncertain for both technical (molecular research is specifically needed on methods that will complement combining ability testing) and economic reasons. In any case, improvements are needed in existing breeding methods to extract and transfer new genetic variability that will contribute to combining ability.

HETEROSIS

In cross-pollinated crops, such as pearl millet, it is not realistic to measure heterosis in relation to the performance of the inbred parents of single crosses because of inbreeding depression. The yield of a top cross (of a line) made with a good local variety compared to the variety parent gives a better value, or the comparison of a hybrid with the best established variety (of the same maturity) of the locality. This will usually show a 20 to 30% yield increase for a good hybrid; however, higher values have been recorded (+59% in Senegal in a variety × variety hybrid) [Iniadi × Souna II] (Lambert, 1982).

Virtually everywhere that sorghum hybrids have been compared with improved and landrace varieties, there has been a yield advantage, commonly on the order of 20 to 60% (Table 35–2) (House 1997). As growing conditions become stressed, the yields of both decline, but the yield differences between hybrids and varieties become proportionately larger, favoring the hybrids.

From a trial to compare the yields of old and new sorghum hybrids in Texas, USA, Miller and Kebede (1981) found that at least 40% yield increase had been realized in 30 years. Doggett (1988) emphasized the importance of yield increase achieved in the parent varieties. He claimed that about one-half of the yield increase can be ascribed to better parents. It is well established in sorghum breeding that good varieties make good parents for hybrids, because heterosis comes primarily from additive gene action (Kambal & Webster, 1965; Miller & Kebede, 1981).

Maunder (1983) summarized benefits and limitations of sorghum hybrids. Benefits were yield, maturity, seedling vigor, disease resistance, insect resistance, adaptation, drought resistance and grain quality. Limitations were stalk quality, cold resistance, grain quality, seed production cost, trueness to type, tillering, and midge resistance.

STRESS TOLERANCE

Both crops are usually grown under stress conditions (particularly moisture and temperature) in semiarid environments. Most cereal breeders acknowledge the benefits of heterosis in providing superior performance of hybrids when grown under stress conditions.

Tomes (1996) used the frequency of corn acres harvested among planted acres as a measure of the superior yield stability of hybrids. During the 'normal' years in the 1920s and 1930s when open-pollinated varieties predominated, about 85% of the corn planted was harvested. During the drought stress years of 1934 and 1936 harvest rates dipped to only 61 and 67%. After widespread adoption of hybrids (1940 and beyond), the proportion of harvested hectares has fluctuated from 85 to 92%, regardless of the environment of any particular year. However, the yield in each of these high stress years was lower than that recorded in more normal years.

Superiority of sorghum hybrids over inbred lines in stress environments also has been reported. They found that the hybrid yield increase was 58% over the best parent under dry-land conditions, but only 22% under irrigation conditions in Texas USA (Quinby et al., 1958). The actual mean yield increases were 612 kg ha^{-1} under dry-land conditions, and 790 kg ha^{-1} under irrigation. From the results of 391 trials carried out in four countries, (Doggett, 1969) concluded that the yield increases of the hybrid sorghum relative to the open-pollinated varieties are constant over a wide range of growing conditions and management levels. Under conditions of low varietal yield levels, hybrids can give more than double the variety yield. In Sudan, three elite hybrids had a combined mean yield of 50% over the best open-pollinated local varieties in 27 yield trials across four seasons (Ejeta,

Table 35–2. Comparative performance of hybrids, improved varieties, and land race cultivars.

Country	Years of testing	Number of trials	No. of entries			Yield kg ha^{-1}			% Increase over		Type of test	Authority
			Hybrid	Var.	Local	Hybrid	Var.	Local	Var.	Local		
India												
Kharif	1985-1990	-	7	4		3665	3189		14.9		Regional	Murty, U.R. 1992
Rabi	1981-1987	13	15-100	1		2400	2900		-17.2		Station	Reddy, B.V.S. 1994
	1983-1987	5	31-100	1		2400	4300		-44			
Zambia	1989-1990		3	3		3977	3238		22.8		Demon.	Verma, B.N.
	1990-1991		3	3		4162	3091		34.6		Demon.	Pers. Com.
	1991-1992		3	3		2741	1928		42.2		Demon.	Pers. Com.
Sudan	1985-irr		1	1		5189	3010		72.3		Regional	Ejeta, G. 1985
	1985-dry	2	1	1		2968	1543		92.3		Regional	Ejeta, G., 1985
	1986	2	1	1		4152	2700		53.7		Regional	Ejeta, G. 1985
		2	1	1		2670	2483		7.5		Regional	
		2	1	1		3891	3113		24.9		Regional	
		9	6	1		4573	3109		47.0		Regional	
Niger	1988-irr		90	33	2	2582	1779	1605	45	61	Station	Kapran, I. 1988
	1988-dry		90	33	2	1799	1081	1204	66	49	Station	
West Central Africa†	1986	87	10-34	1		2970	1770		67.7		Regional	Murty, D.C. Pers. Com.
	1995					4340	3120		39.1		Regional	Pers. Com.
Burkina Faso												
Farako Ba		1	21	2	1	2258-4241	2795-3236	1345	-19 31	67	Regional	Murty, D.S. Pers. Com.
Kamboinse		1	21	2	1	945-1904	0-679	331	180 -33	185	Regional	Pers. Com.
Ouahigouya		1	21	2	1	687-2302	1031-1490	271	54 39	154	Regional	Pers. Com.
India			14	15		3254	2336		39		Intl.	ICRISAT Annual Report
Pakistan			14	15		1663	1535		8		Intl.	
Thailand			14	15		3007	4775		-37		Intl.	
Philippines			14	15		5394	2327		131		Intl.	
El Salvador			14	15		6196	3549		75		Intl.	
Venezuela			14	15		6826	4707		45		Intl.	

† Benin, Burkina Faso, Cameroon, Cote d'Ivoire, Ghana, Mali, Niger, Nigeria, Senegal, Togo.

1983). One of these hybrids was released (HD-1), ATX623 × K1597 as the first sorghum hybrid in sub-Saharan Africa.

In another trial, 126 sorghum genotypes were tested in an international yield trial in 15 different environments in the USA, Africa, and South America (Ibrahim, 1995). Genotypes tested include 30 inbred lines and their hybrids on each of three A-lines, commercial hybrids, and B-lines. The results showed that G × E interaction did not alter the relative rank of cultivar types: hybrids ranked higher than inbred lines in all environments. Differences between mean yield of hybrids and their parental inbred lines were greater in stress environments than in favorable environments. The author concluded that hybrid varieties were superior to inbred lines in all environments and sorghum hybrids appeared to be more reliable than inbred varieties in erratic environments typical of sorghum growing regions in the semi-arid tropics.

SEED INDUSTRY

The paradox is that in spite of the fact that sorghum and pearl millet are usually grown under stress conditions, agricultural policy makers and sorghum researchers in developing countries have been reluctant to adopt sorghum hybrids. They believe that hybrids are adapted to and, therefore, profitable only under high-yielding favorable environments, where modern production practices are employed and production inputs are available. Numerous evaluations (above) show otherwise.

The lack of a viable private-sector seed industry in many countries is the major impediment to the use of hybrids in sorghum and pearl millet. Maunder, at this conference, has reviewed many of the reasons for this predicament. It seems clear that the availability of well-adapted hybrids can be used as a tool to foster the development of a successful seed industry in the private sector with all of the accompanying benefits.

The best example of this is the Indian experience of the last 40 years. A Rockefeller team of 14 scientists laid the groundwork for the development of hybrid maize, sorghum and pearl millet in India. An indigenous seed industry, which now numbers >35 companies, has emerged with significant impact on food production in India. This is an aspect of the green revolution in India, which is not well known outside India, but has important implications for other developing countries, especially in Africa. At times it seems the world has forgotten the importance of seeds and seed technology in delivering agricultural research to the farmers (Douglas, 1980).

SORGHUM HYBRIDS IN NIGER
A CASE HISTORY OF HYBRID DEVELOPMENT IN THE SAHEL

Sorghum is the second most important food crop in the West African country of Niger. The crop is grown as rainfed under 400 to 800 mm of annual precipitation. Annual acreage of sorghum in Niger increased from <500 000 ha in 1961 to two million ha in 1996. In contrast, national average yields declined from 0.6 t ha^{-1} in 1961 to 0.2 t ha^{-1} in 1996. Cultivation into marginal lands with declining fertility and the continued use of low-yielding local cultivars contributed to this limited productivity. Research on sorghum improvement in Niger started in the 1950s with the French Research Institute for Tropical Agriculture (IRAT). Emphasis was on the breeding of open-pollinated varieties up to the establishment of INRAN, the National Institute for Agricultural Research in Niger. In the 1980s, sorghum breeders at INRAN in collaboration with the International Sorghum and Millet Program (INTSORMIL), added a hybrid testing component to the selection of improved cultivars. Experience elsewhere had shown that hybrids have the potential of modernizing the agricultural sector of a country and contributing to national food sufficiency. Our collaborative effort aimed at evaluating the superiority of hybrid sorghum cultivars under the marginal growing conditions prevalent in much of Niger.

Experimental hybrids were synthesized from adapted germplasm provided by the Purdue Sorghum Research Program and tested at several INRAN field research stations. Selected hybrids gradually moved from observation nurseries of preliminary trials to advanced yield trials at several locations for two to three years. The experimental material included at least one open-pollinated local variety as a check. Elite hybrids resulting from these national tests also were evaluated in the West African Sorghum Hybrid Adaptation Trial (WASHAT). Data collected included estimates of maturity, plant height, plant stand, grain yield and grain quality, and reactions to biotic and abiotic stresses. An elite hybrid, ATX623 × MR732, later named NAD-1, emerged as the most promising based on the summary of the cumulative data. Its superior performance has been on demonstration in farmer's fields in several locations. Experimental seed production of NAD-1 has been underway since 1989 to evaluate the feasibility of commercial hybrid seed production.

Agronomic evaluation: a summary of the annual analyses of data reported by INRAN sorghum breeders is presented in Table 35–3. Experimental hybrid yields averaged approximately 2 t ha^{-1}. Best hybrids yielded up to 6.5 t ha^{-1} whereas best variety checks never exceeded 3.8 t ha^{-1}. These results confirm the findings of two graduate thesis projects conducted at Purdue University, which initiated the thrust on sorghum hybrid research in Niger. Kapran (1988) reported heterosis values of 45% under irrigation and 66% in rainfed conditions of Niger. In the same tests, hybrids also out-yielded local checks by 61% with irrigation and 49% under rainfed conditions. Tyler (1988) found that hybrids with parents grouped as exotic, intermediate, or local, were higher yielding than their respective parents by an average of 127, 83, and 66%. Through the results of these early thesis projects and subsequent national tests conducted through INRAN and INTSORMIL collaboration, the agronomic value of heterosis for Niger's agriculture has largely been demonstrated.

NAD-1, a medium-maturing, white-seeded hybrid with good tolerance of mid-season drought (common in Niger), was the most attractive cultivar to breeders and farmers alike. In national trials, its yield average was 3.1 t ha^{-1} between 1986 and 1992. Also in the 1989 WASHAT trial, it ranked third of 20 entries for grain yield at nine locations across West and Central Africa. Estimates of NAD-1 yield potential under farmer management varied between 1.7 t ha^{-1} and

Table 35–3. Yields of sorghum hybrids in research station tests in Niger, 1984-92.

Year	Number of hybrids tested	Grain yield (t ha^{-1})		
		Trial mean	Best hybrid	Local variety
1984	22	2.3	5.6	---
1985	17	---	4.0	---
1986	149	2.5	2.7	1.1
1987	147	2.6	5.5	3.2
1988	81	1.9	4.9	2.7
1989	67	2.8	6.5	3.8
1990	49	2.4	5.0	3.0
1991	78	1.8	2.0	1.0
1992	88	1.9	4.6	---

Table 35–4. Yield of elite sorghum hybrid, NAD-1, under farmer conditions in Niger, 1993 to 1996 (t ha^{-1}).

Year—Activity	NAD-1	Local variety	% Gain from hybrid
1993 Demonstrations	2.4	---	---
1994 Demonstrations	3.3	---	---
1995 NAD-1 vs. MM trial	1.6	1.0	60
1996 NAD-1 vs. MM trial	1.7	0.7	143
Regional network trial	1.8	1.0	80

3.3 t ha^{-1} were obtained starting in 1993 (Table 35–4). Despite severe drought conditions, the hybrid was superior to Mota Maradi (MM), an early-maturing and widely adapted local variety. Also, from on-farm trials of the regional sorghum network, the yield of NAD-1 was 80% higher than the average of farmer checks. The strong interest expressed by farmers in this hybrid over the years is thus easily understandable.

Experimental seed production: NAD-1 seed production by INRAN grew from a 200 m^2 plot in 1989 to several ha today. Under good management, the equivalent of 1.5 t ha^{-1} of hybrid seed has been obtained. Additional hybrid seed production is now in progress by Kapran and House with INTSORMIL, INRAN, ICRISAT and World Bank support.

Seed business activity: Having a hybrid that appeals to farmers and can be produced are key elements for seed marketing. For the first time in Niger, hybrid seed in 1996 was sold at a price eight times that of sorghum grain. However, INRAN, as a public research organization, has no mandate for commercial seed activity. We are actively approaching extension, farmer coops, and individuals that potentially could turn into commercial seed producers.

It is concluded that heterosis can be used to improve agricultural productivity in Niger and similar Sahelian countries. Sorghum hybrid, NAD-1, also demonstrates the value of heterosis in other important crops like millet. The momentum created by this hybrid is today being used by INRAN, INTSORMIL, and the Sahelian Center of ICRISAT, to educate policy makers and private producers on the need for, and advantages of, launching a hybrid seed industry in Niger. The role of the private sector will be paramount, but the involvement of the public sector in gestation and supporting activities is also essential.

CONCLUSION

In both species the exploitation of heterosis has led to sustained improvement in both grain and forage yields. The discovery of cytoplasmic male sterility, in the same decade, was a necessary precursor in both. Though sorghum is predominantly (but no means completely) a self pollinating species, and pearl millet essentially cross pollinating, they both are diploids and show similar useful levels of heterosis over equivalent varieties. Sorghum grain hybrids have been successfully bred, both for productive warm temperate environments in the USA, where yields now average 4.2 t ha^{-1} and for more stressful environments in a number of countries such as India where yields now average 0.88 t ha^{-1}. Pearl millet hybrids for grain, until recently have only been bred for stressful conditions in India, where yields average 0.67 t ha^{-1} (though the same hybrids have reliably shown to give 5 t ha^{-1} in 85 days) yet even at these relatively modest productivity levels sorghum and pearl millet hybrid breeding and production is a multi-million dollar indigenous industry. In all cases, the yield increase trend show no sign of plateauing. With

more powerful breeding techniques now available and with the application of biotechnology, gains in hybrid productivity are assured, firstly through better genetic
control of biotic and abiotic stresses, and secondly through gain in yield potential.
    The huge production areas of sorghum and pearl millet in Africa have yet to
benefit from the use of heterosis. There are great opportunities, as the two instances described in this paper show. Results from hybrid research in many other
countries show that the use of heterosis will be the basis of the dryland (green)
revolution in sorghum and pearl millet production in Africa

REFERENCES

Andrews, D.J., and P.J. Bramel-Cox. 1993. Breeding cultivars for sustainable
    crop production in low input dryland agriculture in the tropics. p. 211–223 *In*
    D.R. Buxton et al. (ed.) International Crop Sci. I.. CSSA, Madison, WI.

Andrews, D.J. and J.F. Rajewski. 1994. Male fertility restoration and attributes of
    the A_4 cytoplasmic-nuclear male sterile system for grain production in pearl
    millet. Int. Sorghum Millets Newsl. 35:64.

Andrews, D.J., G. Ejeta, M. Gilbert, P. Goswamy, Anand Kumar, A.B. Maunder,
    K. Porter, K.N. Rai; J.F Rajewski, Belum Reddy, W. Stegmeier, and B.S. Ta-
    lukdar. 1998. Breeding hybrid parents. p. 173–185 *In* D. Rosenow et al. (ed.)
    Genetic improvement of sorghum and pearl millet. INTSORMIL/UNL.

Burton, G.W. 1958. Cytoplasmic male-sterility in pearl millet *Pennisetum glau-
    cum* (L).R.Br., Agron. J. 50:230.

Burton, G. 1995. Hybrid development in pearl millet. *In* Proc. First Grain Pearl
    Millet Symp. 17–18 Jan. 1995. Tifton, GA.

Burton, G.W., and Forston, J.C. 1966. Inheritance and utilization of five dwarfs in
    pearl millet (*Pennisetum typhoides)* breeding. Crop Sci. 6:69–72.

Dendy, D.A.V. 1995. Sorghum and the millets: Chemistry and technology. Am.
    Assoc. of Cereal Chem., St. Paul, MN.

Doggett, H. 1988. Sorghum. 2nd ed. Longman Scientific & Technical.

Doggett, H. 1969. Yields of hybrid sorghums. Exp. Agric. 5:1.

Douglas, J.E. 1980. Successful seed programs: A planning and management
    guide. Westview Press, Boulder, CO.

Ejeta, G. 1983. Current status of sorghum improvement research and development
    in the Sudan. p. 17. *In* E. Gebisa (ed.). Hybrid Sorghum Seed for Sudan.

Hanna, W.W. 1989. Characteristics and stability of a new nuclear-cytoplasmic
    male-sterile source in pearl millet. Crop Sci. 29:1457–1459.

Hanna, W.W. 1993. Registration of pearl millet parental lines TIFT8677 and
    A1/B1 TIFT90D2E1. Crop Sci. 33:1119.

Hanna, W.W. 1996. Improvement of millets – emerging trends. *In* Proc. 2nd Int.
    Crop Science Congress. Delhi, India.

Hanna, W.W., H.D. Wells, and G.W. Burton. 1985. Dominant gene for rust resistance in pearl millet. J. Hered. 76:134.

House, L.R., B.N. Verma, G. Ejeta, B.S. Rana, I. Kapran, A.B. Obilana, and B.V.S. Reddy. 1997. Developing countries breeding and potential of hybrid sorghum. p. 84–96. *In* Proc. Int. Conf. on Genetic Improvement of Sorghum and Pearl Millet. INTSORMIL Publication 97–5. 23–27 Sept. 1996. Lubbock, TX, Univ. of Nebraska, Lincoln.

Ibrahim, Y. 1995. Genotype × environment interaction and grain yield stability of drought tolerant sorghum inbred lines and hybrids. M.S. thesis. Purdue Univ., West Lafayette, IN.

Kambal A.E., and O.J. Webster. 1965. Estimates of general and specific combining in sorghum. Crop Sci. 5:521–523.

Kapran, I. 1988. Evaluation of the agronomic performance and food quality characteristics of experimental sorghum hybrids in Niger, West Africa. M.S. thesis. Purdue Univ., West Lafayette, IN.

Lambert, C. 1982. IRAT et l'amelioraton de mil. Agron. Trop. 38:78–88.

Maunder, A.B. 1969. Meeting the challenge of sorghum improvement. p. 135–151. *In* J.I. Sutherland and R.J. Falasca (ed.) Proc. 24th Annual Corn and Sorghum Research Conf.. Chicago, IL. 9–11 Dec. 1969. Am. Seed Trade Assoc., Washington, DC.

Maunder, A.B. 1983. Development and perspectives of the hybrid sorghum seed industry in the Americas. p. 39–48. *In* G. Ejeta (ed.) Proceedings of Hybrid Sorghum Seed for Sudan Workshop. Purdue University, W. Lafayette, IN.

Maunder, A.B. 1990. Importance of sorghum on a global scale. p. 8–16. *In* G. Ejeta et al. (ed.). Proc. Int. Conf. on Sorghum Nutritional Quality. Purdue Univ., West Lafayette, IN.

Menkir, A., P.J. Bramel-Cox, and M.S. Witt. 1994. Comparisons of methods for introgressing exotic germplasm into adapted sorghum. Theor. Appl. Genet. 89:233–239.

Miller, F.R., and Y. Kebede. 1981. Genetic contribution to yield gains in sorghum 1950 to 1980. p. 1–14. *In* W.R. Fehr (ed.) Genetic Contributions to Yield Gains of Five Major Crop Plants. CSSA Spec. Publ. 7. CSSA, ASA, Madison, WI.

Quinby, J.R., N.W. Kramer, J.C. Stephen, K.A Lahir, and R.E. Karper. 1958. Sorghum production in Texas. Texas A&M Agric. Exp. St. Bull. 912.

Stephens, J.C., and J.R. Quinby. 1952. Yield of hand produced hybrid of sorghum. Agron. J. 44:231–233.

Tomes, D.T. 1996. Heterosis: Performance stability, adaptability to changing technology, and the foundation of agriculture as a business. p.13–27. *In* K.R. Lamkey, and J.E. Staub (ed.) Concepts and Breeding of Heterosis in Crop Plants. CSSA Spec. Publ. 25. CSSA, Madison, WI.

Tyler, T. 1988. Heterotic pattern and combining ability for agronomic and food grain quality traits in exotic × exotic, exotic × intermediate, and exotic × local sorghum hybrids in Niger. M.S. thesis. Purdue Univ., West Lafayette, IN.

USDA. 1997. Time series data base. Economic Research Service USDA, Washington,

Chapter 36

Heterosis in Vegetable Crops

T. C. Wehner

INTRODUCTION

Darwin's (1877) research on maize (*Zea mays* L.) demonstrated that cross fertilization was generally beneficial, and self fertilization was generally injurious. Later, McClure (1892) reported that hybrids made by crossing maize cultivars were generally superior to the midparent. In vegetable crops, some of the earliest research was by Hayes and Jones (1916) who reported hybrid vigor for cucumber (*Cucumis sativus* L.) fruit size and number. There has been much interest in hybrids in many breeding programs, although hybrids have not been exploited commercially as much in self-pollinated as cross-pollinated crops. The literature includes many estimates of hybrid advantage relative to the midparent, the high parent, or to the best standard (inbred or open pollinated) cultivars. The latter is most interesting to those evaluating the commercial potential of heterosis. In this chapter, I will use heterosis to mean hybrid vigor relative to the better parent, or where inbreeding depression is severe, to the comparable open pollinated cultivars.

Heterosis does not occur universally in vegetable crops. In fact, some self-pollinated vegetable crop species do not express any inbreeding depression or heterosis. Although there is heterosis expressed for yield traits in many of the vegetable crops, a primary advantage of hybrid cultivars is the protection they provide for proprietary lines developed by plant breeders, especially those working in seed companies. Commercial plant breeding programs have flourished in the USA through the use of hybrid cultivars to protect proprietary lines. In the future, the advantage offered by hybrids will be reduced because of patents and plant variety protection (PVP), especially as it becomes economical to distinguish among cultivars using molecular markers (DNA fingerprinting). Other advantages of hybrids include the ability to combine useful dominant genes available in different inbreds lines, optimizing the expression of genes in the heterozygous state, and the production of unique traits such as seedless triploid watermelons [*Citrullus lanatus* (Thunb.) Matsum. & Nakai].

The vegetable crops can be grouped according to how adaptable they are to hybrid production. For purposes of this chapter, I have divided them into self- pollinated crops with few or many seeds produced per cross, and cross-pollinated crops with a low or high rate of natural outcrossing. Important self-pollinated vegetable crops that have been researched extensively include the legumes [bean (*Phaseolus* sp.) and pea (*Pisum sativum* L.)], Solanaceae [eggplant (*Solanum melongena* L.), pepper (*Capsicum nigrum* L.), tomato (*Lycopersicon esculentum* Mill.)], and lettuce (*Lactuca sativa* L.). Important cross-pollinated vegetable crops include the cucurbits [cucumber, melon (*Cucumis melo* L.), squash (*Cucurbita sp.*), and watermelon], the cole crops [broccoli (*Brassica oleracea* L.), cabbage (*Brassica oleracea* L.), and cauliflower (*Brassica oleracea* L.)], root and bulb crops [carrot (*Daucus carota* L.) and onion (*Allium cepa* L.),], asparagus (*Asparagus officianalis* L.), and spinach (*Spinacia oleracea* L.).

SELF-POLLINATED CROPS

Few Seeds Per Cross

Self-pollinated crops that produce few seeds per cross make it difficult to produce hybrids economically. Vegetable crops such as legumes and lettuce are important in the USA and will be used to represent this group.

Fabaceae (Leguminosae)

Phaseolus vulgaris L. is a hermaphroditic annual used for both green beans and dry beans. It is naturally self pollinated, and produces few seeds per cross. Commercially, 70 kg seeds ha^{-1} (with 2900 seeds kg^{-1}) are required to plant a dry bean crop. If hybrid seeds were used, it would take 68 000 crosses to plant a 1 ha crop. Flowers must be emasculated before they open, and the work must be done by hand. Thus, it is not economical to produce commercial hybrids, but crossing could be facilitated by the transfer of genes for extrorse stigma from *Phaseolus coccineus* L. (a cross-pollinated species) to *P. vulgaris*; however, there are no reasons to produce hybrids, since little heterosis for yield has been observed.

Like bean, pea is a hermaphroditic annual with little evidence of inbreeding depression. Although Gritton (1975) reported heterosis of 28% over the high parent for dry seed yield, for most traits there is no evidence of heterosis. Since there is no economical method for hybrid production, future cultivars will likely be inbred lines rather than hybrids.

Lettuce is considered a major U.S. crop with 130 000 ha in the USA in 1996, is a hermaphroditic annual, and is another good example of a crop where commercial use of hybrid cultivars is not feasible (Ryder & Waycott, 1993). Many seeds are required for planting (250 000 seeds ha^{-1}), but few are produced from hand pollinations (20 seeds cross^{-1}). The pollen is sticky, and cannot be carried by the wind-and no insect pollinators are known. Heterosis is generally not important, but may occur for secondary traits such as seedling size (E. Ryder, 1997, personal communication). The use of hybrids for protection of proprietary inbred lines has decreased in importance with the availability of plant variety protection and plant patents and the development of molecular markers.

Many Seeds Per Cross

Many crops in the Solanaceae are self pollinated and are adaptable to hybrid production. Eggplant, pepper, and tomato are examples of successful use of hybrids.

Solanaceae

Heterosis in eggplant is large, with hybrid yield advantage from 33% (Tiwari, 1966) to 97% (Dharme, 1977). Because eggplant flowers are perfect, and normally self pollinate, the natural outcrossing rate is near zero. Since hand pollination is easy and produces many seeds per cross, hybrid production has been commercialized widely. Many hybrid cultivars are now available, with half of the production area in the USA planted to hybrids (Table 36–1).

Table 36–1. Use of hybrids and value of heterosis in 21 vegetable crops for the USA, 1996.

Category *Group* Crop	Crop area†	Crop area in hybrids†	Estimated % F₁ inc. over OPs or inbreds	Amount agric. land saved by heterosis	Additional people fed by heterosis
	ha	%	%	ha	%
Self pollinated, few seeds per cross					
Legumes					
Bean, Green	112 389‡	0	0	0	0
Pea, Green	134 737‡	0	0	0	0
Other					
Lettuce	131 151	0	0	0	0
Self pollinated, many seeds per cross					
Solanaceous					
Eggplant	4 699	50	65	1 527	32
Pepper, Hot	47 429	2	35	332	1
Pepper, Sweet	86 616	36	35	10 914	13
Tomato, Fresh	85 983	56	19	9 149	11
Tomato, Proc.	166 235	74	19	23 373	14
Cross pollinated, low outcrossing rate					
Cucurbits					
Cucumber, Pickle	46 128	100	5	2 306	5
Cucumber, Slicer	43 638	76	5	1 658	4
Melon	85 508	90	8	6 157	7
Squash, Summer	57 655	52	44	13 192	23
Squash, Winter	9 957	35	40	1 394	14
Watermelon	126 802	33	10	4 184	3
Cross pollinated, high outcrossing rate					
Cole crops					
Broccoli	94 480	100	65	61 412	65
Cabbage	82 575	84	14	9 711	12
Cauliflower	35 619	33	10	1 175	3
Root & Bulb					
Carrot	78 462	56	28	12 303	16
Onion	102 665	65§	40	26 693	26
Other					
Asparagus	34 858‡	90	106	33 254	95
Spinach	8 908	100	18	1 603	18
21-crop total or mean	*1 576 494*	-	*27*	*220 337*	*18*

† From Janick (1998).
‡ From USDA (1992).
§ Northern USA is planted to 90% hybrids.

   Pepper is an annual, self-pollinated crop with a natural outcrossing rate of 25%. Dikil et al. (1973), reported yield heterosis of 28 to 47%, with the higher levels occurring when different ecological groups with different growth patterns were crossed. In a study done in Israel where peppers are grown for export, hybrids had a 9% advantage over inbred cultivars for total yield, but a 75% advantage in export-quality yield (Shifriss & Rylski, 1973). Although cytoplasmic male sterility systems exist, they are not reliable for hybrid production. Thus, commercial hybrids are produced through hand emasculation and hand pollination. In the USA, 36% of the production area of sweet pepper is in hybrid cultivars, but only 2% of the hot pepper is in hybrids (Table 36–1).

Tomato is a hermaphroditic annual (or perennial in the tropics). Heterosis was reported 90 years ago by Hedrick and Booth (1907), and many researchers since then have reported heterosis for yield and earliness. Heterosis has been reported in numerous research articles, and hybrid advantage ranges from 0 to 300% over the best inbreds. In one series of studies (Yordanov, 1983), heterosis averaged about 19%, with 50% for early yield (harvests 1–5) and 4% for late yield (harvests 6–15). East and Hayes (1912) suggested that F_1 hybrids could be made easily and have practical value, but hybrid cultivars have become important only since the 1970s when seed companies needed a way to protect their research investment in inbred lines. Besides heterosis, hybrids permit the combination of traits controlled by dominant alleles present in one or the other parent. Several important dominant genes used for tomato improvement include resistance to fusarium wilt, verticillium wilt, root-knot nematode, tobacco mosaic virus, and tomato spotted wilt virus. Also, performance of some genes, such as ripening inhibitor (*rin*) and non ripening (*nor*), is optimum in the heterozygous state: homozygous recessives produce fruits that have poor flavor and no red color, but heterozygotes become red and flavorful while keeping longer. Griffing (1990) showed that heterosis in a cross of two tomato inbreds (one with few large fruits and one with many small fruits) was not due to mock heterosis (a mathematical phenomenon described by Richey, 1942), but rather to faster growth rate in the hybrid.

The natural outcrossing rate in cultivated tomatoes is 0.5 to 4% in temperate areas (Scott & Angell, 1998), but higher in tropical areas. Insects are the main vector, although wind also is a factor. Most hybrids are produced by hand emasculation and hand pollination; however, it may be possible to produce seeds less expensively using male sterility or exserted stigma genes to increase the number of seeds and reduce the time required per cross. Hand emasculation resulted in 27 to 50 seeds per cross depending on year and location, and took 64 to 120 hours to produce 1 kg of seeds (Oba et al., 1945). In field evaluations, Shakya and Scott (1983) obtained 32 to 92 seeds per pollination with hand emasculation, and 118 to 145 seeds per pollination using a male sterile inbred; however, Scott and Angell (1998) made a case for hand emasculation rather than using male sterility because of the additional time required to convert inbred lines relative to the amount of labor saved. Although hybrid seeds are more expensive than seeds of standard (inbred) cultivars, their improved performance, uniformity, and protection of proprietary lines has made them very popular in the USA. Transplants in Florida cost the grower $185 ha^{-1} for hybrid cultivars and $12 ha^{-1} for inbred cultivars (Scott & Angell, 1998). However, the cost of hybrid seed was only 2 to 3 % of the total production and marketing cost for the grower (Smith and Taylor, 1993). Currently in the USA, 56% of fresh market and 74% of processing tomato crop area is planted to hybrid cultivars (Table 36–1).

CROSS-POLLINATED CROPS

Low Outcrossing Rate

Cucurbitaceae

Cucumber is an annual outcrosser with monoecious (first staminate, then pistillate flowers) sex expression. Several genes control sex expression to produce gynoecious (all pistillate flowers), andromonoecious (staminate, then perfect flowers), and hermaphroditic (all perfect flowers) types. No inbreeding depression was found in inbred lines developed from a pickling cucumber population (Rubino &

Wehner, 1986), perhaps because cucumbers are often grown in small groups of plants where there is little outcrossing. Wehner and Jenkins (1985) measured a natural outcrossing rate of 53% (36% between row). Some have reported heterosis from specific hybrid combinations (Ghaderi & Lower, 1979), especially when crossing diverse parents (Hayes & Jones, 1916). However, others have found no heterosis when parents were of similar type (Hayes & Jones, 1916). In a study involving three years, two replications, and eight locations in the USA (Wehner et al., unpublished data), the total once-over harvest yield of the popular gynoecious hybrid 'Calypso' was 60 fruits per plot, while the gynoecious inbred parent Gy 14 had 61 fruits per plot. Where breeders have worked to develop elite hybrids, heterosis may average only a 5% advantage.

Hybrids are produced commercially using a gynoecious inbred as the female parent crossed with a monoecious inbred as the male parent, and honey bees as pollen vectors. Gynoecious inbreds can be developed by self pollination of plants that have been treated with silver nitrate or other ethylene inhibitors (Tolla & Peterson, 1979). One advantage of producing gynoecious by monoecious crosses is that the resulting hybrid will be gynoecious, having pistillate flowers at every node. A monoecious pollenizer must be mixed in with the hybrid if it is not parthenocarpic. Gynoecious cultivars have earlier, and sometimes higher, yield than monoecious cultivars since they have pistillate flowers at every node (Wehner & Miller, 1985).

Melon has no inbreeding depression, but some crosses were reported to express heterosis of 3% for earliness and 8% for yield (Lippert & Legg, 1972). Andromonoecious lines had only 20 to 35% natural outcrossing (Whitaker & Bohn, 1952), so the lack of inbreeding depression might be explained in the same way as for cucumber. Hybrids are produced by hand pollination of emasculated perfect flowers on the female parent using staminate flowers from the male parent. Gynoecious inbreds have been produced, but fruits from pistillate flowers are oval, rather than round like those produced from perfect flowers. Also, gynoecy is controlled by several genes and is complex to work with. The use of cytoplasmic-genic male sterility would reduce hand labor requirements for crossing in hybrid production blocks.

Summer squash (*Cucurbita pepo* L.) has been shown to have heterosis for yield, with 11 to 84% hybrid advantage over open pollinated cultivars in zucchini types and 0 to 82% advantage for yellow types (Elmstrom, 1978). Hybrids are produced by crossing two monoecious inbred lines using honey bees as the pollen vector. One of the inbreds is made gynoecious for the first two to three weeks of flowering by spraying the plants with ethephon at the two and four-leaf stage. Hybrids are used in 52% of the U.S. crop. On the other hand, winter squash (*Cucurbita maxima* Duch and *C. moschata* Duch) is difficult to cross using monoecious inbreds since ethephon does not work well to change sex expression. Heterosis for yield in hybrid cultivars may be 40% higher than for the open pollinated ones, and an increasing proportion of the crop is planted to hybrids (Table 36–1).

Watermelon is a monoecious, annual outcrossing species with little inbreeding depression. Heterosis is expressed for yield in some parental combinations, perhaps averaging 10%; however, the main use of hybrids has been to protect the parental inbreds, and to produce seedless fruits. Seedless hybrids are triploids produced by crossing a tetraploid female parent with a diploid male parent. Tetraploids are produced by doubling the chromosome number of elite diploid inbred lines. A diploid cultivar must be planted in alternating rows with the triploid cultivar to provide the pollen required for fruit set, so seeded watermelons are produced along with the seedless ones.

High Outcrossing Rate

The ideal crops for purposes of hybrid production are those where there is a high rate of natural outcrossing, and methods for economical control of self and cross pollination. The cole crops, root and bulb crops, asparagus, and spinach are good examples.

Brassicaceae (Cruciferae)

Broccoli, cabbage, and cauliflower are hermaphroditic annuals with a high degree of outcrossing enforced by sporophytic self incompatibility. Significant inbreeding depression and heterosis is present. The exception is summer cauliflower (the most recent and modern crop of the species) where there is no inbreeding depression and no self incompatibility. Heterosis of hybrids over open pollinated cultivars may be only 10%. Cabbage hybrids were 12 to 15% better yielding than open pollinated cultivars (Pearson, 1983). In broccoli, hybrid cultivars yielded 40 to 90% better than open pollinated ones (Morelock et al., 1972), but that comparison was not entirely due to heterosis. High rates of natural outcrossing and heterosis in the cole crops has made them ideal for hybrid production, with 84% of cabbage and 100% of broccoli land in hybrid cultivars (Table 36–1). Another significant factor in the adoption of hybrids has been their greater uniformity in yield and quality relative to open pollinated cultivars. Single cross hybrids are produced by crossing two self incompatible inbred lines in an isolation block. Self incompatible inbreds can be self pollinated using bud pollination, since flowers are self compatible before they open. The system permits economical production of hybrid seed, but is "leaky". Occasional self- or sib-pollinations result in seeds of the parental inbreds being sold in the hybrid cultivar. One solution has been to use four-way hybrids (with occasional two-way hybrids mixed in). Cytoplasmic male sterility is being phased in for hybrid production since it is more reliable in protecting the proprietary inbred lines used as parents.

Root and bulb crops

Carrot (*Daucus carota* L.) is a monoecious biennial outcrosser with much inbreeding depression and essentially no naturally self pollination. Heterosis for marketable yield was measured at 25 to 30% over open pollinated cultivars (Bonnet, 1978; Bonnet and Pecaut, 1978). Most cultivars are two- or three-way hybrids, and are produced by cytoplasmic male sterility (either the brown anther or petaloid type). Inbreds are difficult to develop because of inbreeding depression, but can be advanced to S_3 to S_6 by alternating each generation of self pollination with open pollination. Hybrids are produced in isolation blocks with eight rows of the female parent alternating with two rows of the male parent. Hybrids are produced on 56% of the carrot land in the USA.

Onion is a hermaphroditic biennial outcrosser. There is significant inbreeding depression, making the development of inbred lines difficult. Heterosis is large, with hybrids 14 to 67% higher yielding than the best open pollinated cultivars (Dowker & Gordon, 1983). Hybrid production was made economical by the discovery of cytoplasmic male sterility by Jones and Emsweller (1936), and a hybrid production method was described by Jones and Clarke (1943). Commercial hybrids are produced by planting 24 rows of the female parent line alternating with two rows of the male parent line. Bees are used to move pollen from the pollen parent to the male sterile parent for seed production. Europe has a lower proportion of hybrids used (15%) than in the USA (56%), perhaps because of the extra effort required to develop hybrids, and the extra legal protection provided to developers of cultivars in Europe (Dowker & Gordon, 1983).

Other crops

Asparagus is a dioecious perennial outcrosser having genetic control of sex expression through the *m* gene. Male (androecious) cultivars have a 38% higher spear yield than female (gynoecious) cultivars because they do not use photosynthate in seed production each year (Franken, 1970). Androecious (all male) hybrids (*Mm*) have been produced by crossing a super male (*MM*) inbred with a female (*mm*) inbred. Super males are developed by self pollination of occasional perfect flowers on male plants, and test crossing the resulting progeny with a female tester to identify individuals having all male progeny. In the USA, super male hybrids now occupy 45% of the production area, with dioecious hybrids occupying another 40 to 50%. Inbreeding depression results from self pollination, with inbreds yielding 45% that of open pollinated cultivars (Ito & Currence, 1965). Heterosis is important, with the best hybrids yielding 64 to 149% more than open pollinated cultivars (Thevenin & Dore, 1976). The hybrid advantage is expressed mainly in the first four years of production. Asparagus requires two to four years to become established, and has a productive life of 10 to 20 years.

Spinach (*Spinacea oleracea* L.) is normally dioecious, but monoecious flowering also exists. It is an annual outcrosser that is wind pollinated, with a natural outcrossing rate of 60% (near 100% in dioecious populations). Although spinach has been described as having sex chromosomes (Dressler, 1958), sex expression is simply controlled by three alleles at one locus (Janick & Stevenson, 1954, 1955), with XX gynoecious, XY and X^mY androecious, XX^m monoecious (highly pistillate), and X^mX^m monoecious (highly staminate). Inbreeding depression occurs, with a 2 to 6% reduction in yield relative to open pollinated cultivars (Thompson, 1956). Hybrid cultivars outyield open pollinated cultivars by 16 to 20% (Thompson, 1956). Hybrids are produced by planting a monoecious inbred with a high percentage of pistillate flowers (gynoecious inbred) alternating with rows of a monoecious or androecious inbred. The gynoecious inbred is maintained in an isolation block, using the late-appearing staminate flowers for natural self pollination. Hybrid cultivars have been adopted rapidly since their introduction by Jones et al. (1956), occupying 100% of production area in the USA (Table 36–1).

CONCLUSIONS

Benefits of Hybrids

Although many vegetable crops have significant heterosis, some of the major benefits of hybrids relate to other issues. A primary benefit is the protection of parental inbreds used in the production of elite hybrids. This has become more important with increased involvement of private companies in the development of vegetable cultivars. Hybrids can be sold in commerce without making the parental inbred lines available, so the parents are protected under the "trade secrets" provision of the laws of the USA.

A second important benefit of hybrids is uniformity of trait expression among plants of a cultivar. Uniformity was an especially attractive selling point in broccoli and cabbage when they became available as alternatives to open pollinated cultivars. Uniformity of harvest provided earlier and higher yield, and uniformity of the harvested product resulted in higher prices per kg.

Hybrids offer additional advantages for those crops where important traits are controlled by dominant genes. Hybrids make it easier to produce a cultivar with many useful traits by combining dominant alleles from the two parents. A good example is tomato, where single dominant genes control resistance to fusarium wilt, verticillium wilt, root-knot nematode, tobacco mosaic virus, and tomato spotted wilt virus.

There are cases where hybrids may offer additional benefits that are not possible in inbred or open pollinated cultivars. For example, the ripening inhibitor (*rin*) gene in tomato must be in heterozygous condition for it to work partially to allow the fruits to turn red and gain flavor while keeping longer. In homozygous recessive condition, the fruits keep well but stay green and do not develop flavor. Another example of unique traits is the production of seedless triploid watermelons using one tetraploid and one diploid parent. Neither parental inbred has the appropriate chromosome number for seedlessness, but the odd number produced in the triploid hybrid does.

Finally, heterosis may be sufficient by itself to justify the production of hybrids. In spinach, large amounts of heterosis permit higher yielding cultivars to be developed than the open pollinated alternatives. Over the 21 crops summarized here, heterosis averaged 27% improvement for yield of the hybrid cultivars over the inbred or open pollinated ones.

Benefits of Heterosis

The benefit provided to the world by heterosis in the vegetable crops is dependent on the crop. In the case of bean, pea, and lettuce, there has been no effect of heterosis. In other cases, such as tomato, onion, cabbage, and asparagus, there have been significant savings of agricultural land and a large increase in our ability to feed people as a result of heterosis. In the USA, the 21 vegetable crops discussed here occupied 1 576 494 ha of agricultural land, with an average of 63% (not weighted by ha for each crop) of the crop in hybrid cultivars (Table 36–1). Heterosis was estimated to have saved 220 337 ha of agricultural land per year, or to have fed 18% more people without an increase in land use.

REFERENCES

Bonnet, A. 1978. Breeding of carrot F_1 hybrids in France. Biul. Warzywiczy 22:147–150.

Bonnet, A., and P. Pecaut. 1978. Male sterility and production of hybrid varieties of carrot (*Daucus carota*) adapted to French requirements. C.R. Sea. Acad. Agric. France 64:92–100.

Darwin, C. 1877. The effects of cross and self-fertilization in the vegetable kingdom. Appleton, New York.

Dharme, G.M.V. 1977. Genic analysis of yield and yield components in brinjal. Mysore J. Agric. Sci. 11:426.

Dikil, S.P., L.I. Studentsova, and V. S. Anikeenko. 1973. Heterosis in pepper. Trudy po prikladnoi botanike, genetike i selektsii 49:252–269 (Plant Breed. Abstr. 44:161).

Dowker, B.D., and G.H. Gordon. 1983. Heterosis and hybrid cultivars in onions. p. 220–233. *In* R. Frankel (ed.) Heterosis: Reappraisal of theory and practice. Monographs on Theoretical and Applied Genetics 6. Springer-Verlag, Berlin.

Dressler, O. 1958. Cytogenetical investigations on diploid and polyploid spinach (*Spinacia oleracea* L.) with special consideration of the inheritance of sex as the basis of inbred-heterosis breeding. Z. Pflanzenzuecht. 40:385–424.

East, E.M., and H.K. Hayes. 1912. Heterozygosis in evolution and in plant breeding. USDA Bur. Plant Ind. Bull. 243:58.

Elmstrom, G.W. 1978. Evaluation of summer squash varieties for Florida. Proc. Fla. State Hort. Soc. 91:321–324.

Franken, A.A. 1970. Sex characteristics and inheritance of sex in asparagus. Euphytica 19:277–287.

Ghaderi, A., and R.L. Lower. 1979. Heterosis and inbreeding depression for yield in populations derived from six crosses of cucumber. J. Am. Soc. Hort. Sci. 104:564–567.

Griffing, B. 1990. Use of a controlled-nutrient experiment to test heterosis hypotheses. Genetics 126:753–767.

Gritton, E.T. 1975. Heterosis and combining ability in a diallel cross of peas. Crop Sci. 15:453–467.

Hayes, H.K., and D.F. Jones. 1916. First generation crosses in cucumbers. Connecticut. Agric.. Exp. Stn. Ann. Rep. 319–322.

Hedrick, U.P., and N.O. Booth. 1907. Mendelian characters in tomato. Proc. Am. Soc. Hort. Sci. 5:19–24.

Ito, P.J., and T.M. Currence. 1965. Inbreeding and heterosis in asparagus. Proc. Am. Soc. Hort. Sci. 86:338–346.

Janick, J. 1998. Hybrids in horticultural crops. p. 45–56. In K.R. Lamkey, and J.E. Staub (ed.) Concepts and Breeding of Heterosis in Crop Plants. CSSA Spec. Publ. 25. CSSA, Madison, WI.

Janick, J., and E. C. Stevenson. 1954. A genetic study of the heterogametic nature of the staminate plant in spinach (Spinacia oleracea L.). Proc. Am. Soc. Hort. Sci. 63:444–446.

Janick, J., and E. C. Stevenson. 1955. Genetics of the monoecious character in spinach. Genetics 40:429–437.

Jones, H.A., and Clarke. 1943. Inheritance of male sterility in the onion and the production of hybrid seed. Proc. Am. Soc. Hort. Sci. 43:189–194.

Jones, H.A., and S.L. Emsweller. 1936. A male-sterile onion. Proc. Am. Soc. Hort. Sci. 34:582–585.

Jones, H.A., D.M. McLean, and B.A. Perry. 1956. Breeding hybrid spinach resistant to mosaic and downy mildew. Proc. Am. Soc. Hort. Sci. 68:304–308.

Lippert, L.F., and P.D. Legg. 1972. Diallel analysis for yield and maturity characteristics in muskmelon cultivars. J. Am. Soc. Hort. Sci. 97:87–90.

McClure, G.W. 1892. Corn crossing. Illinois Agric. Exp. Stn. Bull. 2:82–101.

Morelock, T.E., M. Peerson, and D. Motes. 1972. Broccoli trials in Arkansas. Arkansas Farm Res. 31:12.

Oba, G. I., M.E. Riner, and D.H. Scott. 1945. Experimental production of hybrid tomato seed. Proc. Am. Soc. Hort. Sci. 46:269–276.

Pearson, O.H. 1983. Heterosis in vegetable crops. p. 138–188. *In* R. Frankel (ed.) Heterosis: Reappraisal of theory and practice. Monographs on Theoretical and Applied Genetics 6. Springer-Verlag, Berlin.

Richey, F.D. 1942. Mock-dominance and hybrid vigor. Science (Washington, DC) 96:280–281.

Rubino, D.B., and T.C. Wehner. 1986. Effect of inbreeding on horticultural performance of lines developed from an open-pollinated pickling cucumber population. Euphytica 35:459–464.

Ryder, E.J., and W. Waycott. 1993. New directions in salad crops: New forms, new tools, and old philosophy. p. 528–532 *In* J. Janick and J.E. Simon, (ed.) New crops: Exploration, research, and commercialization. John Wiley & Sons, New York.

Scott, J.W., and F.F. Angell. 1998. Tomato. p. 450–474. *In* S.S. Banga and S.K. Banga (ed.) Hybrid cultivar development. Narosa Publ. House, New Delhi, India.

Shakya, S.M., and J.W. Scott. 1983. Influence of flower maturity and environment on hybrid and selfed seed production of several tomato genotypes. J. Am. Soc. Hort. Sci. 108:875–878.

Shifriss, C., and I. Rylski. 1973. Comparative performance of F_1 hybrids and open pollinated "Bell" pepper varieties under suboptimal temperature regimes. Euphytica 22:530–534.

Smith, S.A., and T.G. Taylor. 1993. Production cost for selected vegetables in Florida 1991–1992. Florida Coop. Ext. Circ. 1064.

Thevenin, L., and C. Dore. 1976. Asparagus breeding and its principal resource in vitro culture. Ann. Amel. Plantes 26:655–674.

Thompson, A.E. 1956. The extent of hybrid vigor in spinach. Proc. Am. Soc. Hort. Sci. 67:440–444.

Tiwari, R.D. 1966. Studies on hybrid vigor in *Solanum melongena*. J. Indian Bot. Soc. 45:138–149.

Tolla, G.E., and C.E. Peterson. 1979. Comparison of gibberellin A4/A7 and silver nitrate for induction of staminate flowers in a gynoecious cucumber line. HortScience 14:542–544.

U.S. Department of Agriculture. 1992. Agricultural statistics. U.S. Gov. Printing Office, Washington, DC.

Wehner, T.C., and S.F. Jenkins, Jr. 1985. Rate of natural outcrossing in monoecious cucumbers. HortScience 20:211–213.

Wehner, T.C., and C.H. Miller. 1985. Effect of gynoecious expression on yield and earliness of a fresh-market cucumber hybrid. J. Am. Soc. Hort. Sci. 110:464–466.

Whitaker, T.W., and G.W. Bohn. 1952. Natural cross pollination in muskmelon. Proc. Am. Soc. Hort. Sci. 60:391–396.

Yordanov, M. 1983. Heterosis in the tomato. p. 189–219. *In* R. Frankel (ed.) Heterosis: Reappraisal of theory and practice. Monographs on Theoretical and Applied Genetics 6. Springer-Verlag, Berlin.

Chapter 37

Oilseeds and Heterosis

J. F. Miller

INTRODUCTION

Three crop species, soybean [*Glycine max* (L.) Merr.], rapeseed (*Brassica rapa* L. syn. *B. campestris* L. and *Brassica napus* L.), and sunflower (*Helianthus annuus* L.), account for approximately 78% of world vegetable oil production. Heterosis of these crops has been exploited to increase seed yield only over the past few decades. Utilization of heterosis has allowed sunflower to become the major oilseed in many countries of Eastern and Western Europe, Russia, and South America, and is an important crop in the USA, Australia, South Africa, China, India, and Turkey. Of the approximately 16.5 million hectares of sunflower grown in the major producing countries, 11.5 million hectares are planted to hybrids. Hybrid vigor has been the main driving force for acceptance of this oilseed crop. Heterosis is becoming increasingly important in rapeseed. The yield potential of single-cross hybrids has attracted considerable interest from growers in Canada, Australia, Europe, and Argentina. Significant heterosis derived from hybrids of soybean has encouraged researchers to investigate the feasibility of producing hybrid soybean on a commercial scale.

Sunflower

Hybrid sunflower became a reality in the early 1970s with the discovery of cytoplasmic male sterility and an effective genetic male fertility restoration system. In countries that were growing open-pollinated sunflower varieties, comparisons with hybrids clearly indicated the superiority of hybrids for yield and other characteristics. When hybrid technology became commercially available, the switch from varieties to hybrids was swift and complete. However, heterosis was not the only advantage hybrids displayed over the varieties. Dominant genes controlling disease resistance to downy mildew and rust were present in the restorer inbred lines. Resistance to Verticillium wilt was found in female inbred lines. Crossing the male and female lines to produce a hybrid provided multiple resistance to all three diseases.

Single-cross hybrids of sunflower have uniform plant height, flowering date, and seed quality (particularly important in nonoilseed sunflower). Uniform plant height and maturity are very important for mechanical harvest, and uniform flowering is important for chemical control of sunflower insect pests (most notably the sunflower moth and the red sunflower seed weevil) which are effectively controlled when heads are beginning anthesis. Hybrids also have distinctly improved self-fertility. This characteristic increases sunflower yields in areas lacking insect pollinator populations sufficient to facilitate fertilization of open-pollinated varieties.

Few open-pollinated varieties are being developed in countries producing hybrids, making comparisons with hybrids difficult. In trials grown at Casselton, ND from 1994 to 1995, hybrids from France, Turkey, Hungary, Ukraine, Bulgaria, Yugoslavia, Romania, and the USA were compared with the most advanced open-pollinated variety available, Peredovik 92, developed by VNIIMK, Krasnodar, Russia (Table 37–1). The hybrids averaged 2472 kg ha^{-1} whereas Peredovik 92 averaged 1591 kg ha^{-1}, giving the hybrids an average yield advantage of 34%.

Heterosis in oil content of hybrids vs. Peredovik 92 was not evident, with the oil content of Peredovik 92 at 52.2% and the hybrids averaging 50.1%. The supposition that hybrids are taller in height also was not evident in this sunflower trial. Peredovik 92 was 1.9 m in height and the hybrids averaged 1.9 m in height. Present day hybrids grown in the USA have decreased in height by nearly 0.5 m, increasing grower acceptance and decreased lodging susceptibility.

Table 37–1. Hybrid sunflower performance trial organized by the FAO and grown at Casselton, ND from 1994 to 1995.

Hybrid	Country	Yield	Oil	Height
		kg ha^{-1}	%	m
Azur	France	2227	50.2	1.6
Alinka	France	2740	48.7	1.9
Albena	France	2814	47.4	1.7
01480	France	3108	49.7	2.1
94250	France	2242	50.8	1.4
AS440	France	2320	54.2	1.8
AS3211	France	2732	52.2	1.7
XF4113	France	2255	51.7	2.1
Natil	France	2108	52.0	1.9
Sanbro	France	2881	48.0	1.9
Santiago	France	2686	51.3	1.8
RI 933	France	2803	49.4	1.8
AE 401	France	2110	53.0	2.0
Voltasol	France	2835	48.9	1.9
Trakya	Turkey	2204	50.7	1.9
ST2250	USA	2990	47.8	2.0
ST2350	USA	2876	49.0	1.8
Mado	Hungary	2476	48.2	1.8
Sonrisa	Hungary	2265	52.7	1.7
00504	Ukraine	1966	53.1	1.9
Zgoda	Ukraine	2634	49.1	1.9
HB 9405	Bulgaria	2447	48.7	1.8
NSH-111	Yugoslavia	2190	50.7	2.0
NSH-113	Yugoslavia	2452	49.0	1.7
RO 2160	Romania	2334	49.4	1.8
RO 2158	Romania	1591	49.3	1.9
Peredovik 92	Russia	1591	52.2	1.9
Mean of hybirds		2472	50.1	1.9
CV %		14	2.2	7.0
LSD 5%		562	1.8	0.2

Four distinct heterotic groups within sunflower are now being utilized by breeders throughout the world (Vear & Miller, 1993). The open-pollinated varieties developed in Russia are used in deriving female maintainer inbred lines. The USA restorer group, derived by crossing wild annual species of sunflower with cultivated lines, is a distinct source of disease resistance and fertility restorer genes. Romanian female lines, along with their South African derivatives, are used throughout the industry. Also used are the Argentinean INTA open-pollinated cultivars for developing female lines.

Tersac et al. (1993) collected 39 open-pollinated sunflower populations from 10 countries. Using four tester lines, an analysis of combining abilities for seed yield, seed moisture content at physiological maturity, and seed oil content was performed. Results indicated that the between-population structure based on specific combining ability (SCA) distinctly corresponded to the country in which the population originated. This suggested that little introgression of germplasm has taken place between the 10 countries due to the relatively young status of sunflower in the world. Genetic diversity created in sunflower over the last century has remained distinctive.

The genetic diversity of elite sunflower breeding lines was investigated by Cheres and Knapp (personal communication, Oregon State Univ., Corvallis, OR. Coancestries between 106 oilseed and 50 confection inbred lines were determined by cluster and principal coordinate analysis. Unique germplasm groups were identified and lines were subdivided into subgroups, which theoretically reflected unique heterotic groups. Subsequent crosses between selected lines have verified the existence of the heterotic groups.

Rapeseed

Rapeseed hybrids produced by crossing inbred lines have shown heterosis in yield of 30 to 60% over midparents in both spring and winter forms of *Brassica napus* and spring forms of *B. rapa* (Sernyk & Stefansson, 1983; Grant & Beversdorf, 1985). This level of heterosis stimulated efforts to develop commercial hybrids in Canada, Australia, Western Europe, Argentina, and the USA.

Diallel crosses between two Canadian cultivars and four European cultivars of *B. napus* averaged 23% better than the parental average (Grant & Beversdorf, 1985). The more productive hybrids, in terms of absolute seed yield, were those produced from inter-continent crosses (Canadian × European). Only a few hybrids with significant heterosis were observed in crosses of European × European cultivars and either no heterosis or negative heterosis was observed for the hybrid between the two Canadian cultivars grown at two locations of testing in Canada. This observation led the authors to speculate that the European and Canadian germplasm may represent two distinct heterotic groups.

High levels of heterosis for seed yield were found in hybrids involving four cultivars of *B. rapa* and grown in Saskatchewan, Canada (Falk et al., 1994). The level of heterosis varied greatly among the hybrids, but during a three-year period, an average of 13% increase in seed yield was observed. Heterosis for seed yield was greatest in crosses between genetically diverse cultivars; however, heterosis for seed oil content was not found. Also, it was noted that for a hybrid to have zero erucic acid and low glucosinolate, both parents must express these characteristics. Therefore, it will be necessary to widen the genetic base to improve the heterotic potential of future *B. rapa* hybrids in Canada.

The agronomic performance of 15 cultivars and the hybrid Hyola 401 were compared in a trial grown at Prosper, ND in 1995 (Table 37–2). The hybrid produced 1308 kg ha^{-1} as compared to an average of 786 kg ha^{-1} for the cultivars, a 39% increase in seed yield. The hybrid was significantly shorter in height. No

Table 37–2. Agronomic performance of cultivars and a hybrid of *Brassica napus* grown at Prosper, ND in 1995.

	Yield	Height	Oil
	kg ha^{-1}	cm	%
Variety:			
Pioneer (4 cultivars)	841	153	40.1
Svalov (7 cultivars)	956	126	39.8
Limagrain Gen. (3 cultivars)	820	128	39.7
DLF Trifolium (1 cultivar)	836	135	39.8
OAC Guelf (1 cultivar)	734	103	40.0
Legend (Canadian check)	870	110	40.7
Hybrid:			
Hyola 401 (Advanta)	1308	98	40.6

heterosis was observed in oil content in this hybrid. The location was selected for the potential of summer heat stress on rapeseed performance. An advantage of this hybrid appears to be its tolerance to stress conditions.

Parental cultivars of *B. napus* from Europe and Asia were crossed and grown over two years in France (Lefort-Buson et al., 1987). Comparisons between European × European, Asian × Asian, and European × Asian cultivar crosses were observed. Superiority of hybrids over the check cultivar Bienvenu was not expressed in the European × European and Asian × Asian hybrids; however, heterosis of 25 and 8% was expressed by the European × Asian hybrids during the two years of study. The authors concluded that within the European or Asiatic groups of cultivars, variability for yield was essentially additive. The superiority of mixed hybrids emphasized the interest in utilizing parents, which have very different geographic origins.

Using RFLP analysis, 38 of 43 clones tested were polymorphic among cultivars and inbreds of *Brassica napus* (Diers et al., 1996). The 38 clones revealed a total of 90 polymorphic fragments, which were used to calculate genetic distance. Seven major clusters were identified and appeared to be grouped together by geographic origin. This clustering was consistent with heterotic responses observed in crosses between these groups.

Soybean

High-parent heterotic responses for yield of hybrid soybeans have been reported from 3 to 26% with an average of approximately 13% (Burton, 1987; Nelson & Bernard, 1984). In a study investigating hybrid soybean, genetic male sterility was used to produce four female lines and these were crossed with improved cultivars adapted to Illinois, USA (Nelson & Bernard, 1984). Hybrids were grown at two locations during two years. Five hybrids yielded significantly more (13–19%) than their better parent in at least one season. One hybrid exceeded the yield of the best pure line cultivar in the test. The authors also noted that the yield of the hybrids could not be predicted from their pedigrees. These results provide evidence that hybrid soybean can be justified by the significant heterosis for yield and that future research should be directed to feasibility aspects.

The primary barriers to commercial use of hybrid soybean are the lack of an economical method of seed production, limited pollen transfer by insects, and the sheer amount of seed needed to plant one hectare, making the cost of seed high relative to the degree of heterosis (Fehr, 1987). The author indicated it would be difficult to obtain enough hybrid soybean seed from 1 ha to plant 20 ha of commercial production, whereas enough hybrid maize is produced on 1 ha to plant at least 150 commercial hectares. Sunflower hybrid seed production is similar to that of maize with 1 ha producing enough seed to plant 120 to 200 ha of commercial hybrid.

One explanation for the somewhat limited heterosis in soybean may be due to the low degree of genetic diversity among North American soybean cultivars (Sneller, 1994). Coefficient of parentage analyses were performed with 122 lines from the northern and southern regions of the USA. Results indicated that recent efforts of public and proprietary breeders have had little impact on diversity among current elite lines. Only minor differences in genetic diversity were noted between lines from public or proprietary sources.

Genetic diversity patterns of North American soybean cultivars confirmed the widely held view that northern and southern cultivars trace to contrasting genetic bases (Gazlice et al., 1994). However, analyzing coefficient of parentage data with multivariate and regression procedures, the North–South distinction accounted for only 21% of the variability found in soybean pedigrees (Gazlice et al. 1996). Heterotic groups, tracing to introduced cultivars from China, Japan, and Korea (Hymowitz & Bernard, 1991), could be used as parents in breeding programs to increase diversity in soybean cultivar development.

REFERENCES

Burton, J.W. 1987. Quantitative genetics: Results relevant to soybean breeding. p. 211–247. *In* Soybeans: Improvement, production, and uses. Agron. Monogr. 16. ASA, CSSA, and SSSA, Madison, WI, USA.

Diers, B.W., P.B.W. McVetty, and T.C. Osborn. 1996. Relationship between heterosis and genetic distance based on restriction fragment length polymorphism markers in oilseed rape (*Brassica napus* L.). Crop Sci. 36:79–83.

Falk, K.C., G. Rakow, R.K. Downey, and D.T. Spurr. 1994. Performance of intercultivar summer turnip rape hybrids in Saskatchewan. Can. J. Plant Sci. 74:441–445.

Fehr, W.R. 1987. Soybean. p. 533–576. In W.R. Fehr (ed.) Principles of Cultivar Development. Vol. 2. Macmillan Publ. Co., New York.

Gazlice, Z., T.E. Carter, Jr., and J.W. Burton. 1994. Genetic base for North American public soybean cultivars released between 1947 and 1988. Crop Sci. 34:1143–1151.

Gazlice, Z., T.E. Carter, Jr., T.M. Gerig, and J.W. Burton. 1996. Genetic diversity patterns of North American public soybean cultivars based on coefficient of parentage. Crop Sci. 36:753–765.

Grant, I., and W.D. Beversdorf. 1985. Heterosis and combining ability estimates in spring-planted oilseed rape (*Brassica napus* L.). Can. J. Genet. Cytol. 27:472–478.

Hymowitz, T., and R.L. Bernard. 1991. Origin of the soybean and germplasm introduction and development in North America. p. 147–164. *In* Use of Plant Introductions in Cultivar Development. Part 1. CSSA Spec. Publ. 17. CSSA and ASA, Madison, WI.

Lefort-Buson, M., B. Guillot-Lemoine, and Y. Dattee. 1987. Heterosis and genetic distance in rapeseed (*Brassica napus* L.): Crosses between European and Asiatic selfed lines. Genome 29:413–418.

Nelson, R.L., and R.L. Bernard. 1984. Production and performance of hybrid soybeans. Crop Sci. 24:549–553.

Sernyk, J.L., and B.R. Stefansson. 1983. Heterosis in summer rape (Brassica napus L.). Can. J. Plant Sci. 63:407–413.

Sneller, C.H. 1994. Pedigree analysis of elite soybean lines. Crop Sci. 34:1515–1522.

Tersac, M., D. Vares, and P. Vincourt. 1993. Combining groups in cultivated sunflower populations (*Helianthus annuus* L.) and their relationships with the country of origin. Theor. Appl. Genetics. 87:603–608.

Vear, F., and J.F. Miller. 1993. Sunflower. p. 95–111. *In* Traditional Crop Breeding Practices: An historical review to serve as a baseline for assessing the role of modern biotechnology. OEDC, Paris, France.

Chapter 38

Heterosis in Crops I

Discussion Session

QUESTIONS FOR A.R. HALLAUER

P. Nelson, Sensako, Republic of South Africa: How many heterotic groups are generally used in the U.S. Corn Belt? How are these heterotic groups being maintained?

Response: Generally, there is the one genetically broad-based heterotic group (Reid Yellow Dent - Lancaster Sure Crop) considered, but specific heterotic groups are used, based on the specific families of inbred lines emphasized in each breeding program. Heterotic groups are maintained by crossing lines within each group, but this is not strictly adhered to in all instances; an example is the derivation of inbred lines from good performance hybrids.

J. Brewbaker, University of Hawaii, USA: Have the reciprocal recurrent programs at Ames led to comparable increases in heterosis for yield components (e.g., kernels per row and kernel weight)? Can heterosis for the yield components be multiplied to equal heterosis for yield?

Response: The reciprocal recurrent selection programs have not been evaluated to determine the changes for the different yield components. Yield was emphasized in selection, and yield was measured to determine response to selection, whatever the changes due to components of yield. Although we have not measured the components of yield, I suspect the changes are small and that the multiplication of the heterosis of the components of yield would not equal heterosis for yield.

S. Smith, Pioneer Hi-Bred International, USA: What would be the outline of a comprehensive germplasm enhancement program for U.S. maize to provide a broader and more useful germplasm base for the 21st century?

Response: We currently are working with a genetically narrow base of elite germplasm that has a very good level of productivity. To provide germplasm for the 21st century, we will need to build on the current base of elite germplasm by gradually incorporating other elite germplasm to broaden our genetic base of elite germplasm. For continued genetic improvement, long-term selection programs must be conducted to maintain incremental genetic improvement over time. In maize breeding, some form of reciprocal recurrent selection is recommended to enhance heterotic patterns for future line and hybrid development. The long-term selection programs have to be an important component of organizations because individuals conducting the programs will change. The general objectives have to be retained even though personnel changes will occur.

H. Ceballos, Universidad Nacional de Columbia, Colombia: Duvick's data showed that higher productivity of modern hybrids is related to higher yield potential of parental lines rather than increased heterosis. Does this mean that perhaps we should emphasize inbred performance per se rather than the heterotic response of their cross?

Response: Improvements of inbred lines for yield, for other agronomic traits, and for pest resistance have been expressed in their crosses, i.e., improvements in lines are reflected in their crosses. Maize breeders do give consideration to the performance of the lines themselves, but the ultimate use of the lines is in crosses. Selection is based on both line performance per se and in crosses.

P. Nelson, Sensako, Republic of South Africa: Why are heterotic groups being forced into two or three groups? Are we not disrupting favorable blocks of genes and thereby reducing opportunities for exploiting epistasis, in particular.

Response: Heterotic groups were identified from empirical data that involved crosses among lines from different source populations. After specific trends of the performance of lines in crosses were observed, breeding emphasis was given to the lines involved by developing second, third, etc., cycle lines to enhance the observed heterotic patterns. And the recycling of lines involved in the heterotic patterns probably exploits specific gene combinations, or epistasis. To have other heterotic patterns, it will be necessary to have performance levels similar to or better than currently used heterotic groups. Breeding programs will determine heterotic groups. There are other heterotic groups that could be used, but they either have not received the same emphasis in breeding or they are not competitive with the current heterotic groups.

QUESTIONS FOR S.K. VASAL

J. Hawk, University of Delaware, USA: What are your major inbred maize testers for tropical breeding program?

Response: We are using CML 247 and CML 254 for late white and CML 287 and CL 00331 for late yellow germplasm.

M. Goodman, North Carolina State University, USA: What are KS6 and CL00331?

Response: KS6 is a synthetic developed by the Kasetsaart University in Thailand and CL00331 is a yellow endosperm line developed from CIMMYT's tar spot resistant synthetic.

E. Silva, Ecuador: Maize germplasm from the Andean highlands shows very high inbreeding depression. How can we overcome this problem?

Response: Recurrent selection based on selfed progenies should help reduce inbreeding depression in any germplasm.

A. Ortega, CIMMYT, Mexico: Tuxpeno × Tuxpeno continues to be one of the best heterotic patterns for lowland tropics. Could this situation increase genetic vulnerability in tropical maize?

Response: There are plenty of Tuxpeno type inbred lines available. The risk of vulnerability can be minimized if many such lines are used in hybrid development.

G. Granados, CIMMYT, Mexico: Please comment about research being conducted to develop early and extra-early maize hybrids.

Response: CIMMYT has developed many early inbred lines from its population and pools. Beginning in 1996, the Center has also started evaluating early maturity white and yellow endosperm hybrids internationally.

B. Rana, AICSIP, National Research Center for Sorghum, India: What do you mean by "comprehensive integrated strategies" for hybrid maize development?

Response: By comprehensive integrated strategies, I mean incorporation of evaluation and selection for stress traits such as drought, high density, etc., in different stages of line and hybrid development and population improvement.

QUESTIONS FOR J. AXTELL

M. Carena, Argentina: What would be the environmental, nutritional, and breeding impact of replacing sorghum by corn in Africa's acreage?

Response: This has been taking place since the colonialists introduced maize to Africa. Maize will find its place in Africa where it's most appropriate, adapted, viable and competitive in the market place. We must maintain active breeding and improvement in both crops, and let the market and the environment sort out which crop has an advantage where. One thing that contributes to the superiority of maize is its bird resistance because of its husks.

G. Granados, CIMMYT, Mexico: Is there any variability in sorghum species for protein quality?

Response: We discovered the only naturally occurring high lysine sorghum of any of the major cereal crops, in Ethiopia. Even though it yields low, it is being grown by farmers because it tastes better. Since it has higher level of reducing sugars, it caramelizes when roasted. At Purdue, we have developed another source that has a more normal yield. We think that it is important to continue the effort on improving the quality of sorghum. The protein digestibility is also important, and we have observed some genetic variation for this.

B. Rana, AICSIP, National Research Center for Sorghum, India: World wide sorghum area is declining except in Africa. What strategies can be suggested to increase competitiveness of sorghum vis a vis other crops replacing it?

Response: The answer would be more money and more scientists. Sorghum is a crop that's grown under stress. Are we really losing the genotypes that have high yields under optimal conditions? Are we eliminating the genes of higher yield potential? At Purdue, we look for genotypes that better express under optimal conditions, and then test them for stress tolerance. I'd like to see more sorghum breeding done under optimal conditions - to pick out the winners, not the survivors.

E. Preciado, INIFAP, Mexico: Please discuss heterotic patterns in sorghum.

Response: More recently, scientists are using public inbreds with private inbreds which is often very effective. Germplasms from Gambia, Ethiopia, and Sudan have been used as testers. They have provided lines with excellent grain quality, high yield, and good combining ability.

QUESTIONS FOR T.C. WEHNER

P. Sun, Dairyland Research International, USA: What is the relationship of inbreeding depression and hybrid vigor for the vegetable crops with high outcrossing rate?

Response: The amount of heterosis appears to be inversely related to the amount of inbreeding depression, although some crops have a high inbreeding depression associated with moderate heterosis (onion), while others have high heterosis associated with moderate inbreeding depression (asparagus).

S. Banga, Punjab Agricultural University, India: Do you think the presence of self incompatibility will interfere with CMS-line maintenance in the cole crops?

Response: Yes, it could be a problem. It would be necessary to have different self incompatibility alleles from the male sterile line in the restorer line. Alternatively, the breeder could incorporate self fertility alleles in all the lines to be used in hybrid production using male sterility. In cauliflower, there is already much self fertility since that crop is more recently developed than other members of the species (such as cabbage).

K. Rajendra, IARI, India: Is there heterosis for fruit quality in tomato and cucumber? We find that hybrids are highly productive, but lack of quality has resulted in a failure to attract consumers and good prices for hybrids.

Response: I am not aware of any reports for heterosis relative to fruit quality (shape, color, flavor, internal defects) in those crops. In cucumber (and probably tomato), it is necessary for both parents of the hybrid to have good fruit quality since the hybrid is usually intermediate for those traits.

K. Tadmor, ARO, Israel: (Comment) Significant heterosis for brix in tomato was found by D. Zamir et al. with lines derived from the introgression of *L. pennelii* x *L. esculentum*.

L. Nduulu, Purdue University, USA/Kenya: You reported that not much heterosis is expressed in most vegetable crops with regard to yield per se. Has much research been done on identifying heterotic groups?

Response: I am not aware of any formal classification of inbred lines into heterotic groups in the vegetable crops.

J. Chavez B., SNICS-SAGAR, Mexico: Please comment on heterosis in potato.

Response: There is heterosis for yield in potato, and research by Peloquin using 4X by 2X crosses has shown that having three or four different alleles at a locus (maximum heterozygosity) produces better performing lines than having only one or two.

A. Tsaftaris, University of Thessalonike, Greece: Many vegetable crops are allopolyploid. Thus, could exploid built-in heterosis be a reason that those crops show little heterosis?

Response: Built-in heterosis may explain the reduced heterosis shown in the cole crops (which are allotetraploid) as compared to high heterosis in carrot and onion.

A. Barreras, INIFAP, Mexico: In Sinaloa, we cultivate three different types of cucumbers for export which have high resistance to downy mildew. What is the relationship of resistance to heterosis?

Response: Resistance to downy mildew in cucumber is recessive, with one to three genes involved. Resistance must be in both parents for the hybrid to be resistant, and there is no relation of the genes to heterosis.

QUESTIONS FOR J.F. MILLER

K. Virupakshappa, UAS, GKVK, India: Besides the *H. petiolaris* cytoplasm presently used in all sunflower hybrids, is there any other source of cytoplasm which could be used in a heterosis program?

Response: Yes, there are 21 additional sources of cytoplasm which produce male sterility in sunflower. However, few are as effectively restored as the petiolaris cytoplasm. I believe that only two or three could be effectively utilized at the present time.

D. Duvick, Iowa State University, USA: Of the four major heterotic groups in sunflower, are all two-way combinations equal in heterosis? Or are some combinations better than others?

Response: This is a difficult question to answer, because we have not had time to explore all the combinations thoroughly. However, we know that crossing the U.S. restorer heterotic group with lines derived from the other three groups produces very high yielding hybrids, nearly equal, in many environments. Combinations with the Argentinean B-line group would be more adapted to longer season environments, and, therefore, this group has not been tested as extensively in short season environments.

N. Machado, Cargill S.A.C.I. Argentina: Did you detect any improvement in the level of heterosis present in sunflower when you compared the 1980 hybrids vs. the 1990 hybrids, or was the increase in yield only through pure parental line improvement?

Response: The data I presented in my talk compared hybrid yield with open-pollinated variety yield, as it is very difficult to measure heterosis with the inbred lines that we utilize in sunflower. We did a study where we found that inbred line yield was not correlated with the ultimate hybrid yield. I believe this would be true if we tested lines we used in 1997. Therefore, the increase in yields is due to the effort of finding and developing lines with genes for higher yield potential that are expressed in hybrids, not in the yield of the pure line per se.

S. Sanchez, Mexico: Do you believe that the short day and long day requirements in soybean are heterotic groups? I know that soybean has a short day photoperiod requirement, but you tell us that soybean from north and south latitudes are crossed.

Response: The genetic diversity studies reported in soybean indicate that there is considerable diversity between soybean cultivars developed in the southern U.S. latitudes and cultivars developed in the northern U.S. latitudes. This has led to the supposition that the two could be considered as heterotic groups for hybrid soybean programs as well for crossing in a conventional breeding program. One soybean cultivar, Asgrow 3127, is a cross between north and south varieties, Williams and Essex, which has been very successful.

R. Kumar, Indian Agricultural Institute, India: Seed storability is the major problem in soybean as seeds lose viability very fast. Do you find heterosis for storability (longer viability under ambient situations) in soybean?

Response: I did not find any information regarding heterosis for storability in soybean.

K. Lamkey, Iowa State University, USA: Do soybeans exhibit inbreeding depression? If not, what is the genetic basis of heterosis in soybean?

Response: Soybean breeders generally find a decrease in yield when comparing the F_1 generation or hybrid with the variety that is ultimately derived from that F_1 cross. Soybean breeders believe heterosis is due to additive epistasis in soybean.

Chapter 39

Wheat and Heterosis

J. P. Jordaan, S. A. Engelbrecht, J. H. Malan, and H. A. Knobel

INTRODUCTION

Hybrid vigour has been exploited for decades in plant and animal improvement programmes; however, for wheat (*Triticum aestivum* L.), commercialisation of heterosis has been regarded as mostly unsuccessful. The reasons for this have been argued on many platforms, and for almost four decades, but breeders are still seeking answers and discussing plans and technologies that could make hybrids a commercially successful proposition.

History of Hybrid Wheat

The history of hybrid wheat has been well documented (Pickett, 1993) and (Pickett & Galwey, 1997). Research and development have centered around the identification of suitable systems to produce hybrid wheat seed, the expression of heterosis, and the performance of hybrids.

Present Status of Hybrid Wheat

1. Although the majority of companies have withdrawn from breeding wheat hybrids, multinational organisations like Pioneer Hi-Bred International and Hybritech Seeds International as well as national institutions such as Hybrinova (France), Hybrid Wheat Australia (Australia), SENSAKO (South Africa) and CARNIA (South Africa) are still involved in the production of hybrid seed and marketing of hybrids. The only government known to be involved is the Peoples Republic of China. Acreages being grown to hybrid wheat seems to be non-significant in comparison to that being grown to conventional varieties.

2. In general, hybrid development was accomplished as a result of profit driven motives. The key opportunity of hybrid wheat would therefore lie in its ability to draw investment from private enterprise to support research and development of hybrids. Since hybrids offer the best product protection, it is a potential investment with secured returns. Hybrids also provide private companies involved in the field of biotechnology with a vehicle to sell their biotech products or package.

3. The conclusion is made that most wheat breeders have abandoned rather than adopted hybrid technology. Breeders lacked sufficient results from directed breeding for combining ability, and during the first 20 years resources were spent on sorting out sterilising agents, fertility restoration and crossability. Real comit-

ment to hybrids was curtailed because most breeders also were involved in the breeding of conventional varieties. They also were far behind their colleagues in other crops in establishing heterotic patterns. In most cases, additional selection strategies were lacking or nonexistent, while breeding for heterosis without knowledge of heterotic patterns, has proved to be a hit or miss approach.

4. There is concern that the hybrid yield gain, regarded by commercial hybrid wheat breeders to be 10 to 20%, is too low to successfully commercialise hybrid wheat. If selection for heterosis had been successful, improvement in hybrids should have equalled the significant progress that breeders, involved in the development of conventional varieties, have achieved.

5. Hybrid wheat has put seed companies at risk because the difference between seed cost and added profit has been too small or even negative. Likewise the uncertainty of hybrid seed production for any given year is a major problem. In general, the seed multiplication factor is low in comparison with other hybrid crops like corn.

Farmers Acceptance

A free market system seems to be the best environment for the growth and development of hybrid wheat. Selling hybrid seed should be market driven and in general, progressive farmers seem to accept hybrid seed. Sales strategy has been to sell hybrid seed to farmers in environmental niches having high yield potential. In an environment with high yield potential the contribution of higher seed cost in comparison to gross income is small and hybrids become cost effective. Hybrids are also more acceptable in environments where the seeding rates are low (varying from 10 kg ha^{-1} to 50 kg ha^{-1}) like in New South Wales (Australia) and in the Free State (South Africa). While farmers regard hybrids as being a superior product, they expect a quantified added value before pursuing the growing of hybrid wheat.

HYBRID TECHNOLOGY

Animal and maize (*Zea mays* L.) breeders have been successful in exploiting heterosis and developing technology to optimise selection gain. These methods have been adopted by breeders in most other crops by exploiting heterosis, and are worthy of implementation by hybrid wheat breeders.

Maize: The Role Model

Procedures and techniques that have been developed were empirically evaluated and a successful model is regarded to include three distinct phases (Sprague & Eberhart, 1977): (i) "the development of two or more breeding populations from diverse sources so that the population-cross means will be at the highest level possible and the population will have maximum additive genetic variation within each, (ii) continuous population improvement by an effective recurrent selection program, and (iii) the development of superior hybrids from each cycle of selection by an efficient and systematic procedure."

Originally, first cycle inbreds were selected from open pollinated varieties and used to produce double cross hybrids. Lately, breeders have started recycling inbreds by intercrossing them (Hallauer, 1990) to create new populations for further inbreeding and selection.

A Wheat Model

Knowledge regarding the gene action expressing yield, its components and quality in wheat is abundant. This was largely generated by unpublished postgraduate studies but also reported in a large number of reports, summarized by

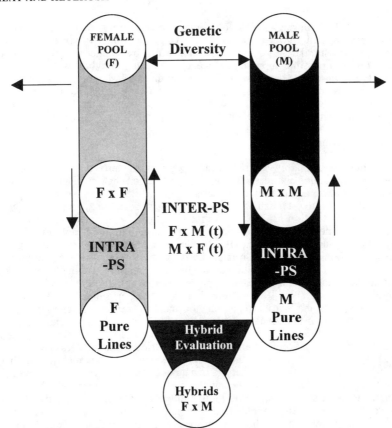

Fig. 39–1. Model for hybrid wheat development including Intra-Population Selection (Intra-PS), Inter-Population Selection (Inter-PS), M(t) and F(t) refer to a specific male or female tester.

Pickett (1993). From these reports gene action can be regarded to be mainly additive, the contribution of dominance significant, and the presence of epistasis acknowledged. In pursuit of the maize model, hybrid wheat should evolve from two different populations (Fig. 39–1). Making provision for: (i) identification of male (M) and female (F) germplasm based on knowledge of heterotic patterns and genetic diversity, (ii) creation of populations within female and male pools to select from (INTRA–PS), (iii) inter-population selection (INTER–PS) to select for combining ability, (iv) recycling of elite lines to improve the average performance of F × F or M × M populations, and (v) final evaluation of hybrids (F × M) for commercialisation.

Identifying Female and Male Populations

The success that was achieved by the introgression of wheat germplasm to enhance genetic variability in conventional wheat breeding programs, contributed negatively to the development of hybrid wheat. This diversion is supported by evidence that hybrids derived from modern, higher yielding varieties appear to show less heterosis compared with the performance of older, lower yielding varieties

(Pickett, 1993). Kronstad (1996), found that winter × spring crosses proved to express higher hybrid vigour than crosses derived from winter × winter or spring × spring germplasm. These gene pools are known to be distinct and Kronstad suggested this as a possible approach to breeders interested in hybrid wheat. Although diversity is regarded as being paramount, heterotic expression is an absolute necessity to optimise cross performance between the two gene pools. Knowledge of heterotic patterns within the available germplasm has been found to be a basic requirement to secure success in hybrid breeding for all crops.

Intra-Population Selection

Interest is mainly centred around populations derived from two parents. This raises questions about: (i) the type of gene action that is important for yield, disease resistance, and quality characteristics in hybrids; (ii) on what basis parents should be chosen to establish source populations for selection; and (iii) the generation in which selection for cross performance should be initiated. Procedures and technology to develop pure lines are well known and are applicable in the development of female (F) and male (M) lines per se. In the presence of additive genetic variance it should be possible to raise the population means to equal the performance of lines derived in conventional breeding programmes. Although some breeders might consider F and M products for release, the emphasis in a well structured hybrid programme should be on selection for cross performance to the opposite pool. Procedures may differ regarding the mechanism used to change the self-fertilising nature of the wheat plant to produce hybrid seed. In the case of nuclear male sterility (NMS) and cytoplasmic male sterility (CMS), breeding emphasis is placed on recombination breeding on the female side while backcrossing to a narrow based male pool would be the appropriate strategy. The availability of chemical hybridising agents (CHA) changes this strategy to one where equal numbers of female and male progeny could be handled in testcross programs to male M(t) and female F(t) testers (t), thus putting more emphasis on selection for cross performance at an earlier F generation of line development. Since male and female inbred lines are only developed within their respective germplasm pools, the testers that are used should be chosen on their specific combining ability to the opposite germplasm pool. While selection for line performance per se enhances the use of additive genetic variance. Laubscher (1984, unpublished data) showed that although female × male interaction accounted for 63.8% of hybrid variation, the genetic variation within the female and male sets were significant. Selection for cross performance in wheat will change this. The expectation is that with elite male and female lines the additive component for yield within heterotic populations will decline, and the ratio of additive to nonadditive variance may shift in the direction of nonadditive genetic variance. Continuous selection progress for line performance per se will be, as it is for conventional breeding, accomplished by making better use of germplasm resources.

Since wheat is mainly used for human consumption the quality of the end product will be decisive in the acceptance of hybrids in industry. Overall evidence (Pickett, 1993) shows variation for milling and baking characteristics to have an additive nature and that hybrid performance could be predicted from parental performance. Selection procedures within the female and male pool should be aimed at end product quality.

Inter-Population Development

In the maize model, inter-population selection starts off as early as possible (S_1, F_2 or S_2, F_3) in the line development cycle. Information on the relative importance of this is lacking in the case of wheat, but it might be reasoned that starting off testcrossing at a later stage, preferably the F_4 or F_5, would be a better procedure to exploit both additive and nonadditive variance and to select for other characters such as disease resistances. A system where testcross evaluation is delayed until a high degree of homozygosity has been attained is satisfactory with limited numbers of lines, but becomes burdensome when large number of lines are available. The testcross technique also is dependent on changing the breeding nature of the wheat plant, either by CMS, NMS, or CHA. In the latter case, reciprocal testcrosses can be made at will, but in the case of CMS, testcrossing females to a tester male should be delayed to follow on backcrosses after cytoplasm substitution. Since line development will be targeting known adaptation phenomena, selection for cross performance between lines within the same heterotic pattern would tend to maintain favourable linked epistatic gene combinations, as was found to be the case in maize (Melchinger et al., 1988).

Homozygosity may be forced on line development at a very early stage to promote recurrent selection procedures. The haploid technique using maize pollen has proved to be effective in creating large numbers of homozygous lines within selected families, reducing generation time and increasing genetic gain (Howes et al., 1997). This technique ensures rapid recycling of selected male or female lines after testcrossing to a reciprocal tester. It also complies with the goal to turn out reproducible hybrids derived from crossing homozygous parents at every cycle of selection.

IDENTIFICATION OF SUCCESS FACTORS

Yield Potential

Comparing hybrids to conventional varieties across environmental sites was done (Jordaan, 1996) by comparing the relative consistency of yield performances (Lin & Bins, 1988). This analyses defines yield potential as the distance mean square between the genotypes response, and the maximum response within an environment, averaged across all testing sites. The consistency of performance of hybrids and comparable conventional varieties tested across a range of localities in South Africa is summarised in Fig. 39–2. This data shows that on the average, hybrids are more consistent in performance than varieties, and when comparing elite cultivars of conventionally derived varieties to hybrids, the hybrids also demonstrate their superiority.

It also is popular to describe genotype × environment interactions as a linear response to environmental yield and the deviations from that response (Eberhart & Russell, 1966). In comparing hybrids with conventional varieties the environmental index was defined as the average performance of the conventional varieties (Peterson et al., 1997). They showed hybrids tested on the Great Plains of the USA to be more responsive to yield potential. Hybrids were significantly higher yielding than pure lines and the yield advantage increased with environmental yield potential showing no crossover at low yield levels, and with comparable deviations from regression.

A similar analyses was run on data from winter wheat hybrids developed and tested in South Africa. These results, compare favourably with results that were reported by Bruns and Peterson (1997) and Peterson et. al. (1997) despite differences in growing conditions and management practices. Hybrids that showed the highest level of heterosis proved to be daylength sensitive with little or no vernalization requirement. The regression slope for yield of the best performing hybrid is compared with those of the newest releases of conventional varieties (Fig. 39–3). The average performance of the conventional varieties was regarded as the environmental index for yield potential at each location. The regression coefficient of pure lines was 1.00, intercepting at 0.0 kg ha^{-1}. A 95% confidence interval was calculated for the hybrids regression using the GLM program (SAS Institute, 1982). Hybrids like this were bred to target stress environments and although the highest

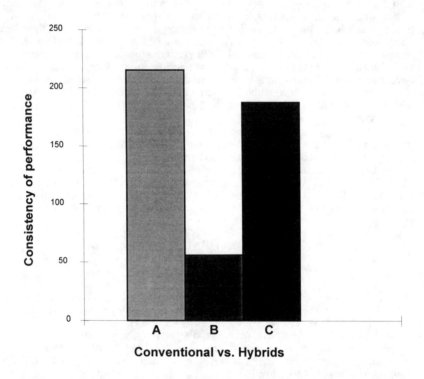

Fig. 39–2. The relative consistency of yield performance (Lin & Bins, 1988) of:
(A) the average of the four latest releases of hybrids with *Duraphius noxia* resistance, (B) the average of the four latest releases of conventional varieties
with resistance to *Duraphius noxia,* (C) the average of all hybrids in test, expressed as a deviation from the average consistency of performance of all conventional varieties in test. Units of consistency are calculated as the distance
mean square between the genotypes response, and the maximum response
within an environment averaged across all testing sites and divided by 1000.

percentage of heterosis was shown under stressed conditions, they also proved to
be responsive to yield potential, showing no crossover at high yield potential.

Effect of Lower Plant Population and Row Width

The sustainability of heterosis at low yield levels is argued when the effect
of the additional cost of hybrid seed is considered. Production practises for growing
dryland winter wheat in South Africa are unique and based on low seeding rates
(15 kg ha^{-1} to 30 kg ha^{-1}) and wide rows (30–60 cm). In a study by Engelbrecht
(1991) the expression of heterosis was specifically adapted to combinations of row
width and plant population. In narrow rows, the expression of heterosis was higher
at the high plant densities and better adapted to high yield potential (high midparent values). In wider rows, the deviation from mid-parent values was the highest

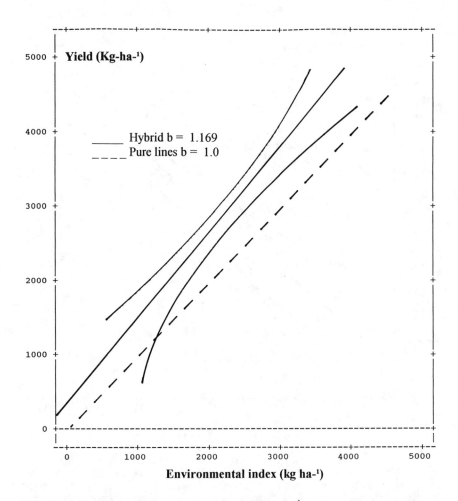

Fig. 39–3. The linear regression of mean yield (kg ha^{-1}) of the hybrid on the latest releases of conventional varieties. The average performance of the conventional varieties was regarded as the environmental index for yield (kg ha^{-1}) at each location.

at low seeding rates when the mid-parent values were low (low yield potential). The hypothesis of wider rows and low seeding rates for low yielding environments seems to hold in practice as well as when conventional pure lines are considered. This study (Engelbrecht, 1991) and also a study involving a Nelder layout (Nelder, 1962, unpublished data) suggest that relative hybrid performance relates to space arrangements of plants at specific row width and plant population and that heterosis can be optimised by agronomic practices. Pickett and Galwey (1997) have reported that higher levels of heterosis were associated with lower plant densities although it also was found that heterosis was the highest at normal seed rates.

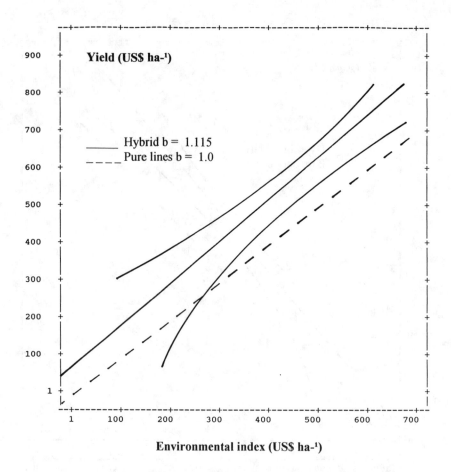

Fig. 39–4. The linear regression of mean yield (US$ ha⁻¹) of the hybrid on the mean
yield of the latest releases of conventional varieties. Hybrid yield was corrected
for hybrid seed cost, taking different seeding rates for differences in yield potential into account. Seeding rates of the conventional varieties were regarded
to be constant at 25 kg ha⁻¹ and their average yield (US$ ha⁻¹) taken to be the
environmental index for yield potential at each location.

Adjusting seeding rates for yield potential has a significant effect on the commercialisation prospect for hybrids (Jordaan, 1996). If the yield data of the hybrid
which was reported in Fig. 39–3 is regarded as US$ ha⁻¹, and corrected for addi-
tional input costs due to the higher price of hybrid seed (1.78 US$ vs. 0.53 US$),
and provision was made for different seeding rates at different yield potentials (5
kg per ha⁻¹ seed at 750 kg ha⁻¹ yield to 25 kg per ha⁻¹ seed at 5000 kg ha⁻¹ yield),
then the yield advantage (US$ ha⁻¹) of the hybrid also holds for the low yielding
potentials (Fig. 39–4). Seeding rate of the conventional varieties were regarded to
be constant at 25 kg ha⁻¹ and their average yield (US$ ha⁻¹) taken to be the envi-
ronmental index for yield potential at each location.

HYBRID WHEAT STRATEGIES

Long Term Commitment

Successful hybrid wheat development is characterised by commitment of the breeders involved to make hybrid wheat work. The authors prefer that this should be done parallel to conventional breeding and the release of conventional cultivars. Releases from the female pool (and where CHA is being used also from the male pool) would allow the breeder to raise the mean of the female (male) population to commercial performance, and heterosis on top of that would immediately result in hybrid advantage.

Availability of Germplasm

In the past, plant breeding has benefited from sharing germplasm. The value of an elite gene pool, where many traits are at an acceptable level, cannot be overestimated (Rasmusson, 1996). Valuable resources have been developed as the result of crossing, selection and genetic recombination. Hybrid development did not share in this evolutionary process, and lacks core germplasm to ensure continuous improvement. Variability to select for combining ability must be virtually untapped, emphasising the need to identify heterotic groups and the need of appropriate technology to optimise recurrent selection procedures. The question whether heterosis can be derived from exploitation of the diversity between spring types, and winter types, hard red winter types, European germplasm, germplasm of an Eastern European origin and the unknown Chinese pool is still not answered. If this information remains proprietary, a large number of wheat breeders will be excluded from the successful exploitation of heterosis.

Role of Biotechnology

"Biotechnology now available allows the wheat geneticist to elucidate and modify the genetic architecture of most characters in terms of the numbers of loci involved, their relative magnitudes, their dominance and epistatic relationship, and their primary and pleiotropic effects" (Snape, 1996). Tagging genes for restoration, pollen shedding, anther extraction, seed set, female receptivity, combining ability, yield, quality and disease resistance could revolutionise hybrid wheat breeding. Marker assisted selection will not only speed up the development of parental lines, but will also allow pyramiding genes from different species that influence the same character. Wheat has already been transformed, enabling the introduction of foreign genes, like the gene for resistance to round-up into wheat. This might facilitate the development of a nuclear encoded male sterility system.

Alternative to Success

In most cases a 10 to 20% of hybrid yield advantage was not enough to commercialise wheat hybrids. If added profit relative to the cost of seed is too low then this ratio can be changed by: (i) Raising the percentage of heterosis to break the 20% ceiling. This can only be achieved by better knowledge of the germplasm involved in hybrid breeding and by improving hybrid breeding technology. (ii) Improve the agronomy of seed production. The higher the female yield the smaller the difference between the cost of hybrid seed and seed of conventional varieties. (iii) Improve female characteristics. Openness of the wheat flower and longevity of the stamen have been found to be under genetic control and can be improved through selection. (iv) Selection for better pollen quality and greater quantities thereof produced by the male. The technology is also available to transfer these characteristics from rye (*Secale cereale* L.) or even triticale (× *Triticosecale* Wittmack) to wheat.

(v) Some breeders consider CHA and CMS to be outdated and are looking at alternatives to improve crossability. A large number of breeders have found the CMS genetic restorer system to be effective and advantageous because it does not add to the cost of seed production like when chemical hybridising agents or other sophisticated methodology are used.

ACKNOWLEDGMENTS

The authors express their appreciation to the following breeders who have shared their views on matters concerning hybrid wheat: Paul Brennan, Queensland Wheat Research Institute, Australia; Ian Edwards, Pioneer Hi-Bred International, USA; Warren Kronstad, Oregon State University, USA; Sid Perry, Goertzen Seed Research, USA; Stephen Sunderwirth, Hybrinova, France; Jim Wilson, Trio Research, USA, and Dr. Peter Wilson, Hybrid Wheat Australia, Australia.

REFERENCES

Bruns, R., and C.J. Peterson. 1997. Yield and stability factors associated with hybrid wheat. *In* Proc. of the 5th Int. Wheat Symp. Ankara, Turkey. CIMMYT, México, DF. México.

Eberhart, S.A. and W.A. Russell 1966. Stability parameters for comparing varieties. Agron. J. 6:36–40.

Engelbrecht, S.A. 1991. Stability of heterosis in hybrids for yield and yield components at different plant populations and yield potential. Ph.D. diss. in Afrikaans. University of the Free State, South Africa.

Hallauer, A.R. 1990. Methods used in developing maize inbreds. Maydica 35:1–16.

Howes, N.K., S.M. Woods, and T.F. Townley-Smith, 1997. Simulations and practical problems of applying multiple marker assisted selection and double haploids to wheat breeding programs. *In* Proc. of the 5th Int. Wheat Symp. Ankara, Turkey. CIMMYT, México, DF. México.

Jordaan, J.P. 1996. Hybrid Wheat: Advances and Challenges. *In* M.P Reynolds et al. (ed.) 1996. Increasing yield potentials in wheat: Breaking the barriers. CIMMYT, México, DF. México.

Kronstad, W.E. 1996. Genetic diversity and the free exchange of germplasm in breaking yield barriers. *In* M.P Reynolds et al. (ed.) 1996. Increasing yield potentials in wheat: Breaking the barriers. CIMMYT, México, DF. México.

Laubscher, M.C. 1984. The evaluation of the cross-performance of probable hybrid breeding parents. M.S. thesis in Afrikaans. Univ. of Stellenbosch, South Africa.

Lin, C.S., and M.R. Bins. 1988. A superiority measure of cultivar performance for cultivars × location data. Can. J. Plant. Sci. 68:193–198.

Melchinger, A.E., W. Schmidt, and H.H. Geiger. 1988. Comparison of testcrosses produced from F_2 and first backcross populations in maize. Crop Sci. 28:743–749.

Nelder, J.A. 1962. New kinds of systematic designs for spacing experiments. Biometrics 18:283–307.

Peterson, C.J., J.M. Moffat, and J.R. Erickson. 1997. Yield stability of hybrids vs. pureline hard winter wheats in regional performance trials. Crop Sci. 37:116–120.

Pickett, A.A. 1993. Hybrid wheat: Results and problems. Plant Breed. 15. Paul Parey Sch. Publ., Berlin.

Pickett, A.A., and N.W. Galwey. 1997. Plant varieties and seeds. 10:15–32.

Rasmusson, D.C. 1996. Germplasm is paramount. *In* M.P Reynolds et al. (ed.) 1996. Increasing yield potentials in wheat: Breaking the barriers. CIMMYT, México, DF. México.

SAS Institute. 1982. SAS user's guide: Statistics. SAS Inst., Cary NC.

Snape, J.W. 1996. The contribution of new biotechnologies to wheat breeding. *In* M.P Reynolds et al. (ed.) 1996. Increasing yield potentials in wheat: Breaking the barriers. CIMMYT, México, DF. México.

Sprague, G.F., and S.A Eberhart. 1977. Corn Breeding. *In* G.F. Sprague (ed.) Corn and corn improvement. Agron. Monogr. 18., ASA, CSSA, and SSSA, Madison, WI.

Chapter 40

Exploitation of Heterosis for Shifting the Yield Frontier in Rice

S. S. Virmani

INTRODUCTION

Rice (*Oryza sativa* L.), the premier food crop in the world was cultivated in 1995 on about 149 million ha producing 553.6 million tons of grains. It is the staple food for 2.4 billion people in the developing world and is grown mostly in humid and subhumid tropics and humid subtropics where land is intensely cultivated. Most rice growing countries, particularly in Asia where 90% of the world rice is produced and consumed, have done remarkably well in meeting their rice needs during the past three decades using the green revolution technologies; however, the future poses even more challenging and ambitious tasks because by 2030 the world must produce 60% more rice than it produced in 1995 to meet the demand created by increasing populations and rising incomes. This production increase must be achieved on less land, with less labor, less water, and less pesticides, and it must be sustainable. Shifting the yield frontier in rice beyond the level of semi-dwarf varieties is considered an important strategy to meet this challenge. Chinese rice scientists have amply demonstrated that heterosis in rice can be exploited commercially to increase rice varietal yields (Yuan et al., 1994). Results from IRRI (Virmani, 1994a, b; Virmani, 1997), India (Siddiq et al., 1997), Philippines (Lara et al., 1994; de Leon et al., 1996), and Vietnam (Luat et al., 1995; Hoan et al., 1997) also have also confirmed that hybrid rice offers an economically viable option to increase varietal yields beyond the level of semi-dwarf rice varieties. This chapter presents the current status and future opportunities of commercial exploitation of heterosis for shifting the yield frontier in rice.

HETEROSIS FOR YIELD AND PRODUCTIVITY

Ever since heterosis in rice was first reported by Jones in 1926, the presence of significant standard heterosis for yield, yield components and several other agronomic traits has been reported by numerous rice researchers (see recent reviews by Kim & Rutger, 1988; Virmani, 1994a; Virmani, 1996). For commercial exploitation of heterosis hybrid rice varieties have been developed in China since 1975 that showed about 20% yield advantage over inbred varieties (Lin & Yuan, 1980). A farm level study in Jiangsu province in China (He et al., 1987a) showed a 16% yield advantage of hybrid rices over farmers varieties. Studies conducted at IRRI comparing the best rice hybrids with best inbred rices during 1986 to 1995 showed about 17% yield advantage (Virmani, 1996). Significant heterobeltiosis and standard heterosis for yield also has been observed in national programs such as those of India, Philippines, Pakistan, and the USA (c.f. Virmani, 1996). The magnitude of heterosis is higher in Indica (I)/Japonica (J) hybrids than in Indica/Indica or Japonica–Japonica hybrids (Maruyama, 1988; IRRI, 1989; Ikehashi, 1991). In a study conducted at IRRI, various groups of hybrids and inbreds yielded in order of Tropical Japonica (TJ)/I (Indica) > I/I > TJ/TJ = > I > TJ (Khush et al., 1997).

Heterotic hybrids can be developed with a range of desired growth duration because growth duration does not correlate with expression of heterosis (Akita et al., 1986). The yield advantage of hybrids over inbred checks was higher in high yielding environments than in low yielding environments (Virmani et al., 1994), although the percentage heterosis may be reduced due to higher mean yields of check varieties in higher yielding environments. Rice hybrids not only yield higher but also flower earlier than parents (Young & Virmani, 1990a); therefore, heterosis for per-day productivity is even higher than heterosis for yield.

Increased yield in heterotic rice hybrids is due to their increased dry matter, resulting from higher leaf area index (LAI) and higher crop growth rate, and increased harvest index, resulting from their increased spikelet number and grain weight (Ponnuthurai et al., 1984; Akita et al., 1986; Agata, 1990; Peng et al., 1997). Kuroda et al. (1995) suggested that heterosis for growth rate of hybrid rice was caused by the acceleration of cell division rate.

PREDICTION OF HETEROSIS

Per-se performance, genetic diversity (measured on the basis of variation of geographic origin, D^2 statistics) and combining ability of parental lines are the criteria commonly used to breed heterotic rice hybrids. Vijaya Kumar et al. (1997) found a positive relationship between better parent heterosis and reproductive phase duration (i.e., number of days from panicle initiation to 50% flowering. Genetic diversity of parental lines can now be better determined by using polymorphism for molecular markers such as RFLP, RAPD, microsatellites (SSLP), and AFLP. Some positive results have been reported on the relationship between heterosis in hybrids and genetic diversity of parental lines as measured through RFLP markers (Kato et al., 1994; Zhang et al., 1994) and microsatellites (Zhang et al., 1994; Saghai-Maroof et al., 1997). The level for correlations between marker distance and hybrid performance was dependent on the germplasm. Cost-effective AFLP markers can now be deployed to assess genetic diversity among potential parental lines for selecting those that are genetically distant to develop heterotic rice hybrids.

GENETIC BASIS OF HETEROSIS

As in other crops, the genetic basis of heterosis in rice has not been understood clearly. Critical studies on gene action on yield and yield components on rice are very few and most suffer from small population sizes, wider spacings than normal, and limited evaluations at one location or for one year (Kim & Rutger, 1988). Combining ability studies (see Kim & Rutger, 1988; Virmani, 1994a) showed significant GCA and SCA effects for yield. Relative proportions of GCA and SCA variances were found to vary in different studies. Young (1987) and Peng and Virmani (1990) reported higher variances for SCA than GCA for yield. The environment did show strong influence on GCA implying that these parameters should be estimated over different environments before a generalization is made. There also is ample evidence of positive and negative cytoplasmic effects on agronomic traits in rice (Sasahara et al., 1986; Chen et al., 1987; Young & Virmani, 1990b; Wang & Wen, 1995). Xiao et al. (1995), using RFLP markers for

QTL analysis, concluded that dominance complementation (including partial dominance) was the major genetic basis of heterosis in rice.

Kuroda et al. (1995) reported that heterosis for cell division rate in rice was controlled by dominance and additive gene effects; the mean degree of dominance was higher than unity, indicating expression of over-dominance. Yang et al. (1996) reported profound alteration of gene expression measured by mRNA amplification in some rice and maize hybrids compared with their parental lines and concluded that   gene expression in hybrids was altered in a variety of ways; some of these alterations were indicative of an over-dominance phenomenon.

From the above information it can be concluded that dominance, apparent over-dominance due to nonallelic interaction, overdominance due to allelic inter-action at the cellular and enzymatic level, linkage disequilibrium and cytoplasmic-nuclear interactions all contributed to heterosis in rice.

GENETIC TOOLS TO BREED RICE HYBRIDS

Rice, being a self-pollinated crop, must involve the use of an effective male sterility system to develop and produce F_1 hybrids. The most effective male steril-ity system in this crop is cytoplasmic-genetic male sterility (CMS) (see reviews by Shinjyo, 1975; Virmani, 1996).

The first set of commercially usable cytoplasmic male sterile lines (Er Jiu Nan 1A, V20A, Zhen Shan 97A) were developed in China in 1973 from a male sterile plant that occurred naturally in a population of wild rice (*Oryza sativa* f. *spontanea*) growing on Hainan Island, China in 1970 (Lin & Yuan, 1980). This plant was designated wild rice with aborted pollen (WA) and the cytoplasm caus-ing this sterility is popularly known as WA cytoplasm.   Since then a number of CMS lines have been developed from this cytoplasm in China (Yuan & Virmani, 1988; Yuan et al., 1994), IRRI (Virmani et al., 1986; Virmani & Wan, 1988; Vir-mani, 1994b; Virmani, 1997) and in India (Siddiq et al., 1994, 1997).   Besides CMS-WA cytoplasm, several other cytoplasms inducing male sterility in rice have been identified among wild and cultivated accessions in China (Li & Zhu, 1988), IRRI (Virmani & Dalmacio, 1987; Dalmacio et al., 1995; IRRI, 1995) and India (Pradhan et al., 1990; Hoan et al., 1997; Siddiq et al., 1997; Rangaswamy & Jaya-mani, 1997).   However, genetic differences among them have not been established in all cases.   It appears that male sterility inducing cytoplasmic factors are widely distributed in wild and cultivated rices to develop CMS lines possessing diverse cytoplasmic and nuclear backgrounds.   Despite this situation, Zhou (1994) ob-served that 88% of the commercial rice hybrids in China are based on CMS-WA cytoplasm. Hybrid rice breeders tend to deploy this CMS system more frequently because it gives stable CMS lines for which restorers are frequently found, and there is no indication so far of its genetic vulnerability.

Fertility Restoration

Most of the available CMS systems in rice have adequate fertility restorers among elite indica rice cultivars, although some CMS systems (CMS-TN, CMS-MS 577, CMS-*O. perennis*, and CMS-*O. glumaepetula*) have no known restorers so their usefulness in developing commercial rice hybrids is limited. The frequency of restorer lines was higher among rice varieties originating in lower latitudes compared to those originating in higher latitudes; also restorer frequency was higher among indica rices, compared with japonicas.   In China, late maturing in-dica rices showed higher frequency of restorers than the early maturing indica rices (Yuan, 1985).   However, outside China, such a correlation has not been noted be-cause the elite lines have been bred by extensive hybridization among late and early rice cultivars (Yuan & Virmani, 1988). Li and Zhu (1988) observed that among the three ecotypic rice cultivars, aman, aus and boro (indica rice cultivated during dry season in the eastern region of the Indian subcontinent); aman and boro cultivars had a higher frequency of restorers compared to aus cultivars. Similarly, among the bulu and tjereh varieties from the Java island of Indonesia, tjereh varieties showed a higher frequency of effective restorers than bulu rices. The latter are now classified as japonicas (Glaszmann, 1987). Effective restorers were found mainly in South and Southeast Asian countries and Southern China, while non-restorers were concentrated in Northern China and far-eastern Asia. Japonica rice hybrids commercialized in China and developed in Japan and the Democratic Peoples Republic of Korea have been bred by incorporating restorer gene(s) from Indica rices into japonica backgrounds. Literature on the genetics of fertility restoration in rice has been reviewed by Virmani (1994a, 1996). The effect of restorer gene(s) for CMS-Bo and CMS-Di cytoplasms was gametophytic whereas the effect of restorer genes for CMS-WA cytoplasm was sporophytic. The inheritance of fertility restoration in CMS-Bo and CMS-Di cytoplasm was monogenic and the two genes were allelic, *Rf 1*, whereas the fertility restoration of CMS-WA cytoplasm was controlled by two genes, one stronger than the other and their mode of action varied with the cross. Govinda Raj and Virmani (1988) found four groups of restorers among six restorer lines possessing different pairs of restorer genes. The existence of a large number of *Rf* genes explains the occurrence of high frequency of restorer lines among elite indica breeding lines for the CMS-WA cytosterility system. Shen et al. (1996) identified a new restorer gene that was different from other restorer genes such as IR24 and Minghui 63.

Although the CMS system has been found to be effective for developing commercial rice hybrids it is cumbersome and its use is restricted to those germplasms in which maintainers and restorers are abundant. Alternatively, the EGMS system in which the expression of a nuclear sterility gene is regulated by environmental factors viz., photoperiod (Shi, 1981; Shi & Deng, 1986) and/or temperature (Zhou et al., 1988, 1991; Wu et al., 1991; Maruyama et al., 1991a;) can be used for the purpose. Several photo-sensitive (PGMS) and thermo-sensitive genic male sterile (TGMS) lines have been developed in China, Japan, the USA, IRRI and India (see Lu et al., 1997; Virmani, 1996). PGMS lines are completely pollen sterile under long day (>13.75 h) conditions and show fertility reversion under short day (daylength <13.5 h) conditions. Similarly, TGMS lines show complete pollen sterility when the maximum day-temperature is above 29°C and partial fertility when the maximum day temperature is 25 to 28°C. Some TGMS lines (e.g., 1VA) developed in China expresses sterility at low temperature (24°C) and fertility at higher temperature. The critical daylength and temperature for expression of sterility–fertility vary depending on the PGMS/TGMS gene and the genetic background in which it has been incorporated. The critical thermo-sensitive stage for fertility alteration in TGMS lines was reported to be 22 and 26 days before heading in Japan (Maruyama et al., 1991a) and 6 to 15 days after panicle initiation at IRRI (Borkakati, 1994). Both PGMS and TGMS are recessive traits controlled by single genes. TGMS genes reported by Sun et al. (1989) and Maruyama et al. (1991a) have been designated as *tms 1* and *tms 2*, respectively (Kinoshita, 1992). The *tms 1* gene has been located on chromosome 8 (Wang et al., 1995) and the TGMS gene reported from IRRI in mutant IR32364 TGMS has been found to be nonallelic to *tms 2* (Borkakati & Virmani, 1996) and located on short arm of chromosome 6. It has been designated as *tms 3* (Subudhi et al., 1997). The PGMS system is useful in temperate regions where striking daylength differences exist during rice growing season(s), however, the TGMS system is more useful under tropical conditions where daylength differences are marginal and temperature differences occur between low and high altitude locations and/or in different seasons at the same locations.

Wide Compatibility Genes

In order to enhance the level of heterosis for yield observed in Indica and Japonica rices, use of Indica/Japonica (including temperate and tropical Japonica) crosses were proposed because the two ecotypes are genetically diverse. However, hybrids between Indica and Japonica rices show a variable degree of hybrid sterility. Ikehashi and Araki (1984) discovered wide compatibility gene(s) to overcome this problem. They showed that gamete abortion by an allelic interaction at a locus (designated as S_5) caused hybrid sterility in S_5^i-S_5^j but not in S_5^n-S_5^i and S_5^n S_5^j (S_5^i representing indica, S_5^j japonica and S_5^n a neutral allele). Thus incorporation of the S_5^n allele into one of the parents overcame sterility in the desired hybrids. A donor of a S_5^n allele was called a wide compatibility variety (WCV). A number of WCVs originating from Indonesia and West Bengal were identified by Ikehashi and Araki (1984). Subsequently, the S_5^n locus was found to be closely linked with marker genes C (chromogen for apiculus pigmentation) and *wx* (waxy endosperm) located in chromosome 6 (Ikehashi & Araki, 1987). Studies at IRRI (Malik & Khush, 1996) have shown a close linkage (4.1 cM) of WC gene with *Amp 3* and *Est 2*. All WCVs have allele 2 and all non-WCVs have allele 1 at *Amp 3*. Such tight linkage with an isozyme marker is of great practical significance for selecting WC individuals in segregating populations. Several WCVs also have been identified in China (Luo et al., 1990) and at IRRI (Vijaya Kumar & Virmani, 1992). The S_5^n allele has been incorporated into japonica types and successfully used for producing indica-japonica hybrids in Japan (Araki et al., 1988; Ikehashi, 1991) and China (Wang et al., 1991). At IRRI, WCV are being identified among elite new plant type (tropical japonica) breeding lines to develop indica/tropical japonica rice hybrids (Khush et al., 1997). In addition to the S_5^n locus for wide compatibility five new loci (viz., S-7, S-8, S-9, S-15, and S-16) located on chromosome 4, 6, 7, 12 and 1, respectively have been identified by Yanagihara et al. (1992) and Wan et al. (1993). Allelic interaction at these loci can cause hybrid sterility in inter-varietal hybrids, independently of each other, and neutral alleles to overcome this problem have been identified in different rice cultivars (Ikehashi & Wan, 1997). These neutral alleles at different loci are extremely important for enhancing the level of heterosis in inter-varietal group hybrids.

Although high cost of hybrid rice seed is over-compensated by the higher yield of rice hybrids, nevertheless, it still discourages resource-poor rice farmers from using hybrid rice technology. The identification and deployment of apomixis in rice will enable even such farmers to use rice hybrids. Besides it will increase the efficiency of hybrid rice breeders in producing many new true-breeding hybrids compared with three- and two-line rice hybrids. The availability of large number of hybrids will help increase the genetic diversity and reduce genetic vulnerability (Khush et al., 1994). Chinese, IRRI, and U.S. scientists are searching for apomixis in rice using various strategies such as analysis of tetraploid *Oryza* spp., induction of mutations, and use of molecular approaches and genetic engineering techniques.

RICE HYBRIDS UNDER BIOTIC AND ABIOTIC STRESSES

Rice hybrids are resistant to biotic stresses if their parents are resistant or if one parent is resistant due to dominant genes. If both parents are susceptible hybrids are also susceptible (IRRI, 1990). With appropriate choice of parents rice hybrids with desired level of resistance to biotic stresses can be bred. Studies in China (Mew et al., 1988) and IRRI (Virmani, 1994a) did not find any evidence to associate any disease or insect susceptibility in rice with CMS-WA cytoplasm which has been deployed extensively in commercial rice hybrids in and outside China.

With regards to abiotic stresses, rice hybrids have shown improved seedling tolerance to low temperature (Kaw & Khush, 1985) and salt tolerance (Senadhira & Virmani, 1987). Ali et al. (1996) developed a salt tolerant rice hybrid: TNRH 16 in Tamil Nadu, India; derived from two moderately salt tolerant parents IR58025A and C-20R. Significant heterosis, commonly observed for vegetative vigor and root characteristics should also make hybrid rice promising for drought and/or submergence prone rainfed lowland environments where transplanting is practiced or direct seeding with reduced seed rate can be done using some seeding equipment.

GRAIN QUALITY CONCERNS IN RICE HYBRIDS

The major determinants of grain quality in rice are: milling and head rice recovery; size, shape and appearance of rice grains and cooking and eating characteristics. Since rice is primarily consumed as whole grains, it is important to consider the effect of F_2 segregation of the grains harvested from commercial F_1 rice hybrids. Khush et al. (1988) critically studied grain quality of F_2 grains of several rice hybrids in comparison to their divergent parents and concluded that genetic heterozygosity of hybrids did not impair their grain quality in terms of physical and chemical characteristics as long as one of the parents was not of poor grain quality. Therefore, it should be possible to develop rice hybrids of acceptable grain quality by using parental lines of the desired grain quality.

Hybrid rices developed in China when introduced in Japan, Korea, and USA, where consumers are highly quality conscious, were rejected on account of their bold grains, excessive chalkiness and low head rice recovery. Even in Vietnam, where consumers are not as quality conscious, farmers have complained about the grain quality of Chinese rice hybrids even though these have yielded 1 to 2 t ha^{-1} higher than check varieties, because these do not fetch the price comparable to check varieties and hence, do not increase farmers' profit proportionally (Luat et al., 1995). Most of the IRRI-bred CMS and restorer lines have grain quality superior to the popular Chinese CMS lines (e.g., V20A, Zhen Shan 97A, V41A, Bo A), and similar to popular check varieties such as PSBRC4, and IR72. Therefore, grain quality of rice hybrids derived from such lines should not pose serious constraints. There are some reports from India that the first set of rice hybrids released for commercial cultivation do have some chalkiness and lower head rice recovery (Ish Kumar, personal communication). Perhaps in a haste to release heterotic rice these hybrids were not critically evaluated for grain quality.

In India and Pakistan where special quality basmati rices are grown for selected local and export markets, research on developing basmati rice hybrids has also been initiated. A major constraint in breeding basmati rice hybrids is the low frequency of restorers among elite basmati rice breeding lines that can be tackled by deploying TGMS basmati rice lines.

OPPORTUNITIES AND CHALLENGES IN
HYBRID RICE SEED PRODUCTION

Rice, being a self pollinated crop, shows very limited outcrossing, however, male sterile cultivated rices show outcrossing rates ranging from 14.6 to 53.1% in hybrid rice seed production plots in China, IRRI and India (Xu and Li, 1988; Anonymous, 1995). Variability in the extent of natural outcrossing on male sterile lines of rice can be attributed to variations in flowering behavior, floral characteristics of male sterile and pollen parents and variation in environmental factors. Some plant characteristics such as plant height, flag leaf length, and angle and panicle exsertion, also affect natural outcrossing (see review by Virmani, 1996).

Seed yield obtained on a male sterile line used in a hybrid rice seed production plot is a function of (i) yielding ability of the male sterile line as determined by the yielding ability of its fertile counterpart, (ii) proportion of the male sterile line in relation to the pollen parent, and (iii) outcrossing rate of male sterile line. Improvement in any of these components can help to increase rice seed yields (Virmani, 1996).

Based on experience in China (Yuan, 1985; Mao, 1988), IRRI (Virmani, 1994a) and Japan (Kato & Namai, 1987) seed production guidelines and practices have been recommended (Yuan, 1985; Mao, 1988; Virmani & Sharma, 1993; Virmani, 1996). These are being used in India, Philippines, and Vietnam with seed yields of 0.2 to 2.5 t ha^{-1} being reported with a median yield of 1.5 to 2 t ha^{-1} (Virmani, 1996). Low seed yields are attributed primarily to nonsynchronous flowering of seed and pollen parents, unfavourable weather conditions and lack of experience of seed growers. On the other hand, high seed yields are obtained through increasing the proportion of female rows, increasing the transplanting density of female parents and decreasing transplanting density of male parents, increasing dosage of GA$_3$ and discontinuing flagleaf clipping etc. (R.C. Yang & C.X. Mao, personal communication). In China, practices of hybrid seed production are continuously modified by seed growers; the highest seed yields reported from China have reached more than 6 t ha^{-1} (Mao et al., 1997).

Challenges in hybrid seed production include: getting consistently high seed yields beyond 2 t ha^{-1}, reducing the cost of production, and mechanization of seed production technology especially, in labor scarce countries. In China and north-western India male sterile lines used in seed production plots have been found to have a higher incidence of seed borne diseases (such as paddy bunt, caused by *Neovassia horinda* and false smut caused by *Ustilagonoides virens*) compared to pollen parents. This problem needs closer attention to prevent serious outbreak of these minor diseases on commercial crops of hybrid rice.

In countries such as the USA, Japan, Korea, and Malaysia having a low labor/land ratio and high wages, hybrid rice seed production needs to be mechanized to make it cost effective and economically viable. In this context, strategies such as use of facultative female sterile line as pollinator (Maruyama & Oono, 1983) and incorporation of a herbicide sensitive gene or phenol reaction (Ph) gene in the pollen parent and practicing mixed planting of seed and pollen parent followed by mechanized harvesting (Maruyama et al., 1991b) have been suggested. For countries lacking a suitable seed industry infra-structure, a self-sustaining hybrid rice seed production system (Virmani et al., 1993b) has been developed at IRRI.

CURRENT STATUS OF INFRA-STRUCTURE FOR HYBRID RICE RESEARCH AND SEED PRODUCTION IN DIFFERENT COUNTRIES

Eighteen rice growing countries and IRRI are currently involved in hybrid rice research mostly in public sector (Table 40–1). In India, the USA, Brazil, Japan, and the Philippines, the private sector also has invested to various degrees in hybrid rice research. Research infra-structure in the public sector is very strong in China, India, and the Philippines, moderately strong in Japan (because of low priority given to this subject) but weak in other countries. The hybrid rice research in the Philippines by itself is not strong but IRRI's presence makes it very strong.

The seed production infra-structure in China is strong and solely under the public sector whereas in Brazil and USA it is solely with the private sector. In India, a strong seed industry exists in public, private and NGO sectors but private and NGO sectors are playing a more important role in hybrid rice seed production than the public sector. In all other countries, the hybrid rice seed industry is still only developing. In Vietnam, hybrid rice technology has been introduced from China to the farmers despite a weak infra-structure for research and seed production. This is not a sustainable arrangement; however, attempts are being made by the Government of Vietnam to strengthen both research and seed production infra-structure within the country.

FUTURE OUTLOOK

Developments in China, India, and Vietnam have encouraged several other countries to invest in development and use of this technology. It is most likely that countries viz., the Philippines, Bangladesh, Sri Lanka and Indonesia also will start growing hybrid rices commercially during the next five years. Perhaps 2.5 to 3 million ha rice area shall be covered with rice hybrids outside China by 2002.

In order to enhance the magnitude of heterosis observed in Indica or Japonica hybrids, molecular markers can be deployed to select genetically diverse parental lines and tag and incorporate heterotic gene blocks into selected parental lines to improve their combining ability. Additionally, Indica × tropical Japonica and tropical Japonica × temperate Japonica crosses would also help to enhance heterosis. For developing Indica × tropical Japonica rice hybrids, CMS, restorer and TGMS lines in indica and tropical japonica background possessing WC genes are being developed at IRRI. In China and Japan some two- and three-line indica × temperate Japonica hybrids have already been developed and are being tested in on-farm trials. Marker aided selection using the isozyme marker (Amp 3, allele 2) found linked with $S\text{-}_5^n$ locus should be extremely useful for developing parental lines possessing the WC gene.

Xiao et al. (1996) identified genes from wild rice *O. rufipogon* which improved yield in cultivated rice. The lines so derived when used as parents may help to improve heterosis in rice. The strategy proposed by Xiao et al. (1996) may also help to improve combining ability of the best available parental lines that in turn would improve heterosis.

Although there is no indication so far of the genetic vulnerability of the widely used CMS-WA cytoplasm, yet there is need to diversify CMS systems and identify these by using characterization based on hybridization with mt DNA specific probes and mt DNA restriction pattern analyses. Tagging of different *Rf* genes with molecular markers should help to deploy marker aided selection for selecting restorer lines. Mutation breeding approach used by Shen et al. (1996) should be tried to develop restorer lines for *O. perennis* and *O. glumepetula* cytoplasms. PGMS and TGMS systems should be investigated thoroughly for their stability under field conditions and a large number of PGMS and TGMS lines should be developed through conventional and non-conventional (anther culture, marker-aided selection) breeding procedures. In japonica and basmati rices which have very low frequency of restorer lines deployment of PGMS and TGMS systems instead of CMS systems, would be more practical. Intensive research must be done to search, breed and/or genetically engineer apomixis in rice. If successful this will place hybrid rice technology within the reach of even resource poor farmers.

Disease–insect resistance of some popular CMS and TGMS lines used in commercial hybrids can be improved by incorporating in them available cloned genes such as *Bt, Xa 21* through transformation procedures. Research should be intensified to develop rice hybrids for the *boro* season, inland salinity prone areas, and rainfed lowland drought and/or submergence prone ecosystem.

Researchers should ensure that heterotic rice hybrids released for commercial cultivation do possess grain quality which is at least comparable if not superior to inbred check varieties grown by rice farmers. Special attention should be paid to traits such as chalkiness and head rice recovery. In India and Pakistan research should be intensified for developing rice hybrids possessing basmati grain quality.

The experience from China has shown that both genetic improvement for flowering behavior and floral traits of seed and pollen parents and modifications in seed production practices are helpful, in increasing seed yields beyond 2 t ha^{-1}. These strategies should be adopted to increase seed yields in other countries. The prohibitive cost of GA$_3$ in many countries requires the identification of a cheaper substitute and/or improvement of male sterile lines for their panicle exsertion, duration of floret opening and stigma exsertion rate.

From the foregoing information it is evident that heterosis in rice has been helpful in shifting the yield frontier in rice. Hybrid rice has already made tremendous impact on rice production in China and it is beginning to contribute to increased rice production in several other countries. It is anticipated that the world would have 30 to 35 million ha under hybrid rice by 2010.

Table 40–1. Level of development of infra-structure for hybrid rice research and seed production in different countries.

| Country | Year started | Level of development | | | | | Cultivated hybrid rice area in 1996 (ha) |
| | | Research | | Seed Production | | | |
		Public	Private	Public	Private	NGO	
China	1970	***	-	***	-	?	17 m
India	1989	***	***	***	***	*	60 000
Vietnam	1992	*	-	*	*	-	80 000†
Philippines							
NARS	1995	*	*	-	*	*	<100
IRRI	1979	***	-	-	-	-	-
Bangladesh	1997	*	-	*	*	*	-
Sri Lanka	1996	*	-	-	-	-	-
Indonesia	-	*	-	*	*	-	-
DPR Korea	1986	*	-	*	-	-	5 000
Rep. of Korea	-	*	-	*	-	-	-
Malaysia	-	*	-	-	-	-	-
Thailand	-	*	-	-	-	-	-
Egypt	1997	*	-	*	-	-	-
Myanmar	1996	*	-	-	-	-	-
Pakistan	1997	*	-	-	-	-	-
Brazil	1993	*	**	-	***	-	-
Colombia	-	*	-	-	-	*	-
Japan	1983	**	*	-	*	*	-
USA	1980	*	***	-	***	-	-

*** Strong; ** fairly strong; * weak.
† Hybrid seed mostly produced in China.

REFERENCES

Agata, W. 1990. Mechanism of high yield achievement in Chinese F_1 rice compared with cultivated rice varieties. Jpn. J. Crop Sci. 59 (Extra 1):270–273. (Japanese).

Akita, S., L. Blanco, and S.S. Virmani. 1986. Physiological analysis of heterosis in rice plant. Jpn. J. Crop Sci. 65 (extra 1):14–15.

Ali, J., M. Rangaswamy, R. Rajagopalan, S.E. Naina Mohamed, and T.S. Mamickam. 1996. TNRH 16 - A first salt tolerant rice hybrid. *In* Proc. of 3rd International Symposium on Hybrid Rice. 14–16 Nov. 1996. DRR and ICAR, Hyderabad, India.

Anonymous. 1995. Development and use of hybrid rice technology. Annual Report 1993–94. Work Plan 1994–95. Directorate of Rice Research, Hyderabad, India.

Araki, H., K. Toya, and H. Ikehashi. 1988. Role of wide compatibility genes in hybrid rice breeding. p. 79–83. *In* Hybrid rice. IRRI, Manila, Philippines.

Borkakati, R. 1994. Genetics of thermo-sensitive male sterility in rice (*Oryza sativa* L.). Ph.D. diss. Assam Agricultural Univ., Jorhat, India (mimeo).

Borkakati, R., and S.S. Virmani. 1996. Genetics of thermosensitive genic male sterility in rice. Euphytica 88:1–7.

Chen, Yinhui, Cai, Junmai, and Lu Haoran. 1987. Effects of male sterile cytoplasmic interaction on the genetic performance of rice. J. Fujian Agric. Coll. 16(3):115–118.

Dalmacio, R.D., D.S. Brar, T. Ishii, L.A. Sitch, S.S. Virmani, and G.S. Khush. 1995. Identification and transfer of a new cytoplasmic male sterility source from *Oryza perennis* into indica rice (*Oryza sativa* L.). Euphytica 82:221–225.

de Leon, J.C., E.D. Redona, I.A. dela Cruz, M.F. Ablaza, F.M. Malabanan, R.J. Lara, and S.R. Obien. 1996. Hybrid rice in the Philippines. *In* Proc. of 3rd Int. Symp. on Hybrid Rice. Hyderabad, India. 14–16 Nov. 1996. DRR and ICAR, IRRI, Manila, Philippines.

Glaszmann, J.C. 1987. Isozymes and classification of Asian cultivated rice varieties. Theor. Appl. Genet. 74:21–30.

Govinda Raj, K., and S.S. Virmani. 1988. Genetics of fertility restoration of WA type cytoplasmic male sterility in rice. Crop Sci. 28:787–792.

He, G.T., X. Zhu, and J.C. Flinn. 1987. A comparative study of economic efficiency of hybrid and conventional rice production in Jiangsu province, China. Oryza 24:285–296.

Hoan, T., N.N. Kinh, B.B. Bong, N.T. Tram, T. Qui, and N.V. Bo. 1997. Current status of hybrid rice research and development in Vietnam. *In* Proc. of 3rd Int. Symp. on Hybrid Rice. Hyderabad, India. 14–16 Nov. 1996. DRR and ICAR, IRRI, Manila, Philippines.

Ikehashi, H. 1991. Genetics of hybrid sterility in wide hybridization in rice (*Oryza sativa* L.). p. 113–127. *In* Y.P.S. Bajaj (ed.) Biotechnology in agriculture and forestry. Springer Verlag, Berlin.

Ikehashi, H., and H. Araki. 1984. Varietal screening of compatibility types revealed in F_1 fertility of distant crosses in rice. Jpn. J. Breed. 34:304–313.

Ikehashi, H., and H. Araki. 1987. Screening and genetic analysis of wide compatibility in F_1 hybrids of distant crosses in rice, *Oryza sativa*. Tech. Bull. 22. Tropical Agric. Res. Center, Japan.

Ikehashi, H., and J. Wan. 1997. Wide compatibility system: Present understanding of its genetics and use for enhanced yield heterosis. *In* Proc. of 3rd Int. Symp. on Hybrid Rice. Hyderabad, India. 14–16 Nov. 1996. DRR and ICAR, IRRI, Manila, Philippines.

IRRI. 1989. Annual Report for 1988. IRRI, Manila, Philippines.

IRRI. 1990. Annual Report for 1989. IRRI, Manila, Philippines.

IRRI. 1995. Program Report for 1994. IRRI, Manila, Philippines.

Jones, J.W. 1926. Hybrid vigour in rice. J. Am. Soc. Agron. 18:424–428.

Kato, H., and H. Namai. 1987. Intervarietal variations of floral characteristics with special reference to F_1 seed production in japonica rice (*Oryza sativa* L.). Jpn. J. Breed. 37:75–87.

Kato, H., K. Tanaka, H. Nakazumi, H. Araki, T. Yoshida, O. Yashuki, S. Yanagihara, N. Kishimoto, and K. Maruyama. 1994. Heterosis of biomass among rice ecospecies and isozyme polymorphism and RFLP. Jpn. J. Breed. 44:271–277.

Kaw, R.N., and G.S. Khush. 1985. Heterosis in traits related to low temperature tolerance in rice. Philipp. J. Crop Sci. 10:93–105.

Khush, G.S., R.C. Aquino, S.S. Virmani, and T.S. Bharaj. 1997. Use of tropical japonica germplasm for enhancing heterosis in rice. *In* Proc. of 3rd Int. Symp. on Hybrid Rice. Hyderabad, India. 14–16 Nov. 1996. DRR and ICAR, IRRI, Manila, Philippines.

Khush, G.S., D.S. Brar, J. Bennett, and S.S. Virmani. 1994. Apomixis for rice improvement. p. 1–21. *In* G.S. Khush (ed.) Apomixis: Exploiting Hybrid Vigor in Rice. IRRI, Manila, Philippines.

Khush, G.S., I. Kumar, and S.S. Virmani. 1988. Grain quality of hybrid rice. p. 201–215. *In* Hybrid rice. IRRI, Manila, Philippines.

Kim, C.H. and J.N. Rutger. 1988. Heterosis in rice. p. 39–54. *In* Hybrid rice. IRRI, Manila, Philippines.

Kinoshita, T. 1992. Report of the committee on gene symbolization nomenclature and linkage groups. Rice Genet. Newslett. 9:2–4.

Kuroda, S., H. Kato, and R. Ikeda. 1995. Heterosis for the growth of hybrid rice plant caused by the acceleration of cell division. *In* K. Oono and F. Takaiwa (ed.) Modification of gene expression and non-Mendelian inheritance. NIAR, Kannondai, Tsukuba, Japan.

Lara, R.J., I.M. Dela Cruz, M.S. Ablaza, H.C. Dela Cruz, and S.R. Obien. 1994. Hybrid rice research in the Philippines. p. 173–186. *In* S.S. Virmani (ed.) Hybrid rice technology: New developments and future prospects. IRRI, Manila, Philippines.

Li, Z., and Y. Zhu. 1988. Rice male sterile cytoplasm and fertility restoration. p. 85–102. *In* Hybrid rice. IRRI, Manila, Philippines.

Lin, S.C., and L.P. Yuan. 1980. Hybrid rice breeding in China. p. 35–51. *In* Innovative Approaches to Rice Breeding. IRRI, Manila, Philippines.

Lu, X.G., S.S. Virmani, and R.C. Yang, 1997. Advances in two-line hybrid rice breeding. *In* Proc. of 3rd Int. Symp. on Hybrid Rice. Hyderabad, India. 14–16 Nov. 1996. DRR and ICAR, IRRI, Manila, Philippines.

Luat, N.V., H.T. Minh, and N.V. Suan. 1994. Hybrid rice research in Vietnam. *In* S.S. Virmani (ed.) Hybrid rice technology: New developments and future prospects. IRRI, Manila, Philippines.

Luat, N.V., N.V. Suan, and S.S. Virmani. 1995. Current status and future outlook on hybrid rice in Vietnam. p. 73–80. *In* G.L. Denning and V.T. Xuan (ed.) Vietnam and IRRI: A Partnership in Rice Research. IRRI, Manila, Philippines, and Ministry of Agriculture and Food Industry, Hanoi, Vietnam.

Luo, L.J., C.S. Ying, and Y.P. Wang. 1990. The use of photoperiod-sensitive genic male sterile line in screening for wide compatibility rice varieties. Chinese J. Rice Sci. 4(3):143–144. (Chinese with English summary).

Malik, S.S., and G.S. Khush. 1996. Identification of wide compatibility varieties (WCVs) and tagging of WC genes with isozyme markers. Rice Gen. Newslett. 13:121–123.

Mao, C.X. 1988. Hybrid rice seed production in China. p. 277–282. *In* Rice seed health. IRRI, Manila, Philippines.

Mao, C.X., Virmani, S.S., Kumar, I., and Haugen, L. 1997. Technological innovations to economize hybrid rice production cost. *In* Proc. of 3rd Int. Symp. on Hybrid Rice. Hyderabad, India. 14–16 Nov. 1996. DRR and ICAR, IRRI, Manila, Philippines.

Maruyama, K. 1988. Strategy and status for developing hybrid rice. Iden (Heredity) 42(5):28–31. (Japanese).

Maruyama, K., H. Kato, and H. Araki. 1991b. Mechanized production of F1 seeds in rice by mixed planting. JARQ 24:343–352.

Maruyama, K., and K. Oono. 1983. Induction of mutation in tissue culture and their use for plant breeding. VII. On the female and male sterile rice plants regenerated from seed calluses. Jpn. J. Breed. 33(Suppl. 1):24–25. (Japanese).

Maruyama, K., H. Araki, and H. Kato. 1991a. Thermosensitive genetic male sterility induced by irradiation. p. 227–232. *In* Rice genetics II. IRRI, Manila, Philippines.

Mew, T.W., F.M. Wang, J.T. Wu, K.R. Lin, and G.S. Khush. 1988. Disease and insect resistance in hybrid rice. p. 189–200. *In* Hybrid rice. IRRI, Manila, Philippines.

Peng, J.Y., and S.S. Virmani. 1990. Combining ability for yield and four related traits in relation to hybrid breeding in rice. Oryza 27:1–10.

Peng, S.B., J. Yang, F.V. Garcia, R.C. Laza, R.M. Visperas, A.L. Sanico, A.Q. Chavez, and S.S. Virmani. 1997. Physiology-based crop management for yield maximization of hybrid rice. *In* Proc. of 3rd Int. Symp. on Hybrid Rice. Hyderabad, India. 14–16 Nov. 1996. DRR and ICAR, IRRI, Manila, Philippines.

Ponnuthurai, S., S.S. Virmani, and B.S. Vergara. 1984. Comparative studies on the growth and grain yield of some F_1 rice (*Oryza sativa* L.) hybrids. Philipp. J. Crop Sci. 9(3):183–193.

Pradhan, S.B., S.N. Ratho, and P.J. Jachuck. 1990. Development and new cytoplasmic genetic male sterile lines through indica × japonica hybridization in rice. Euphytica 51:127–130.

Rangaswamy, M. and P. Jayamani. 1997. Diversification of cytoplasmic sources in rice. *In* Proc. of 3rd Int. Symp. on Hybrid Rice. Hyderabad, India. 14–16 Nov. 1996. DRR and ICAR, IRRI, Manila, Philippines.

Saghai-Maroof, M.A., G.P. Yang, Q. Zhang, and K.A. Gravois. 1997. Correlation between molecular marker distance and hybrid performance in U.S. Southern long grain rice. Crop Sci. 37:145–500.

Sasahara, T., C.H. Cui, and M. Kambayashi. 1986. Cold resistance in rice with some reference to cytoplasmic effects. SABRAO 18(1):69–71.

Senadhira, D. and S.S. Virmani. 1987. Survival of some F_1 rice hybrids and their parents in saline soil. Int. Rice Res. Newslett. 12(1):14–15.

Shen, Y., Q. Cai, M. Gao, and X. Wang. 1996. Isolation and genetic characterization of fertility-restoring revertent induced from cytoplasmic male sterile rice. Euphytica 90:17–23.

Shi, M.S. 1981. Preliminary report of later japonica natural 2-lines and applications. Hubei Agric. Sci. 7.

Shi, M.S., and J.Y. Deng. 1986. The discovery, determination and utilization of the Hubei photosensitive genic male-sterile rice (*Oryza sativa* subsp. *japonica*). Acta Genet. Sin. 13(2):107–112.

Shinjyo, C. 1975. Genetical studies of cytoplasmic male sterility and fertility restoration in rice, *Oryza sativa* L. Bull. Coll. Agric. Univ. Ryukus 22:1–57.

Siddiq, E.A., Ahmed M. Ilyas, M. Rangaswamy, R. Vijaya Kumar, B. Vidya Chandra, B.C. Viraktamath, and S.D. Chatterjee. 1997. Current status and future outlook for hybrid rice technology in India. *In* Proc. of 3rd Int. Symp. on Hybrid Rice. Hyderabad, India. 14–16 Nov. 1996. DRR and ICAR, IRRI, Manila, Philippines.

Siddiq, E.A., P.J. Jachuck, M. Mahadevappa, F.U. Zaman, R. Vijaya Kumar, B. Vidyachandra, G.S. Sidhu, I. Kumar, M.N. Prasad, M. Rangaswamy, M.P. Pandey, D.V.S. Panwar, and I. Ahmed. 1994. Hybrid rice research in India. p. 157–171 *In* S.S. Virmani (ed.) Hybrid rice technology: New developments and future prospects. IRRI, Manila, Philippines.

Subudhi, P.K., R.P. Borkakati, S.S. Virmani, and N. Huang. 1997. Molecular mapping of a thermosensitive genetic male sterility gene in rice using bulked segregant analysis. Genome 40:188–194.

Sun, Z.X., S.K. Min, and Z.M. Xiong. 1989. A temperature-sensitive male-sterile line found in rice. Rice Genet. Newslett. 6:116–117.

Vijaya Kumar, M. Ilyas Ahmed, B.C. Viraktamath, and M.S. Ramesh, 1997. Heterosis: early prediction and relationship with the reproductive phase. IRRN (in press).

Vijaya Kumar, R., and S.S. Virmani. 1992. Wide compatibility in rice (*Oryza sativa* L.). Euphytica 64:71–80.

Virmani, S.S. 1994a. Heterosis and hybrid rice breeding. Springer Verlag, Berlin.

Virmani, S.S. 1994b. Prospects of hybrid rice in the tropics and sub-tropics. p. 7–19. *In* S.S. Virmani (ed.) Hybrid rice technology: New developments and future prospects. IRRI, Manila, Philippines.

Virmani, S.S. 1996. Hybrid rice. Adv. Agron. 57:377–462.

Virmani, S.S. 1997. Hybrid rice research and development in the tropics. *In* Proc. of 3rd Int. Symp. on Hybrid Rice. Hyderabad, India. 14–16 Nov. 1996. DRR and ICAR, IRRI, Manila, Philippines.

Virmani, S.S., and R.D. Dalmacio. 1987. Cytogenic relationship between two cytoplasmic male-sterile lines. Int. Rice Res. Newslett. 12(1):14.

Virmani, S.S., K. Govinda Raj, C. Casal, R.D. Dalmacio, and P.A. Aurin. 1986. Current knowledge of and outlook on cytoplasmic genetic male sterility and fertility restoration in rice. p. 633–647. *In* Rice genetics. IRRI, Manila, Philippines.

Virmani, S.S., G.S. Khush, and P.L. Pingali. 1994. Hybrid rice for tropics: Potentials, research priorities and policy issues. p. 61–86. *In* R.S. Paroda and M. Rai (ed.) Hybrid research and development of major cereals in the Asia pacific region. FAO, Bangkok.

Virmani, S.S., M.N. Kumar and I. Kumar. 1993a. Breaking the yield barrier in rice through exploitation of heterosis. p. 76–85. *In* K. Muralidharan and E. A. Siddiq (ed.) New frontiers in rice research.. Directorate of Rice Res., India.

Virmani, S.S., J. Manalo, and R. Toledo. 1993b. A self-sustaining system for hybrid rice seed production. Int. Rice Res. Newslett. 18:4–5.

Virmani, S.S., and H.L. Sharma. 1993. Manual for hybrid rice seed production. IRRI, Manila, Philippines.

Virmani, S.S., and B.H. Wan. 1988. Development of CMS lines in hybrid rice breeding. p. 103–114. *In* Hybrid rice. IRRI, Manila, Philippines.

Wan, J.M., S. Yanagihara, H. Kato, and H. Ikehashi. 1993. Multiple alleles at a new locus causing hybrid sterility between Korean indica variety and a javanica variety in rice (*Oryza sativa* L.). Jpn. J. Breed. 43:507–516.

Wang, B., W. Xu, J. Wang, W. Wu, H. Zhang, Z. Yang, J.D. Ray, and H. Nguyen. 1995. Tagging and mapping the thermosensitive gene male sterile gene in rice (*Oryza sativa* L.) with molecular markers. Theor. Appl. Genet. 91:1111–1114.

Wang, C.L., J.S. Zou, Z.M. Wang, C.G. Li, and H.B. Li. 1991. Identification of wide compatibility and heterosis in rice, strain 02428 × Xuan. Chinese J. Rice Sci. 5(1):19–24. (Chinese with English summary).

Wang, W.M., and Y.C. Wen. 1995. New cytoplasmic male sterile line with lower negative effects of cytoplasm on some quantitative traits in rice. Int. Rice Res. Newslett. 20(2):7.

Wu, X.J., H.Q. Yin, and H. Yin. 1991. Preliminary study of the temperature effect of Annong S-1 and W6154S. Crop Res. (China) 5(2):4–6.

Xiao, J.H., S. Grandillo, S.N. Ahn, S.R. McCouch, S.D. Tanksley, J. Li, and L.P. Yuan. 1996. Genes from wild rice improve yield. Nature (London) 384:223–224.

Xiao, J., J. Li, L.P. Yuan, and S.D. Tanskley. 1995. Dominance is the major genetic basis of heterosis in rice as revealed by QTL analysis using molecular markers. Genetics 140:745–754.

Xu, S., and B. Li. 1988. Managing hybrid rice seed production. p. 157–163. *In* Hybrid rice. IRRI, Manila, Philippines.

Yanagihara, S., H. Kato, and H. Ikehashi. 1992. A new locus for multiple alleles causing hybrid sterility between an aus variety and javanica varieties in rice (*Oryza sativa* L.). Jpn. J. Breed. 42:793–801.

Yang, J., N. Chen, Y. Gao, M. Xu, K. Ge, and C.C. Tan. 1996. Molecular basis of heterosis in hybrid rice and hybrid maize revealed by mRNA amplification. Int. Rice Res. Newslett. (1):12.

Young, J.B. 1987. Heterosis and combining ability over environments in relation to hybrid rice breeding. Ph.D. diss. Univ. of the Philippines, Los Banos, Laguna, Philippines.

Young, J.B., and S.S. Virmani. 1990a. Heterosis in rice over environments. Euphytica 51:87–93.

Young, J.B., and S.S. Virmani. 1990b. Effects of cytoplasm on heterosis and combining ability for agronomic traits in rice (*Oryza sativa* L.). Euphytica 48:177–188.

Yuan, L.P. 1985. A concise course in hybrid rice. Hunan Technol. Press, China.

Yuan, L.P. and S.S. Virmani. 1988. Organization of a hybrid rice breeding program. p. 33–37. *In* Hybrid rice. IRRI, Manila, Philippines.

Yuan, L.P., Z.Y. Yang, and J.B. Yang. 1994. Hybrid rice in China. p. 143–147. *In* S.S. Virmani, (ed.) Hybrid rice technology: New Developments and future prospects. IRRI, Manila, Philippines.

Zhang, Q., Y.J. Gao, S.H. Yang, R.A. Ragab, M.A. Saghai Maroof, and Z.B. Li. 1994. A diallel analysis of heterosis in elite hybrid rice based on RFLPs and microsatellites. Theor. Appl. Genet. 89:185–192.

Zhou, K. 1994. Breeding of CMS lines in indica hybrid rice. Hybrid Rice J. 3(4):22–26.

Zhou, T.B., H.C. Xiao, D.Y. Lei, and Q.X. Duan. 1988. The breeding of indica photosensitive male sterile line. J. Hunan Agric. Sci. 6:16–18.

Zhou, Y.B., T.Q. Yu, and G.Y. Xiao. 1991. Relationship between the ecological background of photoperiod-sensitive genic male sterile rice and induction effects of fertility transformation. J. Hunan Agric. Coll. 17(2):99–105.

Chapter 41

Hybrid Rye and Heterosis

H. H. Geiger and T. Miedaner

INTRODUCTION

Rye (*Secale cereale* L.) is the only cross-pollinated species among the small grain cereals. Selfing is prevented by an effective gametophytic self-incompatibility mechanism (Lundqvist, 1956). Yet, self-fertile forms have been found in several breeding populations and are being routinely used for developing inbred lines in hybrid breeding programs. As in maize (*Zea mays* L.) and other cross-pollinated crops, selfing results in severe inbreeding depression and hybrids display strong heterosis. These phenomena have been known since the beginning of this century (Fruwirth, 1913), but it was not before the detection of a cytoplasmic-genic male sterility about 60 years later (Geiger & Schnell, 1970) that the use of heterosis in hybrid varieties became possible.

In this chapter, we want to (i) briefly indicate the worldwide distribution and production of rye, (ii) review the genetical and methodological basis of hybrid rye breeding and seed production, and (iii) demonstrate the progress due to hybrid breeding accomplished in Germany during the last two decades.

WORLDWIDE DISTRIBUTION AND PRODUCTION

Rye is a major cereal crop in the cool temperate zones of Europe with major growing areas in Russia, Belorussia, Poland, and Germany (Madej, 1996). Important regions also exist in Asia and North America (Table 41–1).

The European rye growing area decreased by more than one-half during the last three decades to about 9.1 million ha in the mid-1990s. In 1995, the world production amounted to 29.5 million t, 94% of which was harvested in Europe.

Russian rye breeders successfully used dominant dwarfing genes to improve the productivity of open-pollinated varieties. Hybrid breeding was started on a small scale in the seventies at St. Petersburg and later also at Nemchinovka near Moscow. But no hybrid cultivar has been released so far (Madej, 1996).

Poland has been famous for its diploid open-pollinated rye varieties throughout the world. It was second after Germany in initiating a hybrid breeding program. The first promising hybrids are presently being tested in the official trials. Some of them are crosses between Polish pollinator and German seed-parent lines.

In Belorussia, breeding work has concentrated on tetraploid ryes after a successful cultivar had been released at Zhodina in 1987. Presently, more than 90% of the Belorussian rye acreage is grown to tetraploid varieties. Hybrid breeding (at the diploid level) is in its infancy (Madej, 1996).

Table 41–1. (a) Acreages and average grain yields of the four major rye growing
countries and (b) rye distribution across continents in 1995.

(a)   (b)

Country	Acreage†	Yield	Continent	Acreage‡
	1000 ha	t ha⁻¹		1000 ha
Russia	3500	1.54	Europe	9130
Poland	2400	2.56	Asia	840
Belorussia	1000	2.60	North America	310
Germany	860	4.21	Others	205
Total	7760	2.29	World	10485

† Madej (1996)   ‡ FAO Yearbook 1995.

In Germany, rye used to be the major cereal crop until about 1960. The area then continuously declined and eventually stabilized at 0.8 to 0.9 million ha during recent years. Average yields are two to three times higher than in eastern European countries. Almost two thirds of the harvest is used for bread making. Hybrid breeding started in 1970 at the University of Hohenheim. Breeding companies supported this work and later initiated programs of their own. At present, 17 hybrid varieties are on the official variety list, occupying almost 60% of the rye acreage. Some of these hybrids also are registered and widely distributed in neighbouring countries.

Outside Europe, the only hybrid rye breeding program we know to exist is the one at the University of Sydney in Australia. The first hybrid cultivars from that program are expected by 2001 (Darvey et al., 1996).

HETEROSIS

The amount of heterosis in rye is similar to that in maize and thus several times higher than in self-fertilizing small grains such as wheat (*Triticum aestivum* L.), barley (*Hordeum vulgare* L.), and rice (*Oryza sativa* L.; Melchinger & Gumber, 1997). Highest heterosis is found for grain yield and its components kernels per spike and 1000-kernel weight (Müntzing, 1943; Wricke, 1973; Geiger & Wahle, 1978), whereas plant density (spikes m⁻²) usually shows little or even negative heterosis. Significant hybrid vigour also exists for winter hardiness, plant development, pollen production, straw length, and sprouting resistance (Wehmann et al., 1991).

Interestingly, as hybrid breeding progressed the relative contributions of midparent value and heterosis to cross perfomance markedly changed: Line perfomance increased faster than cross performance resulting in a decrease of relative midparent heterosis for grain yield from >200% to <100% in about two decades (Table 41–2). A similar change occurred in maize after breeding had shifted from populations to hybrids (Schnell, 1974).

Rye is specifically adapted to sandy soils of low fertility and displays a higher tolerance to abiotic stresses than other cereal grains (Bushuk, 1976). Yet, modern rye cultivars are highly responsive to better growing conditions and can compete with wheat, barley, and triticale (× *Triticosecale* Wittmack) in many high-productivity areas. In recent experiments, the relative superiority of hybrids over their parent lines was only slightly more pronounced on sandy soil and under nitrogen deficiency than on loamy soil and under normal nitrogen supply; however, heterosis for the yield components kernels per spike and 1000-grain-weight was higher under favourable conditions (Table 41–3).

Systematic search for a suitable heterotic pattern soon revealed that the two widely used germplasm groups Petkus and Carsten were particularly well matching.

Table 41–2. Mean midparent, hybrid, and relative heterosis values of eight early and 24 recent single-cross hybrids for grain yield and yield-component traits.

Material Statistic	Grain yield	Spikes m^{-2}	Kernels $spike^{-1}$	1000-kernel weight
	$t\ ha^{-1}$	no.	no.	g
Early hybrids (1974–1975)†				
Midparent value	2.13	468	23.1	19.4
Hybrid value	6.56	392	56.5	29.8
Relative heterosis [%]	*207*	*-16*	*145*	*54*
Recent hybrids (1997)‡				
Midparent value	4.27	444	35.3	28.3
Hybrid value	8.14	473	46.7	37.2
Relative heterosis [%]	*92*	*7*	*32*	*31*

† Data from Geiger and Wahle (1978).
‡ Data from Wilde et al. (1997, unpublished data).

Table 41–3. Means and relative heterosis values of two sets of inter-pool hybrids for yield and yield components as influenced by soil type (Experiment I) and nitrogensupply (Experiment II), respectively.

Statistic	Exp. I ($n = 24$)†		Exp. II ($n = 18$)‡	
Character	sand	loam	N0§	N1
Mean				
Grain yield [$t\ ha^{-1}$]	7.64	8.64	2.93	6.95
Relative heterosis [%]				
Grain yield	*97*	*86*	*98*	*87*
Spikes m^{-2}	*21*	*-5*	*14*	*-10*
Kernels $spike^{-1}$	*29*	*36*	*23*	*39*
1000-kernel weight	*23*	*41*	*39*	*50*

† Data from Wilde et al. (1997, unpublished data).
‡ Data from Hartmann (1997).
§ N0 = No nitrogen fertilization; N1 = 140 kg N ha^{-1}.

Petkus populations excelled by a high tolerance to abiotic stresses, superior plant density and good kernel development, whereas Carsten materials display large spikeswith excellent seed setting. Crosses between open-pollinated varieties from these two pools surpassed their parents by 15 to 20% for grain yield (Hepting, 1978). Inbred lines from the two pools have successfully been used for the development of hybrids.

To broaden the genetic basis of the two pools and to improve the adaptation to more continental environments, Wilde and Geiger (1987, unpublished data) crossed 15 open-pollinated European cultivars with the varieties Halo and Carokurz representing the Petkus and Carsten pools, respectively. Yield trials of the population testcrosses revealed that it is much easier to enhance the Petkus than the Carsten pool, since several populations nicely matched Carokurz but none surpassed Carokurz in crosses with Halo (Fig. 41–1). Apparently, the Carsten pool is quite unique, and broadening its genetic basis does not seem to be possible without initially loosing combining ability to the Petkus pool.

CORRELATION BETWEEN LINE AND TESTCROSS PERFORMANCE

The efficiency of a hybrid breeding program strongly depends on the correlation between line and testcross performance. For grain yield, this relationship is rather loose (Table 41–4). The improvement of combining ability for yield should therefore be based on cross rather than line performance. In contrast, strong correlations were found for most other traits including yield components and growth characteristics as well as resistance and quality traits. These characters can be preselected effectively in the line stage allowing to concentrate the testcrossing effort to the more promising candidates.

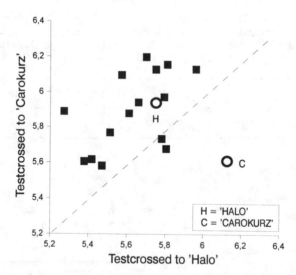

Fig. 41–1. Mean testcross yields [t ha⁻¹] of 15 open-pollinated European cultivars (squares) crossed with 'Carokurz' and 'Halo' representing the two heterotic groups Carsten and Petkus, respectively. For comparison, the trial also comprised reciprocal crosses between 'Carokurz' and 'Halo'(open circles) (data from Wilde & Geiger, 1987, unpublished data).

Table 41–4. Coefficients of genotypic correlation between line and testcross performance of rye in three series of experiments reported in the literature (all estimates larger than twice their standard errors).

Trait	Köhler (1986)	Wilde (1987)	Hartmann (1997)
Grain yield	0.46	0.56	0.56
1000-kernel weight	0.71	0.73	0.76
Plant height	0.86	0.83	0.81
Lodging resistance	0.86	0.90	-
Sprouting resistance	0.73	-	0.90

QUANTITATIVE DISEASE RESISTANCE

Since hybrids are genetically more uniform than open-pollinated populations, they are expected to be more vulnerable to diseases and breeding for resistance requires more attention. The most important diseases of rye in Europe include head blight caused by *Fusarium graminearum* and *F. culmorum*, foot rot caused by *Pseudocercosporella herpotrichoides, Gerlachia nivalis* and *Fusarium* spp., leaf rust caused by *Puccinia recondita*, and powdery mildew caused by *Erysiphe graminis* (EUCARPIA, 1996). Significant quantitative variation for resistance to all of these diseases was observed in various breeding materials (Koch et al., 1985; Miedaner et al., 1993; Gey et al., 1996). In addition, monogenic complete resistances against specific pathogen races occur for leaf rust and powdery mildew (Meyer, 1985; Kobylanski & Solodukhina, 1996). Both types of resistance are presently being used by breeders, but quantitative resistances are considered to be more durable. At Hohenheim, therefore, comprehensive investigations into the genetics of quantitative resistance were taken up in the mid eighties. Briefly, heterosis was low for mildew and foot rot, and medium for head blight resistance. In all cases, general combining ability (GCA) effects were the predominant source of genotypic variation among hybrids. Broad-sense heritability estimates were medium to high in both the inbreds and hybrids, and the resistances of lines and hybrids generally showed good agreement except for head blight resistance (Miedaner et al., 1995; Miedaner & Geiger, 1996). These findings suggest that rapid progress can be expected by selecting within the heterotic pools for quantitative disease resistance in the existing breeding populations. Improvement is necessary in both the seed-parent and pollinator gene pool. Resistance can effectively be pre-selected in the line stage, and in the testcross stage the breeder can largely rely on GCA effects.

CYTOPLASMIC MALE STERILITY AND FERTILITY RESTORATION

Hybrid breeding and seed production in rye is based on cytoplasmic-genic male sterility (CMS) as a hybridizing mechanism. Various CMS sources have been described in the literature (for review see Warzecha & Salak-Warzecha, 1996), but only the 'Pampa' (P) type (Geiger & Schnell, 1970) gained commercial importance. Pampa male sterility is easy to maintain since reliable nonrestorer genotypes occur in all European breeding populations. In contrast, effective fertility restorers are scarce. A first restorer line (L18) for P-CMS was found by Geiger (1972) shortly after the detection of the CMS source; however, this line was never used in breeding programs because of its poor line and testcross performance. Instead, lines with high agronomic performance but moderate restoring ability were employed as pollinator parents. As a result, all presently registered hybrid varieties are only partially restored (Geiger et al., 1995). Under rainy conditions at flowering, this increases the susceptibility to ergot (*Claviceps purpurea*).

To improve the situation, the search for restorers was extented to non-adapted germplasm from Iran and South America. Promising sources are presently being transferred to European breeding materials by repeated backcrossing (Geiger & Miedaner, 1996).

In a recent factorial crossing experiment, the new restorers proved to be universally effective to three genetically diverse CMS tester lines, whereas the European restorers, even line L18, were only acceptable in specific cross combinations (Table 41–5). Molecular marker analyses indicated that restoration is mainly controlled by a dominant gene on chromosome 1R in line L18 and a gene or gene cluster on 4R in the 'exotic' sources (Table 41–6). Ongoing marker projects are aiming at a higher resolution of the respective chromosome segments.

HYBRID BREEDING SCHEME

A simplified scheme of hybrid rye breeding is presented in Fig. 41–2. Seed-parent and pollinator lines are developed from genetically divergent, heterotically matching gene pools. The respective base populations must be continuously improved by recurrent selection and recycling of elite inbred lines. Open-pollinated varieties can be used to broaden the genetic basis but germplasm introductions should be restricted such that an undue increase of the frequency of recessive deleterious genes is avoided and the level of sterility maintenance or fertility restoration, respectively, is not impaired.

Intensive selection for line performance is practiced in selfing generations S_1 and S_2. Major selection criteria at this stage are resistance to disease, lodging, and sprouting. The transfer of the seed-parent lines into the CMS-inducing cytoplasm by repeated backcrossing is generally started in S_2 or S_3. In the same generations, selected lines are first tested for combining ability to the opposite pool. On both the seed-parent and the pollinator side, the testcross seed is produced by means of CMS.

Table 41–5. Male-fertility levels [restorer index, 0-100] of 18 factorial crosses between three CMS seed-parent lines and 6 pollinator lines descending from European (R-L, SFR, L18) and exotic (Iran IX, Pico Gentario, Pastorea Massaux) restorer sources.

| Pollinator line | Seed-parent line | | | |
	L145-P	Lo6-P	L301-P	Mean
R-L	*2.0†*	*2.2*	*69.7*	24.6
SFR-L	49.8	49.1	53.9	50.9
L 18	*19.5*	53.4	73.7	48.7
IRAN IX	78.5	82.8	84.9	82.1
Pico Gentario	79.7	75.5	84.8	80.0
Pastorea Massaux	54.6	79.9	*66.8*	67.1
Mean	47.4	57.2	72.3	59.0

† Figures printed in *italics* refer to significant seed-parent × pollinator line interaction.

Table 41–6. Chromosomal localization of restorer genes and percentage variance explained (R^2) by linked RFLP markers in four rye mapping populations.

Pop.	Restorer source	Chrom. arm	RFLP marker	R^2	Reference
1	L18	1RS	psr596	53.7	Glass et al.
		3RL	psr1077	17.6	(1995)
		4RL	psr119	11.4	
		5RL	psr929	9.8	
2	Iran IX	4RL	mwg573	74.0	Dreyer et al.
3	Pico Gentario	4RL	mwg59	62.5	(1997, unpubl.
4	Pastorea Massaux	4RL	psr899	53.8	data)

For testing the seed-parent lines, topcrosses are made between the CMS versions (backcross generation BC_1) of the candidates and restorer synthetics as testers. The reverse procedure is used for the pollinator lines. In this case the testcross seed is produced on CMS single crosses as testers and the candidates function as males. These crosses are made between plastic walls lined up along the edge of an isolated field. The second stage of combining ability tests is performed likewise with advanced backcross or selfing generations of the seed-parent and pollinator lines, respectively. These tests comprise fewer entries but are grown in a broader range of environments. Attempts to develop doubled haploid lines via anther or microspore culture were successful in specific materials only (Flehinghaus-Roux et al., 1995; Deimling & Geiger, 1996).

Experimental hybrids are produced between CMS single crosses as seed parents and restorer synthetics as pollinator parents. The latter are mostly composed of two restorer lines. A single cross is used to warrant high and stable yields of

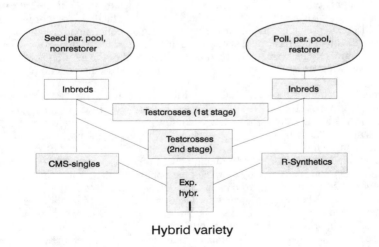

Fig. 41–2. General scheme of a hybrid rye breeding program (CMS = cytoplasmic male sterility, R = restorer).

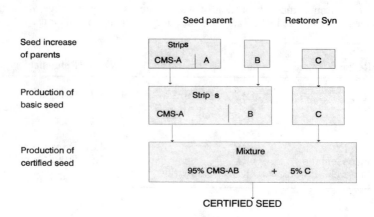

Fig. 41–3. Flow chart of hybrid seed production in rye (CMS = cytoplasmic male sterility, A and B = non-restorer lines, C = restorer synthetic).

vigorous seed. The main advantage of using a synthetic as pollinator parent is an extended pollen-shedding period compared to an inbred line. So, the genetic breadth of most hybrids corresponds to that of a double cross.

The choice of the parent lines for production of experimental hybrids is largely based on GCA since specific combining ability effects seem to be of minor importance even for highly heterotic traits (Wilde et al., 1997, unpubl.). Detailed plans for the development of hybrid rye varieties were suggested by Geiger (1985).

COMMERCIAL HYBRID SEED PRODUCTION

Rye is a highly effective wind pollinator. Seed multiplication fields therefore have to be carefully isolated from other rye fields. This is particularly true for increasing CMS materials. To avoid any risk of genetic contamination, commercial production of pre-basic and basic seed is therefore conducted in regions where no rye is grown otherwise. Only the final certified seed of the hybrid is produced in rye-growing areas. A flow chart of commercial hybrid seed production is given in Fig. 41–3.

For multiplication of the CMS line A and for production of the CMS seed-parent single cross A×B, the female and male parents are grown in alternate strips separated by pathways broad enough to allow roguing and to avoid seed mixtures at harvest. The latter can fully be excluded by removing the pollinator rows after flowering. The first generation of the restorer synthetic (C) should be produced by controlled crossing to exactly obtain equal proportions of the parents. Further multiplication is possible by open-pollination. Strict pollen isolation is necessary since offspring resulting from mispollination are difficult to rogue.

To reduce costs, the certified seed is produced in a mixed stand of about 95% seed-parent and 5% pollinator plants. The pollinator component cannot be separated from the hybrid seed. This may slightly reduce the performance of the hybrid but the reduction is negligible in comparison to the savings in seed production costs.

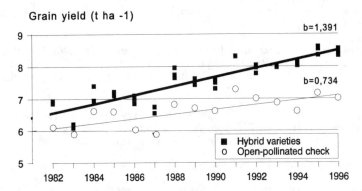

Fig. 41–4. Grain yields of the best hybrids compared to the best open-pollinated check in the official German rye trials 1982–1996 (b = regression coefficient).

ACHIEVEMENTS

During the past fifteen years, substantial progress has been achieved in hybrid rye breeding (Karpenstein-Machan & Maschka, 1996). In the official German trials, the grain yield of the best hybrids surpassed that of the best open-pollinated check by about 10% (=0.6 t ha^{-1}) in 1982 and 20% (=1.4 t ha^{-1}) in 1996 (Fig. 41–4). Progress was also made for other important traits such as lodging resistance, bread making quality, and disease resistance. In Poland, promising results were also obtained in improving the nutritive value of rye as a feed (Rakowska, 1996) and the tolerance to aluminium toxicity (Aniol & Madej, 1996). Compared to population breeding, hybrid breeding proved more flexible in creating varieties for special regions and purposes and more efficient in utilizing the existing genetic variation.

CONCLUSIONS

As a cross-pollinated crop, rye displays high amounts of heterosis for most economically important traits. Hybrids, therefore, soon became more attractive to farmers than open-pollinated varieties. Progress in hybrid breeding steadily increased this superiority, and in the foreseeable future most of the European acreage is likely to be grown to hybrids. Further research is needed for the improvement of fertility restoration, disease resistance, and stress tolerance. For this, new sources of germplasm have to be identified and integrated. Modern biotechnological methods, in particular DNA marker techniques offer great promise in achieving these objectives more efficiently. Concomitantly, biometrical and population genetic research is required to develop optimum breeding plans combining traditional and emerging techniques.

REFERENCES

Aniol, A., and L. Madej. 1996. Genetic variation for aluminium tolerance in rye. p. 201–211. *In* Proc. Intl. Symp. on Rye Breeding and Genetics. Vortr. Pflanzenzüchtg. Vol. 35. (EUCARPIA).

Bushuk, W. 1976. History, world distribution, production, and marketing. p. 1–11. *In* W. Bushuk (ed.) Rye: Production, chemistry, and technology. Am. Assoc. of Cereal Chem., St. Paul, MN.

Darvey, N.L., A. Mohammadkhani, E. Bicar, R. Rodriguez, F. Stoddard, J. Roake, and D.R. Marshall. 1996. Rye developments in Australia. p. 62–63. *In* Proc. Intl. Symp. on Rye Breeding and Genetics. Vortr. Pflanzenzüchtg. Vol. 35. (EUCARPIA).

Deimling, S., and H.H. Geiger. 1996. Anther culture in rye: Methodical improvements and genetic analysis. p. 225–235. *In* Proc. Intl. Symp. on Rye Breeding and Genetics. Vortr. Pflanzenzüchtg. Vol. 35. (EUCARPIA).

FAO Production Yearbook, 1995. FAO, Rome.

Flehinghaus-Roux, T., S. Deimling, and H.H. Geiger. 1995. Anther-culture ability in *Secale cereale* L. Plant Breed. 114:259–261.

Fruwirth, C. 1913. Geschlechtliche Mischung von Roggenformenkreisen. Z. Pflanzenzüchtg. 1:504–507.

Geiger, H.H. 1972. Wiederherstellung der Pollenfertilität in cytoplasmatisch männlich sterilem Roggen. Theor. Appl. Genet. 42:32–33.

Geiger, H.H. 1985. Hybrid breeding in rye (*Secale cereale* L.). p. 237–265. *In* Proc. EUCARPIA Meeting of the Cereal Section on Rye. Svalöv, Sweden.

Geiger, H.H., and T. Miedaner. 1996. Genetic basis and phenotypic stability of male-fertility restoration in rye. *In* Proc. Intl. Symp. on Rye Breeding and Genetics. Vortr. Pflanzenzüchtg. Vol. 35. (EUCARPIA).

Geiger, H.H., and F.W. Schnell. 1970. Cytoplasmic male sterility in rye (*Secale cereale* L.). Crop Sci. 10:590–593.

Geiger, H.H., and G. Wahle. 1978. Struktur der Heterosis von Komplexmerkmalen bei Winterroggen-Einfachhybriden. Z. Pflanzenzüchtg. 80:198–210.

Geiger, H.H., Y. Yuan, T. Miedaner, and P. Wilde. 1995. Environmental sensitivity of cytoplasmic genic male sterility (CMS) in *Secale cereale* L. p. 7–17. *In* U. Kück and G. Wricke (eds.). Genetic mechanisms for hybrid breeding. Blackwell Wissenschafts-Verlag, Berlin.

Gey, A.K.M., T. Miedaner, and H.H. Geiger. 1996. Relationship between leaf rust resistance of inbred lines and test crosses in winter rye. p. 184–185. *In* Proc. Intl. Symp. on Rye Breeding and Genetics. Vortr. Pflanzenzüchtg. Vol. 35. (EUCARPIA).

Glass, C., T. Miedaner, and H.H. Geiger. 1995. Lokalisation von Restorergenen der Winterroggenlinie L18 mit RFLP-Markern. Vortr. Pflanzenzüchtg. 31:52–55.

Hartmann, A. 1997. Untersuchungen zur genetischen Variation von Winterroggen unter ortsüblicher und verringerter Stickstoffversorgung. Ph.D. diss. Univ. of Hohenheim, Stuttgart, Germany.

Hepting, L. 1978. Analyse eines 7×7-Sortendiallels zur Ermittlung geeigneten Ausgangsmaterials für die Hybridzüchtung bei Roggen. Z. Pflanzenzüchtg. 80:188–197.

Karpenstein-Machan, M., and R. Maschka. 1996. Progress in rye breeding. p. 7–13. *In* Proc. Intl. Symp. on Rye Breeding and Genetics. Vortr. Pflanzenzüchtg. Vol. 35. (EUCARPIA).

Kobylanski, V.D., and O.V. Solodukhina. 1996. Genetic basis and practical breeding utilization of heterogenous resistance of rye to brown rust. p. 155–163. *In* Proc. Intl. Symp. on Rye Breeding and Genetics. Vortr. Pflanzenzüchtg. Vol. 35. (EUCARPIA).

Koch, R.J., H.H. Geiger, and W.K. Kast. 1985. Studies on the inheritance of mildew resistance in rye. III. Generation means and components of variation for partial resistance. p. 409–424. *In* Proc. EUCARPIA Meeting of the Cereal Section on Rye. Svalöv, Sweden.

Köhler, K.L. 1986. Experimentelle Untersuchungen zur Aufspaltungsvariation und -kovariation der Linieneigenleistung und Kombinationsfähigkeit bei Roggen. Diss., Univ. of Hohenheim, Stuttgart, Germany.

Lundqvist, A. 1956. Self-incompatibility in rye. I. Genetic control in the diploid. Hereditas 42:293–348.

Madej, L.J. 1996. Worldwide trends in rye growing and breeding. p. 1–6. *In* EUCARPIA (see above).

Melchinger, A.E., and R.K. Gumber. 1997. Overview of heterosis and heterotic groups in agronomic crops. p. 29–44. *In* K.R. Lamkey and J.E. Staub (ed.) Concepts and breeding of heterosis in crop plants. CSSA Spec. Publ. 25. ASA and CSSA, Madison, WI.

Meyer, J. 1985. Breeding for resistance. p. 335–364. *In* Proc. EUCARPIA Meeting of the Cereal Section on Rye. Svalöv, Sweden.

Miedaner, T., D.C. Borchardt, and H.H. Geiger. 1993. Genetic analysis of inbred lines and their crosses for resistance to head blight (*Fusarium culmorum, F. graminearum*) in winter rye. Euphytica 65:123–133.

Miedaner, T., F.J. Fromme, and H.H. Geiger. 1995. Genetic variation for foot-rot and *Fusarium* head-blight resistances among full-sib families of a self-incompatible winter rye (*Secale cereale* L.) population. Theor. Appl. Genet. 91:862–868.

Miedaner, T., and H.H. Geiger. 1996. Estimates of combining ability for resistance of winter rye to *Fusarium culmorum* head blight. Euphytica 89:339–344.

Miedaner, T., W.F. Ludwig, and H.H. Geiger. 1995. Inheritance of foot rot resistance in winter rye. Crop Sci. 35:388–393.

Müntzing, A. 1943. Double crosses of inbred rye. Botaniska Notiser. p. 333–345, Lund.

Rakowska, M. 1996. The nutritive quality of rye. p. 85–95. *In* Proc. Intl. Symp. on Rye Breeding and Genetics. Vortr. Pflanzenzüchtg. Vol. 35. (EUCARPIA).

Schnell, F.W. 1974. Trends and problems in breeding methods for hybrid corn. Manuscript: XVI British Poultry Breeders Roundtable.

Warzecha, R., and K. Salak-Warzecha. 1996. Comparative studies on CMS sources in rye. p. 39–49. *In* Proc. Intl. Symp. on Rye Breeding and Genetics. Vortr. Pflanzenzüchtg. Vol. 35. (EUCARPIA).

Wehmann, F., H.H. Geiger, and A. Loock. 1991. Quantitative-genetic basis of sprouting resistance in rye. Plant Breed. 106:196–203.

Wilde, P. 1987. Schätzung von Populationsparametern und Untersuchungen zur Effizienz verschiedener Verfahren der Rekurrenten Selektion bei Winterroggen. Ph.D. diss. Univ. of Hohenheim, Stuttgart, Germany.

Wricke, G. 1973. Inzuchtdepression und Genwirkung beim Roggen (*Secale cereale*). Theor. Appl. Genet. 43:83–87.

Chapter 42

Cotton and Heterosis

W. R. Meredith, Jr.

INTRODUCTION

Cotton (*Gossypium* spp.) is a major world crop grown primarily for its lint, but also its seed is considered a major oil crop. In 1996, cotton was planted on 33.7 million hectares with a lint production of 19.5 million tons lint (International Cotton Advisory Committee, 1997). Cotton is grown in more than 60 countries. About 96% of world production is planted to upland *G. hirsutum* L., and the remaining 4% to *G. barbadense* L. *G. barbadense* L., is variously called 'Pima', 'Egyptian', or 'Sea Island.' Both of the above species are allotetraploids. A small amount of the diploid species, *G. arboreum* L. and *G. herbaceum* L. are grown in China, India, Myanmas, Pakistan, and Thailand.

In its natural habitat, cotton is a perennial shrub or small tree, but with domestication, it is mostly grown as an annual. The first major success in the USA of using cotton as a crop was in 1833 near Vicksburg, MS, with an introduction, 'Petit Gulf', from the Central Plateau of Mexico (Moore, 1956). Prior to Petit Gulf, the early cottons required a combination of short days, cool nights, and drought stress for fruiting. Their lint was a short, dingy, brown lint and a low lint percentage (Lee, 1984). Petit Gulf's lint was white and longer and its productivity was much greater than other cottons. It was a very popular cultivar and selections and crosses spread over the Cotton Belt in about 15 years. Crosses of Petit Gulf both planned and unplanned, with other grower selections resulted in populations with much genetic variability. Ever since the domestication of Petit Gulf, the Mississippi Delta has been a center of cotton breeding whose germplasm is grown across much of the USA and world.

Early cottons produced large showy flowers with elevated styles that promoted out crossing. Hays et al. (1955) classified cotton as an often cross-pollinated species. Simpson (1948) determined with morphological markers that the expression of heterozygotes was 47% in the F_4 in areas where natural crossing approached 50%. This high level of natural crossing resulted in both a large genetic variance within a cultivar, and opportunities for the expression of heterosis. Simpson (1948) and later Weaver (1986) reported that natural crossing could result in substantial expression of heterosis. The large genetic variance explains why early cotton breeders were frequently successful in improving cultivars by selecting within a cultivar. Presently in the USA, cotton behaves like a self-pollinated crop (Meredith & Bridge, 1973) and most cultivars show little within cultivar variability (Meredith, 1995).

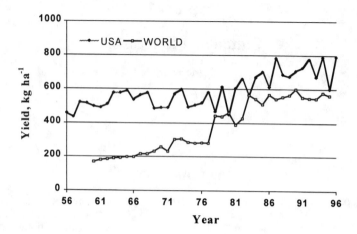

Fig. 42–1. USA and World average yields from 1956 to 1996 (U.S. Agricultural Statistics, 1956 to 1997).

While most crop's yields are increasing, the world's cotton yields seem to be on a plateau. The results in Fig. 42–1 show the average world and USA lint yields over time. Statistics are taken from U.S. Agricultural Statistics (1956 to 1997). In the last decade, there have been only small increases in world average yields. Chaudhry (1997a) reports, that with the exception of India and Turkey, almost all cotton producing countries' yields are on a plateau. The mention of these yield statistics is both to give background on cotton's position as a world crop and to pose the questions of where will increased yields come from and, with an increasing world population and a good economy, how will the world clothe its people?

One possibility is to use manmade synthetic fibers, but presently, cotton has some comfort advantages over synthetics. Increased production comes from two sources; increased hectarage or increased production per ha. It isn't likely that any appreciable increase in world hectarage will occur. Can expanded use of heterosis in cotton be used to address this problem?

Despite much research, there is very little worldwide use of heterosis in cotton. The primary exception is India, where >40% of their cotton is produced from F_1 intraspecific hybrids and 8% interspecific (*G. hirsutum* × *G. barbadense*) hybrids. A small planting of *G. arboreum* × *G. herbaceum* F_1 hybrids also is grown in India (Chaudhry, 1997b). China grows about 330 000 hectares of F_1 and F_2 hybrids (Chaudhry, 1997b).

There are two questions that have to be successfully answered for hybrid cotton to be grown commercially. First, is there sufficient yield heterosis so that productivity of hybrids is significantly better than that of conventional, predominately self-pollinated cultivars? If the answer to the first question is yes, then the next question is, can the production of seed be practical on a commercial economic scale?

The objective of this chapter is to provide the readers with a brief overview of the potential use of heterosis in cotton and to address some of the problems currently limiting heterosis use. This chapter will not cover all cotton heterosis publications, but will use selected publications to demonstrate problems, progress, and potential. The chapter will focus mostly on *G. hirsutum* × *G. hirsutum* research. The term heterosis will refer to the superiority of the F_1 or F_2 over their midparent; and the term useful heterosis will refer to the superiority of the F_1 or F_2 over its highest yielding parent or an established cultivar. In most cases, the highest yielding parent of a specific F_1 or F_2 also is an established cultivar. The term heterosis is useful to explain gene action involved in a specific cross. Useful heterosis is more relevant in explaining the economic importance of a specific F_1 or F_2. Since lint yield is the most important economic determinant of successful cotton culture, this review will focus mostly on lint yield.

PREVIOUS HETEROSIS REVIEWS

Only five of the numerous cotton heterosis reviews are referenced here. Loden and Richmond (1951) reported there was much early skepticism that cotton would show much heterosis. These earlier studies were generally with cultivars that weren't pure lines. Nevertheless, many examples of outstanding heterosis were reported for both intraspecific and interspecific hybrids.

Davis (1977) reported 26 years later that interest in *G. hirsutum* × *G. barbadense* hybrids was growing. He reported interspecific F_1 yield ranged from zero to 80% or 20 to 520 kg ha^{-1} more than the best *G. hirsutum* cultivars. These hybrids had greater seedling vigor, earlier leaf development, earlier maturity, and to some extent, too much vegetative growth than *G. hirsutum* cultivars. The hybrids showed dominance for fiber length, strength, and fineness. Davis (1977) also reported interspecific F_1 yield advantage over their parents ranged up to 138% or 915 kg ha^{-1}; however, intraspecific hybrids showed little dominance for fiber properties.

Table 42–1, taken from a review by Meredith (1984), shows average heterosis [F_1 - midparent (MP)] was 18% and ranged from 3 to 33%. The most influential yield components were number of bolls ha^{-1}, 13.5%, and bollweight, 8.3%. Lint percentage showed little heterosis as its components, lint and seed index, were equal in heterosis magnitude, about 4%, and tended to cancel out any net change in lint percentage.

Basu (1995) reported the first major success of cotton in the world was 'Hybrid 4', which was produced in India by Patel (1971). Parents of Hybrid 4 were 'G67', a commercial Indian cultivar, and Stoneville nectariless (Basu, 1995). Apparently, Stoneville nectariless came from the USDA-ARS in Mississippi. He reported that the average seed cotton yield increase of Hybrid 4 over G67 grown under optimum agronomic practices was 1157 kg ha^{-1} or 171%. Since Hybrid 4's release in 1970, numerous other intraspecific, *G. hirsutum* × *G. barbadense* and a few *G. herbaceum* × *G. arboreum* hybrids have been released in India.

Chaudhry (1997b) indicated almost all of India's hybrids are produced by hand emasculation and pollination. He outlined the steps taken to use cytoplasmic (CMS) and genetic male steriles (GMS) to produce hybrids. He reported on Chembred's attempt to commercialize *G. hirsutum* × *G. hirsutum* F_2 hybrids. He reported that despite much research and testing there are still no commercial F_1 or F_2 hybrids being grown in the USA.

Table 42–1. Heterosis %† for yield, components of yield, and fiber properties (from Meredith, 1984).

		Yield components					Fiber properties		
Yield	Lint %	No. bolls	Boll Weight	Seed weight	Seed boll^{-1}	Lint seed	Length	Strength	Fineness
18.0	1.5	13.5	8.3	34	4.7	4.2	2.0	0.1	0.0

† Heterosis % = 100 × (F_1 - Midparent)/Midparent.

POLLINATION SYSTEMS

As mentioned earlier, cotton was originally often cross-pollinated, but now in the USA is mostly a self-pollinated crop. Unlike maize (*Zea mays* L.) cotton's pollen requires a pollen vector, usually bumble bees (*Bombus* spp.) or honeybees (*Aphis mellifera* L.). There are other native pollinators that vary among regions. There have been three types of pollen control practiced by man: use of male steriles, male gametocide, and hand emasculation followed by hand crossing.

Male Sterile Systems

There are a large number of male sterile systems, many of which are not widely reported; however, Percy and Turcotte (1991) have described 13 genetic male systems (GMS). Five are dominant expressed, four are recessive expressed and four are recessive duplicated loci. The most frequently researched GMS is the MS_{56}. There also are many cytoplasmic male sterile systems (CMS). The most frequently used cytoplasm in the USA was derived from *G. harknessii* Brandagee by Vesta Meyer (1975) and is designated as D_2 CMS. It has an array of restorer factors, which originated from *G. harknessii* and *G. barbadense* and other quantitatively inherited modifiers. This large number of restorer loci required, resulted in slow development of good combiners with good restorer capacity. The D_2 CMS also has about a 5% lower yield than isogenic B fertile lines or CMS hybrids compared with comparable isogenic hybrids (Weaver, 1986). Other CMSs descend from *G. anomalum* wawra and Peyce and *G. arboreum* (Meyer, 1970); however, many of the CMSs are temperature and environmentally sensitive; sometimes resulting in CMS A lines becoming fertile and sometimes resulting in the F_1 not expressing complete fertility.

Another CMS system was developed by Stewart (1992) by crossing *G. trilobum* L. × *G. hirsutum* with this CMS designed D_8. Stewart and Zhang (1996) also developed a single gene restorer factor that also descends from *G. trilobum*. He reports also that F_2s from this system are 100% fertile. Both D_8 and the restorer factors need to be studied in environments other than that of Fayetteville, AR. It's also not reported at this time what effects D_8 has on other agronomic effects.

Male Gametocides

Another method of producing male steriles is the use of male gametocides. The first system reported was FW-450 (sodium a-dichloroisobutyrate). This system requires spray application every two to three weeks being applied to the female rows. These applications also reduce female fertility (Scott, 1961). There was no carryover effect of reducing vigor on growth in the F_1 by this gametocide.

Chembred focused its efforts on producing male steriles by the use of TD-1123, the chemical constituents of which are not made available to the public. This system also required several applications throughout the season and resulted in reduced yields of the female rows. The gametocide also reduced F_1 vigor and had regulatory restrictions on its use. This prompted Chembred to pursue F_2 populations for commercial production. About 12 F_2 cultivars were released but none produced yields significantly better than conventional cultivars. This, coupled with the logistics of seed production, resulted in Chembred closing its operation in 1995.

Hand Emasculation and Crossing

The only successful hybrids produced for commercial use are F_1 and F_2 hybrids produced by hand emasculation and subsequent crossing. India reports about 48% of its cotton is from F_1 hybrids (Chaudhry, 1997b). About 40% of this production is *G. hirsutum* × *G. hirsutum*, 8% *G. hirsutum* × *G. barbadense*, and <1% of *G.*

arboreum × *G. herbaceum*. China is reported by Chaudhry (1997b) to grow about 330,000 hectares of F_1 and F_2 hybrids. In both of these countries the ample supply of hand labor enables the production of these very productive hybrids.

Novel Male Sterile Systems

The use of genetic engineering offers new possibilities for producing male sterility systems. Stewart (1991) describes inducing male sterility by introducing transgenes that degrade the RNA that is normally produced for the plant's pollen development. The second parent, or R line, would be introduced to restore fertility of the F_1 produced by crossing the CMS × R line. Another possibility is to use herbicide resistant cultivars as male parents in a natural crossing field. Susceptible female rows would produce a mixture of selfed- and out-crossed seed. The mixture of F_1 and S_1 seed from female rows would be planted the next season. Upon treatment of the mixture of F_1 and S_1 seed with the herbicide, only resistant F_1 plants will survive. The resulting F_1s could be used to produce F_2 planting seed.

There have long been discussions of using incompatibility and apomixis to utilize heterosis in cotton; however, there is no published research that reinforces the use of either of these two genetic systems. There also have been proposals to produce synthetic cultivars in areas where there is sufficient natural crossing (Duncan et al., 1962).

PHYSIOLOGICAL ASSOCIATIONS WITH HETEROSIS

Wells and Meredith (1986) evaluated the growth rate of four well-adapted Midsouth cultivars and their six F_1s grown in three environments. The average yield of the parents and F_1s was 1453 and 1654 kg ha^{-1}, respectively, a 13.8% heterotic effect. The growth rate of the F_1s was significantly higher than their parents beginning with the early seedling stage and continuing through the late boll development stage (Fig. 42–2). Marani and Avieli (1973) also reported that much of the superiority of F_1s could be explained by the early growth of the hybrids. Wells and Meredith (1986) concluded that the early growth advantage of F_1s formed the basis for the later growth

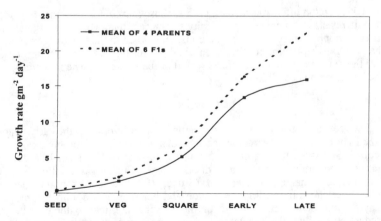

Fig. 42–2. Average growth rate of four parents and the six F_1s over three environments (Wells & Meredith, 1986).

differences. The association of relative growth rate and relative leaf area growth rate was high, $R^2 = 0.99$, for the seedling stage, and $R^2 = 0.85$ for the early vegetative growth stage. The greater early growth of both dry matter and leaf area index seemed to have a compound interest effect on subsequent growth. In another study conducted in two environments, Wells and Meredith (1988) found the F_1s had significantly greater leaf area index and canopy photosynthesis during the early plant stages studied. The relationship of leaf area index and photosynthesis was highly significant in the early development stages ($R^2 = 0.71$), but in later stages after first flowering, no relationship was detected. Early increased growth rates of F_1s were translated into greater first harvest yields in the first study (Wells & Meredith, 1986), which accounted for most of the yield difference between parents and F_1s. The first harvest yield of parents and F_1s averaged 812 and 1002 kg ha^{-1}, respectively. The total yield difference was 201 kg ha^{-1}. In summary, hybrids had increased growth during seedling development, which resulted in greater leaf area indexes. Increased leaf area resulted in greater canopy photosynthesis and ultimately greater first harvest and total yield. At this time the true biochemical and physiological causes of heterosis that are close to the genes' cause and effect are not known.

The Case for F_1s

Chaudhry (1997b) reported that 48% of India's cotton production will be from F_1 hybrids in 1996 to 1997. Hybrids in India result in greater yields per ha, receive a higher price per unit of seed cotton, and their cost of production per kilogram is less than that of conventional cultivars (Basu, 1995). The steady increase in hybrids used ranged from zero in 1970 to 1971 to about two million hectares in 1992 to 1993 (Basu, 1995) and 3.8 million hectares in 1994 to 1995 (Chaudhry, 1997b).

The results in Table 42–2 from eight USA and one Australian study show an average advantage of F_1 vs. nonhybrids of 21.4% and 276 kg ha^{-1}. The range is 11.1% to 57.0% or 111 to 552 kg ha^{-1}. It is evident that even in the USA, which has a long history of improving nonhybrid cultivars, F_1s produce significantly higher yields than the best nonhybrid cultivars. There are no commercial F_1 hybrids grown in the USA. The limiting factor therefore is not due to lack of yield advantage, but is due to the unsolved problems of economically producing sufficient F_1 seed. The success in India is because they have a sufficient labor force to produce F_1 seed by hand emasculation and crossing.

Hybrid cotton offers transgenics, such as those that confer resistance to insects or herbicide resistance, a greater opportunity for success than when those transgenes are placed in nonhybrid cultivars. Some transgenes when in the homozygous condition, have negative pleiotropic effects on yield, maturity, and/or harvesting ease. Much of the negative pleiotropic effects would probably be lost when the transgene is in the heterozygous condition.

The Case for F_2s

The reason for using double-cross rather than F_1 hybrids in the early history of maize hybrids, was to overcome the logistics and economic problems of seed production. Similarly, the reason for trying F_2s instead of F_1s is seed production logistics and economics. The obvious disadvantage of using F_2s is inbreeding depression. Just as in maize and other crops (Stuber, 1996), the gene action model that most closely explains inbreeding depression and heterosis is a dominance model. With any dominance model; complete, partial, or super dominance; the F_2 would be expected to express half the F_1 heterosis. A general scan of reported F_1 and F_2 summary in Table 42–2 shows F_1s average useful heterosis of 21.4%, and F_2s average 10.7%; however, inbreeding depression varies for specific crosses. For example, Meredith and Bridge (1972) reported an average F_1 and F_2 lint yield of 920 and 940

Table 42–2. Review of heterosis and useful heterosis.

Reviews	Env.	F_1s	Heterosis[†] %	kg ha^{-1}	Useful heterosis[‡] %	kg ha^{-1}	Max. yield kg ha^{-1}
	no.	no.					
F_1 references:							
Sheetz and Quisenberry (1986)	3	39	31.5	-	15.1	-	-
Weaver (1986)	2	21	-	-	19.3	200	-
Wells and Meredith (1986)	3	6	13.8	201	21.4	311	1764
Thomson and Luckett (1988)	2	22	15.8	207	20.3 `	308	1827
Meredith (1990)	12	21	11.8	111	11.1	114	1145
Cook and Namken (1993)	1	10	-	-	57.0	552	1375
Cook and Namken (1994)	1	10	-	-	20.5	303	1715
Reid (1995)	2	15	10.0	197	13.0	261	2278
Bowman (unpublished data)	1	15	21.0	210	14.9	157	1302
Mean			17.3	185	21.4	276	1629
F_2 references:							
Olvey (1986)	2	39	-	-	18.0	334	2002
Weaver (1986)	4	2	-	-	7.5	120	1552
Meredith (1990)	12	21	7.6	72	8.0	80	1082
Tang et al. (1993)	4	64	7.4	87	17.9	232	1529
Reid (1995)	7	15	3.0	54	4.7	88	1944
Bauer and Green (1996)	6	10	4.1	35	-	-	-
Meredith and Brown (1998)	3	120	9.3	90	7.9	96	1314
Mean			6.3	68	10.7	158	1570

† Heterosis $= F_1 -$ Midparent.
‡ Useful heterosis $= F_1$ or F_2 - (highest yielding parent or commercial cultivar check)

kg ha^{-1}, respectively from a cross of 'Deltapine 16' × 'Stoneville 603'. This study was conducted at four locations and the parents yields averaged 856 and 862 kg ha^{-1}, respectively. The review of F_2 heterosis ranged in Table 42–2 from 4.7 to 18.0% or 80 to 334 kg ha^{-1}. These results show some F_2s could produce yields significantly higher than popular cultivar checks.

Another use raising F_2 hybrids is that better combination of yield and fiber quality can be obtained. In a 16-parent diallel, Meredith and Brown (1998) reported the regression of yield (y) on fiber strength (T_1) to be: $y = 1866 - 4.46T_1$, with a significant R^2 of 38.6%. The regression for the F_2s was: $y = 1375 - 1.52T_1$, with a nonsignificant R^2 of 4.2%. The mean F_2 yield for the study was 1066 kg ha^{-1} and the T_1 average was 208 kNmkg^{-1}.

In theory, a genetically heterogeneous population, such as an F_2 should have a greater stability of yield when grown across a wide range of environments than that for homozygous pure line cultivars or F_1 hybrids. A few studies, Reid (1995) and Bauer and Green (1996), have shown that F_2s yield relative to their parents is greater under stress environments than under high yield environments. Sheetz (1997, personal communication) reported observing that some F_2s tended to produce yields greater than the best commercial

cultivar when averaged over many environments. At present, there is not sufficient research across numerous environments to give a generalized conclusion on F_2s greater stability. Some have been concerned that the presence of large genetic variability for fiber properties in intraspecific F_2 would present problems in spinning efficiency and quality; however, the variation within a bale from a conventional cultivar due to environmental and developmental factors is great. Also, the textile industry regularly blends bales with widely varying fiber properties without having any difficulties in spinning efficiency or quality; however, lint from interspecific F_2 hybrids such as *G. hirsutum* × *G. barbadense* has much greater fiber variability and can present major spinning efficiency and yarn quality problems.

F_2s can be used experimentally to detect which F_1s would likely produce the highest yields. Data from the study by Meredith (1990) with 21 F_1s and their F_2 hybrids are highly correlated, $r = 0.86$.

SELECTION FOR GOOD COMBINERS

Commercial cultivars have in general, been the parents tested for combining ability. This approach is essentially one of first selecting for useful additive genetic variance. In a previous review (Meredith, 1984), four of eight combining ability studies showed significant general combining ability (GCA) and four showed significant specific combining ability (SCA). Prasad et al. (1984) in India reported that SCA components for F_1 and F_2 hybrids was significant and about five times greater than the GCA component.

In the USA there are four major cotton production regions and breeding for regional adaptation has long been practiced. These regions are the East, Delta, Plains, and West. In 1990, we conducted a half-dialled of 16 parents and 120 F_2s at three locations. The 16 parents consisted of four cultivars from each of the four USA regions. Analyses were made first for all 120 F_2s and then partitioned into 10 combining ability analyses. There were four within region analyses for a total of 24 F_2s (four regions with six crosses each). Six analyses were made among regions, each involving 16 F_2s. The analysis for all 120 F_2s showed significant heterosis with the parental and F_2 average being 976 and 1066 kg ha^{-1}, respectively, or an average heterosis of 9.3%. For all 120 F_2s, GCA and SCA were highly significant, and the SCA component was 2.2 times greater than that for GCA. F_2 dominance is expected to be 0.5 that expressed in the F_1 and the dominance variance is expected to be 0.25 that expressed in F_1 populations. The average F_2 and MP yields within and among regions is shown in Fig. 42–3.

Within regions, heterosis was significant only for the East and West regions; 9.6 and 27.0%, respectively. Among regions, average heterosis was detected from the three groups involving the Western region. The West × East, Delta, and Plains region was 18.7, 11.5, and 15.9%, respectively.

In Fig. 42–3, we see the general relationship of greater statistical heterosis (F_2-MP) occurring with the lowest yielding parents. The West × Delta F_2s produced the best combination of yield and fiber strength; however, the highest yields were from Delta × Delta crosses, although the average increase in yield over parents was only 47

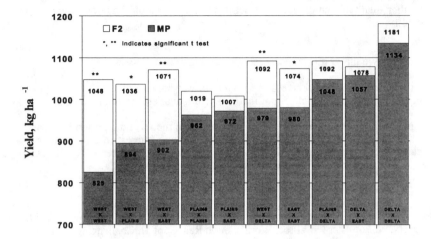

Fig. 42–3. Yield of F₂s and midparent for crosses within four USA cotton growin regions and their six inter-region crosses. Within a region, the mean is from si crosses, and the mean for inter-regions is from 16 crosses. Upper value is F average, and lower value is midparent average.

kg ha^{-1}. Significant heterosis was detected only with one parent, 'MO 344', which averaged 91 kg ha^{-1}. The average heterosis expressed by the three F_2s among the other three Delta cultivars was -1 kg ha^{-1}. RLFP data showed little variation among these three Delta cultivars (Meredith, 1995). The highest yielding parent was 'DES 119' with a yield of 1218 kg ha^{-1}; yield of MO 344 was 1031 kg ha^{-1}. Average F_2 yield was 1314 kg ha^{-1}, which approaches statistical significance over DES 119. Tang et al. (1993) conducted F_2 studies in the Eastern region and reported that the best yielding F_2s average was 1529 kg ha^{-1}. It's parents were MO 344 and Coker 315 whose yields were 1297 and 1272 kg ha^{-1}, respectively.

Where Does One Find Parents to Produce High Yielding Hybrids?

Besides MO 344, good combiners for yield have been discovered within other established Delta cultivars; 'Stoneville 603' has been shown to have excellent combining ability by Meredith and Bridge (1972), and Chembred. While genetic diversity is desired among potential hybrid parents, it does not guarantee heterosis. The weak correlation, $r = 0.08$, between genetic distance (Meredith, 1995) and heterosis among the 16 parents used in that diallel reinforces this generalization.

The positive results with Hybrid 4 in India possibly suggests a generalized approach for choosing parents. Hybrid 4's parents were an established Indian cultivar, G 67, and a USA introduction, Stoneville nectariless. The breeding program in which MO 344 was developed was located in the North Delta, and involved extensive introductions of exotic germplasm. There are no simple rules in selecting good yield producing parents, however, since it is always encouraged, some suggestions follow:

1. At least one parent should be chosen based on its performance and adaptation to the targeted growing region.

2. Using the adapted parents as test parents, conduct test crosses with genetically diverse lines or cultivars.
3. If fiber quality is an objective, at least one parent should have excellent fiber quality.

SUMMARY

The two major questions to using heterosis in cotton were:

1. Is there sufficient useful heterosis for yield to warrant hybrid research?
2. Is there an efficient practical pollination system available to use heterosis?

The answer to the first question is yes. The answer to the second question is yes and no. Yes, the success of heterosis in India and growing interest in other countries indicate, where a sufficient supply of labor is available, use of heterosis can be successful. The answer to the rest of the world is no, not yet.

REFERENCES

Basu, A.K. 1995. Hybrid cotton results and prospects. p. 335-341. *In* G.A. Constable and N.W. Forrester (ed.) Proc. World Cotton Res. Conf.: I. Challenging the future, Brisbane, Australia.

Bauer, P.J., and C.C. Green. 1996. Evaluation of F_2 genotypes of cotton for conservation tillage. Crop Sci. 36:655–658.

Chaudhry, M.R. 1997a. Cotton yields stagnating. The ICAC Recorder XV(1):3–7.

Chaudhry, M.R. 1997b. Commercial cotton hybrids. The ICAC Recorder XV(2):3–4.

Cook, C.G., and L.N. Namken. 1993. Yield and fiber quality of F_1 hybrids. p. 1539–1540. *In* D. Richter (ed.) Proc. Beltwide Cotton Prod. Res. Conf. Natl. Cotton Council of Am., Memphis, TN.

Cook, C.G., and L.N. Namken. 1994. Performance of F_1 hybrids in the lower Rio Grande Valley. p. 674–675. *In* D. Richter (ed.) Proc. Beltwide Cotton Prod. Res. Conf. Natl. Cotton Council of Am., Memphis, TN.

Davis, D.D. 1977. Hybrid cotton: Specific problems and potentials. Adv. Agron. 30:129–157.

Duncan, E.N., J.B. Pate, and D.D. Porter. 1962. The performance of synthetic varieties of cotton. Crop Sci. 2:43–46.

Hays, H.K., F.H. Immes, and D.C. Smith. 1955. Cotton and sorghum breeding. p. 238–266. Methods of plant breeding. 2nd ed. McGraw-Hill Book Co., New York.

International Cotton Advisory Committee. 1997. Cotton world statistics. October 1997. ICAC.

Lee, J.A. 1984. Cotton as a world crop. p. 1–25. *In* R.J. Kohel and C.F. Lewis (ed.) Cotton. Agron. Monogr. 24. ASA, CSSA, and SSSA, Madison, WI.

Loden, H.D., and T.R. Richmond. 1951. Hybrid vigor in cotton: Cytogenetic aspects and practical applications. Econ. Bot. 5:387–408.

Marani, A., and E. Avieli. 1973. Heterosis during the early phases of growth in intraspecific and interspecific crosses of cotton. Crop Sci. 13:15–18.

Meredith, W.R., Jr. 1984. Quantitative genetics. p. 131–150. *In* R.J. Kohel and C.F. Lewis (ed.) Cotton. Agron. Monogr. 24. ASA, CSSA, and SSA, Madison, WI.

Meredith, W.R., Jr. 1990. Yield and fiber-quality potential for second-generation cotton hybrids. Crop Sci. 30:1045–1048.

Meredith, W.R., Jr. 1995. Use of molecular markers in cotton breeding. p. 303–308. *In* G.A. Constable and N.W. Forrester (ed.) Proc. World Cotton Res. Conf.: I. Challenging the future, Brisbane, Australia.

Meredith, W.R., Jr., and R.R. Bridge. 1972. Heterosis and gene action in cotton, *Gossypium hirsutum* L. Crop Sci. 12:304–310.

Meredith, W.R., Jr., and R.R. Bridge. 1973. Natural crossing in cotton (*Gossypium hirsutum* L.) in the Delta of Mississippi. Crop Sci. 13:551–552.

Meredith, W.R., Jr., and S. Brown. 1998. Relationship of regional cotton cultivar origin and F_2 yield. Cotton Sci. (accepted).

Meyer, V.G. 1970. Factors affecting male sterility in cotton. p. 55–56. *In* J.M. Brown (ed.) Proc. Beltwide Cotton Prod. Res. Conf., Natl. Cotton Council of Am., Memphis, TN.

Meyer, V.G. 1975. Male sterility from *Gossypium harknessii*. J. Heredity 66:23–27.

Moore, J.H. 1956. Cotton breeding in the Old South. Agric. Hist. 30:95–104.

Olvey, J.M. 1986. Performance and potential of F_2 hybrids. p. 101–102. *In* T.C. Nelson (ed.) Proc. Beltwide Cotton Prod. Res. Conf., Natl. Cotton Council Am., Memphis, TN.

Patel, C.T. 1971. Hybrid 4, a new hope towards self sufficiency in cotton in India. Cotton Dev. 1:1–6.

Percy, R.G., and E.L. Turcotte. 1991. Inheritance of male sterile mutant MS_{13} in American Pima Cotton. Crop Sci. 31:1520–1521.

Prasad, R.C., J.A. Siddiqui, and R.B. Mehra. 1984. Heterosis and combining ability studies in hirsutum cotton. Indian J. Agric. Res. 18:224–228.

Reid, P.E. 1995. Performance of F_1 and F_2 hybrids between Australian and USA commercial cotton cultivars. p. 346–349. *In* G.A. Constable and N.F. Forrester (ed.) Proc. World Cotton Res. Conf.: I. Challenging the future, Brisbane, Australia.

Scott, R.A. 1961. Mechanism and reversal of gametocide response in cotton. Plant Phys. 36:529–538.

Sheetz, R.H., and J.E. Quisenberry. 1986. Heterosis and combining ability effects in Upland cotton hybrids. p. 94–98. *In* T.C. Nelson (ed.) Proc. Beltwide Cotton Prod. Res. Conf., Natl. Cotton Council of Am., Memphis, TN.

Simpson, D.M. 1948. Hybrid vigor from natural crossing for improving cotton production. J. Am. Soc. of Agron. 40:970–979.

Stewart, J.M. 1991. Biotechnology of cotton: Achievements and perspectives. p. 1–50. ICAC Recorder Rev. Article on Cotton Prod. Res. No. 3. CAB. Int. Wallingford, England.

Stewart, J.M. 1992. A new cytoplasmic male sterile and restorer for cotton. p. 160. *In* T.C. Nelson (ed.) Proc. of Beltwide Cotton Prod. Res. Conf., Natl. Cotton Council of Am., Memphis, TN.

Stewart, J.M., and J. Zhang. 1996. Cytoplasmic influence on the inheritance of the D8 CMS restorer. p. 622. *In* D. Richter (ed.) Proc. Beltwide Cotton Prod. Res. Conf., Natl. Cotton Council of Am., Memphis, TN.

Stuber, C.W. 1996. Heterosis in plant breeding. Plant Breed. Rev. 12:227–251.

Tang, B., J.N. Jenkins, J.C. McCarty, and C.E. Watson. 1993. F_2 hybrids of host plant resistant germplasm and cotton cultivars: I. Heterosis and combining ability for lint yield and yield components. Crop Sci. 33:700–705.

Thomson, N.J., and D.J. Luckett. 1988. Heterosis and combining ability effects in cotton. II. Heterosis. Aust. J. Agric. Res. 39:991–1002.

Weaver, J.B. 1986. Performance of open pollinated cultivars, F_2s and CMS Upland × Upland restorer strains. p. 98–100. *In* T.C. Nelson (ed.) Proc. of Beltwide Cotton Prod. Res. Conf., Natl. Cotton Council of Am., Memphis, TN.

Wells, R., and W.R. Meredith, Jr. 1986. Heterosis in upland cotton: I. Growth and leaf area partitioning. Crop Sci. 26:1119–1123.

Wells, R., and W.R. Meredith, Jr. 1988. Heterosis in upland cotton: II. Relationship of leaf area to plant photosynthesis. Crop Sci. 28:522–525.

Chapter 43

Trees and Heterosis

J. L. Brewbaker and W. G. Sun

INTRODUCTION

Hybrid superiority characterizes one or more traits of most tree hybrids, but trees provide many constraints to the full exploitation and understanding of heterosis. Many tree species are in the early stages of domestication, with limited exploitation of natural variability that extends across wide, often disjunct natural distributions. Population improvement alone has made rapid gains with such species. Trees are by definition woody and long-lived perennials that attain heights of 3 m or more and are usually single-stemmed. They are often slow to flower and fruit, obviating multiple-generation breeding approaches. Most species are outcrossing, often imposed by self incompatibility or dichogamy or dioecy, and the high heterozygosity can lead to severe inbreeding depression in fertility, viability and growth. Production of hybrid seeds is often impractical in trees, where few systems such as cytoplasmic male-sterility have been identified and where emasculation and controlled crossing is impractical or impossible. Vegetative propagation is economically practical for fruit trees and a few forest trees, allowing the exploitation of heterosis through clones; however, genetic improvement for most trees is far from the stage that breeding to capitalize on heterosis is practical.

Forest tree breeders focus increasingly on interspecific hybrids that successfully combine traits leading to improved yields across a wide range of ecosystems. Interspecific fertility exceeds 50% in many tree genera, as in the genus *Leucaena* where 77% of the combinations of 4 polyploid and 12 diploid taxa were fertile (Sorensson & Brewbaker, 1994). Other examples cited by these authors include *Salix* spp. (53% success), *Ulmus* spp. (43% success) and the genera *Erythrina, Liquidambar,* and *Quercus* (>90% success). Coadapted gene complexes are believed to be common in highly outcrossed trees, suggesting an important role of epistatic interactions in hybrid phenomena. Complementarities due to combinations of traits that interact to create superiority in yield, form or environmental adaptability are common. As discussed later, the role of multiplicative effects in such hybrid superiority can be important (Schnell & Cockerham, 1992). Evidence among trees for heterosis based on gene or allelic interactions is not easily obtained and not always distinguishable from complementarity. Despite the challenges in its genetic interpretation, heterosis and general hybridization ability are important targets for many tree breeders. This review will begin with selected major groups of trees, such as poplars, eucalypts and conifers, and conclude with discussions of the roles of complementarity and genetic distance in hybrid superiority.

HYBRID SUPERIORITY IN *POPULUS* SPECIES

The genus *Populus* (Salicaceae) includes the poplars, aspens, and cotton-woods, trees that provided some of the earliest evidence of heterosis in plants. These are the fastest growing northern temperate trees, and are subject of a comprehensive new manual by Stettler et al. (1996). There are 34 species, now spread worldwide, among which hybrids increasingly dominate new plantations. Nearly all species are dioecious, precluding inbreeding by self-fertilization, and they clone easily. The three primary subgroups of the genus *Populus* are Aegiros (black poplars), Populus (aspens and white poplars) and Tacamaca (balsam poplars). The Aegiros group includes the Eastern Cottonwood, *P. deltoides*, and the European Black Poplar, *P. nigra*. The Populus subgroup includes widespread aspens of the species *P. tremuloides* and *P. tremula*. The Sub-group Tacamaca includes *P. trichocarpa* and *P. balsamifera.*

Natural hybrids occur among several *Populus* species and were recognized in Europe as early as 1750 among progeny of the Eastern American Cottonwoods (*P. deltoides*) that had been introduced. The first hybrids involved the European Black poplar (*P. nigra*), and are now referred to as *P. × canadensis* (or *P. × euroamericana*). These include popular modern clones such as 'Robusta.' Other hybrids occurred naturally when varieties such as the columnar Lombardy Poplar (*P. nigra 'italica'*) were introduced worldwide. Among the largest poplars is *P. trichocarpa*, the Western Balsam or Black Cottonwood, that attains heights of 60 m on the Pacific Coast of North America. TT32 is a widely grown hybrid of *P. trichocarpa* × *P. balsamifera* that can exceed 3 m y^{-1} even in temperate climates. Similar interspecific hybrids dominate plantation and afforestation plantings in countries such as Korea, Pakistan, and China.

Intraspecific Hybrids in the Genus *Populus*

The literature on tree hybridization is concerned primarily with intraspecific crosses, as one expects where specific products such as fruits, nuts, or quality timber are of major importance. The genus *Populus,* like that of several to be discussed, provides products such as pulpwood for which the breeder has much less constraint on genetic options. Heterosis research for genera of this type is thus dominated by reports of inter- rather than intra-specific hybrids. Li and Wu (1996) compared heterosis in intra- and inter-specific hybrids of aspens, reviewed below, and emphasized the commercial significance of interspecific hybrids. However, hybrid vigor was evident among their intraspecies hybrids of *Populus tremuloides* for stem growth and volume index, as earlier shown by Foster and Shaw (1988) for intraspecies hybrids in *P. deltoides*. In each case, nonadditive variance explained almost all of the genetic variance, and led authors Li and Wu (1996) to a detailed discussion of the role of pseudo-overdominance, considered later in this review.

Interspecific Hybrids in Poplars

R.F. Stettler provided early evidence for the economic benefits of hybridization in poplars. Yields of hybrids were double those of parents *P. deltoides* and *P. trichocarpa* in six years growth (Stettler et al., 1988, 1996). Stem volumes of hybrids exceeded parent *P. deltoides* by 130% and *P. trichocarpa* by 50%. Bradshaw and Stettler (1995) have mapped QTLs associated with heterotic growth traits in these and other forest trees. Artificial hybrids of *P. trichocarpa* and *P. deltoides* (Ceulemans et al., 1992) achieved impressive growth rates exceeding the parents, with four-year volumes and heights as follows:

	P. trichocarpa	*P. deltoides*	*Hybrid*
Volume	140%	100%	230%
Height	122%	100%	130%

These differences were enhanced by close spacing (1 × 1m) and by inter-clonal competition. Heterosis among these species increased as parents were drawn from different regions. Artificial hybrids of the American cottonwood (*P. deltoides*) and *P. nigra* or *P. simonii* also exceeded their parents in volume. While little is often known about root development in trees, there is evidence that poplar species hybrids show better rooting ability. Natural hybrids characterize many *Populus* populations, and often show no clear evidence of heterosis. Campbell et al. (1993) observed no heterosis among natural hybrids in Alberta among *P. angustifolia, P. Balsamifera*, and *P. deltoides*, concluding that introgressive gene exchange had minimized genetic distances and heterosis. Triploid hybrids occur in the genus *Populus*, among which are several clones in popular use in Europe. Wu (1995) developed quantitative genetic models to estimate the nature of gene action in triploids. Bate et al. (1988) describe high correlations between hybrid vigor in height and contents of endogenous gibberellins in four interspecific poplar hybrids.

Interspecific Hybrids in Aspens

Hybrids among the aspens have long been reported to show impressive vigor. Hybrids of bigtooth aspen *(P. grandidentata)* and quaking aspen *(P. tremuloides)* with other species often outyield parents by 100%. In a definitive paper, Li and Wu (1996) compared intra- and interspecific hybrids of *P tremuloides* and *P. tremula*. They expressed concern that quantitative genetic models underemphasize the role of multiple allelism, and noted its common occurrence in trees. They stressed, along with other authors, that highly outcrossing trees have very high genetic loads of deleterious recessives. Their analytic method revealed heterosis in height to be under control of 6 to 8 QTLs, in stem diameter of 9 to 10 QTLs, and in stem allometry of only 3 to 4. While additive effects were common to all traits, a few major QTL loci were concluded to dominate variation in stem growth and heterosis. Heterozygotes at key QTLs showed improved stem growth over homozygotes and the authors concluded that pseudo-overdominance was a likely cause. This can occur as a result of interaction between deleterious recessive alleles at one locus with beneficial dominant alleles at closely linked loci.

HYBRID SUPERIORITY IN *EUCALYPTUS* SPECIES

The genus *Eucalyptus* (Myrtaceae) contains about 600 Australasian species, of which >200 are now grown abroad (Eldridge et al., 1993). A century ago, McClatchie (1902) illustrated >40 species growing in the Americas and wrote ". . . they have served more aesthetic and utilitarian purposes than any other forest trees . . . planted on this continent." Eucalypts constitute the most important flowering trees in plantations, complementing the conifers worldwide and dominating wood production in the tropics. These are evergreen trees that vary from shrubs to some of the tallest trees known (*E. regnans* to 100 m). Eucalypts are the fastest growing trees in most of the tropics, achieving yields in excess of 100 m3 ha^{-1} yr^{-1} in clonal plantations of Brazil and Congo. One of the most widely cultivated trees in the world is *E. globulus*. Among the most important eucalypts are *E. camaldulensis, E. globulus, E. grandis, E. maculata, E. nitens, E. paniculata, E. robusta, E. saligna, E. tereticornis, E. urophylla,* and *E. viminalis.* Most of these occur in the subgenus Symphomyrtus. Important high-value species include *E. deglupta* for timber and many others as ornamentals.

Genetic improvement is ongoing in more than 20 species of eucalypts, and domestication continues with many more (Eldridge et al., 1993). Eucalypts in natural populations have a mixed mating system with outcrossing in varying proportions. Thus, inbred trees often occur from open-pollinated seeds. Aradhya and Phillips (1995) reported very high heterozygosity among Hawaiian eucalypts that was poorly associated with heterosis for juvenile traits. Inbreeding led to inflated values for heritability and dominance variance in *E. globulus* (Hodge et al., 1996). Inbreeding depression is normally very great in eucalypts, and inbreds are not automatically eliminated in nature nor does nursery culling achieve any gains (Hardner & Potts, 1995). The use of inbreeding to purge deleterious alleles and as a means of identifying superior parents has been propounded by Griffin and Cotterill (1988), since many species can be bred on a three- to five-year cycle.

Intraspecific Hybrids in Eucalypts

Genetic distance based on RAPD markers was highly correlated with heterosis for intraspecific hybrids of *E. globulus* (Vaillancourt et al., 1995). Data based on a 8 × 26-parent factorial showed significant SCA effects with dominance variance comprising 60% of genetic variance, and heterosis accounted for only 5% of the variation in SCA. Within provenance SCA effects were small and unpredictable. Genetic mapping of QTLs is ongoing in several species of *Eucalyptus*, with particular attention to genes affecting clonability and traits such as frost-tolerance.

Interspecific Hybrids in Eucalypts

Hybrids among *Eucalyptus* species have attracted increasing attention for their superiority through apparent complementarity of growth and form traits, and evidently also of heterosis. Natural hybrids occur freely where species are mingled, and artificial hybridization has focussed on fast-grown species for pulp and timber industries. Unlike poplars or leucaenas, hybrids have not segregated polyploids in eucalyptus (Grattapaglia & Bradshaw, 1994). *E. grandis, E. urophylla*, and their hybrid *E. × urograndis* are the most common hardwood plantation trees, covering more than ten million hectares outside of Australia (Eldridge et al., 1993). They can be cloned by cuttings and are favored for both pulp and timber. Seed production is difficult for parents or hybrids, and clonal hybrids dominate many plantations, despite plant costs 3 to 5 times that of seedlings. Vigneron (1995) has initiated reciprocal recurrent selection schemes to exploit dominance variance in *E. × urograndis*, and similar selection is ongoing with hybrids of *E. urophylla* and *E. pellita*. Heritability values were highly correlated with density of spacing in African plantation trials of these interspecific hybrids, but no marked changes occurred between juvenile and mature trees in variances and heritabilities measured (Bouvet & Vigneron, 1995). Trees were 15 m tall at age of four years and narrow-sense heritabilities were quite accurately estimated at this stage.

Significant gains in yield have been achieved in S. Africa by exploiting interspecific heterosis or general hybridizing ability in eucalypt breeding. The most common hybrids involve *E. grandis* with *E. camaldulensis* and *E. urophylla* with *E. tereticornis* (Denison & Kietzka, 1996). These authors note many instances in which hybrids grow vigorously on sites where the pure species is stunted or has failed. *E. grandis* contributes superior form and desirable pulp and timber qualities, but is low in density and is similar to these other species in its tropical–subtropical habit. Controlled cross of *E. grandis* and the cold-tolerant *E. nitens* showed gains up to 25% in four years' growth, with promising superiority in their fitness or adaptability over a wider range of climatic zones (Wex & Denison, 1996).

Hybrids of the tropical *E. tereticornis* and *E. camaldulensis* are widespread in India, where they show heterosis in flowering precocity and vegetative growth. Seed production and seed weights were also heterotic (Venkatesh & Thapliyal,

1993). Hybrids of *E. globulus* and *E. gunnii* were less frost-tolerant than parental midpoint showing partial dominance of tolerance and a high heritability value of 76% (Tibbits et al., 1991).

HYBRID SUPERIORITY IN CONIFERS

Four conifer genera appear most frequently in the heterosis literature-- *Larix, Picea, Pinus,* and *Pseudotsuga*—although interest in hybrids extends to several other conifers. The genus *Larix* (larches) includes nine temperate species similar to cedars (but with deciduous leaves) that provide fine timber, resins and tannins. Major species in breeding research include *L. decidua* (European larch), *L. kaempferi* (Japanese larch), *L. laricina* (tamarack) and their hybrids. The genus *Picea* (spruces) includes 34 temperate trees of which *P. abies* (Norway spruce) is the most studied. The genus *Pinus* (pines) includes about a hundred species of monoecious trees. Among the most prominent in breeding are *P. banksiana* (Jack pine), *P. caribaea* (Caribbean pine), *P. contorta* (Lodgepole pine), *P. elliottii* (Slash pine), *P. nigra* (Austrian pine), *P. parviflora* (Japan pine), *P. pinaster* (Maritime pine), *P. ponderosa* (Ponderosa pine), *P. radiata* (Monterey pine), *P. strobus* (White pine), *P. sylvestris* (Scots pine) and *P. taeda* (Loblolly pine). *Pseudotsuga* is allied to *Picea* and best known through the Douglas fir, *P. menziesii*.

The conifers are highly outcrossed and notably intolerant of inbreeding, with great reduction of filled seeds upon self-pollination. Natural selfs range between 6 and 9% fertility in the Ponderosa pine (*P. ponderosa*). Even the polyploid redwoods (*Sequoia sempervirens*) showed lower survival rates and 60% reduction in height at 14 years after selfing (Libby et al., 1981). Douglas firs (*Pseudotsuga menziesii*) showed up to 25% selfing when grown in solid stands in Switzerland. Fertilization occurs many months after pollination by a haploid megagametophyte, and many seeds carry embryo lethals or seedling lethal genes. Estimated numbers of lethal equivalents can range well above those of man (about 3) to values of 8 to 10 (Williams & Savolainen, 1996), implicating a thousand or more loci. Inbreeding depression is severe in seedlings (mortality, vigor), and appears to have no clear trends for changes with age. Depression in vigor was related linearly to degree of inbreeding in maritime pine and averaged abut 13% in tree volume (Durel et al., 1996) and slash pine, *P. elliottii* (Matheson et al., 1995). The latter authors found no significant change in regression values from two to 13 years of growth. Purging such deleterious genes is being considered as a breeding tool in some tree species, as it also has been in improving tropical maize to increase inbreeding tolerance (several reports, this conference). However, Williams and Savolainen (1996) note that genetic loads are too high in most conifers to make selfing a viable option, recommending instead a slower initial purging by the use of sib- or even random-mating.

Intraspecific Hybrids in Conifers

Heterosis has been rather unimpressive among intraspecific conifer crosses. Heterosis in height was observed by Magnussen and Yeatman (1988) for selected jack pine hybrids among highly dispersed provenances, although it was otherwise uncommon. Inter-provenance crosses in Europe of maritime pine (*P. pinaster*) showed heterosis for most traits, with large SCA values for height and branch angle (Durel & Kremer, 1995). Hybrids among S_1 lines in this species did not show heterosis for vigor traits, where inbreeding depression ranged from 25 to 65%; however, heterosis was observed for stem straightness, for which inbreeding depression was not significant (Durel & Kremer, 1995). Estimates of genetic effects in three pine species revealed similar levels of additive and dominance variances for growth traits (Cotterill et al., 1987). SCA effects were especially significant for stem diameter (at eight years) while heights were more effectively estimated using GCA values.

Interprovenance crosses across a wide geographic range of loblolly pine (*P. taeda*) were evaluated for 10-year growth and tolerance of fusiform rust (*Cronartium quercuum*) by Schmidtling and Nelson (1996). They reported positive heterosis values (not statistically significant) for rust infection in four of six comparisons. Similarly, heteosis values were all positive for height and plot volume, and were statistically significant in most comparisons.

Varietal hybrids in the Douglas fir (*Pseudotsuga menziesii*) were heterotic for growth and frost tolerance (Braun, 1988). Provenance hybrids of the Norway spruce (*Picea abies*) showed significant GCA values but no SCA (or heterosis) for height or diameter growth, and family heritabilities of 40% (Kaya & Lindgren, 1992). Clones of Norway spruce showed intermediacy and no heterosis for plagiotropism, a serious problem with the East European parent (Johnsen & Skroppa, 1992), while year × clone interactions were reduced greatly in the hybrids. Wide interprovenance hybrids in the hoop pine, *Araucaria cunninghamii*, showed significant superiority in volume (average 17%) over controls in Australia. However, they were no better in straightness, reflecting the lack of intensive selection for straightness of the northern, male parent (Nikles & Johnson, 1993).

Interspecific Hybrids in Conifers

Some interspecific pine hybrids are being cloned into plantations, often for obvious complementarity of parental traits such as straightness and vigor. Interfertility characterizes many species, and natural hybrids are often vigorous, as in *P. elliottii* and *P. taeda*. Outstanding are hybrids of *P. elliotii* × *P. caribaea* (Nikles, 1996). Both F_1 and F_2 populations are excellent in performance and replacing the parent species in Australian plantations. Hybrid superiority appears largely to rely on additive effects predictable through GHA (General Hybridization Ability) for growth, straightness and wind firmness (Dieters et al, 1995). These authors conclude that breeding strategies involving reciprocal recurrent selection are not advisable, but they encourage evaluations of GHA.

Hybrids of *P. thunbergii* and *P. massoniana* included backcrosses in Japan that showed significant hybrid vigor in growth and branch traits at age of 11 years. Hybrids of *P. strobus* × *P. griffithii* revealed very low SCA values for many traits, although it was high for stem straightness, where heritability was 38, vs. 85% for diameter (Blada, 1992). Other promising hybrids among pines include *P. cembra* × *P. walliciana*, combining blister rust resistance and fast growth, and *P. halepensis* × *P. brutia*, that combines fast early growth and high timber quality. Shelbourne (1993), Nikles (1993) and Dieters et al. (1995) review schemes for interspecific hybrid breeding in conifers.

Species hybrids in larch (*Larix* spp.) have long been noted for their heterotic growth. The hybrid of the European larch, *L. decidua*, with the Japanese larch, *L. leptolepis* (= *L. kaempferi*) exceeded the parents in stem heterosis by three-fold at age 33 years, but showed no difference in daily carbon or nitrogen uptake, nor in stomatal conductance (Maryssek & Schulze, 1987). Crosses of the Western larch *(L. occidentalis)* and the slower growing but cold-hardy Alpine larch *(L. lyallii)* are occasionally observed in nature and appear heterotic, but controlled hybrid seedlings showed intermediate height growth to that of the parents with, however, greater stem thickness (Carlson, 1994).

Interspecific spruce hybrids are not uncommon in nature, and often difficult to differentiate from parents. Khasa and Dancik (1996) reported that RAPD markers could be useful in identifying hybrids among the 100 million seedlings of the white and Engelmann spruces (*Picea glauca* and *P. engelmannii*) planted annually in British Columbia, Canada.

HYBRID SUPERIORITY IN LEUCAENA

The genus *Leucaena* (Leguminosae; Mimosoideae) is an American genus of woody legumes noted for its agroforestry uses throughout the tropics (Brewbaker, 1987). There are now 22 named species, several of them endangered (Hughes, personal communication). Most are functional diploids, with 2n = 52 and 56. Four species are known polyploids, with 2n = 104 and 112. The diploids and one of the polyploids are self-sterile while the other polyploid species are self-fertile. Outcrossing in the polyploids is rare. These are fast-growing trees that can provide a diversity of products including fuelwood and posts, high-protein fodder and foods.

Leucaena leucocephala has been widely deployed internationally since 1600. It is self-fertilizing and a single genotype, the common or Hawaiian leucaena, covers much of the tropics. Species such as *L. pallida* (2n = 104), *L. diversifolia* (2n = 52, 104) and *L. pulverulenta* (2n = 56) and their hybrids are now entering breeding research and commercial plantings (Brewbaker & Sorensson, 1990). About one-half the thousand accessions grown in Hawaii have been of *L. leucocephala*, that are largely pure lines. Interspecific hybrids of this species are often strikingly heterotic, including seedless triploids with the species *L. esculenta*.

Intraspecific Hybrids in Leucaena

Intraspecific heterosis has been evaluated by Sun (1995) in a partial diallel study of six lines of *L. leucocephala* (Table 43–1). Trees averaged about 7 m in height and 3.6 cm diameter at two years, while annualized wood biomass yields averaged 18.3 Mg ha^{-1} for hybrids and 14.3 for parents. Heterosis values averaged 31.7% and ranged from -17 to +75%. Some significant differences occurred among trees within lines (not shown in table) for lines K584 and K608, suggesting genetic variation among these normally-selfed lines. Intraspecific hybrids among a six-parent diallel were evaluated also by Gupta and Patil (1986), and heterosis noted only in three F_1 hybrids. Overall GCA and SCA values were significant for all measured traits, with GCA dominating. Outstanding hybrids in both studies produced dry wood yields in excess of 20 Mg ha^{-1} yr^{-1}, comparable to superior eucalypts but marked by much higher wood density and coppiceability. An F_2 population created from superior F_1 intraspecific hybrids has been marketed by Hawaii Agricultural Research Center in order to capitalize on heterosis while providing a broad germplasm base to growers. Preliminary yield data indicate a loss of no more than 10% in the heterosis when moving from F_1 to F_2 in this polyploid species.

Interspecific Hybrids in Leucaena

Interspecific hybridization has been practiced among *Leucaena* spp. to exploit trait complementarity, e.g., provide hybrids with the vigor of *L. leucocephala* and the tolerance to cold temperatures and to psyllid insects from *L. pallida*. Interspecific heterosis was evaluated using four lines of *L. pallida* and seven of *L. leucocephala* in a partial factorial under a forage harvest regime of seven harvests in 18 months (Table 43–2). Average total biomass yield (dry matter) of the 19 hybrids was 24.2 Mg ha^{-1} yr^{-1}. Heterosis values averaged 49.3% and ranged from -75 to +160%. Overall GCA and SCA values were significant. Outstanding specific combining abilities occurred among hybrids of K748, a rather ugly shrubby tree that would normally have been discarded by breeders. In 1997 we completed an expanded germplasm collection of this poorly known species, *L. pallida,* in Oaxaca and Puebla, Mexico, in part through assistance of CIMMYT, and trees to 25 m were seen.

Table 43–1. Dry matter yields (Mg ha^{-1}), specific combining ability (SCA), general combining ability (GCA), and heterosis (%) for two-year old hybrids of *Leucaena leucocephala* at Waimanalo, Hawaii. Varietal data on diagonal, hybrid data, and SCA values above diagonal, heterosis percentage below diagonal.

Entry	K397	K565	K584	K608	K636	K638	GCA
K397	13.60	24.39	13.45	24.40	17.43	10.84	
		2.71	-2.35	2.99	-1.80	0.29	-0.20
K565	75%	14.25	18.87	23.80	20.47	----	
			-0.71	-1.39	-2.54	----	3.57
K584	19.0	62.6	8.95	16.65	20.39	10.67	
				-2.67	3.26	2.21	-2.30
K608	73.0	64.0	41.0	14.70	----	----	
							3.31
K636	0.2	16.0	16.0	----	21.20	----	
							1.12
K638	-17.0	----	-0.5	----	----	12.50	
							-7.55

Table 43–2. Dry matter yields (Mg ha^{-1} yr^{-1}), general combining ability (GCA), specific combining ability (SCA), and interspecific heterosis (%) for 18-month old hybrids between *Leucaena leucocephala* and *L. apllida* averaged across seven forage harvests at Waimanalo, Hawaii.

L. pallida parents	*L. leucocephala* parents								GCA
	K8	K481	K584	K608	K636	K865	K997	KX3	
K178	5.47	22.67	----	----	22.07	----	----	28.27	
	----	-1.01	----	----	-1.29	----	----	8.13	-4.55
	56%	17%	----	----	9%	----	----	23%	
K376	26.53	----	----	----	----	----	----	----	
	5.10	----	----	----	----	----	----	----	2.36
	79%	----	----	----	----	----	----	----	
K748	33.20	33.80	39.07	6.27	34.53	35.40	2.67	24.47	
	13.11	8.27	1.45	-7.05	6.88	-5.01	----	-3.13	2.01
	117%	105%	110%	-49%	92%	160%	-75%	45%	
K806	19.03	----	22.07	----	27.13	31.33	23.93	21.33	
	-3.92	----	-8.68	----	-0.62	0.87	11.07	-4.37	-0.03
	26%	----	38%	----	45%	126%	114%	11%	
GCA	-3.11	4.06	6.40	-	3.74	9.20	-	0.52	

Sorensson and colleagues (Sorensson & Brewbaker, 1994) produced over 90 interspecific hybrids in Leucaena and grew to maturity 73 of these. Many showed hybrid superiority over one or both parents. Twelve hybrids were highly or completely sterile, including triploids such as *L. leucocephala* × *L. esculenta* with no fertile pollen. These species are common in Mexico and their natural hybrids were observed by the authors in the 1970s. The hybrids grow to 15 m at a rate of 5 m per year, an unusual vigor that is in part due to sterility, since the fruits constitute a major sink for fixed carbon in legume trees of this type. Interspecific hybrid superiority is widely recognized for ornamental trees, since a long flowering season often accompanies seedlessness. Common examples in Hawaii are hybrids of the genera *Bougainvillea, Cassia, Erythrina*, and *Plumeria.*

MISCELLANEOUS GENERA

Genetic improvement of fruit, nut and industrial trees has first to contend with product quality, often greatly outweighing concern about quantity. Hybrid development in such trees seeks first to combine useful traits affecting product quality, and later to address hybrid vigor for yield per se. Cloning is the rule for most fruit and industrial tree crops, and outcrossing with high interspecific compatibility is common as with forest species. Heterosis for tree growth in the palm, *Astrocaryum mexicanum* was closely linked to isozymic heterozygosity (Eguiarte et al., 1992). The commercial oil palms (*Elaeis oleifera*) of highest oil yield are monogenic hybrids of the Dura type (dd) with a thick shell and the Pisifera type (DD) with no shell and female sterility. Inbreeding leads to great depression in bunch yield, and a modified reciprocal recurrent selection scheme is used on the two parent types for genetic improvement (Hartleg, 1977). Heterosis for latex yield is common among provenance hybrids of rubber (*Hevea brasiliensis*), and mean annual yields show high heritability and additive variance values (Licy et al., 1992). Hybrids of cashew (*Anacardium occidentale*) and other nuts have been selected for improved nut size. Interspecific hybrids are increasingly commercialized among fruits such as the *Annona* spp., *Citrus* spp., and species of peaches, pears, and apricots. The high inter-cross fertility among *Citrus* spp. has led to tangelos, limequats, citranges, and others, but research is more concerned with quality than with heterosis (Herrero et al., 1996).

Birch hybrids based on lines of *Betula pendula* selfed for three cycles showed a high correlation of heterosis with the degree of inbreeding (Wang et al., 1996). Selfed seeds germinated <2%, vs. 25 to 50% for normal seed, and selection during inbreeding was viewed as an intensive and useful eliminator of lethal alleles.

ROLE OF COMPLEMENTARITY IN HYBRID SUPERIORITY

Most trees grow for long durations in polymorphic ecosystems with very limited environmental control. Stability or homeostasis of growth over time and space are thus highly sought traits. Complementarity or combinations of parental traits appears to be a major component of hybrid superiority in trees (Wang et al., 1996). While these combinations may not be examples of heterosis in the strict sense, they assume special importance in expanding the range of adaptation of tree genotypes to diverse environmental challenges over time, and they are not easily distinguished from heterosis per se. As examples, hybrids between Douglas fir varieties from coastal and inland regions combine vigor of the coastal variety with winter hardiness of the inland variety. Similar hybrids are known among South African eucalypts and among North European Norway spruce. Crosses between the highland and lowland Mexican leucaenas perform as well as the best parent in both environments. Complementarity is also evident in many tree hybrids that combine vigor with straightness, tolerance of drought with tolerance of water-logging, and improved wood properties with high yield. A notable example of the combination of straightness and vigor is the hybrid of *Pinus elliotii* and *P. caribaea* in Australia (Nikles, 1996). Nikles (1993) emphasizes that tree breeders are unwise to base a hybrid breeding program solely on heterosis, but programs must be geared to exploit both complementarity and heterosis. While heterozygosity is often important in fitness, it is difficult to attribute this to overdominance since many types of epistasis can also be invoked (Turelli & Ginzburg, 1983; Namkoong & Kang, 1990).

Hybrid superiority in the genus *Larix* (larch) is dramatic. European hybrid larches of *L. decidua* and *L. leptolepis* have three times the biomass of parent species at age of 33 years (Maryssek & Schulze, 1987). The hybrid vigor is apparently due in large part to the combination of long twigs in the *L. leptolepis* parent with high needle density in the *L. decidua* parent. In other examples of heterosis for tree growth or wood yield, complementarity of genes for diameter growth and vertical extension also can be invoked.

Component traits of yield often can be assumed to interact multiplicatively, rather than by simple addition. Schnell and Cockerham (1992) carefully reviewed examples of this type and provided linear parameters for pertinent digenic and trigenic models. They emphasized the distinction between the two multiplicative effects, i.e., that of the multiplicative interaction of non-allelic genes, as in studies of Powers (1944), W. Williams (1959) and Bos and Sparnaaij (1993), and that of the multiplicative accumulation that can occur when a small amount of heterosis occurs in each component of a complex trait, as in studies of Immer (1941) and Grafius (1961). These effects are referred to as multiplicative interaction and component heterosis and are shown to arise independently of dominance. The authors conclude that multiplicative interaction between loci of arbitrary dominance is one of the classic forms of non-additive gene action, and mainly concerns traits that are products of several sub- or component-traits.

Sparnaaij and Bos (1993) point out that non-additive interaction of complex traits (specific combining ability) may be viewed primarily as the result of complementary action, i.e., when a high value of one component in parent A is complemented by a high value in parent B of a heterotic cross, as in the larch example cited above. Selection for parents that are especially contrasting in such traits should maximize this type of recombinative heterosis. The present authors' research with tropical maize provides many examples suggesting the significance of this phenomenon in crosses of tropical with temperate maize, and this review affords other examples from interspecific crosses of tree species. An important conclusion from the analyses of Sparnaaij and Bos (1993) is that log transformation of observed data for components of complex traits should be avoided, at the risk of misinterpretation of such data.

GENETIC DISTANCE AND HETEROSIS

Tree breeders must emphasize strategies that improve their predictability of performance, working as they do with long-rotation crops. The option of using genetic distance (GD) as a correlative predictor of hybrid performance thus has special appeal. Adding difficulty to GD-heterosis correlations in trees is the question of maturity at which correlations are measured. Strauss (1986) did not observe significant correlations among growth traits and GD in *Pinus attenuata*. In *Eucalyptus globulus* Vaillancourt et al. (1995) reported high GD-heterosis correlations based on RAPD markers. Interproveanance birch hybrids have been studied for populations ranging from northern Finland to Austria (Wang et al., 1996). A general trend of improved heterosis with geographic distance was encountered, but complicated by localized inbreeding and by problems of adaptation to local environments.

Significant correlations based on RAPDs characterized several traits and genetic distances in larch species studied by Arcade et al. (1996). Correlation and statistical significance values were as follows for tree heights at ages of two to 10 years in this study: $r = 0.17$ at two years; $r = 0.31$ at four years; $r = 0.55***$ at six years; $r = 0.50***$ at eight years; and $r = 0.50***$ at 10 years The data involved over a thousand hybrid trees in *Larix decidua* (European larch), *L. kaempferi* (Japanese larch), and their widely used interspecific hybrids. GD (Jaccard) values were 0.39 for *L. kaempferi*, 0.45 for *L. decidua* and 0.72 for the hybrids. Correlations and their trends with age were consistent among families. Branch angle also showed significant GD correlations, while specific gravity, stem straightness and other traits did not. The latter were thought to be under control of few QTLs, as opposed to height and yield. Many factors can be cited to explain failure of such correlations, including inadequate genome coverage and competition by RAPDs not linked to important QTLs. Melchinger et al. (1990) and others emphasize the significance in a heterosis breeding program of markers known to be linked to QTLs for the trait in question. The positive larch correlations and their change with age must however encourage expanded research of this type in tree species.

DISCUSSION AND CONCLUSIONS

Tree improvement research has yet to add greatly to our understanding of the underlying mechanisms of heterosis. Many commercial tree species are only recently domesticated, highly intolerant of inbreeding, slow to breed, uneconomic to clone or otherwise difficult to justify breeding solely to capitalize on heterosis. However, the predominance of hybrid superiority among interspecific, in contrast to intraspecific, hybrids of tree species emphasizes the large role of complementarity among components of complex traits like wood yield. Forest scientists also emphasize the significance of interspecific hybrid performance through assessment of GHA (general hybridizing ability). They also emphasize the roles of multiple allelism, co-adapted gene complexes, epistasis, and high heterozygosity as important influences on hybrid superiority data. Although evidence for overdominance is reported, pseudo-overdominance due to complex linkages of deleterious and beneficial alleles is generally believed to be more common in trees. Complementarity of traits undoubtedly involves multiplicative interactions of non-allelic genes and component heterosis with multiplicative effects when heterosis characterizes each component of a complex trait (Schnell & Cockerham, 1992). Quoting Brewbaker's (1964) Chapter 6 on Heterosis, "if anything, the tenuous evidence for overdominance seems to encourage the view that the dominance theory of heterosis is, as biological theories go, an uncommonly reliable one for applied research." To this we must add that hybrid superiority increasingly involves complementarity and multiplicative effects for the many components of complex traits such as the yield of a tree.

REFERENCES

Aradhya, K.M., and V.D. Phillips. 1995. Lack of association between allozyme heterozygosity and juvenile traits in Eucalyptus. New Forests 9:97–110.

Arcade, A., P. Faivre-Rampant, B. Le Guerroue, L.E. Paques, and D. Prat. 1996. Heterozygosity and hybrid performance in larch. Theor. Appl. Genet. 93:1274–1281.

Bate, N.J., S.B. Rood, and T.J. Blake. 1988. Gibberellins and heterosis in poplar. Can. J. Bot. 66:1148–1152.

Blada, I. 1992. Analysis of genetic variation in a *Pinus strobus* × *P. griffithii* F1 hybrid population. Silvae Genet. 1:282–289.

Bos, I., and L.D. Sparnaaij. 1993. Component analysis of complex characters in plant breeding: II. The pursuit of heterosis. Euphytica 70:237–245.

Bouvet, J.M., and P. Vigneron. 1995. Age trends in variances and heritabilities in *Eucalyptus* factorial mating designs. Silvae Genet. 44:206–215.

Bradshaw, H.D. Jr., and R.F. Stettler. 1995. Molecular genetics of growth and development in Populus: IV. Mapping QTLs with large effects on growth, form, and phenology traits in a forest tree. Genetics 139:963–973.

Braun, H. 1988. Results of hybridization in Douglas fir (Pseudotsuga menziesii). Beitrage fur Forstwirtschaft 22:1–7. (In German).

Brewbaker, J.L. 1964. Agricultural Genetics. Prentice-Hall, New Jersey

Brewbaker, J. L. 1987. Leucaena; A genus of multipurpose trees for tropical agroforestry. p. 289–323 In H.A. Steppler and P.K.R. Nair (ed.) Agroforestry: A decade of development. ICRAF, Nairobi, Kenya.

Brewbaker, J.L., and C.T. Sorensson. 1990. Leucaena Bentham; new tree crops from interspecific Leucaena hybrids. p. 283–289. In J. Janick and J. Simon (ed.) Advances in new crops. Timber Press, Portland, OR.

Campbell, J.S., J.M. Mahoney, and S.B. Rood. 1993. A lack of heterosis in natural poplar hybrids from southern Alberta. Can. J. Bot. 71:37–42.

Carlson, C.E. 1994. Germination and early growth of western larch (Larix occidentalis), alpine larch (L. lyalii) and their reciprocal hybrids. Can. J. Forest Res. 24:911–916.

Ceulemans, R., G.T. Scarascia-Mugnozza, B.M. Wiard, J.H. Braatne, T.M. Hinckley, R.F. Stettler, J.G. Isebrands, and P.E. Heilman. 1992. Production physiology and morphology of Populus species and their hybrids grown under short rotation: I. Clonal comparisons of 4-year growth and phenology. Can. J. Forest Res. 22:1937–1948.

Cotterill, P.P., C.A. Dean, and G. van Wyk. 1987. Additive and dominance genetic effects in Pinus pinaster, P. radiata and P. elliottii and some implications for breeding strategy. Silvae Genet. 36:221–232.

Denison, N.P., and J.E. Kietzka. 1996. The use and importance of hybrid intensive forestry in South Africa. S. African Forestry J. 165:55–60.

Dieters, M.J., D.G. Nikles, P.G. Toon, and P. Pomroy. 1995. Hybrid superiority in forest trees - concepts and applications. p. 152–155. In Eucalypt plantations: Improving fibre yield and quality. CRCTHF–IUFRO Conference, Hobart, Australia.

Durel, C.E., and A. Kremer. 1995. Hybridization after self-fertilization; a novel perspective for the maritime pine breeding program. Forest Genet. 2:117–120.

Durel, C.E., P. Bertin, and A. Kremer. 1996. Relationship between inbreeding depression and inbreeding coefficient in maritime pine (Pinus pinaster). Theor. Appl. Genet. 92:347–356.

Eguiarte, L.E., N. Perez-Nasser, D. Pinero. 1992. Genetic structure, outcrossing rate and heterosis in Astrocaryum mexicanum (tropical palm): Implications for evolution and conservation. Heredity 69:217–228.

Eldridge, K., J. Davidson, C. Harwood, and G. van Wyk. 1993. Eucalypt Domestication and Breeding. Oxford Univ. Press, Oxford, England.

Foster, G.S., and D.V. Shaw. 1988. Using clonal replicates to explore genetic variation in a perennial plant species. Theor. Appl. Genet. 76:788–794.

Grattapaglia, D., and H.D. Bradshaw, Jr. 1994. Nuclear DNA content of commercially important *Eucalyptus* species and hybrids. Can. J. Forest Res. 14:1074–1078.

Grafius, J.E. 1961. The complex trait as a geometric construct. Heredity 16:225–238.

Griffin, A.R., and P.P. Cotterill. 1988. Genetic variation in growth of outcrossed, selfed and open-pollinated progenies of *Eucalyptus regnans* and some implications for breeding strategy. Silvae Genet. 37:124–131.

Gupta, V.K., and B.D. Patil. 1986. Heterosis and combining ability in *Leucaena leucocephala* (Lam.) de Wit. J. Tree Sci. 5:67–73.

Hardner, C.M., and B.M. Potts. 1995. Inbreeding depression and changes in variation after selfing in *Eucalyptus globulus* ssp. *globulus*. Silvae Genet. 44:46–54.

Hartleg, C.W.S. 1977. The oil palm. Longmans, London.

Herrero, R., M.J. Asins, E.A. Carbonell, and L. Navarro. 1996. Genetic diversity in the orange subfamily Aurantioideae: 1. Intraspecies and intragenus genetic variability. Theor. Appl. Genet. 92:599–609.

Hodge, G.R., P.W. Volker, B.M. Potts, and V. Owen. 1996. A comparison of genetic information from open-pollinated and control-pollinated progeny tests in two eucalypt species. Theor. Appl. Genet. 92:53–63.

Immer, F.R. 1941. Relation between yielding ability and homozygosis in barley crosses. J. Am. Soc. Agron. 33:200–206.

Johnsen, O., and T. Skroppa. 1992. Genetic variation in plagiotropic growth in a provenance hybrid cross with *Picea abies*. Can. J. Forest Res. 22:355–361.

Kaya, Z., and D. Lindgren. 1992. The genetic variation of inter-provenance hybrids of *Picea abies* and possible breeding consequences. Scand. J. Forest Res. 7:15–26.

Khasa, P.D., and B.P. Dancik. 1996. Rapid identification of white-Englemann spruce species hybrids by RAPD markers. Theor. Appl. Genet. 92:46–52.

Li, B., and R. Wu. 1996. Genetic causes of heterosis in juvenile aspen: A quantitative comparison across intra- and inter-specific hybrids. Theor. Appl. Genet. 93:380–391.

Libby, W.J., B.G. McCutcheon, and C.I. Millar. 1981. Inbreeding depression in selfs of redwood. Silvae Genet. 30:15–25.

Licy, J., A.O.N. Panikkar, D. Premakumari, Y.A. Varghese, and M.A. Nazeer. 1992. Genetic parameters and heterosis in *Hevea brasiliensis*: I. Hybrid clones of RRII 105 × RRIC100. Indian J. Nat. Rubber Res. 5:51–56.

Magnussen, S., and C.W. Yeatman. 1988. Provenance hybrids in jack pine. Silvae Genet. 37:206–218.

Maryssek, R., and E.D. Schulze. 1987. Heterosis in hybrid larch (*Larix decidua* × *L. leptolepis*): II. Growth characteristics. Trees Struct. Function 1(4):225–231.

Matheson, A.C., T.L. White, and G.R. Powell. 1995. Effect of inbreeding on growth, stem form and rust resistance in *Pinus elliotii*. Silvae Genet. 44:37–46.

McClatchie, A.J. 1902. Eucalypts cultivated in the United States. USDA Bur. Forestry Bull. 35.

Melchinger, A.E., M. Lee, K.R. Lamkey, A.R. Hallauer, and W.L. Woodman. 1990. Genetic diversity for restriction fragment length polymorphisms and heterosis for two diallel sets of maize inbreds. Theor. Appl. Genet. 80:488–496.

Namkoong, G., and H. Kang. 1990. Quantitative genetics of forest trees. Plant Breed. Rev. 8:139–188.

Nikles, D.G. 1993. Breeding methods for production of interspecific hybrids in clonal selection and mass propagation programmes in the tropics and subtropics. *In* J. Davidson (ed.) Proc. of Regional Symposium on Recent Advances in Mass Clonal Multiplication of Forest Trees for Plantation Programmes. FAO/UN. Dec. 1992, Bogor, Indonesia.

Nikles, D.G. 1996. The first 50 years of the evolution of forest tree improvement in Queensland. p. 51–64. *In* M.J. Dieters et al. (ed.) Proc. QFRI–IUFRO Conf. on Tree Improvement for Sustainable Tropical Forestry. Nov. 1996. Caloundra, Qld., Australia,

Nikles, D.G., and M.J. Johnson. 1993. Ten year field performance of wide improvenance hybrids of *Araucaria cunninghamii* tested at 3 locations in southeast Queensland and implications for future breeding and propagation. *In* Proc. 12th Meeting IUFRO Research Working Group 1, Forest Genetics. A.N.U., Canberra, Australia.

Powers, L. 1944. An expansion of Jones's theory for the explanation of heterosis. Am. Nat. 78:275–280.

Schmidtling, R.C., and C.D. Nelson. 1996. Interprovenance crosses in loblolly pine using selected parents. Forest Gen. 3:53–66.

Schnell, F.W., and C. C. Cockerham. 1992. Multiplicative vs. arbitrary gene action in heterosis. Genetics 131:461–469.

Shelbourne, C.J.A. 1993. Interspecific hybrid breeding schemes. p. 92–93. *In* Proc. 12th meeting IUFRO Research Working Group 1, Forest Genetics. A.N.U., Canberra, Australia.

Sorensson, C.T., and J.L. Brewbaker. 1994. Interspecific compatibility among 15 Leucaena species via artificial hybridization. Am. J. Bot. 81:240–247.

Sparnaaij, L.D., and I. Bos. 1993. Component analysis of complex characters in plant breeding: I. Proposed method for quantifying the relative contribution of individual components to variation of the complex character. Euphytica 70:225–235.

Strauss, S.H. , and W.J. Libby. 1987. Allozyme heterosis in radiata pine is poorly explained by overdominance. Am. Nat. 130:879–890.

Stettler, R.F., R.C. Fenn, P.E. Heilman, and B.J. Stanton. 1988. *Populus trichocarpa* × *P. deltoides* hybrids for short rotation culture; Variation patterns and four-year field performance. Can. J. Forest Res. 18:745–753.

Stettler, R.F., H.D. Bradshaw, Jr., P.E. Heilman, and T.M. Hinckley. 1996. Biology of *Populus* and its implications for management and conservation. NRC Res. Press, Ottawa.

Strauss, S.H. 1986. Heterosis at allozyme loci under inbreeding and crossbreeding in *Pinus attenuata*. Genetics 113:115–134.

Strauss, S.H., and W.J. Libby. 1987. Allozyme heterosis in radiata pine is poorly explained by overdominance. Am. Nat. 130:879–890.

Sun, W.G. 1995. Genetic improvement of Leucaena and *Acacia koa*. Ph.D. diss. Univ. of Hawaii, Honolulu, HI.

Tibbits, W.N., B.M. Potts, and M.H. Savva. 1991. Inheritance of freezing resistance in interspecific F_1 hybrids of Eucalyptus. Theor. Appl. Genet. 83:126–135.

Turelli, M., and L.R. Ginzburg. 1983. Should individual fitness increase with heterozygosity? Genetics 104:191–209.

Vaillancourt, R.E., B.M. Potts, M. Watson, P.W. Volker, G.R. Hodge, J.B. Reid, and A.K. West. 1995. Detection and prediction of heterosis in *Eucalyptus globulus*. Forest Genet. 2:11–19.

Venkatesh, C.S., and R.C. Thapliyal. 1993. Heterosis and variation in reproductive parameters of reciprocal F_1 hybrids between *Eucalyptus tereticornis* Sm. and *E. camaldulensis* Dehn. Indian Forester 119:714–721.

Vigneron, P. 1995. Production and improvement of Eucalyptus varietal hybrids in the Congo. *In* Actes du Seminaire de Biometrie et Genetique Quantitative. Montpellier, France. Sept. 1994, CIRAD, Montpellier, France.

Wang, T.L., R. Hagqvist, and P.M.A. Tigerstedt. 1996. Growth performance of hybrid families by crossing selfed lines of Betula pendula. Theor. Appl. Genet. 92:471–476.

Wex, L.J., and N.P. Denison. 1996. Promising potential of the hybrid, *E.* grandis × *E. nitens* in cold to temperate regions of South Africa. p. 295–299. *In* IUFRO Conf. on Silviculture and Improvement of Eucalyptus, IUFRO, Vienna.

Williams, C.G., and O. Savolainen. 1996. Inbreeding depression in conifers: implications for breeding strategy. Forest Sci. 42:102–117.

Williams, W. 1959. Heterosis and the genetics of complex characters. Nature (London) 184:527–530.

Wu, R. 1995. A quantitative genetic model for mixed diploid and triploid hybrid progenies in tree breeding and evolution. Theor. Appl. Genet. 90:683–690.

Chapter 44

Heterosis in Crops II

Discussion Session

QUESTIONS FOR J.P. JORDAAN

N. Machado, Cargill Seeds, Argentina: Do you think hybrid breeding may improve the results regarding disease resistance?

Response: Since resistance to most major pathogens is dominant, hybrid breeding creates an ideal opportunity to choose parents to complement each other - broadening the genetic base and stacking of major genes in the F_1 hybrid.

P. Wilson, Hybrid Wheat Australia, Australia: There have been suggestions that hybrids can be sown at substantially lower seeding rates than varieties and still maintain 10 to 20% yield advantage. Comment please?

Response: In the case of winter wheat production, both under dryland and irrigation, we believe that seeding rates can at least be halved and still maintain hybrid advantage.

B. Cudadar, CIMMYT, Mexico: What are the seeding rates of widely grown pure lines in the region where hybrid wheat is grown in South Africa?

Response: 15 to 30 kg ha-[1].

P. Wilson, Hybrid Wheat Australia, Australia: Will wheat hybrids be successful in South Africa at low seeding rates? Do you expect this success to spread to areas of higher seeding rates in South Africa and other countries?

Response: We do you expect this to happen in South Africa, especially under irrigated conditions. Whether it will be adopted in other countries, will depend on many issues. But there is definitely the potential.

A. Gallais, INRA, France: You have given results showing that heterosis is greater in favorable environments than in unfavorable. Generally it is the other way round, can you explain.

Response: Although our hybrids have been selected to target stress environments, they proved to be responsive to higher yielding environments. This was found to be a result of heterosis for number of spikes m☐ and also spikelet fertility.

QUESTIONS FOR S. S. VIRMANI

H. Geners, Quality Seed CL, South Africa: What research is carried out on dryland rice? Give results—is it a serious breeding objective to increase dryland rice production?

Response: Research on hybrid rice for dryland conditions is scanty. About a decade ago, Brazilian rice scientists were exploring prospects of hybrid rice under such conditions, but without success. A major handicap is the high seeding rate and low base yields, which makes hybrid rice technology for dryland conditions uneconomical.

P. Nelson, Sensako, South Africa: What is the relative stability across environments of rice hybrids versus inbred lines—particularly at low yield levels?

Response: This aspect has not yet been critically studied, although preliminary results are mixed; some studies indicate their stability comparable to inbreds, others indicate higher stability and some even lower stability.

R. Singh, B.H.U. Varanasi, India: What is the scope of rainfed lowland hybrids in comparison to irrigated medium group hybrids particularly for eastern parts of India?

Response: Prospects of rice hybrids under rainfed lowland conditions are being explored in India. Considering the presence of heterosis for traits viz., early vegetative vigor, root system, and submergence tolerance, it is quite likely that rice hybrids for certain rainfed lowland ecosystems, in which transplanting is practiced, would be developed.

L. Hernandez, INIFAP, Mexico: I would like to know what is the best way to identify male parents in the two line hybrid rice system especially for TGMS lines?

Response: For developing two line hybrids using TGMS/PGMS system, any rice variety or breeding line can be used as male parent as long as it gives heterotic hybrids and provides enough pollen to produce hybrid rice seeds economically. We do not require restorer lines as in case of three-line hybrids using CMS system.

M. Camacho, INIFAP, Mexico: How much of the area dedicated to rice hybrids goes under transplanting? Is this hand-made transplanting or mechanical?

Response: Almost all the hybrid rice area in China, India and Vietnam is currently hand-transplanted. With very high seed yields (6 to 7 t ha^{-1}) as reported from China, the cost obtained in commercial seed production plots on consistent basis of hybrid seed can be significantly lowered. Then hybrid rice would be economical even under direct seeded and mechanical transplanting conditions in which seeding rate is much higher than transplanted.

O. Reyes, INIFAP, Mexico: What are the highest yields in commercial production of hybrid rice in China that have been obtained?

Response: In Yunnan province in China hybrids have shown as high as 16.5 t ha^{-1} (with 165 days growth duration) in farmers' fields. The inbred varieties under these conditions were reported to yield 14.5 to15.0 t ha^{-1}.

J. Hawk, University of Delaware, USA: What is the mechanism for transferring pollen in the production of hybrid rice?

Response: The mechanism of pollen transfer in hybrid rice seed production plots is primarily wind pollination. If the wind velocity is lower than 3 m per second supplementary pollination is practiced by shaking the pollen parents by sticks, and/or rope pulling.

P. Peterson, Iowa State University, USA: How can you overcome the Japanese taste preference for non-hybrid rice?

Response: Japanese taste preference is very specific (which is satisfied by the most popular rice variety Koshihikari). This variety was released in 1950s. Since then only a few high yielding rice varieties with Koshihikari type of grain quality have been bred in Japan. Because of very limited genetic variability among such rice varieties heterosis for yield in the hybrids derived from them is also low. Hence, it really makes the development of hybrid rices with Japanese taste preference extremely difficult. Besides, labor intensive hybrid rice seed production technology as available today further makes it difficult for hybrid rice technology to be adopted in Japan.

QUESTIONS FOR W.R. MEREDITH

F. Troyer, USA: Would it be possible and desirable to import cotton seed from India, and grow F_2 seed in the USA?

Response: This has been successfully tried. We have grown F_2s that originated from crosses from India. They had excellent yield and excellent quality. Dr. J.B. Weaver is involved in organizing such an effort.

QUESTIONS FOR J.L. BREWBAKER

D. Duvick, Iowa State University, USA: Why is there more heterosis in interspecific than in intraspecific tree hybrids? One would expect genetic differences between species to be greater than within species?

Response: There's certainly genetic diversity. A great deal of these species of trees that are under study, like the *Eucalyptis*, historically occupy large regions. They are highly outcrossing or completely outcrossing. There seems to be a steady and regular gene flow that's not interrupted by ecological or geographic barriers. In the case of the *Leucenas*, we have a rather different situation in that certain leucenas are self fertile, and they do sort themselves out into very distinct populations here in Mexico. So you can go over the hill into a new valley and pick up something that will be genetically quite diverse, and show a fair amount of heterosis. I should have said genetic distance in general is pretty well correlated with heterosis, and the estimates are probably best from some of the pines and eucalyps. And among the Southern American slash pines, for example, there's quite a lot of research at North Carolina that shows that there is genetic diversity available, but not enough to achieve satisfactory heterosis levels for yield per se, that has been put to use in the production of hybrids.

Chapter 45

Heterosis
What Have We Learned?
What Have We Done?
Where Are We Headed?

A. R. Hallauer

WHAT HAVE WE LEARNED?

Observations on the effects crossbreeding and inbreeding have been re-ported for several centuries in different plant and animal species (Zirkle, 1952). Although the genetic basis of crossbreeding and inbreeding was not understood until the 20th century, it was observed that inbreeding tended to be injurious and crossbreeding tended to be beneficial. Experimental methods of making direct comparisons of either individuals or progenies developed by crossbreeding and inbreeding were not used, but the observations were generally consistent that vigor and productivity were greater in crosses, particularly in crosses produced after some level of inbreeding. Koelreuter, Knight, and Burbank observed greater vigor in crosses, but the concept of using inbreeding and crossbreeding on a consistent basis for crop improvement was not proposed (see Zirkle, 1952). The observations reported by the early hybridizers were not based on a concept of heterosis and how heterosis could be exploited to increase crop productivity.

The potential of hybrids to increase maize (*Zea mays* L.) production was first suggested by Beal (1880). Crosses were produced between cultivars of maize, and the crosses tended to have greater yields than either of the parent cultivars. Sanborn (1890) and McClure (1892) reported similar results in maize. Although Beal (1880), Sanborn (1890), and McClure (1892) did not have an understanding of Mendelian genetics, their interpretations of the necessity of reproducing the hybrid seed for each growing season were correct. The potential of cultivar crosses for increasing maize production in the USA continued to be studied, but the use of cultivar crosses and their impact on U.S. maize production were not important factors for future maize production. Because maize is a cross-pollinated species, a maize cultivar is a heterogeneous mixture of genotypes. Isolation of maize culti-vars was not strickly adhered to, and, consequently, the hybrid vigor observed in cultivar crosses may not have been consistent among growing seasons because of possible contamination. Appropriate experimental techniques for making valid comparisons among crosses and between crosses and parents were either not avail-able or used to make definitive conclusions. Richey (1922) summarized data for 244 variety crosses, and he emphasized that the greatest hybrid vigor was ex-pressed in cultivar crosses of extreme types; i.e., in crosses that probably had the greatest differences in allele frequencies of the parent cultivars. Richey (1922) concluded that the chances were about equal to have a cultivar cross that was or was not better than the better parent. The future course of crossbreeding and in-breeding changed dramatically after the rediscovery of Mendel's paper in 1900.

WHAT HAVE WE DONE?

The rediscovery of Mendel's laws of inheritance stimulated research in gaining an understanding of the genetics for the inheritance of traits and how the genetic information could be used in plant improvement. The genetic information also provided a genetic interpretation of the effects of inbreeding and crossbreeding. The studies reported by East (1908) and Shull (1908) on the effects of inbreeding in maize suggested the future direction of maize breeding. Shull (1908, 1909, 1910) correctly predicted the breeding methods used in present-day maize breeding, and the early beginnings of the heterosis concept were reviewed by Shull (1952).

The inbred-hybrid concept was developed by 1910, but the realization of hybrid maize was not fully in place until the 1930s. The concept suggested by Shull (1910) was of interest, but it seemed the procedure was not practical. After Jones (1918) suggested the use of double-cross hybrids rather than the use of single-cross hybrids by the growers, greater interest was given to the potential of hybrid maize. Intensive and extensive research was initiated in 1922 to exploit the potential of hybrid maize to increase maize yields in the USA. The first task was the development of inbred lines. Different source populations (generally open-pollinated cultivars) were sampled to initiate inbreeding. Research was conducted to determine the relations of traits between inbred and their hybrids, methods to evaluate lines in crosses, generation of inbreeding to evaluate lines in hybrids, combinations of lines to use in double-cross hybrids, relative importance of general and specific combining abilities, and methods to predict hybrid performance of lines that had preliminary evaluation in single crosses and testcrosses. Hybrids were produced, evaluated, and seed of the better double-cross hybrids was being made available to the growers in the 1930s. An experimental basis of the inbred-hybrid concept had been developed (Richey, 1946; Sprague, 1946a). Because it was necessary to reproduce hybrid seed for each growing season, commercial seed organizations were developed to produce and market a dependable supply of high-quality seed for the growers. The model for hybrid maize was extended to include other crop species and many vegetable, horticultural, tree, and ornamental species. Large multinational commercial organizations currently develop and market hybrids for a broad array of plant species, which have contributed significantly to the genetic advances made in plant improvement during the 20th century.

Genetic Effects and Variance

Our knowledge of genetics, of the inheritance of quantitative traits, and of breeding and selection methods has increased significantly since the last heterosis conference (Gowen, 1952). The molecular structure of DNA, the significance of transposable elements, the development of mating designs to study genetic variation of quantitative traits in plant populations and recurrent selection methods to improve germplasm resources, the partition of genetic variation among different types of hybrids, the relative importance of general and specific combining ability, and the genetic basis of heterosis were either recently introduced concepts or had limited use in 1950. Progress was reported at this conference in our understanding of the different aspects of genetics and their use in plant improvement for several crop species with one notable exception: the genetic basis of heterosis. The explanations for the expression of heterosis in crosses discussed by Richey (1950) have relevance to our current discussions except for the greater emphasis given to possible interactions of alleles among loci, or epistasis.

The term, heterosis, was coined by Shull (Shull, 1952) because it seemed hybrid vigor was the result of heterozygosis in the crosses of different parents. The genetic basis of heterosis had received extensive study and discussion after inbreeding and crossbreeding became common in maize. Heterozygosity itself, physiological stimulation in crosses, additive effects of alleles with partial to complete dominance, overdominance, and interactions among loci (epistasis) all have been suggested as explanations for the expression of heterosis (Sprague, 1949; Richey, 1950). Most traits of economic importance, particularly yield, are inherited in a quantitative manner, and different methods were needed to study the genetic basis of their inheritance. Methods for the estimation of genetic variation and the types of genetic effects important in the inheritance of quantitative traits were developed by Fisher (1918) and Wright (1921), but the methods had not been extensively used to study the inheritance of quantitative traits in plant species. It was not until Mather (1949) and Comstock and Robinson (1948) suggested methods for studying the inheritance of quantitative traits in plant populations that the types of genetic effects of primary importance in plant populations and hybrids were studied.

Evidence of the types of genetic effects of greater importance in plant populations and crosses was reported (e.g., Hallauer & Miranda, 1988). It was found that additive genetic effects were of greater importance, and that selection should be effective for the improvement of quantitative traits. Genetic and selection studies were conducted to determine the types of genetic effects of greater importance in the expression of heterosis in crosses. Depending on the plant population and the mating design used, the estimates ranged from additive effects with partial to complete dominance to overdominance. Dominance effects were usually detected via generation mean analyses, whereas level of dominance estimated via components of variance estimated from mating designs was usually in the range of partial to complete dominance. One exception was the estimates of level of dominance for F_2 populations developed from crosses of inbred lines. Estimates of levels of dominance in F_2 populations were frequently in the overdominant range, but subsequent estimates in the same F_2 populations after intermating were usually in the partial to complete dominant range. Because of the effects of linkage, pseudo-overdominance was detected rather than true overdominance.

The relative importance of the genetic effects important in the expression of heterosis of quantitative traits has been difficult because of the complexity of their inheritance. There is certainly additivity of effects among loci, but the interactions of alleles within loci and among loci also must be important. Examples of the possible interactions that can be expressed in crosses were reported by Russell and Eberhart (1970) for inbred B14 and by Russell (1971) for inbred Hy. Both studies included isolines for B14 and Hy that differed only for three gene loci and short segments of chromosomes linked to the three genetic markers. The 27 genotypes (all combinations of the homozygous and heterozygous genotypes) were evaluated, and the analyses partitioned for the linear comparisons (additive effects), deviations from regression (dominance effects), and interactions among loci (epistatic effects). In both studies, highly significant differences were detected among genotypes for nine traits and epistasis, averaged over all traits, accounted for 41% of total variability for B14 and 29% of the total variability for Hy. The frequency of significant epistatic effects was greater in Hy than in B14, but epistasis accounted for a greater portion of the total sums of squares in B14 than in Hy. And the frequency for significant dominance effects was greater for Hy than for B14. These studies included only two inbred lines, B14 and Hy, but the information does illustrate the complexity of the possible genetic effects that can occur in crosses of lines having different genetic backgrounds. It seems that the frequency and the multiplicity of possible interactions that can occur in crosses will make it exceedingly difficult to determine the exact genetic basis of heterosis; the types of genetic effects will be specific for each cross.

Reciprocal Recurrent Selection

Recurrent selection methods were introduced in the decade before the previous heterosis symposium (Gowen, 1952), but only Sprague (1952) and Hull (1952) discussed their possible uses. The common goal of recurrent selection methods is to increase the frequencies of favorable alleles of traits inherited in a quantitative manner. The types of testers suggested for use in recurrent selection depended on the relative levels of dominance (Jenkins, 1940; Hull, 1945). The relative importance of the level of dominance of alleles affecting yield in maize hybrids led Comstock et al. (1949) to develop the method of reciprocal recurrent selection. They demonstrated theoretically that reciprocal recurrent selection was as effective as methods that either emphasized selection for alleles with partial to complete dominance (Jenkins, 1940) or emphasized selection for alleles with over-dominant effects (Hull, 1945). Reciprocal recurrent selection would be more effective than either method if there were additive and nonadditive genetic effects.

Recurrent selection studies were initiated to determine the relative importance of different genetic effects and to determine their relative effectiveness for the genetic improvement of germplasm for breeding programs (Pandey & Gardner, 1990; Hallauer, 1992). In most instances, the proposed recurrent selection methods were effective for increasing the frequencies of favorable alleles in maize (Hallauer & Miranda, 1988). Most of the evidence from recurrent selection suggested that the additive effects of alleles with partial to complete dominance were of greater importance, but dominant and epistatic interaction effects could not be discounted. Similar response to recurrent selection methods was realized in other crop species (Hallauer, 1985).

Reciprocal recurrent selection studies conducted during the past 50 years were effective for improving the cross performance (direct response) of the two populations included. Reciprocal recurrent selection seems to be the appropriate choice to either improve or develop germplasm resources for hybrid breeding programs because both additive and nonadditive effects are included in selection.

Hybrid Prediction

Double-cross hybrids were used almost exclusively until the 1960s. Prediction methods were used to predict performance of double crosses based on preliminary combining ability data (Jenkins, 1934). Because pedigree selection methods were effective in developing improved recycled inbred lines from elite line crosses, the potential of producing single-cross hybrids at acceptable costs was enhanced. Cockerham (1961) also demonstrated that the effectiveness of selection among single-cross hybrids was greater than among three-way crosses and double crosses, particularly if nonadditive effects were important. He showed that the relative advantages will always be a minimum of 1.00 (single crosses) to 0.75 (three-way crosses) to 0.50 (double crosses) when all of the genetic variance is additive. The relative advantages among the three types of crosses will increase in favor of the single crosses when the genetic variance due to dominant and epistatic effects increases. Cockerham (1961) emphasized that the need for inbreeding of the parent lines used in hybrids increases with the deviations from an additive model, and that the presence of nonadditive variance is the primary justification for conducting hybrid breeding programs. Single-cross hybrids presently are the predominant types of hybrids made available to producers. Hence, the prediction methods used for double-cross hybrids are not appropriate for single-cross hybrids. Because of the increases in number and in magnitude of breeding programs, methods have been developed to predict the performance of non-tested single-cross hybrids, based on the performance of lines in crosses (Bernardo, 1996) and from use of molecular markers and related information (Bernardo, 1994).

Heterotic Groups

Parents of the early double-cross hybrids included lines that had good combining ability, had acceptable agronomic traits, and had acceptable levels of tolerance to important pests. Limited attention was given to the origin of the lines included in the early double-cross hybrids. Eckhart and Bryan (1940) presented data that suggested that double-cross hybrids had greater yields and were more uniform if they were produced from single crosses that included lines having similar origin; i.e., (A × A')(B × B') rather than (A × B)(A' × B'), (A × B)(B' × B''), and (B × B')(B'' × B'''), where A and B represent different origins. Breeding groups were established that included lines having similar parentage. It was suggested that recycled lines developed via pedigree selection methods should be from crosses of lines (e.g., B × B') from the same breeding group.

Heterotic groups are identified by plant breeders for hybrid breeding purposes. Heterotic groups do not evolve naturally, except for being genetically dissimilar for allele frequencies. We have broad heterotic groups within different areas because of the original germplasm sources used by the early maize breeders. For the U.S. Corn Belt, the more frequent heterotic group is designated as the Reid Yellow Dent-Lancaster Sure Crop. But if greater emphasis had originally been given to Midland-Leaming, we may have had a different heterotic group. Heterotic groups evolved by trial and error from the genetic materials that were available for producing hybrids.

Heterotic groups tend to be genetically broad-based classifications. Because of the extensive development of breeding programs, each will have smaller heterotic groupings for specific families of inbred lines unique to each breeding program. Each specific heterotic group will emphasize selection for specific sets of alleles that will enhance the heterosis expressed in hybrids. Specific genetic effects (dominance and epistasis) will be emphasized, and the genetic effects will be unique to each heterotic group. Heterotic groups are important components of hybrid breeding programs, both in the broad and narrow context. Different specific heterotic groups will arise within the broad heterotic groups.

Interactions

Interactions, genetically and environmentally, are always important components of plant breeding. Any deviations from an additive model will always dampen either expected or predicted response to selection. Interactions can occur at all levels of plant breeding and cause difficulties at all levels. Interactions (genotype × environment, dominance, epistasis, and all possible combinations) are difficult to analyze and describe explicitly. Interactions are difficult because they involve two variables—genotypes and environments—that are not predictable. Plant breeders know the importance of environmental effects on genotype expression and have been in the forefront in the development and use of experimental and mating designs and statistical analyses to estimate the relative importance of interactions. Progress has been made to identify genotypes for specific environments (e.g., maturity, heat and drought, salt tolerance) and to identify specific heterotic groups to take advantage of dominant and epistatic genetic effects expressed in hybrids. Evaluation trials, however, continue to be important to determine which genotypes have good, stable performance across environments. It seems this will continue to be important in the future because of the possible sources of interactions.

Early Testing

Stage of testing new lines in hybrids was an important issue in 1950 (Gowen, 1952). Early testing was suggested because of the large number of lines being advanced in breeding nurseries for which no information was available for their relative performance in crosses. The main goal of early testing was to identify lines that are either above, or below, average in combining ability (Sprague, 1946b). The proponents of early testing did not suggest that there would be an exact ranking of lines in crosses at early and later generations of inbreeding. The main goal for use of early generation testing was that further breeding efforts (inbreeding, selection, and testing) would be given only to those lines that exhibited above average combining ability. Early testing also was an important aspect of the recurrent selection methods suggested by Jenkins (1940), Hull (1945), and Comstock et al. (1949) to identify progenies for intermating that have the best performance in crosses. Opponents to early testing argued correctly that plant breeders could practice effective selection for several traits before conducting the expensive evaluation trials. Both arguments were valid. During the past 40 years, however, one of the more significant changes made in maize breeding has been the increased use of testing at earlier generations of inbreeding. It is not uncommon for maize breeders to test individual plants at the S_0, S_1, and S_2 generations, which is similar to the generations used in recurrent selection. Combining ability is the trait of greatest importance because the ultimate use of lines is in hybrids. The present trend is for maize breeders to continue selection in advanced generations of inbreeding for lines known to have above average combining ability, identified in early generations of selfing.

Experimental Methods

One consistent theme throughout the symposium was the genetic improvements made in the hybrids developed for the different plant species. There is evidence that hybrids have been improved genetically, but several ancillary developments have been very important (and perhaps as important) in the genetic advances made during the past 40 years. These include the development of plot equipment to plant and harvest experimental plots; the development of computer hardware and software to collect, record, and analyze data; the development of experimental designs and analyses to increase the precision of comparisons among hybrids; the availability of and access of off-season nurseries to either advance progenies by inbreeding or to develop testcrosses for evaluation; the development of prediction methods for single crosses; the development of laboratory equipment and techniques to clarify parentage of lines used in breeding crosses and to follow the transfer of genes and chromosome segments in backcrossing programs; the development of laboratory equipment and techniques to monitor grain quality techniques; and there are others. These developments are nongenetic, but they have increased our effectiveness and efficiency of genetic gain. The change to greater use of early testing in maize breeding, for example, was because modern, dependable plot equipment permitted a greater number of experimental plots to be planted and harvested in a timely manner. The equipment and techniques mentioned were not available in 1950. The advances made in the development of equipment and techniques parallel similar genetic advances made since 1950. Similar advances will continue to be made in the future, and they will impact on the breeding skills and knowledge of genetics we use in plant improvement.

WHERE ARE WE HEADED?

Heterosis in many plant species has been used successfully. Heterosis is an important component in plant improvement, and efforts will be continued in many plant species in which hybrids are either not currently used or widely used. Heterosis has been used successfully even though its genetic basis has not been for the most part determined. It seems it will remain so in the near future.

Although the genetic basis may not be elucidated to specific types of genetic effects, newer techniques have been developed that will provide additional information on the types of genetic effects important in the expression of heterosis and refine our breeding and selection methods to enhance the expression of heterosis in crosses. Molecular genetic studies have developed more extensive genetic maps for use for our important crop species (e.g., Stuber, 1995, 1998). Restriction fragment length polymorphisms (RFLPs) have been used to determine changes in allele frequencies with selection and how the changes in allele frequencies were reflected in the inbred lines derived from original and selected populations (Neuhausen, 1989; Messmer et al., 1991); to assign inbred lines to heterotic groups (Lee et al., 1989); to characterize genetic diversity among inbred lines (Melchinger et al., 1991); and to determine if the genotype of the inbred lines can be used to predict hybrid performance (Godshalk et al., 1990; Melchinger et al., 1990). It has been demonstrated that RFLPs can monitor genetic diversisty and assign inbred lines to heterotic groups, but the relation between RFLP-based genetic distance and hybrid performance was not predictable. It seems that a greater number of probe-enzyme combinations are needed to obtain reliable estimates of genetic distance. Smith (1984, 1988) and Stuber and Goodman (1983) used isozyme data to characterize inbred lines and to determine genetic diversity among inbred lines and hybrids of maize. At this time, valuable and useful information has been reported, but none of the research provides a genetic basis to explain the expression of heterosis of crosses among inbred lines. Future research may provide greater precision than the previous studies, but the complexity of the inheritance of the traits and resultant interactions within and among loci make it difficult to determine the genetic basis of heterosis of specific crosses.

Genetic progress has been realized in nearly all instances for the improvement of germplasm resources and development of inbred lines and hybrids. Although definitive genetic information was not available for the breeding and selection methods used during the past 45 years, accumulated information provided parameters that increased the effectiveness of breeding and selection methods. Genetic information is increasing at a rapid pace, and genetic progress should continue when we have a greater understanding of the inheritance of quantitative traits, and how they can be manipulated in breeding and selection methods for plant improvement. The integration of all facets of plant breeding, both genetic and non-genetic, will continue to be necessary to determine the optimum assemblage of genetic factors (genotypes) that have consistent performance across environments. Plant breeding methods will become more complex, but it is necessary that plant breeders integrate all of the available information and techniques for continued genetic advance.

ACKNOWLEDGMENTS

Contribution from the Department of Agronomy, Iowa State Univ. and Journal Paper J-17640 of the Iowa Agric. and Home Econ. Exp. Stn. Project 3082.

REFERENCES

Beal, W.J. 1880. Indian corn. p. 279–289. *In* Michigan Board Agriculture.

Bernardo, R. 1994. Prediction of maize single-cross performance using RFLPs and information from related hybrids. Crop Sci. 34:20–25.

Bernardo, R. 1996. Best linear unbiased prediction of maize single-cross performance. Crop Sci. 36:50–56.

Cockerham, C.C. 1961. Implications of genetic variances in a hybrid breeding program. Crop Sci. 1:47–52.

Comstock, R.E., and H.F. Robinson. 1948. The components of genetic variance in populations of biparental progenies and their use in estimating the average degree of dominance. Biometrics 4:254–266.

Comstock, R.E., H.F. Robinson, and P.H. Harvey. 1949. A breeding procedure designed to make maximum use of both general and specific combining ability. Agron. J. 41:360–367.

East, E.M. 1908. Inbreeding in corn. p. 419–428. *In* Rep. Connecticut Agric. Exp. Stn. for 1907.

Eckhardt, R.C., and A.A. Bryan. 1940. The effect of the method of combining the four inbred lines of a double cross upon the yield and variability of the resulting hybrid. J. Am. Soc. Agron. 32:347–353.

Fisher, R.A. 1918. The correlation between relatives on the supposition of Mendelian inheritance. Trans. R. Soc. Edinburgh 52:399–433.

Godshalk, E.B., M. Lee, and K.R. Lamkey. 1990. Relationship of restriction length fragment polymorphisms to single-cross hybrid performance of maize. Theor. Appl. Genet. 80:273–280.

Gowen, J.W. 1952. Heterosis. Iowa State Univ. Press, Ames.

Hallauer, A.R. 1985. Compendium of recurrent selection methods and their application. Crit. Rev. Plant Sci. 3:1–33.

Hallauer, A.R. 1992. Recurrent selection in maize. p. 115–179. *In* J. Janick (ed.) Plant breeding reviews. Vol. 9. John Wiley & Sons, New York.

Hallauer, A.R., and J.B. Miranda, Fo. 1988. Quantitative Genetics in Maize Breeding. 2nd ed. Iowa State Univ. Press, Ames.

Hull, H.F. 1945. Recurrent selection and specific combining ability in maize. J. Am. Soc. Agron. 37:134–145.

Hull, H.F. 1952. Recurrent selection and overdominance. p. 451–473. *In* J.W. Gowen (ed.) Heterosis. Iowa State Univ. Press. Ames.

Jenkins, M.T. 1934. Methods of estimating the performance of double crosses in corn. J. Am. Soc. Agron. 26:199–204.

Jenkins, M.T. 1940. Segregation of genes affecting yield of grain in maize. J. Am. Soc. Agron. 32:55–63.

Jones, D.F. 1918. The effect of inbreeding and crossbreeding upon development. Connecticut Agric. Exp. Stn. Bull. 207:5–100.

Lee, M., E.B. Godshalk, K.R. Lamkey, and W.L. Woodman. 1989. Association of restriction fragment length polymorphisms among maize inbreds with agronomic performance of their crosses. Crop Sci. 29:1067–1071.

Mather, K. 1949. Biometrical Genetics. Methuen, London.

McClure, G.W. 1892. Corn crossing. Illinois Agric. Exp. Stn. Bull. 21:82–101.

Melchinger, A.E., M. Lee, K.R. Lamkey, and W.L. Woodman. 1990. Genetic diversity for restriction fragment length polymorphisms: relation to estimated genetic effects in maize inbred. Crop Sci. 30:1033–1040.

Melchinger, A.E., M.M. Messmer, M. Lee, W.L. Woodman, and K.R. Lamkey. 1991. Diversity and relationships among U.S. maize inbreds revealed by restriction fragment length polymorphisms. Crop Sci. 31:669–678.

Messmer, M.M., A.E. Melchinger, M. Lee, W.L. Woodman, E.A. Lee, and K.R. Lamkey. 1991. Genetic diversity among progenitors and elite line from the Iowa Stiff Stalk Synthetic (BSSS) maize population: comparison of allozyme and RFLP data. Theor. Appl. Genet. 83:97–107.

Neuhausen, S.L. 1989. A survey of Iowa Stiff Stalk Synthetic parents, derived inbreds, and BSSS(HT)C5 using RFLP analysis. Maize Genet. Coop. Newsl. 63:110–111.

Pandey, S., and C.O. Gardner. 1990. Recurrent selection for population, variety, and hybrid improvement in tropical maize. Adv. Agron. 48:1–86.

Richey, F.D. 1922. The experimental basis for the present status of corn breeding. J. Am. Soc. Agron. 14:1–17.

Richey, F.D. 1946. Hybrid vigor and corn breeding. J. Am. Soc. Agron. 38:833–841.

Richey, F.D. 1950. Corn breeding. Adv. Genet. 3:159–192.

Russell, W.A. 1971. Types of gene action at three gene loci in sublines of a maize inbred. Canadian J. Genet. Cytol. 13:322–334.

Russell, W.A., and S.A. Eberhart. 1970. Effects of three gene loci in the inheritance of quantitative characters in maize. Crop Sci. 10:165–169.

Sanborn, J.W. 1890. Indian corn. Rep. Main Dep. Agric. 33:54–121.

Shull, G.H. 1908. The composition of a field of maize. Rep. Am. Breeders' Assoc. 4:296–301.

Shull, G.H. 1909. A pure-line in corn breeding. Rep. Am. Breeders' Assoc. 5:51–59.

Shull, G.H. 1910. Hybridization methods in corn breeding. Am. Breeders' Mag. 1:98–107.

Shull, G.H. 1952. Beginnings of the heterosis concept. p. 14–48. *In* J.W. Gowen (ed.) Heterosis. Iowa State Univ. Press. Ames.

Smith, J.S.C. 1984. Genetic variability within U.S. hybrid maize multivariate analysis of isozyme data. Crop Sci. 24:1041–1046.

Smith, J.S.C. 1988. Diversity of United States hybrid maize germplasm: isozyme and chromatographic evidence. Crop Sci. 28:63–69.

Sprague, G.F. 1946a. The experimental basis for hybrid maize. Biol. Rev. 21:101–120.

Sprague, G.F. 1946b. Early testing of inbred lines of corn. J. Am. Soc. Agron. 38:108–117.

Sprague, G.F. 1949. Heterosis. p. 113–136. *In* W.E. Loomis (ed.) Growth and differentiation in plants. Iowa State Univ. Press, Ames.

Sprague, G.F. 1952. Early testing and recurrent selection. p. 400–417. *In* J.W. Gowen (ed.) Heterosis. Iowa State Univ. Press. Ames.

Stuber, C.W. 1995. Mapping and manipulating quantitative traits in maize. TIG:477–481.

Stuber, C.W. 1998. Case history in crop improvement: yield heterosis in corn. p. 197–206. *In* G.H. Paterson (ed.) Molecular dissection of complex traits. CRS Press, Boca Raton, FL.

Stuber, C.W., and M.M. Goodman. 1983. Allozyme genotypes for popular and historically important inbred lines of corn, *Zea mays* L. USDA-ARS, ARR–16. USDA, New Orleans, LA.

Wright, S. 1921. Systems of mating. I. The biometric relation between parent and offspring. Genetics 6:111–123.

Zirkle, C. 1952. Early ideas on inbreeding and crossbreeding. p. 1–13. *In* J.W. Gowen (ed.) Heterosis. Iowa State Univ. Press. Ames.

Chapter 46

Role of Heterosis in Meeting World Cereal Demand in the 21st Century

P. L. Pingali

INTRODUCTION

As we look ahead to 2020, cereal supply situation in developing countries does not lend itself to complacency. Developing country demand for the basic cereals, rice (*Oryza sativa* L.), wheat (*Triticum aestivum* L.) and maize (*Zea mays* L.), is anticipated to grow at least at the rate of 2% per annum. Even if cereal crop productivity growth continued at current levels of 1.5% per annum, developing countries are expected to import some 200 million tons of cereals annually by the year 2020. In the absence of a significant increase in cereal crop productivity, beyond current levels, the global availability of such import volumes at an affordable price is questionable. The current down turn in research and infrastructural investments, along with reduced farm level profitability of cereal production, make the problem of future imports and food security even more daunting.

There is an urgent need to increase cereal crop productivity growth in developing countries in order to meet future cereal food and feed requirements. A substantial shift in the yield frontier for cereal crops through the exploitation of heterosis is likely to be the most cost-effective strategy for increasing productivity in the short to medium term. A shift in the yield frontier also is likely to increase the competitiveness of cereal crop production by reducing the cost per ton of output produced. The rapid development and dissemination of cereal crop hybrids ought to be a high priority for the public and private sector research systems in developing countries.

This chapter provides an assessment of countries and production environments where high returns can be expected from investments in hybrid research and development. Particular attention is paid to hybrid development for the three major cereals, rice, wheat and maize. The returns to hybrid research and development are assessed both for traditional food importing countries as well as food exporting countries. In the case of maize, returns are assessed for food as well as feed demand. The final section of the paper discusses the distribution of responsibilities between the public and private sector in the development and dissemination of hybrids.

CEREAL SECTOR IN THE DEVELOPING WORLD: LOOKING TOWARDS 2020

Recent projections by IFPRI indicate that, by 2020, 96% of the world's rice consumption, two-thirds of the world's wheat consumption and 57% of the world's maize consumption will occur in developing countries. Relative to their 1993 levels, the global demand in 2020 for rice wheat and maize is expected to rise by 36, 40, and 47%, respectively. The global annual demand for rice in 2020 is expected to be around 490 million tons, while that of wheat and maize is expected to be 775 million tons each (Rosegrant et al., 1997). While per capita rice consumption is expected to decline due to income induced diversification of Asian diets (Pingali et al, 1997), the total demand for rice is expected to continue rising due to the growing absolute number of people in the region. In the case of wheat, the expected increase in demand is, in addition to population growth, the result of substitution out of rice and coarse grain cereals as incomes rise and as populations become increasingly urban based. While per capita demand for food maize is expected to decline in all regions except sub-Saharan Africa and Central America, the demand for feed maize is expected to rise dramatically as the demand for livestock products increases.

Virtually all future cereal output growth, in the case of rice and wheat, must come from increased yield per unit of land since the opportunities for further area expansion are exhausted in Asia and extremely limited elsewhere. Yet the opportunities for further yield growth are limited, for the two crops, due to the narrowness of the economically exploitable gap between the technology frontier and farmer performance (Pingali et al, 1997; Pingali & Heisey, 1996). Given current technology and relative price levels it is not profitable for cereal crop farmers, in the favorable as well as the unfavorable environments, to further bridge this gap. In the case of maize, the yield gap is still quite large, exploiting that gap, however, requires high levels of investments in road, transport and market infrastructure. Productivity growth in developing country maize could be expected to rise with the anticipated growth in feed demand for the rapidly expanding livestock sector, particularly in Asia. As the demand for feed maize rises, the relative price of maize can be expected to increase, hence leading to a supply response and increased productivity growth. Increasing maize imports into Asia could trigger maize productivity growth in other regions, particularly Latin America and Africa, as they compete to supply the Asian market. There continue to be opportunities for area expansion in maize, however, such expansion would be increasingly into the more marginal production zones and could come with high environmental costs. Therefore, the focus in maize, as in the case of rice and wheat, ought to be on increasing farm level yields as the primary source of productivity growth.

The developing world is currently under going unprecedented policy reforms that could have significant effects on cereal crop productivity growth over the next several decades (Pingali, 1997). The reforms fall into two broad categories, liberalization of the agricultural sector, and increased global economic integration. The primary impact of the above reforms is to move developing country agriculture away from its traditional focus on food self sufficiency. Crop choice and land use decisions will be increasingly made on the principles of comparative advantage rather than on the imperatives of domestic food needs. Production inputs, even those available on the farm, such as family labor, will be increasingly competed for by the non-agricultural sector and will be valued at their true opportunity.

Liberalization of the agricultural sector could imply an almost total removal of policy protection and support to the cereal crops sector. Output price supports, input subsidies and preferential access to credit will all be gradually phased out, or at least substantially reduced. Infrastructure and research support to the cereals sector could also be expected to be reduced as governments diversify their agricultural portfolios to include crops that are competitive on the global market. The anticipated liberalization of developing country economies and their increased global integration will have significant consequences for the organization and management of agricultural production (Pingali, 1997). The anticipated withdrawal of labor from the agricultural sector will lead to an increase in the opportunity cost of labor, and make small holder-intensive cereal production systems less profitable relative to other income earning and livelihood opportunities. Land and water will face similar competitive pressures from the nonagricultural sectors. While provisioning food for the growing urban conglomerates is expected to be a major challenge of the 21st century, domestic sources of supply would have to compete with, the often cheaper, international sources of supply, especially in the case of coastal megacities (Nailor, 1994).

With the progression towards global integration, the competitiveness of domestic cereal agriculture can only be maintained through dramatic reductions in the cost per ton of production. New technologies designed to significantly reduce the cost per unit of output produced, either through a shift in the yield frontier or through an increase in input efficiencies, would substantially enhance farm level profitability of cereal crop production systems. Exploitation of heterosis in the major cereals can potentially lead to dramatic reductions in unit production costs by shifting the yield frontier while at the same time enhancing input use efficiency. Unit cost reductions can be achieved either through a parallel shift in the yield frontier, (i.e., a percentage increase in output using hybrids with the same input level as for varieties), or through a change in the slope of the input frontier, making it more steeper and hence more input responsive. A shift in the yield frontier for the major cereals, through heterosis or otherwise, could lead to savings in land, primarily by reducing the need for further intensification, and savings in other inputs, through increased utilization efficiencies. To achieve labor use efficiency, hybrid technology would have to be coupled with the adoption of other labor-saving techniques.

TARGETING CEREAL CROP HYBRID RESEARCH AND DEVELOPMENT

Hybrid technology is well advanced and is being widely adopted in maize (Morris, 1998), it is in the final stages of development in rice (Virmani, 1996), and in the initial stages in wheat (Pingali & Rajaram, 1997). There is no doubt, however, that even in the latter two crops, the short to medium term prospects for hybrids are good. In the first two decades of the 21st century, cereal crop hybrids can be expected to be the main sources of yield growth. Yet, the returns to hybrid research and development are not expected to be uniformly high, there will be substantial variability between crops, between countries, and between production environments. Some of the factors that ought to be taken into account in determining the returns to hybrid research and development for a particular crop are: (i) the size of the domestic market; (ii) export potential; and (iii) the proportion of high potential area under the crop. The hybrid research and development that is contemplated here is a combination of public and private sector efforts, the former working more on the upstream research end and the latter on product development and dissemination.

The size of the domestic market is determined by the aggregate population projections as well as by the prospects for income growth. Countries with large populations, such as India and China, would want to invest in hybrid technologies, in order to buffer the domestic consumer against the vagaries of the international market. As countries move from low income to mid-income status, there is a fall in per capita maize and rice consumption for food, while the per capita consumption of wheat tends to rise. At very high income levels, per capita wheat consumption also tends to fall. China for example, with rapidly rising incomes is expected to see a drop in per capita maize consumption from its 1993 level of 26 kg per year to 18 kg in 2020. During the same period, rice consumption per capita is expected to drop marginally from 96 kg per annum to 92 kg per annum. While wheat consumption is expected to rise from 83 kg per capita per annum in 1993 to 88 kg per capita per annum in 2020 (Rosegrant et al, 1997). India on the other hand is expected to see a rise in per capita wheat consumption from 55 to 64 kg per annum, while its rice consumption remains relatively stable at 80 kg per capita. The movement of the poorer sections of the population from coarse grains to rice explains its relative stability, in the case of both India and China.

As countries move from low to middle income status, per capita consumption of livestock and poultry products increases substantially, and hence the drop in per capita maize consumption for food is more than compensated for by the rise in demand for maize as feed grains. For instance, aggregate demand for maize in China is expected to rise from 92 million tons in 1993 to 170 million tons in 2020 (Rosegrant et al., 1997). Therefore, while the returns to rice and wheat hybrid research and development are largely population driven, the returns to maize research and development are driven by aggregate population size as well as by income growth in all regions of the world.

The returns to hybrid research and development are also expected to be high in cereal exporting countries, both current and prospective. The use of hybrids for exports is well established in the case of maize, although much of the supply now comes from the developed world. With the explosion in feed demand, especially in Asia, one could anticipate several countries in Latin America and even some in Africa investing in production for exports. The additional 250 million tons of maize required by 2020 cannot be met exclusively through the current sources of supply. Food exporting developing countries, such as Thailand for rice and Argentina for wheat, and countries with potential for such exports, may find invest in hybrids to be attractive provided that their grain quality is comparable to varieties. Countries with high land to labor ratios with good market infrastructure and suitable agroclimatic conditions would be suitable candidates for expansion in future exports. Myanmar, Brazil, and South Africa are countries to watch in this regard, the first for rice and possibly maize, the second two for maize and possibly wheat and rice.

The returns to hybrid research and development are of course conditioned by agroclimatic conditions and by the extent of high potential environments for each of the crops. The demand for hybrid technology is generally expected to be the greatest in the high potential production environments, especially in countries with large domestic demand for cereals. In the case of wheat and rice, the irrigated environments were the primary beneficiaries of the Green Revolution, and sustaining productivity growth, in the short to medium term, in these environments depends largely on the successful exploitation of heterosis in these crops.

Although hybrids have been shown to out perform inbreds in low potential environments, especially drought prone environments, the demand for them is likely to be low in these environments for the foreseeable future. The switch from a low yield, low input, traditional cereal crop production system to a commercial oriented, high input, hybrid production system is not likely to be profitable, especially in the absence of good market infrastructure. The one significant exception to the above is the movement of hybrid maize into traditional upland rice areas in Southeast Asia, the switch occurring because of the rapid rise in demand for feed maize in the region. The uplands of China also are witnessing the rapid spread of hybrid maize for feed.

From a research and development perspective, the concentration of hybrid technology development on the high potential environments makes economic sense. The relative homogeneity of the high potential environments worldwide leads to high spillover benefits across borders and hence high returns to research investments. The costs of developing hybrid technology for the low potential areas will always be higher and the returns to research investments, especially private sector investments, lower due to the relative heterogeneity of the production environment. Small yield gains in the high potential environments may make the adoption of hybrids profitable while large gains in the low potential environments may not.

Given the above, where should one expect high returns to investment in rice wheat and maize hybrid research and development? For each of the crops, countries or regions with high returns to meeting domestic needs, and those with high returns to expanding exports are identified.

In the case of rice, the returns to investing in hybrids for meeting domestic demand requirements in 2020 are going to be the greatest for China, India, Bangladesh, Indonesia, and Philippines. Asian countries with potential for sustaining or expanding rice exports that will likely see high returns to hybrid invest are: Thailand, Vietnam and Myanmar. Of course, substantial improvements have to be made in hybrid rice quality before these countries would find hybrid rice R&D to be attractive. Over the long term rice exports into Asia can be expected from Latin America and possibly Africa (Pingali, 1997). Countries to watch in this regard are Brazil in Latin America and Mozambique in Africa.

China and India also are the top countries for anticipated high returns to investment in hybrid wheat, primarily directed towards meeting domestic demands. Mexico, South Africa, and Egypt also are likely to find hybrid wheat investments, directed to the domestic market, to be attractive, the latter through spillover benefits gained from other regions of the world. Among wheat exporting countries, the largest gains through investment in hybrids are likely to be in Argentina. The open question in the case of wheat is the potential of the countries in the Former Soviet Union to respond to global wheat demands through increased exports.

The potential for returns to maize hybrid research and development are to be seen separately for food maize vs. feed maize. Sub-Saharan Africa, Mexico and Central America will continue to be the predominant maize consuming areas of the world. Substantial public and private sector investment in maize hybrids would be needed to sustain food supplies. Domestic investments are required in these regions since much of the export maize is yellow maize targeted to the feed sector rather than the food grain sector. South Africa and Zimbabwe are likely to have the highest returns to investment in food maize exports, primarily within Africa. In the case of feed maize, China, followed by Indonesia and the Philippines are likely to see high returns to investing in hybrids targeted specifically towards meeting domestic requirements. Asian countries with potentially high returns to expanding feed maize exports are: Thailand, India, Vietnam, and Myanmar. Non-Asian countries that could expand their exports, specifically targeting the Asian market, are: Brazil, Argentina and Chile. Once again, the capacity of the Former Soviet Union countries in this regard is an open question.

The above discussion is not meant to imply that countries with smaller areas under the above crops will not find their production profitable or that they will not be able to benefit from improved hybrid technologies. Several of these countries may find that with modest investments and by maximizing spillover benefits from other countries with similar agroclimatic environments they will see productivity gains similar to the countries mentioned above. In the case of hybrids, the private sector can play a major role in the dissemination of information across national boundaries in much the same way that the international system played with conventional varieties.

Are we ignoring the marginal environments? Specific targeting of hybrid research and development for the marginal environments, such as the drought prone environments or the problem soil environments, is not likely to be profitable, especially for the private sector. Given the heterogeneous nature of the marginal environments, research costs are likely to be relatively higher than those for the favorable environments, research lags are likely to be longer and the risk of failure higher. Where spillovers are possible from the favorable environments, hybrids do move into the marginal environments under appropriate market conditions. There is some evidence that hybrids, that spillover from the favorable environments, perform better than conventional varieties in the marginal environments (Virmani, 1996). Where spillovers are not possible, and where a shift in the yield frontier is potentially the only source of productivity growth, public research investment would be needed to develop appropriate hybrids. The private sector cannot be expected to take the high risks of investments for these environments, yet when the hybrids are available they may be the main conduit for widespread dissemination of the technology.

PUBLIC–PRIVATE SECTOR LINKAGES IN HYBRID DEVELOPMENT AND DISSEMINATION

The successful development and dissemination of cereal crop hybrids depends on the development of strong collaborative linkages between public and private sector research organizations. The development of the U.S. maize seed industry attests to the crucial importance of public-private sector linkages (Duvick, 1998). The public sector, both national and international, can be expected to be active in pretechnology and technology development phases, while the private sector tends to be more active in product development and dissemination activities. The private sector would be reluctant to invest in public good research that is high risk and with long time lags, while being extremely motivated to develop and disseminate a product that has proven profit potential. The private sector takes on an increasingly important role as hybrids move from the pretechnology stage to the product development and dissemination stages. Of course, there is substantial crossover between the distribution of responsibilities between the public and private sector. In advanced countries, such as the USA, large private sector companies are actively involved in pre-technology and technology development activities, especially with the advent of intellectual property protection. On the other hand, in several less developed countries, such as India, the public sector is active in product development and dissemination.

Of the three cereals considered in this paper, heterosis in wheat requires the most pre-technology development research. It is not yet clear that the exploitation of heterosis in wheat is technically feasible and economically profitable (Pingali & Rajaram, 1997). Both the private sector and the national public sector would be reluctant to invest in hybrid wheat research and development at this early stage, the international public sector is the only player with sufficient resources to take on such a high risk venture. Similar high-risk ventures where the international public sector is active are: the development of apomictic maize and rice, and the development of maize hybrids for stress environments.

Both public and private sector institutions tend to be active in technology development activities, these include the development of stable heterotic lines, incorporation of resistance and tolerance to stresses, quality enhancement, and improved agronomic and crop management practices. Tropical rice hybrids are currently in the technology development phase, with active involvement of both the public and the private sector, especially in India. In the case of maize, technology development activities are concentrated on the development of hybrids for the tropical lowlands and for the mid-altitude environments. Product development and dissemination activities, such as the development of marketable heterotic combinations, seed production and distribution, tend to be concentrated in private hands. Public institutions generally do not have a comparative advantage at the market end of technology dissemination. The one significant exception has been the widespread promotion of hybrid rice in China done exclusively by the public sector (Lin & Pingali, 1994), although with the liberalization of the Chinese economy one is beginning to observe significant divestment of this activity to the private sector.

Public-private sector collaborative linkages in hybrid research and development are likely to be affected by the increasing movement towards intellectual property protection and plant patenting. Enhanced proprietary protection would encourage the private sector to move up stream and invest in technology development activities, while still relying on the public sector to invest in pre-technology research. On the other hand, under restricted funding situations, the public sector may be encouraged to move downstream and develop its own marketable products rather than invest in public good research (Pingali & Traxler, 1997). In the absence of strong legal structures, many developing countries are likely to find that the movement to intellectual property protection is likely to constrain the movement of hybrids internationally and thereby reduce the spillover benefits of research and technology generation.

Product dissemination activities depend crucial on private sector investments in the seed industry. The development of the seed industry is often seen as a constraint to hybrid adoption, a more thorough assessment would indicate in most cases that there are broader economic factors, including transport and market infrastructure, that have constrained adoption. In the absence of policy impediments, the seed industry has generally established itself in countries and regions where the underlying economic environment has been conducive to hybrid adoption. The lack of interest of the private sector in developing and disseminating hybrids for the marginal environments may be an exception to the above generalization (Morris, 1998). In targeting the geopolitical and agroecological domains for the promotion of hybrids, it is important to identify, and try to alleviate, the infrastructural and institutional constraints to the development of the seed industry. The most important reforms needed to encourage private sector investment in the seed industry are: the liberalization of trade regimes, including international capital movement, and a legal system that protects property and patent rights.

REFERENCES

Duvick, D.N. 1998. United States. *In* M.L. Morris (ed.) Maize seed industries in developing countries. Lynne Rienner and CIMMYT. Boulder, CO.

Lin, J.Y., and P.L. Pingali. 1994. An economic assessment of the potential for hybrid rice in tropical Asia: lessons from the Chinese experience. *In* S.S. Virmani (ed.) Hybrid Rice Technology: New Developments and Future Prospects. IRRI, Los Baños, Laguna, Philippines.

Morris, M.L. 1998. *In* M.L. Morris (ed.) Maize seed industries in developing countries. Lynne Rienner and CIMMYT. Boulder, CO.

Nailor, R. 1994. Provisioning the cities in the 21st century. IIS: Stanford.

Pingali, P.L. 1997. From subsistence to commercial production systems: The transformation of Asian agriculture. Am. J. Agric. Econ. August.

Pingali, P., and P. Heisey. 1996. Cereal crop productivity in developing countries: past trends and future prospects. p. 51–94. *In* Conf. Proc. Global Agricultural Science Policy for the Twenty-First Century, 26–28 August 1996, Melbourne, Australia,

Pingali, P.L., M. Hossain, and R.V. Gerpacio. 1997. Asian rice bowls: The returning crisis? IRRI/CAB International.

Pingali, P.L., and S. Rajaram. 1997. Technological opportunities for sustaining wheat productivity growth. *In* Illinois World Food and Sustainable Agriculture Program Conf. Proc. of Meeting the Demand for the Food in the 21st Century: Challenges and Opportunities for Illinois Agriculture. 27 May, 1997.

Pingali, P., and G. Traxler. 1997. International linkages in wheat varietal improvement: will the incentives for co-operation persist in the 21st Century? Proc. 41st Ann. Conf. of the Australian Agricultural and Resources Economics Society, Queensland, 20–25 Jan. 1997.

Rosegrant, M.W., M.A. Sombilla, R.V. Gerpacio, and C. Ringler. 1997. Global food markets and US exports in the twenty-first century. *In* Illinois World Food and Sustainable Agriculture Program Conf. Proc. of Meeting the Demand for the Food in the 21st Century: Challenges and Opportunities for Illinois Agriculture. 27 May, 1997.

Virmani, S.S. 1996. Hybrid rice. Adv. Agron. 57: 377–462.

Chapter 47

Research Needs in Heterosis

R. L. Phillips

SETTING THE STAGE

This "Genetics and Exploitation of Heterosis in Crops" international symposium has stimulated many ideas regarding possible mechanisms to account for the important biological phenomenon of heterosis. Each hypothesized mechanism needs to be tested; this presentation is an attempt to suggest research needs for the future and summarize the ideas that are researchable with today's tools.

Even though heterosis has been in the forefront of our thinking for many years, the phenomenon is not much better understood today than it was when Gowen's famous book on heterosis was published 45 years ago (Barabas, 1992). Barabas (1992) said "One of the greatest achievements in the 20th century was the utilization of the effect of heterosis . . . obviously, a fantastic chain-reaction was widely expected: hybrids in all species of plants. However, the process came to an abrupt halt." This symposium shows that there has been a resurgence of interest in hybrids. However, one question still timely today that relates to the mechanism of heterosis is: Why does heterosis exist in some crops but not in others?

Is it possible that crops that are pure breeding have built-in heterosis? Are polyploid crops such as oat (*Avena sativa* L.) and wheat (*Triticum aestivum* L.) actually improved by maximizing heterosis among the genomes? Does barley (*Hordeum vulgare* L.) have sufficient duplication among its DNA sequences to allow pure breeding heterosis due to heterozygosity between duplicate loci (AA, aa) as opposed to between alleles at individual loci? How does it happen that maize (*Zea mays* L.) and rice (*Oryza sativa* L.) have different degrees of expression of heterosis? Could it be related to the recent findings that at least two genomic equivalents of rice can be found among the chromosomes of maize?

We tend to think of heterosis as due to heterozygosity of many individual genes. Can the heterotic effect be more of a global (pan-genomic) nature? Are there global changes to the genome (chromatin) that occur when different genotypes are brought together in a hybrid? Molecular biological technologies now allow the investigation of certain global considerations. DNA methylation, DNA content changes, silencing–cosuppression, paramutation, amplification, unequal crossing over, and other genetic phenomena can occur in the prezygotic, zygotic or early developmental stages and alter the phenotypic outcome not based strictly on the genotype of the parents. Elevated levels of epistasis also might be expected from such *de novo* changes. We need to go beyond the thought that heterozygosity of a collection of individual genes accounts for heterosis. In this meeting, we have been searching for leads such as whether global gene expression changes occur when different genotypes are combined and for new examples where a gene behaves differently in the heterozygous versus homozygous states. We are looking for a path to follow in order to solve the heterosis puzzle.

ASSUMPTIONS

Many of our assumptions regarding heterosis have as their basis the theoretical foundation of quantitative genetics. These assumptions have served well but do not clearly lead us in a defined direction when considering the molecular basis of heterosis. One assumption, for example, is that quantitative traits are controlled by polygenes - that is, many genes each with small effects. Although this assumption is common in our thought process, we know it is not the case. With the advent of molecular genetic markers, such as restriction fragment length polymorphisms (RFLPs), the evidence clearly shows the existence of genomic regions with major effects controlling quantitative traits along with regions with minor efforts (Stuber, 1994).

Another assumption is that selection leads to fixation of alleles and a subsequent plateau in the selection response. Long-term selection experiments do not support this view either (Dudley & Lambert, 1992). One can argue that the number of segregating loci were underestimated in the original source population and all that is needed to explain the results is to raise the estimate of the number of loci differing in the original population; however, this view may limit our thinking as to other explanations.

A third assumption is that genetic gain is the result of the accumulation of favorable alleles. This would appear logical from a Mendelian genetics point of view but also may not properly assess the situation. Changes in yield could be due to new chromatin configurations which may influence the expression of large domains of genes. In tissue culture, lines derived which are several generations removed from the culture step may be significantly different for a number of quantitative traits (Dahleen et al., 1991). That the tissue culture process would mutate a sufficient number of individual genes to cause such variation in all these quantitative traits is difficult to imagine. Such trait changes may reflect chromatin alterations, such as resulting from DNA methylation modifications, and affect the expression of many genes at once.

A fourth assumption is that genes available for selection in the progeny of a cross must have been represented in the two parents by different alleles; however, the continued genetic gain achieved even as pedigrees become narrower may be due to the occurrence of *de novo* variation (Rasmussen & Phillips, 1997). We now know that in addition to standard point mutations, variants can arise via transposable elements, unequal crossing over, intragenic recombination, DNA methylation, paramutation, gene amplification, and other causes.

Other assumptions include our tendency to think that inbreds are homozygous. Again, we know this is not the case but generally make the assumption anyway. Inbreds maintained for a long time often have become genetically different (Russell et al., 1963). We also know that doubled haploids, which initially are presumably homozygous at all loci, gain variability over a short period of time. Sprague et al. (1960) pointed out that the occurrence of variability in doubled haploid maize was too fast to be explained by classical mutation. Tsaftaris (1999, this publication) presents information indicating that the DNA of inbreds are more methylated than that of hybrids. Since doubled haploids represent the most severe form of inbreeding, perhaps doubled haploid lines are highly methylated. The deamination of methylated cytosine results in a thymine; a mutation can be the result depending on the repair mechanism that occurs (Phillips et al., 1994). Under these circumstances doubled haploids would be expected to have a high mutation rate. Goodnight (1999, this publication) indicated that additive genetic variance is increased with inbreeding. Could an increased mutation rate also account for this observation?

We should not let the assumptions made unduly limit our thinking about the mechanism of heterosis. Even though some of the assumptions are consistent with the data, they may not be molecularly correct.

THINKING ABOUT HETEROSIS

In regard to heterosis, we need to look for features that will lead to certain predictions as a means of testing the concept. Perhaps we should come back to the idea that heterosis and inbreeding depression are flip sides of the same phenomenon. In Falconer's *Introduction to Quantitative Genetics*, 1989 edition, he states that: "Complementary to the phenomenon of inbreeding depression is its opposite 'hybrid vigor' or 'heterosis'. . . That the phenomenon of heterosis is simply inbreeding depression in reverse can be seen by consideration of how the population mean depends on the coefficient of inbreeding." One might argue, therefore, that a molecular explanation of heterosis also should explain inbreeding depression, perhaps by some kind of opposing molecular effect.

Although the dominance hypothesis or the pseudo-overdominance hypothesis may fit the results in many instances, such classical explanations have not allowed much progress in understanding or improving heterosis. Duvick (1999, this publication) has studied maize hybrids developed over the decades and found that heterosis per se has not been improved. The inbreds were improved and hybrid performance followed, but the hybrid performance as a function of the inbred did not improve. In other words, selection for hybrid performance over the years did not improve heterosis. Intuitively, this seems quite unexpected. Non-conventional explanations for heterosis/inbreeding depression now seem appropriate to explore since molecular genetic techniques allow increasingly exquisite genomic tests.

Timothy Reeves (1999, this publication) indicated that "Existing technology and knowledge will not permit the necessary expansion in food production to meet needs". How do we gain understanding of a phenomenon like heterosis which is so important to food production yet alludes a definitive explanation? When attempting to understand a genetic phenomenon, I like to think back to the adage "Treasure your exceptions". In genetics, advances are often made by examining exceptions to the general rule. Mutants have always been a mainstay of understanding genetics, for example. There are at least three different sets of results mentioned at this symposium that defy explanation at this time. The first is Duvick's results mentioned above that hybrid heterosis as a proportion of inbred productivity has not improved, or perhaps it even decreased, even though materials have been selected carefully for hybrid performance. Few, if any, traits have not been amenable to improvement by breeding. This result is corroborated by Geiger's results in hybrid rye (1999, this publication).

The second unexpected result deals with predictions based on genetic distance measures and has been discussed by Melchinger (1999, this publication) and others. Genetic distance measures assist in assigning genetic materials to heterotic groups and are a good predictor of heterosis within groups; however, genetic distance is not a good predictor of heterosis in crosses between heterotic groups. This result has been a disappointment but again is in need of explanation.

The third result that would not have been predicted by most breeders is that genetic gain can be made within a narrow germplasm base. Some of the more mature breeding programs are crossing highly related materials and still making admirable gains (Rasmusson & Phillips, 1997). This is making us take a closer look at *de novo* variation as generated by a number of genetic mechanisms, including transposable elements and intragenic recombination (Peterson, 1999, this publication) as well as several other mechanisms. All three of these "Treasure your exceptions" cases are trying to tell us something. Experiments to elucidate these observations should help in explaining the genetic basis of heterosis.

RESEARCHABLE IDEAS

Is it possible that autogamous crops have a form of "pure breeding" heterosis due to inherent duplications in the genome? Comparative mapping is revealing a much greater than expected degree of synteny across broadly related species, such as the grasses (Bennetzen & Freeling, 1993). This recent revelation leads to the possibility that the heterotic mechanism may not be fundamentally different among species but rather a function of the species' genetic structure which leads to achieving heterosis by different approaches. For example, oat (*Avena sativa* L.) and wheat are allohexaploids with three genomes. Based on genomic in situ hybridization, two of the genomes of oat are probably quite closely related, perhaps deriving from the same diploid progenitor (see Kianian et al., 1997). In both wheat and oat, considerable intergenomic duplication or triplication is present. If different alleles are present at homoeologous loci, a heterotic expression could occur due to heterogeneity between or among these loci while each locus is individually homozygous. This would lead to a pure breeding heterotic pattern. Perhaps selection for performance in such species is, in effect, selection for heterogeneity (pure breeding heterozygosity) among homoeologous loci. RFLP analyses often reveal three bands when oat or wheat DNA is hybridized with a single probe. One could test how the frequencies of 3-banded versus 2-banded vs. 1-banded sequences are correlated with performance. Sequencing might reveal how the polymorphisms relate to activity.

Even crop species that have been considered as diploids have considerable duplication of DNA segments. Maize has been shown to contain two subgenomes with most regions represented twice. Perhaps maximizing inter-subgenome heterogeneity is necessary to project maximum heterosis. This could be the basis of the different heterotic groups in maize; that is, the groups might be based on which lines in hybrid combinations maximize inter-subgenomic heterozygosity. Barley is another diploid crop but still contains duplications at the DNA level. Perhaps performance in barley relates to pure breeding intragenomic heterogeneity among duplicated loci.

Melchinger (1993) and others "suggest potentially successful utilization of RFLP-based genetic distance measures for predicting F_1 performance, mid-parent heterosis, or specific combining ability when hybrids are produced between lines that do not belong to well-established, genetically divergent heterotic groups . . .; however, with respect to the typical situation of hybrid breeding, in which crosses are produced between lines from genetically divergent heterotic groups, the results indicate that an unselected set of RFLP markers are of no value for predicting hybrid performance. It seems necessary to identify specific marker genotypes for those chromosomal segments that determine the expression of the traits of interest." Why would this be the case? Could it be related to the two ways heterozygosity is achieved in maize - that is, allelic heterozygosity at a locus versus heterozygosity between the subgenomes?

Can some species achieve greater heterozygosity than others? Peloquin has proposed that maximum heterozygosity at a locus should be the goal in autotetraploids such as potato (*Solanum tuberosum* L.; Peloquin & Ortiz, 1992). He and colleagues believe that achieving tetraploidy by doubling the chromosome complement of a diploid potato with colchicine will never result in maximum yield because it would not be possible to have the maximum number of alleles at a locus. Perhaps outcrossing may, but not always, lead to heterosis because heterozygosity is achieved by two means - heterozygosity (whether based on dominance, overdominance, or some other model) at a locus and heterozygosity among duplicated loci. Again the idea is that maximum heterozygosity leads to maximum performance in general. In rice, which is perhaps as true a diploid as any cereal species, Yu et al. (1997) found a low correlation between overall heterozygosity and yield. This result was in contrast with the finding of a relationship between marker distance and hybrid performance in diallel crosses (Zhang et al., 1994, 1995). Could this suggest that heterozygosity at a locus is less important than pure breeding heterozygosity involving duplicated loci?

Probes often detect multiple sequences. These sequences, which are highly homologous, may map to several different chromosomes. Such sequences are somewhat different and, therefore, represent cases of multiple alleles. Peloquin would argue that maximizing the number of alleles at a locus is desirable. Because inbreds may differ in the number of loci detected by a single probe, the hybrid would likely have more alleles present than either inbred. In addition, the hybrid might have certain loci in a hemizygous condition, a situation not encountered in inbreds.

Is gene dosage within the complex genomes of crops an important aspect of heterosis? In recent years, the effects of gene dosage alterations have been shown to be much more complex than previously thought (Guo & Birchler, 1994). An increase in dosage at a locus can result in an increase, a decrease, or no change in the level of gene product. Epistatic gene interactions clearly also could be affected by the dosage of the respective genes. Bowen (1996) indicated that, "In hybrids, the combination of high and low dose alleles at appropriate loci could result in a more optimal stoichiometry of interactive components and increase levels of downstream gene expression, the manifestation of which would be hybrid vigor." Understanding heterosis likely will require the elucidation of the complex genomic structure of our crop plants. New molecular biological techniques will be of value to analyze gene expression, such as expressed sequence tags (ESTs), RNA Differential Display, and others. Lee (1999, this publication) pointed out that the emerging genome projects will provide an enabling infrastructure such as extensive sequence data and molecular genetic maps. Functional genomics will become powerful where genes will be associated with a particular phenotype perhaps by reverse genetics. In addition, thousands of genes across the genome will be simultaneously assessed using some form of chip technology (Schena et al., 1996; Lockhart et al., 1996). He also pointed out the value of bioinformatics in making matches of structure and function across species. Such technology for genome-wide assessments of genetic function should assist in the elucidation of heterosis.

Is heterosis the result of chromatin alterations in hybrid combinations that change the expression of large domains of genes? Heterozygosity may impose a global effect on the genome. Lines of maize that carry active Mutator (*Mu*) elements usually become inactive upon inbreeding. This inactivity appears to correlate with methylation of the *Mu* elements (Chandler & Walbot, 1986). Outcrossing restores *Mu* activity and also results in *Mu* elements being less methylated. Large domains of genes, conceivably across much of the genome, could be altered in expression by global DNA methylation effects. Since gene expression is generally negatively correlated with DNA methylation, heterosis also could be related to methylation. If selfing leads to increased methylation and less gene expression, inbreeding depression might be expected. Lamkey (these proceedings) said that "The genetics of inbreeding depression may be the key to understanding heterosis. Crops that do not show inbreeding depression may not show much heterosis".

Tsaftaris' (1999, this publication) data imply that inbreeding results in increased DNA methylation. An important experiment would be to measure methylation as inbreeding is performed. Inbreeding depression could be caused by the shut down of genes by methylation as well as the expression of deleterious recessive genes.

Should near-isogenic lines be produced for studying heterosis? Encouraging the development of isogenic lines useful in heterosis studies seems quite appropriate at this time. Stuber (1999, this publication) indicated that isogenic lines were being developed in maize, through the use of molecular genetic markers, which would represent the substitution of 10–20 cM from the donor parent into the recurrent parent on a segment by segment basis for the whole genome. These lines should be quite useful for assessing the effect of one genetic segment at a time. Another approach is to look for markers linked to heterotic effects but which are still heterozygous in a

Recombinant Inbred Line. Selfing allows the recovery of near-isogenic lines that represent the two homozygous marker classes at that locus. These lines would presumably carry different forms of the quantitative trait locus (QTL) for the heterotic response.

An effective breeding system for studying heterosis might be derived from tissue culture studies. Maize lines obtained from the tissue culturing of an inbred line vary greatly in yield. Crosses among these lines yield more than either derived line and normally achieve nearly normal yields—that is, the yield of the non-cultured (control) inbred line. Kaeppler and Phillips (1993) showed that this was probably a reflection of recessive mutations that were induced by the tissue culture process. In some cases, the cross yielded more than the control but not significantly in their studies (Kaeppler, 1992). If this were a consistent result when making the hybrid, such near-isogenic lines could reveal specific genetic regions important for heterosis.

HETEROSIS MODEL

Heterosis is likely caused by a combination of genetic mechanisms, not only one. Lamkey (1999, this publication) presented the idea that heterosis should be divided into two components—baseline heterosis and functional heterosis. Perhaps progress in our thinking would accrue if we considered baseline heterosis as resulting from those mechanisms that could be captured in an inbred line (homozygous condition). This would include performance improvements due to dominance, pseudo-overdominance, and certain aspects of epistasis. Functional (or value added) heterosis would be those mechanisms not likely to be captured in an inbred and would be found mainly in the hybrid. Loci with dosage divergence between inbreds leading to unpredictable interactions in the hybrid and the result of maximizing multiple alleles might be in the functional heterosis category. In addition, hemizygosity for certain loci would be present only in the hybrid. Overlaid on these considerations is the possibility that the DNA may be differentially methylated in inbreds versus hybrids. The hybrid could have a chromatin structure different from the inbreds and thereby have large domains of genes differentially expressed. A simplified model is presented in Fig. 47–1 where C and D illustrate a pseudo-over-dominance situation, B and E illustrate masking of deleterious alleles in the hybrid, and X illustrates homologous loci (detected with a single probe) with dosage divergence among lines giving multiple alleles (X^1, X^2, X^3, X^4, X^5) and hemizygosity (X^5) in the hybrid. Epistasis among various loci also would be expected.

CONCLUSION

Heterosis is the basis of a multi-billion dollar business in agriculture. Much of our ability to meet the world's food needs depends on heterosis. Yet we do not know the molecular genetic basis of this important biological phenomenon. Let's hope that ideas from this symposium will lead to breakthroughs toward achieving an understanding of heterosis.

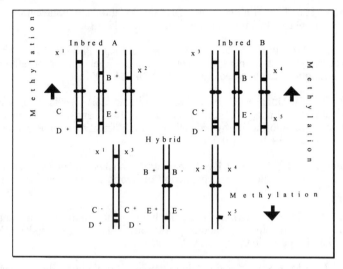

Fig. 47-1. Heterosis model. Combination of genetic mechanisms contributing to baseline and functional (value added) heterosis.

REFERENCES

Barabas, Z. 1992. A new era in the production of hybrid varieties? Hungarian Agric. Res. 1:17–21,

Bennetzen, J.L., and M. Freeling. 1993. Grasses as a single genetic system: Genome composition, colinearity and compatibility. Trends Genet. 9:259–261.

Bowen, B. 1996. Thinking about the molecular basis for heterosis in corn. *In* Workshop on Heterosis. ASA, CSSA, and Am. Soc. Hort. Sci., Indianapolis, IN.

Chandler, V.L., and V. Walbot. 1986. DNA modification of a maize transposable element correlates with loss of activity. Proc. Natl. Acad. Sci. USA 83:1767–1771.

Dahleen, L.S., D.D. Stuthman, and H.W. Rines. 1991. Agronomic trait variation in oat lines derived from tissue culture. Crop Sci. 31:90–94.

Dudley, J.W., and R.J. Lambert. 1992. Ninety generations of selection for oil and protein in maize. Maydica 37:81–87.

Falconer, D.S. 1989. Introduction to quantitative genetics. 3rd ed. John Wiley & Sons, New York.

Guo, M., and J.A. Birchler. 1994. Trans-acting dosage effects on the expression of model gene systems in maize aneuploids. Science (Washington, DC) 266:1999–2002.

Kaeppler, S.M. 1992. Molecular and genetic studies of tissue culture-induced variation in maize. Ph.D. diss., Univ. of Minnesota. St. Paul.

Kaeppler, S.M., and R.L. Phillips. 1993. DNA methylation and tissue culture-induced variation in plants. In Vitro Cell. Dev. Biol. 29P:125–130.

Kianian, S.F., B.-C. Wu, S.L. Fox, H.W. Rines, and R.L. Phillips. 1997. Aneuploid marker assignment in hexaploid oat with the C genome as a reference for determining remnant homoeology. Genome 40:386–396.

Lockhart, D.J., H. Dong, M.C. Byrne, M.T. Follettie, M.V. Gallo, M.S. Chee, M. Mittmann, C. Wang, M. Kobayashi, H. Horton, and E.L. Brown. 1996. Expression monitoring by hybridization to high-density oligonucleotide arrays. Nature Biotech. 14:1675–1680.

Melchinger, A.E. 1993. Use of RFLP markers for analysis of genetic relationships among breeding materials and prediction of hybrid performance. Int. Crop Sci. Congr., Ames, IA.

Peloquin, S.J., and R. Ortiz. 1992. Techniques for introgressing unadapted germplasm to breeding populations. p. 485–507 In H.T. Stalker and J.P. Murphy (ed.) Plant Breeding in the 1990s. CAB Int.

Phillips, R.L., S.M. Kaeppler, and P. Olhoft. 1994. Genetic instability of plant tissue cultures:Breakdown of normal controls. Proc. Natl. Acad. Sci. USA. 91:5222–5226.

Rasmusson, D.C., and R.L. Phillips. 1997. Plant breeding progress and genetic diversity from de novo variation and elevated epistasis. Crop Sci. 37:303–310.

Russell, W.A., G.F. Sprague, and L.H. Penny. 1963. Mutations affecting quantitative characters in long-time inbred lines of maize. Crop Sci. 3:175–178.

Schena, M., D. Shalon, R. Heller, A. Chai, P.O. Brown, and R.W. Davis. 1996. Parallel human genome analysis: Microarray-based expression monitoring of 1000 genes. Proc. Natl. Acad. Sci. USA. 93:10614–10619.

Sprague, G.F., W.A. Russell, and L.H. Penny. 1960. Mutations affecting quantitative traits in selfed progeny of doubled monoploid maize stocks. Genetics 45:855–865.

Stuber, C.W. 1994. Breeding multigenic traits. In R.L. Phillips and I.K. Vasil (ed.) DNA-Based markers in plants. Kluwer Acad. Publ., Dordrecht, The Netherlands.

Yu, S.B., J.X. Li, C.G. Xu, Y.F. Tan, Y.J. Gao, X.H. Li, Qifa Zhang, and M.A. Saghai Maroof. 1997. Importance of epistasis as the genetic basis of heterosis in an elite rice hybrid. Proc. Natl. Acad. Sci. USA. 94:9226–9231.

Zhang, Q., Y.J. Gao, S. Yang, R. Ragab, M.A. Saghai Maroof, et al. 1994. A diallel analysis of heterosis in elite hybrid rice based on RFLPs and microsatellites. Theor. Appl. Genet. 89:185–192.

Zhang, Q., Y.J. Gao, M.A. Saghai Maroof, S.H. Yang, and J.X. Li. 1995. Molecular divergence and hybrid performance in rice. Molec. Breed. 1:133–142.

Chapter 48

How to Feed the 21st Century?

The Answer is Science and Technology

N. E. Borlaug

INTRODUCTION

It is a pleasure to participate in this International Symposium on the Genetics and Exploitation of Heterosis in Crops, the first meeting of its kind, I understand, in the last 47 years.

I am now in my 53rd year of continuous involvement in food production programs in developing nations. During this period, I have seen much progress in increasing the yield and production of various crops, especially the cereals, in many food-deficit countries. Clearly, the research that backstopped this progress has produced huge returns. Yet, despite a more than tripling in the world food supply during the past three decades, the so-called Green Revolution in cereal production has not solved the problem of chronic under nutrition for hundreds of millions of poverty-stricken people around the world who are unable to purchase the food they need, despite its abundance in world markets, due to unemployment or under-employment.

The invention of agriculture, some 10 000 to 12 000 years ago, heralded the dawn of civilization. It began with rainfed, hand-hoe agriculture, which evolved into an animal-powered, scratch-tooled agriculture, and finally into an irrigated agriculture along the Euphrates and Tigris rivers that for the first time allowed humankind to produce food surpluses. This permitted the establishment of permanent settlements and urban societies which, in turn, engendered culture, science and technology. The rise and fall of ancient civilizations in the Middle East and Meso-America were directly tied to agricultural successes and failures, and it behooves us to remember that this axiom still remains valid today.

ORIGINS OF FOOD CROP SPECIES

We will never know with certainty when nature first began inducing genetic diversity, making recombinations, and exerting selection pressure on the progenitors of the plant species that would be chosen much later by man as his food crop species. But as the Mesolithic Age gave way to the Neolithic there suddenly appeared, in widely dispersed regions, the most highly successful group of plant and animal breeders that the world has ever seen—the Neolithic domesticators. Within a relatively short geological period, apparently only 20 to 30 centuries, Neolithic man, or more probably woman, domesticated all of the major cereals, grain legumes, root crops, and animal species that remain to this day as humankind's principal sources of food.

Agriculture and animal husbandry spread rapidly from their cradles of origin across vast areas of Asia, Africa, Europe, and the Americas. These migratory diffusions were in large part possible because of the tremendous genetic diversity that existed in the original land races and populations of the domesticated crop plants. This genetic variability permitted—with the aid of continued mutations, natural hybridizations, and recombination of genes—the spinning off of new genotypes that were suitable for cultivation in many environments.

The groundwork for genetic improvement of crop plant species by scientific man was laid by Darwin in his writings on the variation of life species (published in 1859) and through Mendel's discovery of the laws of inheritance (reported in 1865). Darwin's book immediately generated a great deal of interest, discussion, and controversy; however, Mendel's discovery was largely ignored at first. Nearly 40 years transpired before these two strands of scientific though were joined by Carl Correns, Erich Von Tschermak, and Hugo De Vries, in independent studies. This rediscovery of Mendel's laws in 1900 provoked a tremendous scientific interest in genetics. The fact that Mendel had worked out his principals on a plant [the sweet pea (*Pisum sativum* L.)] encouraged many to prepare themselves for a career in applied plant genetics.

It was recognized early on that inbreeding leads to reduced vigor in the following generation and that vigor can be restored by crossing. Darwin noted this phenomenon in The Vegetable Kingdom, published in 1876. The first organized attempt to exploit hybrid vigor in maize (*Zea mays* L.) was initiated by W.J. Beal at Michigan State College in 1875. Beal's work and that of other stimulated little interest for 25 years until Edward East and George Shull proved conclusively that although maize lost vigor on inbreeding, when inbred lines were crossed, the progeny of the next generation exhibited an explosive recovery of vigor called heterosis. A few years after I was born, Donald Jones, an associate of Edward East, figured out a solution to the high seed cost of hybrids, with the development of the double-cross hybrid.

But there was no stampede to exploit this potential until the mid 1920s when H.A. Wallace, later to become Secretary of Agriculture and Vice President under Franklin Roosevelt, founded Pioneer Hi-Bred, the first private hybrid seed company. Because of the disastrous economic depression of the 1930s, the use of hybrids did not really take off until the early 1940s. But by the mid-1950s, hybrids dominated U.S. maize production and the use of open-pollinated varieties had virtually disappeared.

Since the commercial introduction of the first hybrids in the USA some 50 plus years ago, many improved elite, double-, three-way and single-cross hybrids have been developed with continually higher yields, improved disease and insect resistance, and shorter and stronger stalks. During this period, average U.S. yields have increased from 1.8 to 8.2 t ha^{-1}, a phenomenal 4.5-fold increase.

Tremendous genetic improvement for yield and yield dependability also has been achieved in wheat (*Triticum aestivum* L.), rice (*Oryza sativa* L.), barley *(Hordeum vulgare* L.), and other cereals, grain legumes, roots and tubers, sugarcane, (*Saccharum officinarum* L.) fruits and vegetables, and the fiber and tree crops.

During the past seven decades, conventional breeding has produced a vast number of varieties and hybrids that have contributed immensely to higher grain yield, stability of harvests, and farm income. Surprisingly, there has been no major increase in the maximum genetic yield potential of the high-yielding semidwarf wheat and rice varieties commercially being grown since those that served to launch the so-called Green Revolution of the 1960s and 1970s. There have been, however, important improvements in resistance to diseases and insects, and in tolerance to a range of abiotic stresses, especially soil toxicities. But we must also find new and appropriate technology to raise genetic yield levels higher, if we are to cope with the food production challenges before us. I'll say more about this later in this chapter.

Of course, plant breeding, or genetic improvement, has been only one component of the total research effort to improve—in the case mentioned above —maize production. Research and development efforts in soil fertility, weed science, and pest management have also led to the development of high- analysis mineral fertilizers and effective crop protection chemicals, which have permitted the new management-responsive varieties to express their genetic yield potential. And research and development in farm machinery has allowed the farmer to increase enormously his labor productivity.

OUR WORLD FOOD SUPPY

In 1994, global food production of all types stood at 4.74 billion metric tons of gross tonnage and 2.45 billion tons of edible dry matter (Table 48–1). Of this total, 99% was produced on the land. Only about 1% came from the oceans and inland waters, even though 70 percent of the earth's surface is covered with water. Plant products constituted 93% of the human diet, with about 30 crops species providing most of the world's calories and protein, including eight species of cereals, which collectively accounted for 66% of the world food supply. Animal products, constituting 7% of the world's diet, also come indirectly from plants.

Had the world's food supply been distributed evenly, it would have provided and adequate diet in 1994 (2350 calories, principally from grain) for 6.4 billion people—about 800 million more than the actual population. However, had

Table 48–1. World food production, 1994. Source: 1994 FAO Production Yearbook.

Commodity	Production, million metric tons			
	Gross tonnage	Edible matter†	Dry protein†	Increase, % 1980–1992‡
Cereals	1950	1623	162	22
Maize	570	501	52	19
Wheat	528	465	55	22
Rice	535	363	31	35
Barley	161	141	14	7
Sorghum/millet	87	78	7	7
Roots & Tubers	583	156	10	9
Potato	265	58	6	3
Sweet potato	124	37	2	7
Cassava	152	56	1	24
Legumes, oilseeds, oil nuts	387	263	88	48
Sugarcane and sugarbeet §	133	133	0	21
Vegetables and melons	486	57	5	32
Fruits	388	53	2	28
Animal products	858	170	75	25
Milk, meat, eggs	760	143	57	21
Fish	98	25	18	29
All Food	4743	2456	343	24

† At zero moisture content, excluding inedible hulls and shells.
‡ 1979–1981 and 1989- 1991 averages used to calculate changes.
§ Sugar content only.

people in Third World countries attempted to obtain 30% of their calories from animal products—as in the USA, Canada, or EEC countries—a world population of only 2.6 billion people could have been sustained—less than one-half of the present world population.

These statistics point out two key problems of feeding the world's people. The first is the complex task of producing sufficient quantities of the desired foods to satisfy needs in environmentally and economically sustainable ways. The second task, equally or even more daunting than the first, is to distribute food equitably. Here, poverty is the main impediment to equitable food distribution, which, in turn, is made more severe by rapid population growth.

PROJECTED WORLD FOOD DEMAND

During the 1990s, world population will grow by nearly one billion people and then again by another one billion people during the first decade of the 21st Century. A medium projection is for world population to reach 6.2 billion by the year 2000, 8.3 billion by 2025, 10 billion by 2045, before hopefully stabilizing at 11 to 12 billion toward the end of the 21st Century.

At least in the foreseeable future we will continue to relay on plants and especially the cereals, to supply virtually all of our increased food demand. Even if current per capita food consumption stays constant, population growth will require that world food production increases by 2.6 billion gross tons between 1994 and 2025. However, if diets improve among the destitute, estimated to be 1 billion people living mainly in Asia and Africa, world food demand could increase by 100 percent—from 4.7 to nearly 9 billion gross tons—during this period. It will have to increase by another 5 billion gross tons between 2025 and 2045. Let me summarize. World food production has to double over the next 30 years and triple over the next 50 years.

Raising Yield Levels on Existing Agricultural Lands

While there are still some vast areas to bring into production in South America and Africa, much of the projected increases in food supply will have to come from land currently in production. Fortunately, there are many improved agricultural technologies—already available or well advanced in the research pipeline—that can be employed in future years to raise crop yields, especially in the low-income food deficit countries where most of the hunger and poverty exist.

Yields can still be increased by 50 to 100% in much of the Indian sub-Continent, Latin America, the former USSR and Eastern Europe, and by 100 to 200% in much of sub-Saharan Africa, providing political stability is maintained, bureaucracies that destroy entrepreneurial initiative are reigned in, and researchers and extension workers devote more energy to putting science and technology to work at the farm level. Yield gains in China and industrialized North America and Western Europe will be much harder to achieve, since they are already at very high levels. Still, I am hopeful that scientific breakthroughs—particularly from genetic engineering, will permit another 50% increase in yields over the next 25 years.

The most frightening prospect of food insecurity is found in sub-Saharan Africa, where the number of chronically undernourished could rise to several hundred million people if current trends of declining per capita production are not reversed. Sub-Saharan Africa's increasing population pressures and extreme poverty; the presence of many human diseases, e.g., malaria, tuberculosis, river blindness, trypanosomiasis, guinea worm aids, etc.; poor soils and uncertain rainfall; changing ownership patterns for land and cattle; inadequacies of education and public health systems; poorly developed physical infrastructure; weaknesses in research and technology delivery systems will all make the task of agricultural development very difficult.

Despite these formidable challenges, many of the elements that worked to bring a Green Revolution to parts of Asia and Latin America during the 1960s and 1970s also will work in sub-Saharan Africa. An effective system to deliver modern inputs—seeds, fertilizers, crop protection chemical—and to market output must be established. If this is done, Africa can make great strides toward improving the nutritional and economic well-being of Africa's downtrodden farmers, who constitute more than 70 percent of the populations in most countries.

Since 1986, I have been involved in food crop production technology transfer projects in sub-Saharan Africa spearheaded by the Nippon Foundation and its former Chairman, the late Mr. Ryoichi Sasakawa, and enthusiastically supported by former U.S. President Jimmy Carter. Our joint program is known as Sasakawa-Global 2000, and it currently operates in 12 African countries: Ghana, Benin, Togo, Nigeria, Tanzania, Ethiopia, and most recently, in Mozambique, Guinea, Burkina Faso, Mali, Uganda, and Eritrea. Previously, we also operated similar projects in Sudan and Zambia.

The heart of these projects are dynamic field testing and demonstration programs for major food crops. Although improved technology—developed by national and international research organizations—had been available for more than a decade, for various reasons it was not being adequately disseminated among farmers. Working in concert with national extension services during the past 10 years, more than 600 000 demonstration plots (usually from 0.25 to 0.5 ha) have been grown by small-scale farmers. Most of these plots have been concerned with demonstrating improved basic food crops production technology for maize, sorghum, wheat, cassava, and grain legumes. The packages of recommended production technology include: (i) the use of the best available commercial varieties or hybrids, (ii) proper land preparation and seeding dates and rates to achieve good stand establishment, (iii) proper application of the appropriate fertilizers, including green manure and animal dung, when available, (iv) timely weed control and, when needed, crop protection chemicals, (v) moisture conservation and/or better water use, if under irrigation.

Virtually without exception, the yields obtained by participating farmers on these demonstration plots are two to three times higher—and occasionally four times higher—than the control plots employing traditional methods. Only rarely have plot yields failed to double that of the control. Hundreds of field days attended by tens of thousands of farmers have been organized to demonstrate and explain the components of the production package. In project areas, farmers' enthusiasm is high, and political leaders are now taking much interest in the program. From our experiences over the past decade, I am convinced that if there is political stability, and if effective input supply and output marketing systems are developed including a viable agricultural credit system, the nations of sub-Saharan Africa can make great strides in improving the nutritional and economic well-being of their desperately poor populations.

Bringing New Lands into Production: the Remaining Frontiers

Most of the opportunities for opening new agricultural land to cultivation have already been exploited (Table 48–2). This is certainly true for densely populated Asia and Europe. Only in sub-Saharan Africa and South America do large unexploited tracts exist, and only some of this land should eventually come into agricultural production. But in populous Asia, home to half of the world's people, there is very little uncultivated land left to bring under the plow. Apparently in West Asia there are already some 21 million hectares being cultivated that shouldn't be. Most likely, such lands are either too arid, or, because of topography, so vulnerable to erosion that they should be removed from cultivation.

Table 48–2. Potential cropland (million ha) in the less developed countries. Source: Calculated from Buringh and Dudal (1987) Table 2.6, p. 22, World Bank.

	Africa	West Asia	S/SE Asia	East Asia	South America	Central America	Total
Potentially cultivated	789	48	297	127	819	75	2155
Presently cultivated	168	69	274	113	124	36	784
Uncultivated	621	0	23	14	695	39	1392
% of region	79	0	8	11	85	52	NA
% all regions	29	0	1	0.5	32	2	65

One of the last major land frontiers are the vast acid-soils areas found in the Brazilian *cerrado* and *llanos* of Colombia and Venezuela, central and southern Africa, and in Indonesia. Historically bringing these unexploited potentially arable lands into agricultural production posed what were thought to be insurmountable challenges. But thanks to the determination of interdisciplinary teams in Brazil and international research centers, the prospects of making many acid soil savanna areas into productive agricultural areas has become a viable reality.

Let's look briefly at the Brazilian *cerrado*. The central block, with 175 million hectares in one contiguous area, forms the bulk of the savanna lands. Approximately 112 million hectares of this block is considered potentially arable. Most of the remainder has potential value for forest plantations and improved pastures for animal production. The soils of this area are mostly various types of deep loam to clay-loam latosols (oxisols, ultisols), with good physical properties, but highly leached of nutrients by Mother Nature in geologic time, long before humankind appeared on the planet. These soils are strongly acidic, have toxic levels of soluble aluminum, with most of the phosphate fixed and unavailable.

In pre-colonial times, the area was sparsely inhabited by a number of Amerindian tribes dependent on a culture based on hunting and gathering of wild plants. During the colonial period, and continuing from Independence up until about 35 years ago, the *cerrado* was considered to be essentially worthless for agriculture (except for the strips of alluvial soils along the margins of streams, which were less acidic and where there had been an accumulation of nutrients). The natural savanna/brush flora of poor digestibility and nutritive quality—resulting in low carrying capacity—was used for extensive cattle production.

Through a slow painful process over the past 50 years, involving some outstanding scientists, bits and pieces of research information and new types of crop varieties have been assembled. Only during the past 20 years have these pieces been put together into viable technologies and applied by pioneering farmers. By the end of the 1980s Brazil's national research corporation, EMBRAPA, and several international agricultural research centers (especially CIMMYT and CIAT), have developed a third generation of crop varieties combining tolerance to aluminum toxicity with high yield, better resistance to major diseases, and better agronomic type. These included rice, maize, soybeans (*Glycine max* Merr.), wheat, and several species of pasture grasses, including the panicums, pangola, and brachiaria. Triticale (× *Triticosecale* Wittmack) is an interesting man-made cereal that has a very high level of aluminum tolerance, although it has not been used much yet either for forage or for grain production.

Table 48–3. Potential food production if available technology is adopted on *Cerrado* area already in production. Source: Prospects for the Rational Use of the Brazilian **Cerrado** for Food Production by Jamil Macedo, CPAC, EMBRAPA, 1995.

Land use	Area	Productivity	Production
	million ha	t ha^{-1} y^{-1}	million T
Crops (rainfed)	20.0	3.2	64
Crops (irrigated)	5.0	6.0	30
Meat (pasture)	20.0	0.2	4
Total	45.0		98

Improved crop management systems were also developed, built around liming, fertilizer to restore nutrients, crop rotations and minimum-tillage that leave crop residues on the surface to facilitate moisture penetration and reduce run off and erosion. However, with conservation tillage coming into widespread use, it will be absolutely necessary to work out better crop rotations to minimize the foliar disease infections that result from inoculum in the plant crop residues left on the surface from previous seasons.

In 1990, roughly 10 million hectares of rainfed crops were grown in the *cerrado*, with an average yield of 2 t ha^{-1} and a total production of 20 million tons. The irrigated areas are still relatively small with an average yield of 3 t ha^{-1} and a total production of 900 000 tons. There are also 35 million hectares of improved pasture supporting an annual meat production of 1.7 million tons. During the 1990s the area using improved technology has expanded greatly. If it continues to spread, farmers could attain 3.2 t ha^{-1} in rainfed crops and 64 million tons of production. If the irrigation potential is developed, which can add another 30 million tons of food production, it is likely that by 2010 *cerrado* food production will have increased to 98 million tons—or a four-fold increase over 1990 (Table 48–3).

WHAT CAN WE EXPECT FROM BIOTECHNOLOGY?

I am now convinced that what began as a biotechnology bandwagon some 15 years ago has developed some invaluable new scientific methodologies and products which need active financial and organizational support to bring them to fruition in food and fiber production systems. So far, biotechnology has had the greatest impact in medicine and public health; however, there are a number of fascinating developments that are approaching commercial applications in agriculture. In animal biotechnology, we have bovine somatatropin (BST), now widely used to increase milk production, and porcine somatatropin (PST) waiting in the wings for approval.

Transgenic varieties and hybrids of cotton, maize, potatoes, containing genes from *Bacillus thuringiensis*, which effectively control a number of serious insect pests, are now being successfully introduced commercially in the USA. The use of such varieties will greatly reduce the need for insecticide sprays and dusts. Considerable progress also has been made in the development of transgenic plants of cotton, maize, oilseed rape, soybeans, sugar beet, and wheat, with tolerance to a number of herbicides. This can lead to a reduction in herbicide use by much more specific dosages and interventions.

The development of transgenic plants for the potential control of viral and fungal diseases is also picking up considerable speed. There are some very promising examples of specific virus coat genes in transgenic varieties of potatoes, rice, and vegetables that confer considerable protection. Other promising genes for disease resistance are being incorporated into other transgenic crop species.

Until recently, it has been generally assumed that increases in genetic yield potential in plants (and animals) are controlled by a large number of genes, each with small additive effects; however, the work of recent years shows that there may also be a few genes that are sort of master genes that affect the interaction, either directly or indirectly, of several physiological processes that influence yield. For example, BST and PST are apparently such master genes. They not only affect the total production of milk or meat, but also the efficiency of production per unit of feed intake. It now appears that the dwarfing genes, Rht1 and Rht2, used to develop the high-yielding Mexican wheats that launched the Green Revolution, also acted as master genes, for at the same time that they reduced plant height and improved standability, they also increased tillering and the number of fertile florets and the number of grain per spike (harvest index). Biotechnology may be a new window through which to search for new master genes for high yield potential by eliminating the confounding effects of other genes.

CAN AGRICULTURAL SCIENCE STAY AHEAD OF WORLD POPULATION GROWTH?

So far, agricultural research and production advances—and the efforts of the world's farmers—have kept food production ahead of aggregate world population changes. However, there can be no lasting solution to the world food-hunger-poverty problem until a more reasonable balance is struck between food production/distribution and human population growth. The efforts of those on the food-production front are, at best, a holding operation which can permit others on the educational, medical, family planning, and political fronts to launch an effective, sustainable, and humane attack to tame the population monster.

Still, there is a crying need today for creative pragmatism in research and extension organizations in many parts of the developing world. In particular, we need more venturesome young scientists who are willing to dedicate their lives to helping to solve the production problems facing several billion small-scale farmers. In seeking to push forward the frontiers of scientific knowledge, some researchers often lose sight of the most pressing concerns of farmers and cease to develop products that extension workers can promote successfully. In the low-income developing countries, impact on farmers' fields should be the primary measure by which to judge the value of this research work, rather than by a flood of publications that often serve to enhance the position of the scientist but do little to alleviate hunger.

A growing number of agricultural scientists, myself included, anticipate great benefits from biotechnology in meeting our future food and fiber needs. Since most of this research is being done by the private sector, which patents its inventions, those of us concerned with agricultural policy must face up to a potentially serious conundrum. Most of those being born into this world are among the abject poor, most of whom live in rural areas of the developing world, and who depend on low-yielding agricultural production systems to eke out a meager existence. How will these resource-poor farmers be able to afford the products of biotechnology research? What will be the position of these transnational agribusiness towards this enormous section of humanity that still live largely outside the commercial market economy? This issue goes far beyond economics; it is also a matter for deep ethical consideration. Fundamentally, the question is do small-scale farmers of the developing world also have a right to share in the benefits of biotechnology? If the answer is yes, then what is the role of international and national governments to ensure that this right is met? I believe we must give this matter serious thought.

STANDING UP TO THE ANTISCIENCE CROWD

Science and technology are under growing attack in the affluent nations where misinformed environmentalists claim that the consumer is being poisoned out of existence by the current high-yielding systems of agricultural production. While I contend this isn't so, I ask myself how it is that so many people believe to the contrary? First, there seems to be a growing fear of science, per se, as the pace of technological change increases. The late British physicist and philosopher-writer C.P. Snow first wrote about the split between scientists and humanists in his little book, *The Two Cultures*, published in 1962. It wasn't that the two groups necessarily disliked each other, rather they just didn't know how to talk to each other. The rift has continued to grow since then. The breaking of the atom and the prospects of a nuclear holocaust added to people's fear, and drove a bigger wedge between the scientist and the layman. The world was becoming increasingly unnatural, and science, technology and industry were seen as the culprits.

Rachel Carson's *Silent Spring*—which reported that poisons were everywhere, killing the birds first and then us—struck a very sensitive nerve. Of course, this perception was not totally unfounded. As pointed out in Bittman's little book, *The Good Old Days: They Were Terrible,* about environmental quality in America (and the United Kingdom and other industrialized nations), in the late 19th and early 20th century air and water quality had been seriously damaged through wasteful industrial production systems that pushed effluents often literally into "our own backyards." Over the past 30 years, we all owe a debt of gratitude to environmental movement in the industrialized nations, which has led to legislation to improve air and water quality, protect wildlife, control the disposal of toxic wastes, protect the soils, and reduce the loss of biodiversity. In almost every environmental category far more progress is being made than most commentators in the media are willing to admit. Why? I believe that it's because apocalypse sells. Sadly, all too many scientists, many of whom should (and do) know better, have jumped on the environmental bandwagon in search of research funds. When scientists align themselves with anti-science political movements, like Rifkin's anti-biotechnology crowd, what are we to think? When scientists lend their names and credibility to unscientific propositions, what are we to think? Is it any wonder that science is losing its constituency? We must be on guard against politically opportunistic, charlatan scientists like T.D. Lysenko, whose pseudo-science in agriculture and vicious persecution of anyone who disagreed with him, contributed greatly to the collapse of the former USSR.

Recently a science writer named Gregg Easterbrook wrote an article about me in the U.S. magazine, *Atlantic Monthly*. While I have never met him, I was intrigued by the brief biographical sketch provided about him, which labeled him an eco-realist. I was prompted to buy his new book, *A Moment on the Earth*, published by Penguin in 1996, which I am now reading. It's a fascinating and provocative book, based upon a massive amount of research. Calling himself an "eco-realist" Easterbrook's central appeal is that the worthy inclinations of environmentalism must become grounded in rationality. "Logic, not sentiment," he contends, "best serves the interests of nature." Easterbrook provides much documentation that in the Western world, "the Age of Pollution is nearly over" asserting that "aside from weapons, technology is not growing more dangerous and wasteful but cleaner and more resource-efficient."

However, Easterbrook correctly points out that as positive as trends are in the First World, they are negative in the Third World, where environmental degradation is occurring at an alarming rate, with poverty and rapid population growth the underlying causes. While it is certainly technically feasible for plant Earth to support a human population several times larger than at present without ecological harm, the point is that our current social, political, and economic systems cannot support rapidly growing populations at an adequate standard of living. Thus short-term global population stabilization is of paramount importance.

In sharp contrast to the rich countries, where most environmental problems have been urban, industrial, and a consequence of high incomes, the critical environmental problems in most of the low-income developing countries remain rural, agricultural, and poverty-based. More than one-half of the world's very poor live on lands that are environmentally fragile, and they rely on natural resources over which they have little legal control. Land-hungry farmers resort to cultivating unsuitable areas, such as erosion-prone hillsides, semiarid areas where soil degradation is rapid, and tropical forests, where crop yields on cleared fields drop sharply after just a few years.

I often ask the critics of modern agricultural technology what the world would have been like without the technological advances that have occurred? For those whose main concern is protecting the environment, let's look at the positive impacts of science-based technology on the land. Had 1961 yields still prevailed today, three times more land in India would be needed to equal 1992 cereal production. Obviously, such a surplus of land of the same quality is not available, and especially not in populous China and India.

I have calculated that if the USA attempted to produce the 1990 harvest of the 17 most important crops with the technology and yields that prevailed in 1940 it would have required an additional 188 million hectares of land of similar quantity. This theoretically could have been achieved either by plowing up 73% of the nation's permanent pastures and range lands, or by converting 61% of the forest and woodland area to cropland. In actuality, since many of these lands are of much lower productive potential than the land now in crops, it really would have been necessary to convert an even larger portion of the range lands or forests and woodlands to crop production. Had this been done, imagine the additional havoc from wind and water erosion, the obliteration of forests, extinction of wildlife habitats, and the enormous reduction of outdoor recreational opportunities.

In his writings, Professor Robert Paarlberg, who teaches at Wellesley College and Harvard University in the USA, has sounded the alarm about the consequences of the debilitating debate between agriculturalists and environmentalists about what constitutes so-called sustainable agriculture in the Third World. This debate has confused—if not paralyzed—policy makers in the international donor community who, afraid of antagonizing powerful environmental lobbying groups, have turned away from supporting science-based agricultural modernization projects, so urgently needed in sub-Saharan Africa and parts of Latin America and Asia. The result has been increasing misery in smallholder agriculture and accelerating environmental degradation.

This policy deadlock must be broken. In doing so, we cannot lose sight of the enormous job before us to feed 10 billion people, most of whom will be born into abject poverty in low-income, food-deficit nations. And we must also recognize the vastly different circumstances faced by farmers in different parts of the Third World, and assume different policy postures. For example, in Europe or the U.S. Corn Belt, the application of 300 to 400 kg of fertilizer nutrients per hectare of arable land can cause some environmental problems. But surely, increasing fertilizer use on food crops in sub-Saharan Africa from around 5 kg of nutrients per hectare of arable land to 30 to 40 kg is not an environmental problem but rather a central component in Africa's environmental solution.

CLOSING COMMENTS

Twenty-seven years ago, in my acceptance speech for the Nobel Peace Prize, I said that the Green Revolution had won a temporary success in man's war against hunger, which if fully implemented, could provide sufficient food for humankind through the end of the 20th century. But I warned that unless the frightening power of human reproduction was curbed, the success of the Green Revolution would only be ephemeral.

I now say that the world has the technology—either available or well-advanced in the research pipeline—to feed a population of 10 billion people. The more pertinent question today is whether farmers and ranchers will be permitted to use this new technology. Extremists in the environmental movement from the rich nations seem to be trying to stop scientific progress in its tracks. Small, vociferous, well-funded, and highly effective anti-science and technology groups are slowing the application of new technology, whether it be developed from biotechnology or more conventional scientific methods.

I am particularly alarmed by those who seek to deny small-scale farmers of the Third World—and especially those in sub-Saharan Africa—access to the improved seeds, fertilizers, and crop protection chemicals that have allowed the affluent nations the luxury of plentiful and inexpensive foodstuffs which, in turn, has accelerated their economic development. While consumers in the affluent nations can certainly afford to pay more for the so-called organically produced foods, the one billion chronically undernourished people of the low-income, food-deficit nations cannot. As Richard Leakey likes to remind his environmental supporters, "you have to have at least one square meal a day to be conservationist or environmentalist".

At the closure of the Earth Summit in 1992 at Rio de Janeiro, 425 members of the scientific and intellectual community presented to the Heads of State and Government what is now being called the Heidelberg Appeal. Since then, some 3000 scientists have signed this document, including myself. Permit me to quote the last paragraph of the Appeal:

> "The greatest evils which stalk our Earth are ignorance and oppression, and not science, technology, and industry, whose instruments, when adequately managed, are indispensable tools of a future shaped by Humanity, by itself and for itself, in overcoming major problems like overpopulation, starvation, and worldwide diseases."

Agricultural scientists, agribusiness leaders, and policy makers have a moral obligation to warn the political, educational, and religious leaders about the magnitude and seriousness of the arable land, food and population problems that lie ahead. Let us all remember that world peace will not be built on empty stomachs and human misery. Deny the small-scale, resource-poor farmers of the developing world access to modern factors of production—such as improved varieties, fertilizers and crop protection chemicals—and the world will be doomed, not from poisoning, as some say, but from starvation and social and political chaos.

Chapter 49

A Symposium Overview

J. F. Crow

I have been asked to give a short overview in this final chapter, mainly I suppose because of being on the program of both the 1950 Heterosis Conference in Ames, Iowa, and this one. I am pleased and honored to be invited.

What a contrast there has been between the two conferences! Let me count the ways.

First, the number of attendees was much smaller in 1950. I don't have any numbers—only my memory—but I would guess between one and two hundred in contrast to some 500 here. There were 30 speakers then, although fewer than half dealt directly with heterosis. This time there are 185 listed as speaking or presenting posters. And the organization was very different. In 1950 there was one talk each morning followed by an extended discussion in the afternoon, and the conference was spread over a five-week period; this time the talks were crowded into a very busy four days.

In 1950 the emphasis was almost entirely on maize. In this symposium we have heard talks on sorghum [*Sorghum bicolor* (L.) Moench], millets [*Pennisetum glaucum* (L.) R. Br.], rapeseed (*Brassica napa* L and *B. rapa* L.), wheat (*Triticum aestivum* L.), rice (*Oryza sativa* L.), rye (*Secale cereale* L.), cotton (*Gossypium* sp.), and a number of trees and vegetable crops—including a spectacular success story in sunflowers (*Helianthus annuus* L.). The 1950 conference also included swine and poultry; the organizers of this symposium wisely limited the subject to crop plants.

In 1950 the speakers were almost entirely from the academia. This time there has been a large representation from the private sector. The program has been greatly enriched by the cooperation and complementation between the groups.

Finally, there was a major contrast in the emphasis. In 1950 the discussions were all biological, with emphasis on increasing yield in good environments. At this symposium there has been a deep concern for three areas hardly mentioned in the earlier conference. One of these is ecology—how to increase food supply while doing minimum damage to the present and future environment and leaving as many pristine areas as possible undisturbed. Several speakers have noted the crop area reduction made possible by increased yields per hectare. Second has been an emphasis on developing strains that grow under stressed conditions, such as drought and reduced nutrients. This is an important, indeed urgent consideration for much of the world. A third difference has been a recognition of economic issues, particularly in getting capital and appropriate infrastructure for developing and employing better crops in the tropics.

By the time of the 1950 Conference, the concepts of general and specific combining ability were well established. Selection for combining ability had succeeded in producing much better maize (*Zea mays* L.) hybrids, although inbred yields were low enough that double crosses were still the norm. Dickerson along with Comstock and Robinson presented breeding systems designed to maximize productivity increases with overdominance.

I was particularly struck by two subjects of which we heard more at the present conference. One was the talk by Henderson presenting the beginnings of what is now called Best Linear Unbiased Prediction or BLUP. I vividly remember having a hard time understanding Henderson's statistics. The practical applications had to wait a long time for sufficiently powerful computers, which fortunately are now readily available. I think it is a safe prediction that this method will play an increasing role in plant breeding, as it already is doing in dairy cattle selection. The other subject was Comstock and Robinson's presentation of Reciprocal Recurrent Selection (RRS). This was a system designed to produce improvement with any level of dominance. We have seen in this conference that in fact it works very well.

A WORD ABOUT OVERDOMINANCE

1950 and the next few years was the zenith of overdominance. Then, for reasons given in my talk earlier in this symposium, the opinion shifted and complete or partial dominance became more popular. In looking at the continued upward slope in maize yields, I notice no blip or curvature as opinions changed. The upward trend has been oblivious to changing opinion. Of course there is still uncertainty as to how much seeming overdominance is pseudo-overdominance caused by repulsion linkages. Close linkages can persist for many generations, particularly with inbreeding, so repulsion linkages are effectively overdominant. I find it interesting that one of the seeming overdominant QTLs found by Stuber and his colleagues has been resolved into a repulsion linkage.

A comment from a fellow octogenarian went this way: "They tell you that you lose your mind when you grow older, but what they don't tell you is that you won't miss it very much." I would paraphrase this by saying: "They tell us that we don't understand heterosis, but what they don't tell us is that we can get along very well without understanding it." Let me second the remark of Jim Coors, "'Selection works," echoing the sentiments of Charles Darwin in another century. Selection can work very well even when the developmental details are a black box.

We can expect that heterosis is as complicated as any other aspect of gene action. So we should expect to find all sorts of actions and interactions. The problem is to sort them out. Variance components in populations are mainly additive and dominance. But individual inbred lines and crosses may well show other complications that are lost in the averages over multiple loci. We learned in this symposium from Todd Wehner of a nice instance of overdominance (or codominance?) in tomatoes. It may not increase fitness or yield, but the Rin/rin heterozygote combines desirable properties of both homozygotes. The place where I would expect to find overdominance is in such qualitative traits.

I should like to comment on two particular points that came up in the symposium: mutation and epistasis.

MUTATION

Information on mutation rates since 1950 make it almost certain that mutation rates, especially for genes with small effects, are higher than was previously thought. Sprague's doubled haploids produced 4.5 mutations per trait per 100 gametes and inbred lines gave 2.8. These are strikingly high. Experiments on

Drosophila show a homozygous viability decrease due to mutations of individually minor effects of about 1% per generation. We should expect that, if unopposed by selection, inbred lines should decrease in yield by some such amount.

Of course, mutation is the ultimate source of variability on which all progress depends. But the great majority of mutations are harmful with respect to fitness, and since yield is almost the same as fitness in highly selected plants, the great majority of mutations can be expected to decrease performance. One might argue that mutations are needed to produce variability for future selection. I suggest, however, that it is much better to look for new genes in other plants, where they have been preselected and the more deleterious ones weeded out.

Since we don't want to lose what we have gotten through so much effort, anything that will reduce the mutation rate will, in my view, be good. In this conference we have heard about two controllable causes of mutation in maize.

The first is transposable elements, mostly retrotransposons. At the time of the 1950 conference, McClintock was already widely known, but it was for her work on such things as the breakage-fusion-bridge cycle. Her work on mutable genes was hidden in reports of the Carnegie Institution, which were not widely read and even less widely understood. In this symposium, Peterson has emphasized the large role that these elements play in maize variability. If, as I believe, the variants are almost entirely deleterious, anything that could be done to excise active transposons should be beneficial. Many of the variants induced by transposition will have phenotypes that can be recognized and eliminated, but not all will. So one useful task ahead is devising ways of lowering the mutability of important inbred lines. Perhaps active transposons can somehow be excised. Or, less farfetched, lines with smaller numbers of active transposons could be chosen. (I would hazard the guess that maize would be better off if most of the repeat-sequence space between genes could be eliminated. This "junk DNA" has no known useful effect for the plant and the energy expended in its replication could be put to better use. But I may be way off base in this conjecture.)

The second controllable source of mutations is methylation. Maize lines differ considerably in the amount of methylation. If maize is like other organisms in this regard, and I have no doubt that it is, methylation is correlated with increased mutation at CpG and other sites. These changes are likely to be largely base substitutions, and hence to produce mainly minor effects. But a number of minor effects can combine to produce a major effect. So I suggest that, whenever feasible, one choose inbred lines with low levels of methylation. Whether this will turn out to be of practical value is of course only a guess at this time.

EPISTASIS

We have heard a great deal about epistasis during this conference. Goodnight has emphasized that with epistasis additive within-line variance can increase with inbreeding, similar to what Alan Robertson showed for dominance back in 1952. Hallauer has told us about three introduced genes that showed significant interactions. Developmental biologists tell us that epistasis is everything; every process seems to interact with everything else.

In contrast, animal and plant breeders have almost always ignored epistasis. Usually they can't find statistically significant epistatic components. Why this contrast? How can epistasis be present and yet have so little influence in selection experiments?

I think there are several reasons. The first is that Mother Nature understands least squares (perhaps she was taught by R.A. Fisher). We know, thanks to Fisher, that selection acts mainly on the additive component. But least squares puts much of the epistatic contribution into the additive component. The procedure pulls out as much additivity as it can, and this is what selection acts on.

Second, the coefficients for the epistatic variance components are small; for parent-offspring correlation the coefficient of the additive by additive component is 1/4 and the epistatic components involving dominance do not contribute. For more than two loci the coefficients are still smaller. Third, small amounts of anything are additive, on the same principle that taking only one or two terms in a Taylor expansion often gives a good approximation in a wide variety of physical and biological situations. Fourth, even with the powerful BLUP technique, as we learned from Bernardo, there is very little gain from including epistasis.

Finally, let me mention another possibility. In 1965 Kimura discovered "quasi-linkage equilibrium." Epistasis is known to create linkage disequilibrium (a misnomer since it applies also to unlinked loci). Kimura showed that with loose linkage, after several generations the population arrives at a state in which the linkage disequilibrium is nearly constant. In this circumstance, the rate of change in fitness under selection is determined almost entirely by the additive variance component. The epistatic variance is approximately canceled by the opposite effect of the linkage disequilibrium. This argues that in long-term selection programs a better prediction can be made by ignoring the epistatic variance than by including it. I have no idea whether this is an important consideration in plant breeding, but I think it is worth looking into. For those who may wish to follow this up, it is discussed in my book (Crow & Kimura, 1970), pages 195-225.

To summarize:

Is there epistasis? Of course.

Can we safely ignore it? Often, yes.

But it is there, and perhaps ways can be found to exploit it.

WHAT OF THE FUTURE?

We can be confident that existing methods will continue to yield better performance. The rate of improvement may be lessening, as some have suggested in this symposium, but the slope is still positive, and still steep. In addition, I am confident of the ultimate benefit from molecular trickery.

The application of molecular methods to plant breeding is in its infancy, and the infant will surely grow into a productive adult—soon, I hope. Meanwhile, we must not let ourselves be so dazzled by the excitement of molecular methods that we relax our efforts along traditional lines (administrators and review panels are particularly prone to this proclivity). Our best tool for improving performance is still selection, and it may remain so for some time to come.

John Axtell has brought us a sobering picture of third world demographics. Even with a substantial reduction in age-specific birth rates, the age pyramid is such that the population will continue to grow in the next two decades as those already born enter the high-reproduction and low-death ages. So we must accept the challenge that Timothy Reeves has issued in his opening address. The current rate of growth of crop productivity is simply not enough.

Your job as plant breeders is to find ways to improve performance. But all of us, as world citizens, must share a concern for birth rates. Increasing production in the near future will buy time but the long-time future is uncertain. The nature of exponential growth is such that unless we can bring the world birth rate into some sort of balance with food supply and economic well being, the future for most of the world's people is dismal indeed.

REFERENCE

Crow, J. F., and Kimura, M., 1970. An introduction to population genetics theory. Burgess Int. Edina, MN.